Transportation and Infrastructure for Sustainability

Transportation and Infrastructure for Sustainability

Guest Editors

Larysa Neduzha
Iryna Bondarenko

MDPI

Basel • Beijing • Wuhan • Barcelona • Belgrade • Novi Sad • Cluj • Manchester

Guest Editors

Larysa Neduzha
Department of Technical Mechanics
Ukrainian State University of Science and Technologies
Dnipro
Ukraine

Iryna Bondarenko
Department of Railway Transport
Lviv Polytechnic National University
Lviv
Ukraine

Editorial Office
MDPI AG
Grosspeteranlage 5
4052 Basel, Switzerland

This is a reprint of the Special Issue, published open access by the journal *Sustainability* (ISSN 2071-1050), freely accessible at: https://www.mdpi.com/journal/sustainability/special_issues/2U2081075M.

For citation purposes, cite each article independently as indicated on the article page online and as indicated below:

Lastname, A.A.; Lastname, B.B. Article Title. *Journal Name* **Year**, *Volume Number*, Page Range.

ISBN 978-3-7258-7987-8 (Hbk)
ISBN 978-3-7258-7988-5 (PDF)
https://doi.org/10.3390/books978-3-7258-7988-5

Contents

sustainability

MDPI

Editorial

Transportation and Infrastructure for Sustainability

Iryna Bondarenko [1] and Larysa Neduzha [2,*]

1 Department of Railway Transport, Institute of Mechanical Engineering and Transport, Lviv Polytechnic National University, 79000 Lviv, Ukraine; iryna.o.bondarenko@lpnu.ua
2 Department of Technical Mechanics, Ukrainian State University of Science and Technologies, Lazaryan 2, 49010 Dnipro, Ukraine
* Correspondence: nlorhen@i.ua

The transition towards sustainable development requires the integration of technological innovation, environmental responsibility, and effective governance. In this context, transport and infrastructure systems play a key role, as they determine not only the efficiency of mobility but also the environmental and socio-economic impact of modern societies. Achieving sustainability in this domain requires the ability to measure progress, adapt to emerging challenges, and transform existing systems into more resilient and future-oriented pathways. This transition is closely aligned with the broader global sustainability agenda emphasizing low-carbon development, energy transformation, and long-term environmental resilience [1,2].

This Special Issue, "Transportation and Infrastructure for Sustainability", was conceived as a platform to link knowledge with practical solutions and to foster cooperation among researchers working across different transport modes and infrastructure systems. The collected contributions show that sustainability is not limited to the development of entirely new solutions but also involves the reinterpretation and integration of existing approaches in new contexts and combinations.

This Special Issue brings together 15 published articles that demonstrate the multidimensional nature of sustainability in transport and infrastructure. Several studies focus on the environmental and energy transformation of transport systems, including the use of solar energy in maritime transport, the integration of ecological principles into transport operations, and the development of sustainable port systems. These studies highlight the growing importance of reducing the environmental impact while maintaining operational efficiency.

Another group of contributions addresses infrastructure and engineering solutions, emphasizing the role of materials, technologies, and system design. These include the use of mining waste in road construction, the application of BIM technologies in infrastructure planning, the development of sustainable railway systems, and approaches to enhancing the resilience of large-scale transport networks. Additional studies focus on safety and technical performance, such as tire configurations in freight transport, tractor safety, and navigation systems for inland waterways.

Finally, several articles explore governance, decision-making, and socio-economic aspects of transport and infrastructure systems. These include models linking transport development with quality of life, frameworks for evaluating transport alternatives based on passenger behavior, airport classification systems for better planning, and innovative financing mechanisms such as crowdfunding for sustainable transport projects.

Taken together, the contributions illustrate a clear shift from isolated technical solutions towards a more integrated understanding of transport and infrastructure sustainability, where environmental, technological, and socio-economic dimensions are closely interconnected. This reflects an evolving paradigm in which sustainable transport is viewed as a complex system requiring coordinated action across multiple domains.

To provide a structured overview of the contributions, the articles in this Special Issue were organized into three thematic groups based on their primary research focus. This classification reflects the dominant perspective adopted in each study and allows for a more coherent presentation of the published work.

Several contributions to this Special Issue focus on the environmental dimension of transport and infrastructure, particularly addressing the transition towards low-emission and resource-efficient systems. These studies emphasize the urgent need to decarbonize transport while maintaining operational efficiency and technological feasibility.

The study by Petrychenko et al. explores the potential of solar energy as an alternative to conventional fuels in maritime transport, identifying both technological opportunities and existing barriers, such as storage and economic constraints. The paper highlights promising directions for integrating renewable energy solutions into shipping, contributing to the broader goal of reducing greenhouse gas emissions.

Corazza advances the discussion by proposing a comprehensive framework for integrating ecological principles into transport operations. The study demonstrates that environmental considerations—such as mitigating wildlife–vehicle collisions—can simultaneously generate economic benefits, thereby reinforcing the practical value of sustainability-oriented approaches.

The role of innovation in enhancing sustainability is further examined in the context of port infrastructure. Vaca-Cabrero et al. analyze how technological advancements and collaborative strategies can improve operational efficiency while reducing environmental impact in port systems. Similarly, Paulauskas et al. propose a methodology for assessing port sustainability, focusing on optimizing port operations, reducing emissions, and ensuring safe and efficient vessel handling.

Together, these contributions highlight that the environmental transformation of transport and infrastructure is not limited to energy substitution but also requires a comprehensive approach, integrating technology, operational practices, and regulatory frameworks.

Another group of articles addresses the technological and engineering foundations of sustainable transport and infrastructure. These studies demonstrate how innovations in materials, design, and digital tools can significantly enhance system performance, safety, and resilience. Contemporary infrastructure research increasingly emphasizes resilience, lifecycle efficiency, and sustainability-oriented performance metrics as key criteria for long-term transport system development [3,4].

Galhardo et al. investigate the reuse of mining waste in road construction, showing that such materials can improve mechanical properties while reducing the environmental burdens associated with waste disposal. This approach represents a practical contribution to circular economy principles in infrastructure development.

Digital transformation is addressed by Castañeda et al., who examine the application of Building Information Modeling (BIM) in road construction planning. Their findings demonstrate how BIM enhances coordination, reduces scheduling errors, and supports more efficient project management, thereby contributing to both economic and environmental sustainability.

Large-scale infrastructure resilience is the focus of Rehak et al., who propose a systematic framework for assessing and developing the sustainability of Trans-European Transport Network (TEN-T) infrastructure. Their work emphasizes the importance of integrating legislative, technical, and technological dimensions to ensure long-term system robustness.

https://doi.org/10.3390/su18104760

Railway infrastructure modernization is explored by Bondarenko, who addresses the challenges of integrating different railway gauge systems. The study proposes innovative conceptual approaches to harmonizing technical parameters, thereby contributing to the long-term sustainability and interoperability of railway systems.

Several contributions focus specifically on safety as a critical component of sustainable infrastructure. Žuraulis et al. analyze the impact of tire configurations on freight transport safety, demonstrating how optimized technical choices can reduce both the accident risks and environmental impacts. Simović et al. examine tractor safety in rural transport systems, showing that relatively simple visibility improvements can significantly reduce crash risks and associated resource losses. In the context of inland waterways, Pieron et al. propose improvements to navigation marking systems, contributing to enhanced safety and operational efficiency.

Overall, these studies demonstrate that engineering innovation plays a key role in achieving sustainable transport and infrastructure, enabling both performance improvements and risk reduction.

The third thematic group focuses on the governance and socio-economic aspects of transport and infrastructure systems, emphasizing that sustainability depends not only on technological solutions but also on effective decision-making and strategic planning.

Sivilevičius and Žuraulis develop a multi-criteria model that links transport system characteristics with quality-of-life indicators. Their findings highlight that infrastructure quality, safety, and environmental factors have the most significant impact on societal well-being, providing valuable insights for policy development.

Decision-making processes in public transport systems are examined by Şimşek et al., who propose an integrated framework combining multiple analytical methods to evaluate transport alternatives. The study demonstrates how incorporating passenger behavior into planning can lead to more effective and sustainable mobility solutions.

Strategic planning at the infrastructure level is addressed by Khanjanasthiti et al., who propose a new classification system for airports. By introducing a structured framework based on functional and operational criteria, the study supports more coordinated and sustainable infrastructure development.

The issue of financing sustainable transport innovations is explored by Mańkowska et al., who analyze the role of crowdfunding as an emerging mechanism for supporting high-risk and early-stage projects. Their findings highlight the potential of digital finance to complement traditional funding sources and accelerate innovation in the transport sector.

Collectively, these contributions underline that achieving sustainability in transport and infrastructure requires a holistic approach, where technological advancements are complemented by effective governance, informed decision-making, and innovative economic mechanisms. In this context, governance mechanisms, strategic planning, and stakeholder coordination are increasingly recognized as essential components for implementing sustainable mobility policies and achieving long-term development goals [5,6].

The contributions presented in this Special Issue highlight the evolving nature of sustainability in transport and infrastructure, demonstrating that its achievement requires the integration of environmental, technological, and socio-economic dimensions. Rather than addressing isolated challenges, the collected studies emphasize the importance of coordinated and system-oriented approaches that link innovation, infrastructure development, and governance mechanisms.

A key insight emerging from this Special Issue is that sustainable transport and infrastructure cannot be achieved solely through technological advancements or policy interventions taken separately. Instead, progress depends on the interaction between

https://doi.org/10.3390/su18104760

engineering solutions, environmental strategies, and decision-making frameworks that together shape resilient and adaptive systems.

The findings presented across the articles also underline the growing role of digitalization, alternative energy sources, and advanced analytical methods in addressing current and future challenges. At the same time, they point to the need for continued interdisciplinary collaboration and the development of integrated methodologies capable of responding to the complexity of transport systems.

This Special Issue contributes to the ongoing scientific discourse by providing both theoretical insights and practical solutions, and it is expected to support further research and informed decision-making in the field of sustainable transport and infrastructure.

Ultimately, the transition towards sustainable transport and infrastructure should be understood not as a set of isolated improvements but as a continuous and systemic transformation shaping the future of mobility and society as a whole. We are confident that the insights gleaned from these studies will inspire further research and accelerate the transition towards knowledgeable, sustainable, and equitable transportation systems. We extend our sincere gratitude to all the authors, reviewers, and the editorial team of *Sustainability* for their invaluable contributions to this successful endeavor.

Author Contributions: Conceptualization, L.N. and I.B.; writing—original draft preparation, I.B.; writing—review and editing, I.B. and L.N.; project administration, L.N. All authors have read and agreed to the published version of the manuscript.

Funding: This research received no external funding.

Acknowledgments: We thank the authors of all the papers published in this Special Issue for submitting their excellent work. Additionally, we would like to thank the reviewers for their recommendations and comments, which have improved the quality of this Special Issue.

Conflicts of Interest: The authors declare no conflicts of interest.

List of Contributions:

1. Petrychenko, O.; Levinskyi, M.; Goolak, S.; Lukoševičius, V. Prospects of Solar Energy in the Context of Greening Maritime Transport. *Sustainability* **2025**, *17*, 2141. https://doi.org/10.3390/su17052141.
2. Corazza, M.V. A Comprehensive Research Agenda for Integrating Ecological Principles into the Transportation Sector. *Sustainability* **2024**, *16*, 7081. https://doi.org/10.3390/su16167081.
3. Galhardo, D.C.; Guimarães, A.C.R.; Martins, C.A.; Narciso, M.M.; Monteiro, S.N.; Coelho, L.M. Influence of Iron Mining Waste Addition as a Sustainable Alternative on the Resilient and Physical Properties of Soils for Pavement Design. *Sustainability* **2024**, *16*, 10211. https://doi.org/10.3390/su162310211.
4. Žuraulis, V.; Pečeliūnas, R.; Misevičius, T. Assessment of Safe and Sustainable Operation for Freight Transportation Company Based on Tire Set Configurations Used in Its Trucks' Fleet. *Sustainability* **2025**, *17*, 1500. https://doi.org/10.3390/su17041500.
5. Sivilevičius, H.; Žuraulis, V. Modeling the Impact of Interaction Factors for Transport System Elements on Quality of Life Using Multi-Criteria Decision-Making and Applied Statistical Methods. *Sustainability* **2025**, *17*, 1784. https://doi.org/10.3390/su17051784.
6. Rehak, D.; Vlkovsky, M.; Manas, P.; Apeltauer, J.; Apeltauer, T.; Hromada, M. Sustainability of the Trans-European Transport Networks Land Infrastructure to Address Large-Scale Disasters: A Case Study in the Czech Republic. *Sustainability* **2025**, *17*, 2509. https://doi.org/10.3390/su17062509.
7. Vaca-Cabrero, J.; Domínguez Rastrojo, J.; González-Cancelas, N.; Camarero-Orive, A. Enhancing Sustainability in Port Infrastructure Through Innovation: A Case Study of the Spanish Port System. *Sustainability* **2025**, *17*, 2593. https://doi.org/10.3390/su17062593.

https://doi.org/10.3390/su18104760

8. Castañeda, K.; Sánchez, O.; Peña, C.A.; Herrera, R.F.; Mejía, G. BIM Uses for Mitigating Deficiencies in Road Scheduling Planning. *Sustainability* **2025**, *17*, 2729. https://doi.org/10.3390/su17062729.
9. Khanjanasthiti, I.; Bajracharya, B.; O'Hare, D. Towards More Sustainable Planning Decisions Around Airports: Investigating Global Airport Classifications and Proposing a Four-Tiered System for Australia. *Sustainability* **2025**, *17*, 5259. https://doi.org/10.3390/su17125259.
10. Bondarenko, I. Sustainable Railway Infrastructure: Modernization Strategies for Integrating 1520 mm and 1435 mm Gauge Systems. *Sustainability* **2025**, *17*, 5768. https://doi.org/10.3390/su17135768.
11. Şimşek, P.B.; Özdemir, A.; Kosunalp, S.; Iliev, T. Choosing Sustainable and Traditional Public Transportation Alternatives Using a Novel Decision-Making Framework Considering Passengers' Travel Behaviors: A Case Study of Istanbul. *Sustainability* **2025**, *17*, 5904. https://doi.org/10.3390/su17135904.
12. Simović, S.; Trifunović, A.; Ivanišević, T.; Lukoševičius, V.; Neduzha, L. Assessing Tractors' Active Safety in Serbia: A Driving Simulator Study. *Sustainability* **2025**, *17*, 6144. https://doi.org/10.3390/su17136144.
13. Mańkowska, M.; Kordela, D.; Pettersen-Sobczyk, M. Crowdfunding in Transport Innovation and Sustainability: A Literature Review and Future Directions. *Sustainability* **2026**, *18*, 576. https://doi.org/10.3390/su18020576.
14. Paulauskas, V.; Paulauskas, D.; Paulauskas, V. Achieving Port Sustainability by Harnessing the Potential of Port Operations, Ships and Tugboats. *Sustainability* **2026**, *18*, 1916. https://doi.org/10.3390/su18041916.
15. Pieron, Ł.; Jędrzychowski, K.; Iwicki, S.; Durajczyk, P. Harmonisation of Navigation Signs on Inland Waterways in Poland with European Regulations to Improve Navigation Safety. *Sustainability* **2026**, *18*, 3844. https://doi.org/10.3390/su18083844.

References

1. World Commission on Environment and Development. *Our Common Future*; Oxford University Press: Oxford, UK, 1987. Available online: https://sustainabledevelopment.un.org/content/documents/5987our-common-future.pdf (accessed on 4 May 2026).
2. International Energy Agency. *Net Zero by 2050: A Roadmap for the Global Energy Sector*; IEA: Paris, France, 2021. Available online: https://www.iea.org/reports/net-zero-by-2050 (accessed on 4 May 2026).
3. Jeon, C.M.; Amekudzi, A. Addressing sustainability in transportation systems: Definitions, indicators, and metrics. *J. Infrastruct. Syst.* **2005**, *11*, 31–50. [CrossRef]
4. World Bank. *Sustainable Infrastructure for Development*; World Bank: Washington, DC, USA, 2021. Available online: https://www.worldbank.org/en/topic/infrastructure (accessed on 4 May 2026).
5. United Nations Economic Commission for Europe. *Transport for Sustainable Development: The Case of Inland Transport*; UNECE: Geneva, Switzerland, 2021. Available online: https://unece.org/transport/publications/transport-sustainable-development-case-inland-transport (accessed on 4 May 2026).
6. United Nations. *Transforming Our World: The 2030 Agenda for Sustainable Development*; United Nations: New York, NY, USA, 2015. Available online: https://sdgs.un.org/2030agenda (accessed on 4 May 2026).

https://doi.org/10.3390/su18104760

sustainability

MDPI

Article

A Comprehensive Research Agenda for Integrating Ecological Principles into the Transportation Sector

Maria Vittoria Corazza

Department of Civil, Environmental and Building Engineering—DICEA, Sapienza University of Rome, 00185 Rome, Italy; mariavittoria.corazza@uniroma1.it

Abstract: The paper examines the integration of novel Transportation Ecology principles into transit operations, aiming to address the environmental impacts associated with surface services in urban areas and with the purpose of creating a comprehensive agenda for integrating ecological principles into transit planning and management. The research problem is to quantify the tangible benefits for transit operators, particularly in the context of mitigating wildlife-vehicle collisions and improving overall operational efficiency as a motivator for transit managers to adopt Transportation Ecology principles. The study design, after analyzing the regulatory requirements, implements scenario-based methodology, utilizing data from an average Italian bus fleet to estimate potential monetary savings and benefits. Key parameters, such as maintenance costs, insurance premiums, and collision-related expenses, are analyzed to provide a realistic assessment of the economic advantages of implementing Transportation Ecology measures. The findings reveal that significant cost reductions can be achieved by minimizing accidents involving wildlife, alongside other operational improvements. The scenario demonstrates that even a small fleet, when adopting these principles, can yield substantial financial benefits, thereby making a compelling case for broader implementation. The paper concludes that while the qualitative nature of the analysis necessitates conservative estimates, the results underscore the value of incorporating ecological considerations into transit planning and management. These insights are vital for transit operators and policymakers seeking to balance environmental sustainability with operational profitability and protect urban ecosystems. This also implies the need for a more holistic and interdisciplinary approach to transportation planning and management.

Keywords: transportation ecology; road ecology; transit

1. Introduction to Transportation Ecology: Distant Roots for a Contemporary Study Field

The term "Road Ecology" encompasses the interactions between living organisms and their environment with vehicles and road infrastructure, including structures and artworks [1]. Given that an ecosystem comprises all organisms in an area and their interactions with the physical environment, fostering diversity and cyclicity [2], Road Ecology can be more comprehensively defined as the relationship between road infrastructure, the traffic it generates, and the encompassing ecosystem. This relationship is intrinsic to the concept of mobility infrastructure, which broadly refers to artificial interventions in natural or built environments aimed at fulfilling connectivity needs for people and goods across different locations. Historically, road construction has been viewed as humanity's triumph over nature, achieved through effort (like the Etruscan cuts made in rock to reach necropolises in Tuscia in Central Italy, potentially dating back to the Roman Republican era) or ingenuity, becoming a significant element that generates "place" [3]. Since the rise of ecological thinking in the early 1970s, this dominance has been increasingly questioned, highlighting the adverse impacts of road infrastructure on natural and social environments [4–6]. This shift prompted the need to begin and consolidate research on these impacts [7], develop initial technical specifications for environmental conservation following infrastructure construction [8], and generate numerous studies, many on specific case studies. These studies emerged from diverse investigative fields, such as ecosystem protection focusing on faunal components [9–11], water systems [12], traffic pollutant quantification [13,14], road maintenance [15,16], and public health [17,18], showcasing the multidisciplinary nature of the emerging field of Road Ecology. However, the roots of Road Ecology date back much earlier, addressing issues like mud removal from road surfaces or erosion phenomena, as addressed by John Loudon McAdam in the late 18th and early 19th centuries. Additionally, the focus on roadkill, the phenomenon of animals being struck by vehicles, gained attention in North America following the construction of parkways and scenic routes during the New Deal era [19]. This eventually led to the construction of the first highway overpass in Utah in the late 1970s to facilitate deer migration [1]. Initially, roadkill and Road Ecology were considered synonymous until the term "Straßen-Ökologie" was coined by Ellenberg et al. [20], establishing Road Ecology as a discipline studying the impact of road infrastructure on landscapes. This study was significant not only for coining the term and linking landscape protection with the impacts of road infrastructure but also for introducing other key notions: nature management (Naturhaushalt) and ecosystem (including plant and animal components) integration—major influencing factors (climate, noise, pollutant emissions into the air and infiltration into groundwater and soils, use of deicing salts)—and the development of regulations and design specifications to protect the ecosystem. The analysis also included a case study on the green cover along the Rhoenline, the highway section between Würzburg and Bad Hersfeld, Germany.

The English translation of the term into "Road Ecology" came later, in the 1990s [21], when the scientific literature on the subject was consolidating into several research streams. For example, the negative impacts of road infrastructure are analyzed concerning wildlife mortality, behavioral changes, movement along corridors, and migrations, with areas along roads acting as "ecological traps" for species attracted there in search of food [22]. Regarding the landscape, emphasis is placed on fragmentation and interaction effects in change dynamics [23]. The effects on individual abiotic components of the landscape—such as water, microclimate, light, noise, topography, and soil sedimentation mechanisms—are also highlighted [1,24].

Interestingly, in the 1980s, a related concept, "Natural Disturbance Ecology", emerged in response to the need to preserve natural areas in land-use decision-making processes [25,26]. Based on Odum's definition of an ecosystem, specific characteristics of heterogeneity and dynamism in time and space are subject to disturbance regimes, referring to any event capable of disrupting a natural community, its structure, the resource system it relies on, or the physical environment that hosts it, for natural causes [26,27]. The affinity between the two concepts lies in recognizing the mutability of natural ecosystems and distinguishing between endogenous (Natural Disturbance Ecology) and exogenous—specifically anthropogenic—factors related to road infrastructure construction (Road Ecology) that cause it. Additionally, natural disturbances may not necessarily be harmful [27], while the mutual interaction between nature and roads generates degradation, leading to social costs [1].

In the name and vision of Road Ecology, the disruptive element is the road. However, it would be erroneous to think that other types of road infrastructure are exempt from such a role. Introducing railway connections and their operation can similarly have negative effects, as can any infrastructure related to other modes and forms of transportation, especially if they are high-capacity modes. The shared characteristic of linear terrestrial infrastructure between roads and railways led to the definition of Railway Ecology as an extension of Road Ecology in the early 2000s. However, railway infrastructure impacts on flora and fauna have been studied since the 1980s, in the form of specific case studies, similar to the early stages of Road Ecology. Initially, the focus was on specific structures such as tunnels and underground channels allowing the passage of small mammals [28,29], but the advent of high-speed rail increased interest, highlighting the risks associated with habitat fragmentation [30–32]. However, the provision of passageways is not without risks, especially for smaller mammals that become easy prey due to the size or location of these structures, introducing a central theme in the ecology of road infrastructure: the provision of mitigation or protection measures, addressed later. Despite continuous interest over time, Railway Ecology remains relatively underexplored in the literature, with the predominantly European sources [33] being focused on train–animal collisions, while impacts on habitat (loss and fragmentation), pollution (chemical, noise, light), and the potential for creating suitable environments in areas degraded by railway infrastructure are still relatively under-studied compared to Road Ecology [34].

However, comparing Road Ecology and Railway Ecology introduces numerous points for reflection. Both involve studying the effects on natural ecosystems resulting from the introduction of linear terrestrial infrastructure, prompting the subsequent discussion in this chapter to jointly address rubber and rail modes—their methods of introduction into a given area (earth movement, construction, "sealing" of the area and adjacent strips) are similar, as are the effects of severance and fragmentation of territory and landscape continuity.

1.1. Developing the Transportation Ecology Concept

In both Road Ecology and Railway Ecology, research is addressed in a multidisciplinary manner, but currently, interest within transport disciplines remains marginal and driven by regulatory requirements, mainly in terms of environmental impact assessment. Additionally, railways are considered a less polluting mode compared to roads, but this is not true from the perspective of wildlife impacts, where the presence of catenaries and higher travel speeds can introduce additional hazards [35]. This is compounded by the fact that road accidents, a much more publicized topic than rail accidents, contribute to greater visibility and awareness of the phenomenon. Both Road Ecology and Railway Ecology, which will henceforth be collectively referred to as the broader, and novel, concept of Transportation Ecology, integrate environmental requirements, ecology, and transportation [36], emphasizing territorial preservation and sustainable development. As such, it becomes pivotal not only in quantifying the impacts of roads and railways but also in assessing a territory's sustainability in accommodating these infrastructures and the associated traffic volumes without causing loss or fragmentation. This involves evaluating not only the impact of construction but also the ongoing effects of road and rail operations [19], recognizing that habitat damage is a primary cause of flora and fauna degradation [33].

Therefore, Transportation Ecology and traditional Road Ecology can be considered related fields but have distinct focuses and research agendas. A clear demarcation of their boundaries relies in four main areas—scope, interdisciplinary approach, research focus, and environments considered—is synthesized in Table 1.

Table 1 shows the clear demarcation of the boundaries between Road and Transportation Ecology, and understanding such boundaries between these two areas and their research agendas helps illuminate the critical gaps they aim to fill in the existing literature.

Traditional Road Ecology primarily zeroes in on the environmental effects of road networks, focusing on how roads impact wildlife and ecosystems [1,19,21]. This field has traditionally examined localized effects such as animal mortality due to vehicle collisions, changes in animal behavior, and direct alterations to habitats caused by the presence of roads. However, the research in this domain has often been limited in several ways. One major gap in traditional Road Ecology has been the lack of comprehensive solutions for habitat fragmentation caused by road networks. Studies moved from highlighting the issue to provide infrastructure-based strategies for mitigating these effects, with more recent research beginning to address this by developing and assessing wildlife corridors, overpasses, and underpasses, which aim to enhance habitat connectivity and reduce wildlife mortality, e.g., [23,24]. Another significant gap has been the limited duration of many studies, which often focused on short-term impacts without capturing the long-term effects of roads on wildlife and ecosystems. The field has started to establish long-term ecological monitoring programs to gain a deeper understanding of the chronic impacts of road infrastructure [37,38].

Table 1. Differences between conventional Road Ecology and novel Transportation Ecology.

Areas of Differentiation	Road Ecology	Transportation Ecology
Scope	Primarily focuses on the environmental impacts of roads and highways. It examines the effects of road networks on wildlife, ecosystems, and landscapes. Key topics include wildlife–vehicle collisions, roadkill, habitat fragmentation, and the spread of invasive species along road corridors.	Encompasses a broader range of transportation modes beyond just roads and highways and includes railways and urban transit systems, with a focus on operations in urban environments. It studies the environmental impacts of these various transportation systems on ecosystems and biodiversity. Although marginal, aviation and maritime transportation, if operating in urban areas, can be included.
Interdisciplinary approach	Thusfar specialized, involving ecologists, biologists, and conservationists and focusing on terrestrial ecosystems affected by road infrastructure.	Highly interdisciplinary, involving not only ecologists but also engineers, urban planners, economists, and social scientists. This field considers the full spectrum of ecological impacts from the considered transportation modes.
Research focus	Concentrates on localized effects such as animal mortality, changes in animal behavior, and direct habitat alterations due to road presence.	Looks at broader systemic impacts including air and noise pollution, climate change contributions, landscape connectivity, and the sustainability of transportation networks.
Areas/Environments	Mostly rural, non-urban	Urban, primarily

Furthermore, traditional Road Ecology has not extensively examined how road networks affect ecosystem services, such as water filtration, pollination, and carbon sequestration. This gap is beginning to close as researchers recognize the importance of assessing how roads and operations influence these critical services provided by ecosystems. This has been thus far acknowledged in few cases [39–41].

In contrast, moving from Table 1, Transportation Ecology encompasses a wider range of urban transportation modes, from private vehicles to the whole transit supply. This field adopts a highly interdisciplinary approach, integrating insights from ecologists, engineers, urban planners, economists, and social scientists. Transportation Ecology seeks to understand the full spectrum of ecological impacts arising from all urban transportation systems and operations, addressing several key gaps in the literature, such as:

- The lack of integration of multiple transportation modes into traditional ecological studies. Road ecology has primarily focused on the impacts of roads, often neglecting how different transportation systems interact and their combined effects on the environment. Transportation ecology aims to fill this gap by studying the interactions and cumulative impacts of various transportation modes.
- The limited focus on climate change and energy consumption in Road Ecology. While road ecology has examined localized environmental impacts, it has not fully addressed the broader implications of transportation emissions and energy use. Transportation Ecology can actively research how transportation systems contribute to climate change and exploring sustainable transportation solutions to reduce carbon footprints.
- The lack of integration with urban and regional planning (an area where traditional road ecology has fallen short). Transportation Ecology typically does not address the integration of transportation systems within urban planning frameworks. Transportation Ecology, however, includes the role of transportation in urban sprawl, land use, and regional development, promoting eco-friendly urban and mobility planning practice that consider environmental sustainability.
- Human health and socioeconomic impacts of transportation systems have also been underrepresented in road ecology. Traditional studies rarely consider how transportation affects air quality, noise levels, and overall human well-being. Transportation

ecology can fill this gap by investigating these impacts and addressing socioeconomic disparities in transportation access.
- Lastly, technological innovations in transportation, such as electric vehicles and autonomous transportation, have not been extensively explored in traditional road ecology. Transportation Ecology can help evaluating the environmental benefits and potential drawbacks of these new technologies, aiming to understand their implications for ecological sustainability.

One more gap to consider is that the current knowledge, and scientific and gray literature all reflect the approach from Road Ecology, i.e., how to cope with damages to flora and fauna derived from a given infrastructure, with no attention paid to those generated by traffic (which can be even more disruptive). The proposed novel Transportation Ecology approach considers these gaps and addresses them from the transportation operators' point of view (with a focus on urban transit managers), which differs from that of road planner and managers, since they act on the supply and thus the traffic volumes generated, and not just the infrastructure, per se.

However, the issues above mentioned are not meant to claim Transportation Ecology's superiority over Road Ecology but the need of the transition of the latter towards the former. Both fields together provide a comprehensive understanding of the ecological consequences of human transportation infrastructure and operations, paving the way for more sustainable practice and policies.

1.2. Rationale of the Paper

Moving forward, the paper will briefly describe the impacts of both roads and railways on flora and fauna to highlight their disruptive potential and the need for mitigation with several countermeasures (Section 2), which is partly addressed at the regulatory level (Section 3). However, from the perspective of transport research, transportation ecology remains largely unexplored in impact assessment processes [36,37,41–43]. This is partly due to the lack of a clear definition of which effects to evaluate from the transport operators' point of view. Many studies focus on individual species and isolated case studies [11,28,32,44,45], leading to a selective approach to identifying negative effects, affected territories, and methods for detecting, measuring, and evaluating infrastructure impacts on biodiversity. However, by learning from these case studies, it is possible to transfer Road Ecology criteria and visions to Transportation Ecology and to quantify tangible benefits for key actors in this sector, i.e., transit operators (Section 4). This transfer methodology relies on an exploratory scenario based on assumptions and hypotheses collected partly from the Road Ecology literature and partly from typical operational data from transit management. Results are qualitative and designed to demonstrate that by including the Transportation Ecology criteria in the transit management, monetary savings can be obtained.

The scenario building approach moves from the observation that, although under consolidation, Transportation Ecology is still far from being considered a "regular fixture" in current transportation practice when planning and designing infrastructure, and even less so when operating high-capacity transit systems. This raises a fundamental research question: "How can Transportation Ecology be made integral to the planning and management of roads and rails services?", which also paves the way for a new vision of the relevance of flora and fauna, reversing the current approach that typically seeks to remove natural components (animals and vegetation) as disturbances to operations, as epitomized by efforts to mitigate bird strikes at airports.

To address this, a digression is eventually presented (Section 5) with the goal of establishing a feasible program for transit operators to safeguard local environments from traffic's disruptive effects. Finally, concluding remarks (Section 6) are drawn with the ultimate goal of advancing knowledge in this field further. A flowchart of the work is presented in Figure 1.

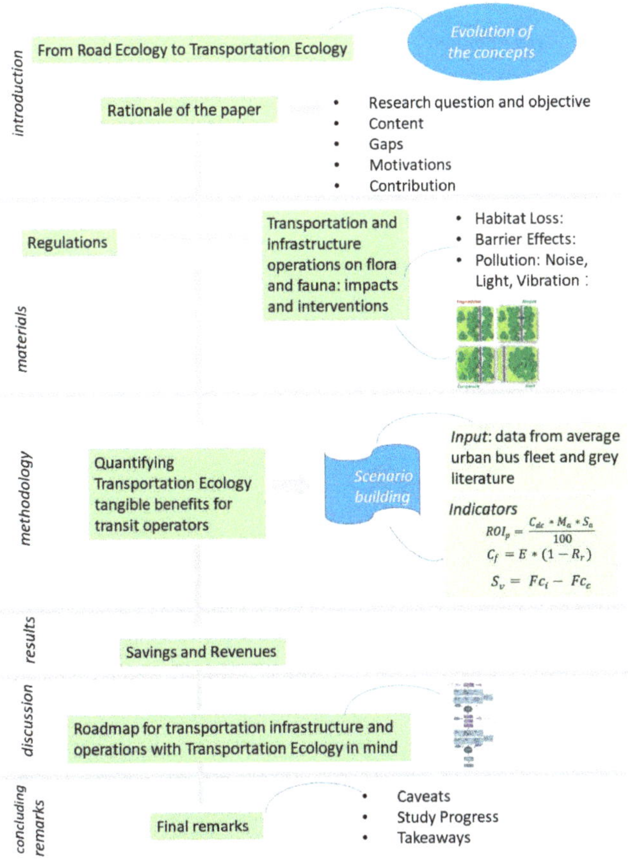

Figure 1. Flowchart of the study.

This study's motivation, thus, stems from the recognition that current approaches to integrating ecological principles within urban transportation systems are limited, particularly in how they address broader environmental impacts beyond localized issues like wildlife mortality and habitat fragmentation generated merely by the infrastructure. While Road Ecology has provided a foundation, the need for a more expansive framework that considers various transportation modes and urban environments has, therefore, become evident. This realization drives the study towards developing the innovative concept of Transportation Ecology, which encompasses not just the infrastructure-based approach of conventional Road Ecology, but also the need to consider operations when assessing the ecological impacts of mobility, especially in urban ecosystems (thus integrating multiple disciplines, including ecology, engineering, and urban planning).

The steps in Figure 1 introduce this novel framework and contribute to the transportation study field. Transportation Ecology not only expands the scope of ecological integration in transportation but also offers practical tools for transit operators. By proposing a scenario-based methodology in Section 4, this study highlights how ecological considerations can lead to economic benefits, such as reduced maintenance costs and lower insurance premiums, thereby aligning environmental sustainability with operational efficiency. This contribution is crucial for advancing both theoretical understanding and practical application in urban transit systems.

2. Transportation and Infrastructure Operations' Impacts on Flora and Fauna: Materials and Methods to Describe the Phenomena

The negative effects due to road or rail infrastructure and operations to consider derive from the well-known concept of "negative externalities" [37]. However, each of these externalities includes further specifics to consider, namely the actual ecological effects already defined by Iuell et al. [38] and summarized in Table 2.

Table 2. Effects on fauna and flora (adapted from [38]).

Phenomena Associated with Transportation Infrastructure and Operations		Effects on:	
		Fauna	**Flora**
Wildlife habitat loss	Physical change of soil surface	Changes in species distribution	Destruction of vegetated surfaces and conversion to paved areas or embankments
Barrier effect	Caesura between portions of territory	Isolation, potentially threatening survival	Modification in the continuity of vegetated surfaces
Risk of fatal events	Mortality due to accidents	Population reduction	
Disturbance and pollution	Changes in hydrological setup	Changes in behavior for seeking water sources	Modification in continuity of vegetated surfaces
			Variation in water resources and water regimes
			Modification of wetland and riparian habitats
		Damage to vital and behavioral functions	
	Noise pollution	Population reduction	
	Vibrations	Changes in behavior	
	Light pollution		Modification in growth
Changes in protective strips (e.g., road shoulders) functions	Alterations of functions along protective strips	Creation of corridors	
		Creation of new habitats	
Land use changes	Physical change of soil surface with increased sealed surfaces (concrete, asphalt, etc.), generation of impervious surfaces	Increased risk due to human presence	
New Human settlements resulting from the opening of infrastructure			

At the root of all the negative effects on biodiversity summarized in Table 2 is the habitat fragmentation caused by the insertion and operation of road and rail infrastructure. It is not merely the introduction of a dividing element within a territory but rather the resultant parceling of the area into smaller zones where the infrastructure becomes a barrier to the continuity of the landscape. Moreover, fragmentation introduces issues of size and mobility. If parceling creates smaller areas unable to support the same amount of wildlife populations, these will tend to move toward new territories. However, if the barrier effect is insurmountable, such mobility will not be possible anymore, making the population more vulnerable in a smaller territory. Furthermore, a territory composed of smaller and less permeable habitable areas generates a different and uneven distribution of species, exacerbated by the presence of other land uses (especially urban areas), which can pose additional risks and conflict with the quest for suitable living spaces by migrating populations.

The capacity for movement within a territory (in search of food or shelter) and of dispersal is a pivotal factor in the survival of a species. This introduces the concept of road permeability, which is related to the type of barrier, the traffic it generates, and animal behavior. As evidenced in the literature, large animals may still utilize areas in close proximity to the infrastructure but may not fully do that [1,26,34]. Conversely, small mammals or avian fauna tend to avoid areas in direct proportion to the extent of the infrastructure. For invertebrates, the road surface itself or the less hospitable substrate of the road shoulders may act as a barrier [1,26].

The most evident phenomenon, mortality, is not only associated with traffic intensity but also with the behaviors induced in fauna by the infrastructure as a barrier. The risk of collision with vehicles, whether cars or trains, is a significant hazard for species attempting to cross roads. However, a high number of fatal events may simply indicate a large population of a given species, such as rodents. It is evident that temporal and climatic factors influence behaviors, which is reflected in observations of accident levels and the interaction with human activities, including seasonal livestock migrations, hunting, and seasonal agricultural activities. In a survey conducted in the Czech Republic, a "scale factor" that connects accident events with various species and their surrounding road catchment areas was also observed. For example, at a landscape scale characterized by a catchment area with a 1000 m radius, the highest mortality rates were noted for foxes, martens, and polecats, particularly near agricultural lands and settlements. Intermediate scales were also defined, down to the local scale, with a catchment area radius of 50 m, where mortality included not only landscape-scale species but also otters, badgers, stoats, and weasels, particularly along forested strips adjacent to infrastructure [46].

Anthropogenic activities introduce various disturbance factors, notably pollution. Although railway infrastructure does not impact the environment through harmful air emissions like road infrastructure, it generates similar issues through noise pollution. This noise pollution significantly affects avian fauna, including bats, which can be disturbed even at a distance of 20 m from the noise source [47]. Raptors are particularly sensitive during reproduction and nesting to noise levels above 40 dB(A) in Leq24h [48], and this threshold is also harmful to some terrestrial mammals [49,50]. Chronic and frequent noise disrupts animals' ability to detect essential sounds, while intermittent and unpredictable noise is often perceived as a threat [51]. Additionally, artificial light significantly impacts nocturnal fauna behaviors, particularly in predatory activities [52].

Finally, when vegetated surfaces are converted into impervious surfaces, whether by asphalt or rail, the result is soil consumption and behavioral changes in fauna and flora. It is well-documented in the literature that green areas along roads become new hunting grounds for predators or new propagation surfaces for non-native and/or invasive plant species. Similarly, the creation of drainage channels generates micro-wetlands with the same attractive power.

2.1. Mitigating Impacts Due to Transportation Operations and Infrastructure

The scientific literature offers numerous examples of strategies to reduce the impact of transportation operations and infrastructure on flora and fauna [1,38], and many case studies demonstrate their successful application. The overarching approach to addressing habitat fragmentation and anthropogenic threats involves three main actions—mitigation, avoidance, and compensation (Figure 2)—especially when preventive measures are not feasible.

In this framework, prioritizing actions to avoid fragmentation is essential. When avoidance is not possible, mitigation measures become necessary, though they cannot completely eliminate negative impacts. Mitigation is the most common approach and can be implemented either ex ante, incorporated into the infrastructure's planning and design once avoidance is deemed impossible, or ex post [53]. The consideration of not proceeding with the project remains critical, as emphasized by Forman et al. [1], who instead suggest the following drastic avoidance measures:

- Not constructing the infrastructure at all.
- Modifying the route.
- Building the infrastructure underground.

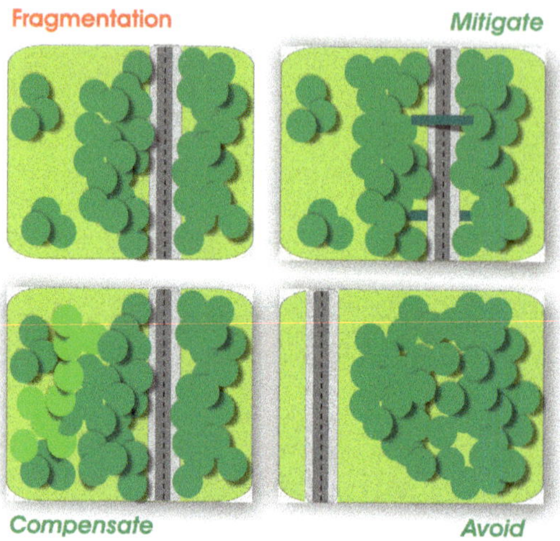

Figure 2. Key actions to counteract fragmentation.

Ultimately, the implementation of compensation measures should be regarded as a last resort, primarily applicable to existing infrastructure. The conceptualization of new infrastructure should, in fact, be founded upon preventive criteria that effectively minimize the necessity for compensation.

Thus, the discussion of Transportation Ecology measures is divided into "preventive" measures for the design phase of new infrastructure and operations, and measures to mitigate the damage from habitat fragmentation caused by existing routes. For new infrastructure, while it is understood that any human intervention alters the natural environment, integrating preventive measures into the design can be categorized into three main areas: (i) route alignment, (ii) earthworks, and (iii) the design of structures as part of the landscaping [38].

For existing infrastructure, the design criteria primarily involve mitigative measures aimed at both fauna and human components [1], typical of transportation operations. Identifying the appropriate locations for these measures requires examining the catchment area, or road-effect zone, which defines the road's interaction with surrounding flora and fauna [1,54]. The extent of this area of influence depends on several factors, including vegetation, terrain morphology, the direction of water and air flows, traffic volumes (particularly in peri-urban areas with frequent railway and road operations), and the characteristics of the road. Determining this area of influence is also crucial for monitoring the effectiveness of mitigation measures and tracking accident occurrences. In addition, the approach based on the three key actions (mitigate, avoid, and compensate) continues to be applicable in this case. The term "avoid" refers to interventions such as reducing railway traffic or closing the road [1]. It is reasonable to conclude that these are drastic actions, difficult to implement in areas where, conversely, it seems more feasible to introduce mitigative actions that, as previously mentioned, influence the behavior of animals or humans. Regarding fauna, interventions fall into two primary categories:

- Creation of structures to reduce risk and safeguard zones or alert animals: This includes crossing structures, barriers, ramps, reflectors, and lighting.
- Habitat modifications: For instance, planting unattractive or unpleasant vegetation along the road.

Studies in the literature address these types of interventions [35,55], highlighting common mitigation criteria that can be summarized as follows:

- Enhancing road permeability: This involves the installation of overhead or underground crossings.

- Modifying the roadbed: Lowering or raising the roadbed relative to the surrounding terrain to reduce traffic disturbances, particularly noise.
- Implementing noise mitigation devices: These can be used both on vehicles and infrastructure.
- Adopting less polluting traction systems: This is essential for reducing environmental impact [56].

Compensatory measures are designed to provide environmental improvements that offset the impacts of infrastructure and habitat fragmentation that cannot be mitigated or avoided. When dealing with habitat loss, degradation, or isolation, the following approach is recommended:

- Creation of compensation areas larger than the impacted areas: These areas should be forested, wetland, etc., rather than making improvements in areas equal in size to the impacted ones.
- Proximity to impacted areas: These compensation areas should be located as close as possible to the impacted areas but outside the zones of influence.
- Re-creation of pre-existing ecological conditions: It is preferable to recreate the same ecological conditions that existed before, rather than introducing different conditions.
- Quality improvement: The aim should be to enhance the quality of ecological conditions compared to pre-existing ones rather than simply restoring the same quality level.

In this context, effective actions include the introduction of larger areas with high-quality avian presence to offset zones affected by traffic noise; further expansion of large areas with natural vegetation; restoration of watercourses or wetlands that have been altered to pre-infrastructure conditions; recreation of corridors or pathways to allow wildlife movement; and eventually, improvement of habitat conditions for rare species and in high biodiversity sites [38].

2.2. Additional Issues

The aforementioned measures are exemplified in well-documented case studies available in the literature thus far reported and are collated in Table 3 for convenient reference.

Table 3. A synthesis of criteria and measures to counteract fragmentation.

<table>
<tr><th rowspan="2">Criterion</th><th colspan="2">Type of Infrastructure/Operations</th></tr>
<tr><th>New</th><th>Already Operational</th></tr>
<tr><td rowspan="2">Avoid</td><td rowspan="2">Do not build/operate at all</td><td>Reduce traffic flows</td></tr>
<tr><td>Operate less polluting vehicles</td></tr>
<tr><td rowspan="7">Mitigate</td><td rowspan="2">Adapt the layout to local morphology</td><td>Define road effect zones</td></tr>
<tr><td>Introduce stuctures and devices to reduce the risk for wildlife</td></tr>
<tr><td rowspan="2">Design elevated or underground layouts</td><td>Modify the habitat (also Landscape the infrastructure area)</td></tr>
<tr><td>Design permeable infrastructure (also Create passageways for wildlife)</td></tr>
<tr><td>Reduce earthworks</td><td></td></tr>
<tr><td>Create passageways for wildlife</td><td></td></tr>
<tr><td>Landscape the infrastructure area</td><td></td></tr>
<tr><td rowspan="4">Compensate</td><td colspan="2">Create compensation areas larger than the impacted ones</td></tr>
<tr><td colspan="2">Locate the compensation areas as closer as possible to the impacted ones</td></tr>
<tr><td colspan="2">Recreate the previous ecological conditions</td></tr>
<tr><td colspan="2">Improve the quality of the previous ecological conditions</td></tr>
</table>

Despite the fact that a considerable number of these measures have become standard practice, there are still instances where issues arise. For example, a common practice involves the installation of transparent noise barriers along roads with bird silhouettes to prevent bird collisions, typically conducted retrospectively. However, bird fatalities resulting from collisions with transparent surfaces have been largely documented, with identifiable occurrences [57]. This further substantiates the necessity for accurate, multi-step decision-making processes, as discussed in Section 5.

It bears noting that the aforementioned considerations collectively raise a caveat: the underlying premise is that the majority of interventions pertain to infrastructure, with traffic operations constituting a secondary consideration. In urban environments, however, impacts are predominantly attributable to the latter, thereby reinforcing the initial research question (how to integrate Transportation Ecology into the planning and management of road and rail services) and the relevant research objective to quantify benefits for key actors in this sector, specifically transit operators.

3. Regulatory Requirements, the European Vision

After highlighting the critical role of Transportation Ecology in aligning infrastructure projects with environmental and urban considerations, it is equally important to focus on the regulatory frameworks (taking the European framework as an example) that should guide its enforcement, ensuring compliance and ecological integrity. In fact, as evidenced in the preceding Sections, Transportation Ecology is of paramount importance in the comprehensive assessment of the environmental consequences of new infrastructure and transportation operations. At the same time, environmental regulations mandate the implementation of comprehensive impact assessments. While this role may be self-evident in the context of decision-making processes pertaining to the construction or management of public infrastructure, given the existence of specific regulations pertaining to strategic assessment, it is equally important in the context of transportation services and especially transit management.

3.1. Regulations on Strategic Assessment for Transport Infrastructures

In Europe, supranational regulatory tools such as the Strategic Environmental Assessment (SEA) and the Environmental Impact Assessment (EIA) promote sustainable development by incorporating environmental considerations into the decision-making process for various interventions. Both SEA and EIA must align with other supranational instruments, like the EU Habitats Directive (92/43/EEC), which focuses on the conservation of natural and semi-natural habitats and wild flora and fauna, and the Birds Directive (2009/147/EC), which is dedicated to the conservation of wild birds.

A key component of the Habitats Directive is the Natura 2000 network, which connects Special Areas of Conservation (SACs), as defined by Directive 2001/42/EC and Directive 85/337/EEC, and Special Protection Areas (SPAs) dedicated to wild bird protection in accordance with the Birds Directive. The Habitats Directive sets conservation targets for natural habitat types and species of Community interest within these areas. When initial screening suggests potential impacts on local ecosystems, a focused study is required to assess the significance of a plan or project on any SACs or SPAs. If such a plan or project must proceed for reasons of public interest, compensatory measures must be implemented. These measures, consistent with the approach stressed in Section 2, can be:

- Conservative: Aimed at maintaining or restoring natural habitats and populations of wild species.
- Contractual: Ensuring compliance through agreements.
- Preventive: Designed to avoid degradation and disturbances around the affected sites.

These measures must align with the essential abiotic and biotic factors necessary for the conservation of different habitat types and species, including their interactions with the physical environment (air, water, soil, vegetation, etc.).

Applying these regulatory criteria in the realm of transportation raises a fundamental query: How should the areas of influence for impact assessment and their implications be defined? Case studies demonstrate that a single road infrastructure hosts animal species with diverse habits and movement ranges, while the disruptive impacts of associated human activities (in this case, traffic) vary accordingly. Moreover, Natura 2000 sites exhibit considerable diversity in size and number. For instance, in the Latium region of Italy, the network encompasses approximately one hundred sites, comprising over 200 SACs and SPAs, with about 90 distinct habitats (some of them within consolidated urban environments) [58].

As a result, determining the influence area for assessing impacts from regulatory tools concerning transportation surface infrastructures and operations (such as municipal transit plans, masterplans, and sustainable urban mobility plans) necessitates a multidisciplinary approach. This approach should integrate technical planning with pertinent scientific knowledge to elucidate behavioral traits of fauna and flora modifications, underscoring the pivotal role of Transportation Ecology in this process.

3.2. Additional Environmental Regulatory Requirements

While the above-mentioned regulations appear to focus on the physical environment, with road and rail infrastructure as the main areas of intervention rather than transportation services, additional regulatory tools require the full involvement of transportation operators, particularly in the fields of transit planning and management. Transportation operators face significant environmental challenges primarily stemming from strict regulatory demands aimed at reducing emissions. These challenges are particularly salient in light of international obligations such as the Nationally Determined Contributions (NDCs) established under the 2015 Paris Agreement. The NDCs mandate each participating nation to implement domestic measures for mitigating emissions, including the establishment of broad-based reduction objectives. Despite their global policy scope and emphasis on emission control, supplementary regulations impose heightened responsibilities on transit operators and infrastructure administrators to evaluate environmental repercussions effectively.

In continuity with the NDCs, the 2024 European Commission Directive on Corporate Sustainability Due Diligence (COM/2022/71 final) mandates that companies identify and mitigate adverse environmental impacts resulting from their operations. Concurrently, the 2024 European Nature Restoration Law (COM/2022/304 final) focuses on enhancing urban ecosystems by addressing the depletion of green urban spaces and tree cover. Together, these directives are poised to synergistically contribute to the preservation of urban flora and fauna. For transit operators, compliance may involve a range of actions, such as integrating more green spaces into infrastructure developments, incorporating vegetation along railway lines and bus routes, implementing water management systems to mitigate runoff pollution harmful to local ecosystems, constructing wildlife passages, restoring indigenous flora along transit corridors, and conducting rigorous environmental impact assessments. These measures emphasize the critical role of both SEA and EIA, ensuring that flora and fauna preservation remains central to transit projects.

- Evidently, a multidisciplinary approach is essential for the effective application of these regulatory tools, SEA and EIA included, to Transportation Ecology. However, the lack of established practice and awareness suggests these regulations may be overlooked due to several factors:
- Firstly, their comprehensive scope spans multiple sectors and industries, potentially diluting specific focus on Transportation Ecology.
- Secondly, sector-specific regulations that are already in force may take precedence over broader regulations such as the Due Diligence Directive and the Nature Restoration Law.
- Lastly, corporate strategies often prioritize compliance with regulations that directly and immediately affect operational aspects, such as emission reduction targets, co-

herently with NDCs and national regulations, and waste management, rather than long-term ecological considerations like Transportation Ecology.

- To address these gaps, targeted efforts are required, with increased advocacy and awareness being of particular importance. In addition, the development of practical solutions is essential, including the formulation of specific guidelines for Transportation Ecology actions, the provision of financial and technical support for projects in this field to encourage implementation by transit operators, and the undertaking of research and data collection to highlight the impacts of transit operations and infrastructure on biodiversity, thereby demonstrating the benefits of integrating Transportation Ecology into broader sustainability efforts and within regular traffic planning and infrastructure management practice.

4. Quantifying Tangible Benefits for Transit Operators

As stressed in the previous Section, the imperative to integrate Transportation Ecology into traffic planning and infrastructure management is mandated by environmental legislation. The crucial question centers on how to quantify the advantages of Transportation Ecology, transforming it from a legal obligation into an appealing practice for transportation managers, especially those operating transit companies, and decision-makers. Key areas of focus might involve quantifiable monetary benefits, serving as foundational elements for a roadmap that facilitates the adoption of Transportation Ecology in planning and management practices.

The literature already offers several studies that assess cost-benefit analyses related to measures discussed in Section 2 [59,60]. However, these typically focus on quantifying expenditures for road managers versus benefits to communities and the environment, primarily through reduced social costs. Specifically, many studies emphasize cost analyses traditionally used to evaluate infrastructure impacts [60,61] yet neglect operational effects.

Nevertheless, potential monetary savings can also be computed for operations, particularly benefiting key stakeholders such as transit managers. In this context, an initial scenario is proposed to quantify monetary benefits specifically for public transport operators, starting from basic indicators.

4.1. A Scenario of Potential Monetary Benefits

Monetary benefits can be associated with a number of parameters, which, in turn, give rise to savings and revenues. While maintenance and insurance costs are the most immediately apparent, there are numerous other potential expenses to consider. It is noteworthy that in the United States, collisions between vehicles and large mammals (e.g., deer and bears) result in damages amounting to 8.4 billion USD and account for nearly 2 million accidents annually. A comparable analysis in the United Kingdom reveals that these incidents result in 42,500 to 74,000 accidents with associated costs of approximately 25 million Euros [62].

To this end, a qualitative scenario is built by considering an average Italian bus fleet (Table 4) a small part of which—20 vehicles—is the scenario fleet potentially involved in collisions. Table 4 also presents the relevant parameters in transit management and planning, used to build this scenario.

Table 4. The scenario input data.

Features	Scenario Input	Source
Average fleet composition (units)	20 as part of the average Italian bus fleet of 457 vehicles	[63]
Average yearly vehicle mileage (km)	55,000	[64]
Average yearly maintenance cost (Euro × km)	0.35	[64]
Average collision cost per vehicle (Euro)	1400	[65]
Estimated basic insurance premium (Euro)	10,077	[66]

It is noteworthy that although qualitative, the Table 4 parameters feeding the scenario reference data describing typical fleet consistency and operations as well as cost parameters and market rates as reported in the literature to describe general real-life operations [63,64,66] or specific case studies [65] on Road Ecology. Given the novelty of Transportation Ecology, no specific inputs are available in this field. For what concerns the scenario settings, the fleet size of 457 buses, with a focus on a subset of 20 vehicles allows for a manageable yet representative sample size, reflecting typical urban transit operations and supply in Italy (but also elsewhere in Europe, for example) and ensures that the findings are scalable and relevant to real-world conditions. In turn, the yearly vehicle mileage, set at 55,000 km, is based on operational standards [64] and therefore, reflects typical usage patterns and provides a solid foundation for cost and savings calculations. Likewise, the maintenance cost data, which are sourced from detailed operational cost analyses of similar fleets [64], and insurance premium rates, which are derived from a major Italian insurance company, ensure accuracy and relevance [66]. Thus, the scenario settings are carefully chosen to provide a realistic and conservative analysis of the economic benefits of wildlife collision mitigation and other operational improvements. In the same way, for the basic equations used in the coming Sections 4.1.1 and 4.1.2 to estimate potential savings and revenues, the assumptions suggested by the cited literature are more prudent to avoid overstated figures and ensure the robustness of the estimates. Therefore, the resulting scenario, although qualitative, is corroborated by representative data and assumptions as provided by the industrial and scientific literature and uses conservative figures for collision reduction and fuel efficiency improvements to avoid overestimating benefits and instead provide a cautious estimate of potential savings that is reliable and tailored on a subset of 20 vehicles, which allows for detailed insights that can be scaled up to larger fleets, making the equations' results thorough, practical, and transferable to a large set of different operational case studies. The scenario's resulting figures presented in Sections 4.1.1 and 4.1.2 can represent qualitative—but convincing—arguments for transit operators willing to adopt the Road Ecology principles in their everyday practice.

4.1.1. Benefits Associated with Improvements

Areas for improvement include reducing accident costs and increasing the attractiveness of the service. Studies have shown that mitigation measures can significantly reduce wildlife–vehicle collisions. For example, the installation of wildlife crossings and fencing along the Trans-Canada Highway led to an 80% reduction in wildlife–vehicle collisions at Banff National Park [67]. Similarly, an animal-detection system on a section of State Route 206 in Arizona resulted in a 97% reduction in collisions with elk [68]. These findings demonstrate that incorporating road biodiversity-conscious design elements, such as wildlife crossings or habitat preservation, can potentially reduce collisions between vehicles and animals. In turn, this can lead to lower vehicle maintenance costs, as they are less likely to be involved in accidents caused by wildlife crossings. Assuming a conservative (if compared to those in [67,68]) reduction rate of 40% and an estimated 20 collisions per year for the total fleet (a prudent estimate given that such incidents are often underreported), the reduced collision frequency is calculated using Equation (1):

$$C_f = E * (1 - R_r) \tag{1}$$

where:

C_f is the new collision frequency per year (unit)
E is the number of events (unit)
R_r is the reduction rate (%)

Equation (1) calculates overall savings of EUR 16,800 for a fleet of 20 vehicles based on the average collision cost of EUR 1400 per vehicle [65], consistent with other sources [69]. According to Table 4 [64], this reduction translates to yearly general maintenance costs per vehicle decreasing from EUR 19,250 to EUR 18,410. Additionally, further savings can be achieved by reducing accidents involving wildlife, which helps public transport operators avoid legal liabilities and associated costs. Collisions with animals may lead to legal actions, compensation claims, and other expenses that can be mitigated through ecology-conscious operations. Similarly, the infrastructure itself benefits: wildlife crossings and other road ecology measures help prevent damage to transportation infrastructure caused by collisions with large animals, thereby reducing the need for costly repairs and maintenance.

According to Bil et al. [62], following a collision with wildlife, drivers' immediate second concern after calling the police is to contact their insurance company, a fact which highlights the significance of these costs. A decrease in wildlife-related accidents can lead to lower insurance premiums for public transport operators; in other words, insurance companies consider risk factors when determining premiums, and fewer wildlife collisions can contribute to a safer (and less expensive) operating environment for transit managers. Online simulations of insurance premiums for buses operating in urban areas in central Italy show varying rates. However, reducing just one accident per year can result in savings of around EUR 1000 per bus, decreasing from EUR 10,077 (with one accident per year) to EUR 8918 (with zero accidents), according to rates provided by one of the major insurance companies in Italy [66]. This means that if the overall basic insurance premium for a fleet of 20 buses was EUR 201,540, applying the same reduced frequency from Equation (1) would lower the new insurance premium to EUR 187,632, resulting in an average savings of approximately EUR 696 per bus.

Additional areas of savings to consider are those associated with strategic operations that consider wildlife habitats when planning and designing the services. Once implemented, they can lead to smoother traffic flow and reduced congestion. This can result in improved fuel efficiency for public transport vehicles and increased overall operational savings. Also in this case, assuming very prudent values and transferring experience from other fields of transportation planning and management, it is possible to quantify the potential fuel savings. This can be achieved by estimating the improved average fuel cost and the savings per vehicle via Equations (2) and (3), respectively, i.e.,

$$Fc_i = Fc_c * \left(1 - \frac{P}{100}\right) \quad (2)$$

$$S_v = Fc_i - Fc_c \quad (3)$$

where:

Fc_i is the improved average fuel cost (EUR)
Fc_c is the current average fuel cost (EUR)
P is the expected improvement in fuel efficiency (%)
S_v are the savings per vehicle (EUR)

In the literature, numerous sources provide different values for P, depending on the type of traffic management. For example, eco-driving has been shown to reduce hard-braking events by 7% and collisions by 4% in a test case in Canada [70]. Similarly, the enforcement of low speed limits, combined with cruise control, can save around 5% to 7% in fuel use [71,72]. Conversely, urban environments with frequent intersections that impose changes in vehicular flow might lead to higher fuel consumption [73]. It is challenging, therefore, to assign a specific value to P for reduced speeds to safeguard fauna in urban environments due to the lack of specific studies. Based on the lessons learned from the aforementioned literature, a very conservative value of P as 1% can again be assumed. Considering an average annual fuel cost of around EUR 3500 per vehicle [74], the savings per vehicle—according to Equations (2) and (3)—are approximately EUR 35.

4.1.2. Potential Revenues by Increasing the Attractiveness of the Service

There is also potential for increasing the attractiveness of the service and generating additional revenues. A commitment to environmental safeguards can enhance a company's brand reputation, making it more attractive to investors, partners, and customers. For a transit company, such a commitment could lead to increased patronage. While specific figures on the propensity to ride from companies committed to environmental issues are hard to find, a holistic approach combining policy, infrastructure, and public awareness is essential to increase bus patronage while addressing environmental challenges. In this, investments in innovation, particularly in greener technologies for the bus sector, can attract more passengers and generate more revenue. A study in the United Kingdom indicates that for each GBP 1 spent on innovating transit operations, an economic return of GBP 4.4 to 8 can be expected [75,76]. Public transport services that demonstrate environmental consciousness and a commitment to wildlife preservation are likely to enhance their reputation similarly. This positive image may attract more passengers, leading to increased ridership and revenue. This is also consistent with customer surveys, where respondents are more likely to purchase goods and services from companies engaged in innovation and inclusivity, with a likelihood of up to 59% [77].

Accordingly, by assuming: (i) a large transit company budget for marketing of around EUR 650,000 per year (as in the case of one of the largest transit company in Italy, operating around 2000 vehicles [74]); (ii) a prudent 5% of consumers who are more likely to travel using companies that are committed to sustainability and innovation, and (iii) an average increase in sales per customer estimated to be about 5%, a return on investment generated by a commitment to achieve more transportation ecology friendly operations for a given company can be estimated by Equation (4):

$$ROI_p = \frac{C_{dc} * M_a * S_a}{100} \tag{4}$$

where:

ROI_p is the potential return on investment a transit company can expect to see as a result of achieving more transportation ecology consciousness (EUR)
C_{dc} is the number of conscious consumers (%)
M_a is the average marketing budget (EUR)
S_a is the average increase in sales per customer (%)

According to all of the above, the potential increase in marketing ROI_p for a transit company that achieves 5% more sales per customer would be about EUR 162,500, corresponding to around EUR 81 per vehicle. Additional returns can be associated with public transport systems operating in ecologically sensitive areas; by preserving natural habitats and promoting eco-friendly transportation options, they can attract more environmentally conscious tourists, thus boosting local economies. Thus, it is also to stress again that incorporating road ecology-conscious practices can ultimately contribute to the long-term sustainability of public transport operations. Although upfront investments may be required, avoiding future costs related to accidents, legal issues, and environmental damage can lead to significant savings over time.

5. Discussion around a Prospective Road Map for Transportation Infrastructure and Operations with Transportation Ecology in Mind

Given the findings above, economic potential is certainly a driver for including Transportation Ecology in transit planning and management. The incorporation of scientific knowledge into Transportation Ecology is crucial for complying with regulations that necessitate intricate assessments such as SEA or EIA, or any infrastructural or operational assessment in the field of transport, and more specifically transit. This integration facilitates the development of foundational knowledge beneficial to transit planners and designers, guiding regulatory decisions that align with the conservation requirements of wildlife and plants coherently with the Habitat Directive, which underscores the importance of preserving human activities in harmony with nature. Such an approach could mitigate the need for extensive plan revisions should disturbances to habitats or species be identified during initial or preliminary assessments. Additionally, integrating this knowledge into professional practice could serve as a valuable asset in decision-making and project development for transit. Establishing databases as a result could further aid in monitoring and controlling impacts post-implementation. The literature proposes various methodologies to assess the effectiveness of consolidated Road Ecology measures [38,53,70,78,79], employing similar multi-step approaches, but coherent with the Road Ecology vision, they focus solely on the infrastructure. These methods emphasize the importance of acquiring detailed knowledge about habitat characteristics, wildlife behavior, flora attributes, and both natural and human-induced environmental conditions (such as local weather patterns and land use dynamics). Additionally, it is recognized that assessment phases should extend beyond simple comparisons before and after infrastructure and services implementation, influencing the design of all phases within the decision-making process [80], as synthesized in Table 5, and where adding transit operations assessment increases the level of complexity.

Table 5. Decision-making process (adapted from [78,80]).

Phases	Intervention Areas	Issues	Regulations	Tools
Scoping	Transport policies	Transportation modes and operations (especially where multimodal supply is missing; conflicts analysis with the areas to safeguard	Strategic Environmental Assessment	Transit plans, landscape plans, any safeguard plan, also including Natura 2000 network requirements
	Identification of corridors	Traffic counting and quantification of conflicts with the local fauna		
Planning	Route and operations identification	Evaluation of planning variants, preliminary study of the mitigation measures (e.g., corridors and survey of the habitat's main features		Surveys and counting of fauna; preliminary study on migration effects, economic analyses
Design	Route and operations design	Location and design of mitigation measures	Environmental Impact Assessment Building/Operations permits	Monitoring plan; Focus on mitigation effects; plans updated versions including mitigation measures; ex ante monitoring/specific habitat safeguard plans associated with the building phases
Construction	Prevention of wildlife in the building sites; operations meeting the habitat requirements	Ecological supervision	Infrastructure and operational plans	Monitoring during the building phases
Operations	Operations/Infrastructure-generated impacts assessment and evaluation of maintenance impacts on fauna; mitigation measures effectiveness (including roadkill)		Business plan Management plan	Operation and maintenance monitoring; ex post evaluation

Moreover, this integration could spur the development of innovative landscaping and streetscaping approaches, advancing research into the impacts of infrastructure networks at the landscape or regional scale (a field still in its early stages [38]), largely due to challenges in gathering comprehensive data. Aligned with the above points, specific analytical fields must be developed. These include studying impact zones and compensation areas, determining the selection and placement of mitigation measures, and establishing procedures for monitoring and control plans. Similar to conventional evaluation processes in transportation studies, clear objectives must be defined in line with the conservation regulatory requirements outlined in Section 3. The evaluation methodology should involve using multi-scope indicators, constructing scenarios to assess various time horizons for effects and setting up monitoring networks and techniques as part of a comprehensive monitoring plan, all tailored to meet specific goals and scenarios [80].

Coherently with all of the above, a roadmap is suggested (Figure 3) in which the calculation of monetary benefits and their subsequent achievement, presented in Section 4, can represent an added value in the adoption of Transportation Ecology criteria in the process. In this scheme, a fundamental requirement involves clearly defining zones for analysis, such as impact areas and sites requiring mitigation, determining the specific locations for mitigation measures, and establishing protocols for monitoring and supervision. Similar to how assessments are conducted in transportation planning, this process involves setting explicit goals that adhere to conservation standards, particularly those outlined in environmental impact assessments and the specifics of potentially affected Natura 2000 sites. The evaluation process employs adaptable metrics, anticipates scenarios over different timeframes to gauge impacts and concludes by implementing a monitoring framework and appropriate techniques that align with the monitoring objectives and scenarios. The indicators must be diverse, encompassing multiple assessment categories and capable of capturing various impacts, facilitated by innovative tools. These tools include advancements in digital mapping for data retrieval, community-involved collaborative mapping, and participatory processes aimed at protecting territories. As these methods become integral to evaluating new transportation projects; their importance will continue to grow with the maturation of knowledge and expertise in this field, shaping decision-making processes accordingly.

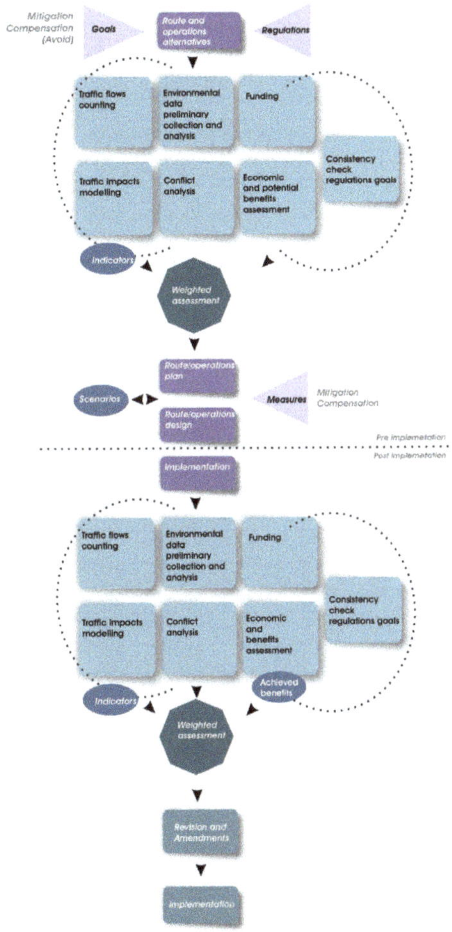

Figure 3. The Transportation Ecology assessment roadmap.

Evaluation should not only focus on environmental aspects but also on economic factors, assessing how human behaviors change with the introduction of measures to mitigate or compensate for impacts. Additionally, it should consider the potential benefits of integrating Transportation Ecology into operations management. This discussion also underscores the financial implications involved. While these measures may not always be expensive (as demonstrated by [1,55]), their implementation and the systems for monitoring and analyzing data should rely on consistent, sustainable funding that aligns with the evolving needs of natural habitats.

5.1. Adapting the Roadmap to Diverse Transportation Contexts

The proposed roadmap is general, but its policy strategies can be applied to different situations and easily integrate Transportation Ecology into transit planning and management, specifically in urban environments (Table 6), starting from the approach of Table 5, i.e., phases and intervention areas.

Table 6 provides a structured approach to integrating Transportation Ecology strategies into urban transportation systems, focusing on rail, bus, and marine networks, with each phase of implementation (from scoping to operations) being detailed to ensure that ecological considerations are embedded throughout the operations lifecycle. More specifically, urban transportation projects should begin with a comprehensive scoping phase, utilizing tools like Strategic Environmental Assessments (SEA) and Geographic Information Systems (GIS), which help map existing urban green spaces and wildlife corridors, ensuring that initial planning stages are informed by robust ecological data. The compliance with urban planning policies, such as Natura 2000, is to underscore the commitment to biodiversity conservation. The Route and Operations Identification phase involves meticulous planning to minimize ecological disruption. Wildlife surveys and the design of green corridors and overpasses are essential strategies to maintain urban biodiversity, when designing surface transit supply. For marine transportation, planning shipping routes to avoid sensitive areas is crucial to protect marine ecosystems. The design phase integrates green infrastructure, such as green roofs and vertical gardens, into urban transit systems. This phase also involves developing detailed mitigation plans to address potential ecological impacts. For marine supply, eco-friendly ship designs are to be incorporated to reduce noise and physical disturbances to marine life. During the construction phase, noise and pollution control measures need to be implemented to minimize environmental impacts, as must ecological supervision to ensure that construction activities are monitored and adjusted as necessary to protect urban wildlife. For marine projects, using silt curtains prevents sediment dispersion, protecting aquatic habitats. When operating, transit supply long-term monitoring of urban ecosystems affected by transportation projects are necessary. Regular updates to management plans ensure adaptive management based on ongoing monitoring data. This phase also includes assessing the impact of ferry operations on marine life, making necessary adjustments to minimize ecological footprints. All the interventions mentioned above (GIS, urban planning policies, regular wildlife surveys, and green infrastructure) are fundamental to implementing these strategies effectively and complement each other, as is adherence to regulations, such as Natura 2000 and marine conservation laws with the final goal of ensuring that urban transportation projects are aligned with broader environmental protection goals.

Table 6. Roadmap policy strategies implementation in urban environments.

Phase	Intervention Area	Policy Strategy	Implementation Examples
Scoping	Transport Policies	Conduct SEA, use GIS for habitat mapping, comply with urban planning policies	Assess impacts on city parks and urban wildlife corridors—e.g., mapping impacts on Berlin tram line expansions on urban parks [81].
	Identification of Corridors	Map green spaces, conduct traffic and wildlife surveys	Identify and map urban wildlife corridors, such as Central Park bird migration paths in New York [82].
Route and Operations Identification	Route and Operations planning	Design green corridors/overpasses, conduct wildlife surveys, plan mitigation measures	Implement wildlife crossings in high-density areas—e.g., wildlife safeguard in Melbourne's urban fabric [83].
		Plan shipping routes to avoid sensitive marine areas	Ensure ferry routes minimize disturbances to marine life—e.g., rerouting ferries in San Francisco Bay to avoid sensitive ecosystems [84].
Design	Route and Operations Design	Integrate green infrastructure, develop detailed mitigation plans	Use green roofs and vertical gardens on transit stations—e.g., green roofs on Utrecht bus stops [85].
		Integrate eco-friendly ship designs	Use quieter propellers and hull designs for ferries—e.g., implementing quiet ship technology in the ferry system of the Vancouver area [86].
Construction	Infrastructure and Operations Plans	Implement noise and pollution control measures, ensure ecological supervision	Limiting impacts—e.g., avoid measures while building the Paris Metro Line 17 [87].
		Implement silt curtains for marine projects	Use silt curtains and other mitigation measures to prevent sediment dispersion during port construction—e.g., several cases worldwide [88].
Operations	Operations/Infrastructure-generated impacts assessment	Monitor long-term impacts on urban ecosystems, update management plans regularly	Ongoing monitoring of urban biodiversity and green space management—e.g., checking birds' vocalization after noise events due to air and surface traffic in San Carlos de Bariloche, Argentina [89].
		Monitor ferry operations' impact on marine life	Assess and mitigate vessels' impacts on marine ecosystems—e.g., continuous monitoring of anthropogenic sounds in 36 locations in the Baltic Sea to ensure minimal impact on marine biodiversity [90].
General Tools		GIS, urban planning policies, regular wildlife surveys, green infrastructure, noise reduction technology	Use GIS to map and protect urban green spaces and wildlife—e.g., GIS mapping of Los Angeles Metropolitan Area urban wildlife corridors [91].
Regulations		Urban planning policies, Natura 2000 requirements, marine conservation laws	Ensure compliance with urban and marine conservation regulations e.g., adhering to Natura 2000 in urban development projects in European cities [92].

The case studies reported as implementation examples demonstrate how various ecological strategies can be effectively integrated into urban transportation planning and management, even when transit is not a priority in the land use management. Moreover, they also evidence the interdisciplinary nature of Transportation Ecology since they originated in other study fields. Transportation Ecology needs different disciplinary perspectives—from economic to scientific, regulatory, and ecological—because it addresses the complex interplay between transportation systems and the urban environment, which involves multiple facets that cannot be understood or managed through a single disciplinary lens. For example, economic insights drive the inclusion of transportation ecology in transit planning by underscoring its potential benefits, making it a key motivator for stakeholders, in the line of the urban ecology studies [93,94]. At the same time, the scientific knowledge, crucial for regulatory compliance, ensures that assessments like the SEA or the EIA are thorough, providing a strong foundation for transit planning that aligns with wildlife and habitat conservation directives [95]. This integration helps mitigate extensive plan revisions by identifying habitat disturbances early on, thus facilitating smoother project development and decision-making processes. Furthermore, the case studies underscore the importance of innovative methodologies and tools, such as digital mapping [96] and community-involved processes [97], to gather comprehensive data and monitor impacts effectively. This holistic approach, combining detailed habitat studies, economic analyses, and strategic regulatory compliance, fosters a robust transportation ecology framework that balances infrastructure development (thus differentiating from the Road Ecology approach) with environmental sustainability, ultimately shaping more informed and responsible decision-making in transportation projects.

5.2. *The Role of the Stakeholders: From Management to Awareness to Education*

The policy implications reported above, even if qualitatively corroborated by the scenario results to motivate transit operators to adopt Transportation Ecology in their practice, need acceptance and awareness. The benefits and savings described in Section 4 can be drivers in fostering acceptance among transit operators, but raising awareness and providing education on Transportation Ecology are critical steps in ensuring the successful integration of ecological principles into the transportation sector. This process requires the active involvement of key stakeholders, including transit managers, local administrators, and transit patrons. Each group plays a distinct role in fostering a culture of environmental responsibility, and their combined efforts can significantly influence the sustainability of transportation systems.

Transit managers are at the forefront of implementing ecologically responsible practices within transportation systems. For them, education should begin with comprehensive training programs that cover the principles of Transportation Ecology. These programs should emphasize best practices for reducing environmental impacts, such as optimizing transit routes to lower emissions, incorporating green infrastructure into operations, and managing noise pollution effectively. Understanding these practices is crucial for transit managers, who must balance operational efficiency with environmental stewardship. However, to ensure that these principles are put into practice, it is essential to introduce ecological performance metrics as part of the management process. These metrics should be treated as key performance indicators (KPIs) that transit managers are held accountable for alongside traditional operational targets. By linking these KPIs to broader organizational goals, transit managers can better appreciate the importance of ecological sustainability in their decision-making processes. Furthermore, providing incentives for sustainable practices (such as recognition programs or bonuses tied to ecological performance) can motivate transit managers to prioritize environmental considerations in their daily operations.

Local administrators also have a pivotal role in promoting Transportation Ecology through public awareness campaigns and policy development. Within their communities, local administrators are well positioned to educate the public about the environmental impacts of transportation and the benefits of adopting more sustainable practices. Public awareness campaigns can be instrumental in this regard, using community workshops, informational brochures, and public service announcements to disseminate knowledge about ecological transportation.

Collaboration with educational institutions is another effective strategy for local administrators. By partnering with schools, universities, and research centers, they can help integrate the concepts of Transportation Ecology into educational curricula. This collaboration not only raises awareness among younger generations but also helps cultivate a new cadre of environmentally conscious professionals who will carry these principles into their future careers.

Eventually, transit patrons also have a role to play in the adoption of Transportation Ecology principles. Educating patrons about the environmental benefits of eco-friendly transit options is essential. Public campaigns can emphasize the positive impact of adopting Transportation Ecology, and engaging patrons in the feedback process can be vital for the continuous improvement of transit services in the "green" direction. By creating platforms for patrons to share their experiences and concerns regarding the ecological aspects of transit, operators can gather valuable insights and adjust their practices accordingly. The success of Transportation Ecology depends on collaboration between transit managers, local administrators, and patrons. Organizing joint workshops where these stakeholders can come together to discuss and learn about ecological transportation can be a powerful tool for fostering mutual understanding and cooperation. Transparent communication about the environmental impacts of transit operations and the steps being taken to mitigate these impacts is essential for maintaining stakeholder engagement.

By aligning their goals towards a shared vision of sustainable and ecologically responsible transportation, these stakeholders can collectively drive meaningful change. Through education and awareness, transit managers, local administrators, and patrons can become active participants in creating a more sustainable transportation system, ensuring that ecological considerations are at the heart of transportation planning and operations. All these above-mentioned managerial implications are synthesized in Table 7.

Table 7. Managerial implications.

Areas	Action
Corporate Education and Training	Transit managers must receive comprehensive education on the principles of Transportation Ecology, including best practices for reducing environmental impacts. This should encompass optimizing transit routes, integrating green infrastructure, and managing noise pollution effectively.
Performance Metrics	It is essential to introduce ecological performance metrics as key performance indicators (KPIs) alongside traditional operational targets. By linking these KPIs to broader organizational goals, transit managers can better understand the importance of ecological sustainability in their decision-making processes.
Incentives for Sustainability	Offering incentives such as recognition programs or bonuses tied to ecological performance can motivate transit managers to prioritize environmental considerations in their operations.
Quantifying Financial Benefits	Transit operators should focus on identifying and quantifying the financial benefits of integrating Transportation Ecology into their operations. This includes potential savings in maintenance, insurance, and other operational costs as well as avoiding expenses related to environmental damage.
Public Awareness and Education	Effective implementation of Transportation Ecology principles requires collaboration with local administrators and transit patrons. This involves public awareness campaigns, educational initiatives, and active engagement.

6. Concluding Remarks

Transportation Ecology is still in its early stages of development and will continue to flourish if able to incorporate contributions from an increasing number of academic disciplines, as just mentioned. This study identified three primary reasons for integrating Transportation Ecology into contemporary transit operation management practices. Firstly, there is a pressing need to mitigate the adverse impacts of anthropogenic mobility on urban ecosystems, particularly when high-capacity operations are involved. Secondly, the environmental regulatory landscape is becoming increasingly stringent, underscoring the necessity for transport operators to align their operations with the principles of Transportation Ecology. Thirdly, there are tangible financial benefits that can be achieved by adopting Transportation Ecology criteria in operational management procedures. As is the case with any exploratory study, there are numerous caveats to be considered. For example, this study is based on the existing literature on road ecology, which is primarily concerned with extra-urban environments and infrastructure supply rather than the operational aspects of vehicular traffic, including transit. A review of the literature reveals that the majority of case studies place greater emphasis on wildlife than flora. The dominant approach is empirical, whereby best practices are identified and applied. In contrast, vehicular traffic and infrastructure constitute a binomial within the urban environment, which is the actual realm of transportation ecology. The approach and vision remain consistent, yet the actors involved have shifted. In the context of transportation ecology, for instance, the responsibility for managing and regulating infrastructure, as well as planning and designing transportation systems, falls upon a diverse set of stakeholders, including drivers, passengers, transit operators, logistics operators, transit authorities, mobility planners, and land use planners. This suggests the need for further investigation into the reasons behind the apparent lack of interest in Transportation Ecology in urban environments and among the aforementioned urban actors, even in the context of strict regulations. There is also a need for more urban case studies to facilitate the creation of best practices and the transferability of excellence as well as for more discussions, such as those highlighted in [98] about considering the ethical dimensions of road construction and maintenance, particularly in terms of the environmental, social, and economic impacts that must be taken into account before a project can be deemed unethical. This highlights the necessity to integrate ecological liability into road construction and maintenance practices to ensure environmental protection.

As with any empirical study, the many limitations can pave the way for future research directions. A primary limitation concerns the qualitative scenario aimed at quantifying monetary benefits for transit operators. Although this is the first such assessment in the literature, it would certainly benefit from more data from consolidated operations and statistics (currently non-existent), which could provide more indicators and enhance evaluation by developing sensitive analyses that consider alternative maintenance, insurance, and fuel cost parameters to assess their consolidated impact on results. Likewise, the availability of consolidated data would enable the development of scenarios according to different time horizons, regulation and policy changes, and the inclusion of different modes of transportation. Although the focus of this paper is to explore how ecologically conscious considerations can affect public transport, the implications for private vehicles and commercial operations are equally important. For example, promoting the use of green spaces, installing green infrastructure like green roofs and walls, and improving urban tree canopies to safeguard flora and fauna can reduce the impervious surfaces dedicated to private cars and promote non-motorized modes. Similarly, the use of quieter engines to avoid disturbing fauna can improve the street soundscape and mitigate the health impacts associated with private vehicle use. For freight transport, the implementation of mitigation measures can optimize logistics and routing to minimize energy use and emissions through advanced logistics software functions, the reallocation of consolidation centers (often located in urban fringe areas where fauna is a significant presence), and efficient route planning. This consideration extends to the integration of advanced transportation planning methodologies like deep learning [99,100], where traffic flows could be managed by incorporating transportation ecology requirements, or algorithms for transit network design problems [101,102] adapted for the same purpose. Maritime and air traffic can also be significantly affected but are not central within Transport Ecology. For instance, urban waterways often intersect with sensitive ecosystems, and transportation ecology emphasizes their protection through measures such as regulating water traffic to prevent habitat disruption, installing pollution control systems, and conducting regular environmental impact assessments to ensure compliance with ecological standards and using quieter vessel engines [82,84,86]. Airports and air traffic, long criticized for being highly polluting, especially during take-off and landing operations [96], are adopting cleaner technologies for engines and fuels [103,104], which can be further expanded by enhancing green spaces within and around airport premises, using renewable energy sources [104], and improving waste management systems. This not only helps in conserving local flora and fauna but also contributes to a more pleasant environment for passengers and staff. Moreover, implementing noise abatement procedures, such as modifying flight paths to avoid residential areas and scheduling flights to minimize nighttime noise, can significantly reduce the impact on urban populations and wildlife.

To conclude, a clear takeaway is the need for a shift from Road Ecology to Transportation Ecology to better address the cumulative environmental impacts of transportation systems in urban areas. This study recommends that transit operators and policymakers integrate more and more ecological principles into the planning and management of transportation systems to achieve both environmental sustainability and economic benefits. This involves considering the full spectrum of ecological impacts, from infrastructure development to daily operations, and emphasizes the importance of long-term ecological monitoring and the inclusion of transportation in urban planning frameworks.

Funding: This research received no external funding.

Institutional Review Board Statement: Not applicable.

Informed Consent Statement: Not applicable.

Data Availability Statement: No new data were created.

Acknowledgments: The author wishes to thank all the members of the UITP Working Group on Transportation Ecology within the Research in Mobility Committee for inspiring this paper.

Conflicts of Interest: The author declares no conflicts of interest.

References

1. Forman, R.T.; Sperling, D.; Bissonette, J.A.; Clevenger, A.P.; Cutshall, C.D.; Dale, V.H.; Fahrig, L.; France, R.L.; Goldman, C.R.; Heanue, K. *Road Ecology—Science and Solutions*; Island Press: Washington, DC, USA, 2003; pp. 12–35.
2. Odum, E.P. *Fundamentals of Ecology*; W.B. Saunders: Philadelphia, PA, USA, 1971.
3. von Herrmann, F.W. (Ed.) *Martin Heidegger. Vorträge und Aufsätze (1936–1953)*; Klostermann: Frankfurt am Main, Germany, 2000; Volume 8.
4. Takeda, F. An approach to make regional developmental effect of road construction compatible with environmental conservation. *Stud. Reg. Sci.* **1971**, *2*, 37–45. [CrossRef]
5. Scheerer, H. *Die Auswirkungen des Fernstrassenbaus und des Fernverkehrs auf Die Natuerliche Umwelt*; Report N. 23; Deutscher Rat für Landespflege: Bonn, Germany, 1974.
6. Mathe, P. Ökologische Gesichtspunkte beim Ausbau und bei der Asphaltierung von Wegen und Strassen mit Verdichtung der Verkehrsnetze auch im landlichen Raum. *Forum Stadte-Hyg.* **1977**, *28*, 189–193.
7. Fielenbach, R. Strasse und Ökologie. *Str. Autob.* **1975**, *26*, 52–54.
8. Bellport, B.P. US Bureau of Reclamation -USBR specifications focus on environment. *Civ. Eng.* **1971**, *41*, 50–52.
9. Muskett, C.J.; Jones, M.P. The dispersal of lead, cadmium and nickel from motor vehicles and effects on roadside invertebrate macrofauna. *Environ. Pollut. A* **1980**, *23*, 231–242. [CrossRef]
10. Ofomata, G.E.K. Impact of road building, urbanisation and general infrastructural development on the Nigerian rainforest ecosystem. *Landsc. Plan.* **1981**, *8*, 21–29. [CrossRef]
11. Davis, J.B.; George, J.J. Invertebrates as indicators of urban and motorway discharges. *Sci. Total Environ.* **1987**, *59*, 291–302. [CrossRef] [PubMed]

12. Dyment, R. Pure waters project includes restoration. *Water Sew. Work.* **1974**, *121*, 52–53.
13. Bacsó, J.; Kis-Varga, M.; Kovács, P.; Kalinka, G. Investigation on the accumulation of lead and other metals in plants caused by motor-traffic and smelting. *J. Radioanal. Nucl. Chem.* **1984**, *81*, 59–65. [CrossRef]
14. Tam, N.F.Y.; Liu, W.K.; Wong, M.H.; Wong, Y.S. Heavy metal pollution in roadside urban parks and gardens in Hong Kong. *Sci. Total Environ.* **1987**, *59*, 325–328. [CrossRef]
15. Bischofsberger, W. Ueber den Einfluss der Tausalze auf Grund- und Oberflaechenwasser. *Str. Tiefbau* **1985**, *39*, 6–10.
16. Scott, N.E.; Davison, A.W. The distribution and ecology of coastal species on roadsides. *Vegetation* **1985**, *62*, 433–440. [CrossRef]
17. Morikawa, T. The influence of road dust on asthmatic children. *Jpn. J. Allergol.* **1985**, *34*, 297–304.
18. Healy, M.A.; Aslam, M. The distribution of lead in a roadside environment and its consequences for health. *Public Health* **1980**, *94*, 78–88. [CrossRef]
19. van der Ree, R.; Jaeger, J.A.G.; van der Grift, E.A.; Clevenger, A.P. Effects of roads and traffic on wildlife populations and landscape function: Road ecology is moving towards larger scales. *Ecol. Soc.* **2011**, *16*, 48. [CrossRef]
20. Ellenberg, H.; Müller, K.; Stottele, T.; Walper, K.H. Straßen-Ökologie: Auswirkungen von Autobahnen und Straßen auf Ökosysteme deutscher Landschaften. *Okol. Straße* **1981**, *3*, 19–122.
21. Forman, R.T.T. Road ecology: A solution for the giant embracing us. *Landsc. Ecol.* **1998**, *13*, III–V. [CrossRef]
22. Harris, L.D.; Scheck, J. From implications to applications: The dispersal corridor principle applied to the conservation of biological diversity. In *Nature Conservation 2: The Role of Corridors*; Saunders, D.A., Hobbs, R.J., Eds.; Surrey Beatty and Sons: Chipping Norton, UK, 1991; pp. 189–220.
23. Coffin, A.W. From roadkill to road ecology: A review of the ecological effects of roads. *J. Transp. Geogr.* **2007**, *15*, 396–406. [CrossRef]
24. Forman, R.T.T.; Alexander, L.E. Roads and their major ecological effects. *Ann. Rev. Ecol. Syst.* **1998**, *29*, 207–232. [CrossRef]
25. Voller, J.; Harrison, S. (Eds.) Connectivity. In *Conservation Biology Principles for Forested Landscapes*; UBC Press: Vancouver, BC, Canada, 1998; pp. 76–97.
26. Parminter, J.V. Natural disturbance ecology. In *Conservation Biology Principles for Forested Landscapes*; Voller, J., Harrison, S., Eds.; UBC Press: Vancouver, BC, Canada, 1998; pp. 3–41.
27. Sousa, W.P. The Role of Disturbance in Natural Communities. *Ann. Rev. Ecol. Syst.* **1984**, *15*, 353–391. [CrossRef]
28. Hunt, A.; Dickens, H.J.; Whelan, R.J. Movement of mammals through tunnels under railway lines. *Aust. Zool.* **1987**, *24*, 89–93. [CrossRef]
29. Yanes, M.; Velasco, J.M.; Suárez, F. Permeability of roads and railways to vertebrates: The importance of culverts. *Biol. Conserv.* **1995**, *71*, 217–222. [CrossRef]
30. De Santo, R.S.; Smith, D.G. An introduction to issues of habitat fragmentation relative to transportation corridors with special reference to high-speed rail (HSR). *Environ. Manag.* **1993**, *17*, 111–114. [CrossRef]
31. Rodriguez, A.; Crema, G.; Delibes, M. Use of non-wildlife passages across a high speed railway by terrestrial vertebrates. *J. Appl. Ecol.* **1996**, *33*, 1527–1540. [CrossRef]
32. Kaczensky, P.; Knauer, F.; Krze, B.; Jonozovic, M.; Adamic, M.; Gossow, H. The impact of high speed, high volume traffic axes on brown bears in Slovenia. *Biol. Conserv.* **2003**, *111*, 191–204. [CrossRef]
33. Popp, J.N.; Boyle, S.P. Railway ecology: Underrepresented in science? *Basic Appl. Geol.* **2017**, *19*, 84–89. [CrossRef]
34. Barrientos, R.; Ascensão, F.; Beja, P.; Pereira, H.M.; Borda-de-Água, L. Railway ecology vs. road ecology: Similarities and differences. *Eur. J. Wildl. Res.* **2019**, *65*, 12. [CrossRef]
35. Borda-de-Água, L.; Barrientos, R.; Beja, P.; Pereira, H.M. *Railway Ecology*; Springer: Cham, Switzerland, 2017.
36. Karlson, M.; Mortberg, U.; Balfors, B. Road ecology in environmental impact assessment. *Eur. J. Wildl. Res.* **2014**, *48*, 10–19. [CrossRef]
37. Santos, G.; Behrendt, H.; Maconi, L.; Shirvani, T.; Teytelboym, A. Part I: Externalities and economic policies in road transport. *Res. Transp. Econ.* **2010**, *28*, 2–45. [CrossRef]
38. Iuell, B. *Wildlife and Traffic: A European Handbook for Identifying Conflicts and Designing Solutions*; KNNV Publishers: BM Zeist, The Netherlands, 2003.
39. Togridou, A. Rail: A test bed for biodiversity innovations. *TCR South—Rev.* **2023**, *26*, 60–63.
40. Goldfarb, B. City of Glass.31.10.2023. Available online: https://www.biographic.com/city-of-glass/ (accessed on 29 March 2024).
41. Transport Infrastructure Ireland. *Design and Delivery of Soft Landscape Treatments in Urban Transport Environments*; Technical Report; TIE: Dublin, Ireland, 2003; pp. 111–113.
42. Naumann, S.; Davis, M.; Kaphengst, T.; Pieterse, M.; Rayment, M. *Design, Implementation and Cost Elements of Green Infrastructure Projects*; Ecologic Institute and GHK Consulting: Berlin, Germany, 2011; pp. 63–109.
43. Biodiversity Andiinfrastructure Synergies and Opportunities for European Transport Networks—BISON. BISON Policy Brief. Available online: https://bison-transport.eu/2023/10/16/bison-policy-briefs/ (accessed on 29 March 2024).
44. Wilansky, J.; Jaeger, J.A.G. Predicting the effectiveness of wildlife fencing along roads using an individual-based model: How do fence-following distances influence the fence-end effect? *Ecol. Model.* **2024**, *495*, 110784. [CrossRef]
45. Wall, A.E.; Biffi, D.; Ackel, A.; Moody, R.W.; Stevens, T.K.; Williams, D.A. Small towns limit dispersal and reduce genetic diversity in populations of Texas horned lizards. *Ecol. Evol.* **2024**, *14*, e70112. [CrossRef] [PubMed]

46. Červinka, J.; Riegert, J.; Grill, S.; Šálek, M. Large-scale evaluation of carnivore road mortality: The effect of landscape and local scale characteristics. *Mammal Res.* **2015**, *60*, 233–243. [CrossRef]
47. Finch, D.; Schofield, H.; Mathews, F. Traffic noise playback reduces the activity and feeding behaviour of free-living bats. *Environ. Pollut.* **2020**, *263*, 114405. [CrossRef]
48. Ortiz-Urbina, E.; Diaz-Balteiro, L.; Iglesias-Merchan, C. Influence of anthropogenic noise for predicting cinereous vulture nest distribution. *Sustainability* **2020**, *12*, 503. [CrossRef]
49. Madadi, H.; Moradi, H.; Soffianian, A.; Salmanmahiny, A.; Senn, J.; Geneletti, D. Degradation of natural habitats by roads: Comparing land-take and noise effect zone. *Environ. Impact Assess. Rev.* **2017**, *65*, 147–155. [CrossRef]
50. Ghadirian, O.; Moradi, H.; Madadi, H.; Lotfi, A.; Senn, J. Identifying noise disturbance by roads on wildlife: A case study in central Iran. *SN Appl. Sci.* **2019**, *1*, 808. [CrossRef]
51. Francis, C.D.; Barber, J.R. A framework for understanding noise impacts on wildlife: An urgent conservation priority. *Front. Ecol. Environ.* **2013**, *11*, 305–313. [CrossRef] [PubMed]
52. Blackwell, B.F.; Devault, T.L.; Seamans, T.W. Understanding and Mitigating the Negative Effects of Road Lighting on Ecosystems. In *Handbook of Road Ecology*; van der Ree, R., Smith, D.J., Grilo, C., Eds.; Wiley: Chichester, UK, 2015; pp. 143–150.
53. Roedenbeck, I.A.; Fahrig, L.; Findlay, C.S.; Houlahan, J.E.; Jaeger, J.A.; Klar, N.; Kramer-Schadt, S.; van der Grift, E.A. The Rauischholzhausen agenda for road ecology. *Ecol. Soc.* **2007**, *12*, 11. [CrossRef]
54. Forman, R.T.T.; Deblinger, R.D. The ecological road-effect zone of a Massachusetts (USA) suburban highway. *Conserv. Biol.* **2000**, *14*, 36–46. [CrossRef]
55. Ciabò, S.; Fabrizio, M.; Ricci, S.; Merten, A. *Strade—Life*; Regione Umbria: Perugia, Italy, 2015.
56. Selva, N.; Kreft, S.; Kati, V.; Schluck, M.; Jonsson, B.G.; Mihok, B.; Okarma, H. Roadless and Low-Traffic Areas as Conservation Targets in Europe. *Environ. Manag.* **2011**, *48*, 865–877. [CrossRef] [PubMed]
57. Loss, S.R.; Will, T.; Loss, S.S.; Marra, P.P. Bird–building collisions in the United States: Estimates of annual mortality and species vulnerability. *Condor: Ornithol. Appl.* **2014**, *116*, 8–23. [CrossRef]
58. Calvario, E.; Sebasti, S.; Copiz, R.; Salomone, F.; Brunelli, M.; Tallone, G.; Blasi, C. *Habitat e Specie di Interesse Comunitario nel Lazio*; Regione Lazio—ARP: Rome, Italy, 2009.
59. Bissonette, J.A.; Kassar, C.A.; Cook, L.J. Assessment of Costs Associated with Deer–Vehicle Collisions: Human Death and Injury, Vehicle Damage, and Deer Loss. *Hum.-Wildl. Interact.* **2008**, *2*, 9.
60. Huijser, M.P.; Duffield, J.W.; Clevenger, A.P.; Ament, R.J.; McGowen, P.T. Cost–benefit analyses of mitigation measures aimed at reducing collisions with large ungulates in the United States and Canada; a decision support tool. *Ecol. Soc.* **2009**, *14*, 15. [CrossRef]
61. Conover, M.R.; Pitt, W.C.; Kessler, K.K.; DuBow, T.J.; Sanborn, W.A. Review of human injuries, illnesses, and economic losses caused by wildlife in the United States. *Wildl. Soc. Bull.* **1995**, *23*, 407–414.
62. Bíl, M.; Grilo, C.; Kubeček, J.; Sedoník, J.; Andrášik, R.; Cícha, V.; Favilli, F.; Stauder, J.; Schwingshackl, F.; Michael, K.; et al. Wildlife Vehicle Collisions: Road Ecology, Monitoring and Mitigation, Citizen Science, Pedagogical and Socioeconomic Aspects. Report ERASMUS+ "EnVeROS" Report, Intellectual Output 1. 2019. Available online: http://www.enveros.eu/ (accessed on 28 June 2024).
63. Automobil Club d'Italia—ACI, 2022. Distribuzione Parco Veicoli per Anno, Comune Capoluogo e Categoria—Open Parco Veicoli Database. 2022. Available online: https://opv.aci.it/WEBDMCircolante/ (accessed on 28 June 2024).
64. Baccelli, O. Scenari e Prospettive Dell'elettrificazione del Trasporto Pubblico su Strada. Un'innovativa Analisi di Benchmark: Il TCRO—Total Cost and Revenues of Ownership. Metodologia e Risultati per l'Italia, Presentation at Bocconi University Milan. 16 November 2021. Available online: https://www.enelfoundation.org/content/dam/enel-foundation/topics/2021/11/211116%20GREEN%20per%20evento%20su%20TCRO%20caso%20Italia_per%20evento%20v7def.pdf (accessed on 29 June 2024).
65. Seiler, A. The Toll of the Automobile—Wildlife and Roads in Sweden. Ph.D. Thesis, Department of Conservation Biology, Swedish University of Agricutural Sciences, Uppsala, Sweden, 2003.
66. Intesa SanPaolo Preventivo Assicurazione Autobus, Autocarri, Caravan, Furgoni, Natanti, Mezzi Agricoli 2024. Available online: https://www.intesasanpaoloassicura.com/preventivatore-altriveicoli (accessed on 29 June 2024).
67. Clevenger, A.P.; Chruszcz, B.; Gunson, K.E. Highway mitigation fencing reduces wildlife-vehicle collisions. *Wildl. Soc. Bull.* **2001**, *29*, 646–653.
68. Gagnon, J.W.; Dodd, N.L.; Sprague, S.C.; Ogren, K.S.; Loberger, C.D.; Schweinsburg, R.E. Animal-activated highway crosswalk: Long-term impact on elk-vehicle collisions, vehicle speeds, and motorist braking response. *Hum. Dimens. Wildl.* **2019**, *24*, 132–147. [CrossRef]
69. Guccione, M.; Gori, M.; Bajo, N. *Tutela della Connettività Ecologica del Territorio e Infrastrutture Lineari*; ISPRA: Rome, Italy, 2008.
70. Robertson, R.D.; Delavary, M.; Lyon, C.; Vanlaar, W. Effect of eco-driving on commercial motor vehicle driver collision risk. *J. Saf. Res.* **2024**, *89*, 190–196. [CrossRef] [PubMed]
71. Litman, T. *Traffic Calming: Benefits, Costs and Equity Impacts*; Victoria Transport Policy Institute: Victoria, BC, Canada, 1999.
72. Haworth, N.; Symmons, M. Driving to Reduce Fuel Consumption and Improve Road Safety. In Proceedings of the Road Safety Research, Policing and Education Conference 2001, Melbourne, VIC, Australia, 18–20 November 2001.
73. Romero, C.A.; Correa, P.; Ariza Echeverri, E.A.; Vergara, D. Strategies for Reducing Automobile Fuel Consumption. *Appl. Sci.* **2024**, *14*, 910. [CrossRef]

74. ATAC Bilancio di Esercizio al. 31 December 2022. Available online: https://trasparenza.atac.roma.it/moduli/downloadFile.php?file=oggetto_allegati/232111122320O__OBilancio+ATAC+al+31.12.2022.pdf (accessed on 29 June 2024).
75. KPMG Greener-Journeys-Maximising-the-Benefits-of-Local-Bus-Services. 2023. Available online: https://greener-vision.com/wp-content/uploads/2020/07/Greener-Journeys-Maximising-the-benefits-of-local-bus-services.pdf (accessed on 29 June 2024).
76. Sustainable Bus. UK Funding Streams for Zero Emission Buses, Well Explained. Available online: https://www.sustainable-bus.com/news/uk-funding-zero-emission-buses-well-explained/ (accessed on 29 June 2024).
77. Piaggio, V. The Global Inclusion Imperative. Available online: https://www.kantar.com/inspiration/brands/the-global-inclusion-imperative (accessed on 22 December 2023).
78. Hlaváč, V.; Anděl, P.; Matoušová, J.; Dostál, I.; Strnad, M. *Wildlife and Traffic in the Carpathians. Guidelines How to Minimize Impact of Transport Infrastructure Development on Nature in the Carpathian Countries. Danube Transnational Programme TRANSGREEN Project*; The State Nature Conservancy of the Slovak Republic: Banská Bystrica, Slovakia, 2019.
79. van der Grift, E.A.; van der Ree, R.; Fahrig, L.; Findlay, S.; Houlahan, J.; Jaeger, J.A.G.; Klar, N.; Madriñan, L.F. Evaluating the effectiveness of road mitigation measures. *Biodivers. Conserv.* **2013**, *22*, 425–448. [CrossRef]
80. Corazza, M.V. Ecologia delle Infrastrutture Viarie a tutela dell'ambiente: Concetti e criteri per una ricerca allargata alle discipline trasportistiche. *Ing. Ferrov.* **2022**, *10*, 787–808.
81. Mihai, F.C.; Schneider, P.; Eva, M. Engineering and green infrastructure in mitigating emerging urban environmental threats. In *Handbook of Ecological and Ecosystem Engineering*; Narasimha, M., Prasad, V., Eds.; Wiley: Chichester, UK, 2021; pp. 95–121.
82. Herrera, D.J.; Fidino, M.; Luther, D.; Mullinax, J.M.; Gallo, T. Historical park planning is associated with modern-day patterns of bird diversity in cities. *Landsc. Urban Plan.* **2024**, *249*, 105132. [CrossRef]
83. Kirk, H.; Soanes, K.; Amati, M.; Bekessy, S.; Harrison, L.; Parris, K.; Ramalho, C.; Van de Ree, R.; Threlfall, C. Ecological connectivity as a planning tool for the conservation of wildlife in cities. *MethodsX* **2023**, *10*, 101989. [CrossRef] [PubMed]
84. Smythe, T.; McCann, J.A.; Fox, C. Spatial Planning for busy waterways: A case study of innovative waterways management in the San Francisco Bay region. In *Case Studies of Marine Spatial Planning Report Series*; McCann, J., Ed.; Coastal Resources Center and Rhode Island Sea Grant College Program, URI Graduate School of Oceanography: Narragansett, RI, USA, 2016; pp. 56–62.
85. Utrecht Municipality. Green-Roofed Bus Shelters in Utrecht. Available online: https://www.utrecht.nl/city-of-utrecht/green-roofed-bus-shelters-in-utrecht (accessed on 1 August 2024).
86. Granger, B.; Burnham, R.; Vagle, S. Soundscape additions from vessels related to transit speed, direction and manoeuvres. *Can. Acoust.* **2023**, *51*, 39–48.
87. Grand Paris Express. Protecting the Environment during the Build. Available online: https://view.genially.com/606574ffda2fe30ce2f43bf5 (accessed on 1 August 2024).
88. Erftemeijer, P.L.A.; Lewis, R.R.R., III. Environmental impacts of dredging on seagrasses: A review. *Mar. Poll. Bull.* **2006**, *52*, 1553–1572. [CrossRef] [PubMed]
89. Bahía, R.; Lambertucci, S.A.; Speziale, K.L. Anthropogenic city noise affects the vocalizations of key forest birds. *Biodivers. Conserv.* **2024**, *33*, 2405–2421. [CrossRef]
90. Nikolopoulos, A.; Sigray, P.; Andersson, M.; Carlström, J.; Lalander, E. 2016: BIAS Implementation. Plan—Monitoring and Assessment Guidance for Continuous Low Frequency Sound in the Baltic Sea, Technical Report. 2016. Available online: www.bias-project.eu (accessed on 1 August 2024).
91. Zellmer, A.J.; Goto, B.S. Urban wildlife corridors: Building bridges for wildlife and people. *Front. Sustain. Cities* **2022**, *4*, 954089. [CrossRef]
92. Ledoux, L.; Crooks, S.; Jordan, A.; Turner, R.K. Implementing EU biodiversity policy: UK experiences. *Land Use Policy* **2000**, *17*, 257–268. [CrossRef]
93. Rautjärvi, S.; MacGregor-Fors, I. Where economic parity meets ecology: Neither biodiversity nor ecosystem integrity values relate to wealth in the context of a medium-sized Finnish city. *Ambio* **2024**, *53*, 292–298. [CrossRef]
94. Dasgupta, P. *The Economics of Biodiversity*; Cambridge University Press: Cambridge, UK, 2024; pp. 107–113.
95. Li, T.; Tian, W.; Zhang, S.; Wang, S. Environmental regulation, high-quality economic development and ecological capital utilization. *Front. Environ. Sci.* **2024**, *12*, 1325289. [CrossRef]
96. Mudalige, U.; Carver, S. Unveiling Sri Lanka's Wilderness: GIS-Based Modelling of Wilderness Attributes. *Land* **2024**, *13*, 402. [CrossRef]
97. Andersson, E.; McPhearson, T.; Pickett, S.T.A. From urban ecology to urban enquiry: How to build cumulative and context-sensitive understandings. *Ambio* **2024**, *53*, 813–825. [CrossRef] [PubMed]
98. Moore, L.J.; Arietta, A.A.; Spencer, D.T.; Huijser, M.P.; Walder, B.L.; Abra, F.D. On the road without a map: Why we need an "Ethic of Road Ecology". *Front. Ecol. Evol.* **2021**, *9*, 774286. [CrossRef]
99. Owais, M. Deep Learning for Integrated Origin–Destination Estimation and Traffic Sensor Location Problems. *IEEE Trans. Intell. Transp. Syst.* **2024**, *25*, 6501–6513. [CrossRef]
100. Owais, M.; Moussa, G.S.; Hussain, K.F. Robust deep learning architecture for traffic flow estimation from a subset of link sensors. *J. Transp. Eng. A* **2020**, *146*, 04019055. [CrossRef]
101. Owais, M.; Osman, M.K. Complete hierarchical multi-objective genetic algorithm for transit network design problem. *Expert Syst. Appl.* **2018**, *114*, 143–154. [CrossRef]

102. Owais, M.; Osman, M.K.; Moussa, G.S. Multi-Objective Transit Route Network Design as Set Covering Problem. *IEEE Trans. Intell. Transp. Syst.* **2016**, *17*, 670–679. [CrossRef]
103. Corazza, M.V.; Di Mascio, P.; Esposito, G. Airports as Sensitive Areas to Mitigate Air Pollution: Evidence from a Case Study in Rome. *Environments* **2022**, *9*, 108. [CrossRef]
104. Di Mascio, P.; Corazza, M.V.; Rosa, N.R.; Moretti, L. Optimization of Aircraft Taxiing Strategies to Reduce the Impacts of Landing and Take-Off Cycle at Airports. *Sustainability* **2022**, *14*, 9692. [CrossRef]

sustainability

MDPI

Article

Influence of Iron Mining Waste Addition as a Sustainable Alternative on the Resilient and Physical Properties of Soils for Pavement Design

Daniel Corrêa Galhardo [1], Antônio Carlos Rodrigues Guimarães [1], Camila Antunes Martins [1], Murilo Miguel Narciso [1], Sergio Neves Monteiro [2] and Lisley Madeira Coelho [1,*]

1 Department of Fortification and Construction, Military Institute of Engineering-IME, Praça General Tibúrcio, 80, Urca, Rio de Janeiro 22290-270, Brazil; danielgalhardoo@hotmail.com (D.C.G.); guimaraes@ime.eb.br (A.C.R.G.); camilaantunes@ime.eb.br (C.A.M.); murilomn58@ime.eb.br (M.M.N.)

2 Department of Materials Science, Military Institute of Engineering-IME, Praça General Tibúrcio, 80, Urca, Rio de Janeiro 22290-270, Brazil; snevesmonteiro@gmail.com

* Correspondence: madeiralisley@gmail.com

Abstract: Mining activities generate large volumes of waste, posing environmental and economic challenges, particularly in Brazil's Quadrilátero Ferrífero region. This study assesses the potential reuse of iron ore waste from Samarco Mineração S.A. in road pavement layers by blending it with phyllite residual soil (PRS) and lateritic clayey soil (LCS). The addition of 50% waste to PRS led to substantial improvements, increasing the resilient modulus (RM) by up to 130% under medium stress and reducing expansibility from 6.1% to 1%, meeting Brazilian standards for sub-base applications. These enhancements make the PRS-waste blend a viable and sustainable option for reinforcing subgrade and sub-base layers. In contrast, the LCS with 20% waste showed moderate RM improvements under high-stress conditions, while higher waste contents reduced stiffness, indicating that higher dosages may adversely affect performance. This study highlights the potential of inert, non-hazardous mining waste as a safe and efficient solution for pavement applications, promoting the sustainable use of discarded materials.

Keywords: sustainable materials; iron mining waste; soils; resilient mechanical behavior

1. Introduction

In Brazil, highways play a critical role in transportation logistics, accounting for 61.1% of freight movement and 96% of passenger transportation [1]. This scenario highlights the importance of highways for the flow of agricultural and industrial production, often representing the only option for moving goods between production centers, consumers, ports, and airports.

For road pavement, it is essential that the materials used, including soils in their natural state, meet specific characteristics. Pavement layer materials must comply with minimum standards and possess properties that ensure resistance to local traffic loads and guarantee project durability [2]. In this context, soil stabilization emerges as a widely adopted solution, although the scarcity of standardized natural materials presents a significant challenge for road infrastructure. To overcome this limitation, soil improvement techniques, such as the application of lime and cement as stabilizing agents, have been traditionally employed [3]. Additionally, various alternative materials have been incorporated to enhance the properties of underperforming soils, enabling their effective use in road construction projects [4]. Examples include RAP-soil blends [5–9], soil–slag mixtures [10], soil–aggregate combinations [11], and soils stabilized with asphalt emulsion [12].

Brazil's vast territory and the predominance of road-based transportation require sustainable approaches to infrastructure development. The shortage of materials in certain regions, combined with high transportation costs and the need for waste reuse, has led professionals to explore alternative materials. This search aims to innovate in low-cost pavements, both in asphalt mixtures [13–15] and granular layers [6,16–18]. This trend is growing not only in Brazil but also in other countries [19–23], driven by the environmental and economic benefits associated with such practices.

Recent studies have highlighted the technical feasibility of using recycled aggregates from mining processes in infrastructure projects [24–26]. Apaza et al. [25] evaluated the incorporation of sandy waste composed of quartz minerals (88%) and hematite (9%) as aggregates in cold slurry seal mixtures for pavement surfaces. The results were satisfactory, showing that the iron ore waste presented no chemical, environmental, mineralogical, or physical restrictions that could limit their use as aggregate in slurry seals. Similarly, Friber et al. [24] emphasized that using mining waste to produce calcined aggregates for paving not only improves the mechanical performance of mixtures but also promotes sustainable practices by mitigating the environmental impacts of raw material extraction. Additionally, Silveira et al. [26] investigated the feasibility of recycling iron ore waste through cost-effective pavement techniques, demonstrating significant performance improvements in stabilized mixtures with the addition of dust-control treatments. In parallel, Guimarães et al. [27] evaluated the physical and mechanical behavior of asphalt mixtures incorporating iron ore residue as a fine aggregate replacement and found it viable for use in local roads without compromising pavement load capacity. These studies demonstrate the potential of iron ore waste to foster more sustainable road infrastructure practices, contributing to more efficient resource management and reduced environmental impact.

Given the growth of Brazil's mining sector and its reliance on road infrastructure, research on the use of waste in pavement structural layers becomes particularly relevant. In addition to reducing the environmental impact caused by waste dispoLCS, such practices contribute to preserving raw materials and improving the efficiency of road construction [28]. This study aims to assess the technical feasibility of using soil mixtures with sandy waste from iron ore processing for application in road pavement layers.

2. Materials and Methods

This section details the methodologies, procedures, and information required to perform the physical, chemical, and mechanical characterization tests of the materials used. The experimental work was carried out at the soil and science and materials laboratories of the Military Institute of Engineering (IME), Rio de Janeiro, Brazil, as well as at the Laboratory of the Mineral Analysis Coordination at the Mineral Technology Center (COAM/CETEM), Rio de Janeiro, Brazil.

2.1. Materials

Two types of soils were used, supplemented by iron mining waste (IMW). The phyllite residual soil was sourced from an abundant area within the perimeter of the Alegria Mine, located in Mariana-MG. The lateritic clay soil was collected from the western region of Maranhão, in deposits mapped for the expansion of the Carajás Railway. The iron ore waste used in this study was provided by Samarco Mineração S.A., situated in the municipality of Mariana-MG, within the Iron Quadrangle region, and was generated during the flotation stage of iron ore processing at the Germano Mine. Figure 1 presents the visual appearance of the soils and waste used in this study.

The proportions of waste in the mixtures with phyllite residual soil and lateritic clay soil were empirically determined, based on a comprehensive literature review conducted for this study. The selected values reflect those frequently reported in specialized literature.

The nomenclature of the mixtures is defined as follows:

- **PRS**: 100% phyllite residual soil
- **LCS**: 100% lateritic clay soil
- **LCS80/IMW20**: 80% lateritic clay soil and 20% iron mining waste
- **PRS80/IMW20**: 80% phyllite residual soil and 20% iron mining waste

- **PRS60/IMW40**: 60% lateritic clay soil and 40% iron mining waste
- **PRS50/IMW50**: 50% phyllite residual soil and 50% iron mining waste

Figure 1. Visual aspect of the soils and the waste used in this study. (**a**) PRS. (**b**) LCS. (**c**) IMW.

2.2. Physical Characterization.

For the physical characterization of the samples, the guidelines established by the Brazilian Association of Technical Standards (ABNT) were followed, including NBR 6457 [29], which addresses sample preparation, compaction, and characterization; NBR 6458 [30], which specifies the determination of apparent specific mass; NBR 6459 [31], which establishes the method for determining the liquid limit; NBR 7180 [32], which discusses the determination of the plastic limit; and NBR 7181 [33], which defines procedures for particle size analysis. The particle size analysis was performed through sieving for the coarse fraction (particles with a diameter > 0.074 mm) and sedimentation for the fine fraction (particles with a diameter < 0.074 mm), using sodium hexametaphosphate as a deflocculating agent, as described in NBR 7181 [33].

The compaction tests for all materials involved, including soils and their mixtures, were conducted according to the guidelines established by NBR 7182 [34]. The compaction of the samples for obtaining the compaction curves and subsequent densities was carried out manually using a Proctor cylinder, employing intermediate compaction energy and without reusing material. The samples had a diameter of 10 cm and a height of 12 cm.

To evaluate the expansion abilities of the PRS and LCS, as well as their mixtures with the IMW, a test was carried out in accordance with the NBR 9895 standard [35], which establishes guidelines for determining the soil expansion in the laboratory. For each type of soil and its mixtures, a sample was molded with the moisture contents obtained during the compaction test, under intermediate energy. After saturation, the samples were monitored to check for expansion over a period of 96 h.

2.3. Geotechnical Characterization

The geotechnical characterization of this study is directed towards the MCT (Miniature, Compacted, Tropical) methodology, recognized for its suitability in addressing compacted tropical soils. This approach will be contrasted with traditional classification methods, such as the SUCS (Soil Use and Classification System) and the guidelines from the TRB (Transportation Research Board) through the AASHTO (American Association of State Highway and Transportation Officials) system [36]. This will enable a more comprehensive and specific analysis of the properties of the soils under investigation.

2.3.1. MCT Methodology

The MCT methodology (DNIT 259 [37]) is based on mechanical and hydraulic properties obtained through the miniature compaction of test specimens. The main tests for determining these indices are the mini-MCV compaction test (DNIT 258 [38]) and the mass loss test by water immersion (DNER-ME 256 [39]).

The MCT classification separates soils into two main groups: (i) lateritic behavior soils, indicating the class "L"; (ii) and non-lateritic behavior soils, represented by class

"N", as described by Nogami and Villibor [40]. The testing procedure essentially involves compacting the soil fraction passing through a 2.0 mm sieve according to the number of blows (n) specified in the standard, interrupting this compaction procedure until the difference between the height of the compacted specimen with "4n" blows and the corresponding height for "n" blows is less than 2.0 mm. At the end of the test, families of compaction curves and curves called Mini-MCV will be obtained, from which data can be extracted for classification purposes. For the mass loss test by water immersion, the standard specifies that this parameter will be obtained after quantifying the mass detached from the previously compacted specimen when submerged in water for a minimum of 20 h, positioned horizontally with its top protruding 1.0 cm above the metallic compaction mold. Figure 2 illustrates the MCT equipment used for compacting miniature specimens and the immersion procedure of the samples immediately after compaction. The classification tests were conducted only on the fine fraction of the soil that passes through the 2 mm sieve, following the MCT methodology for each of the materials studied.

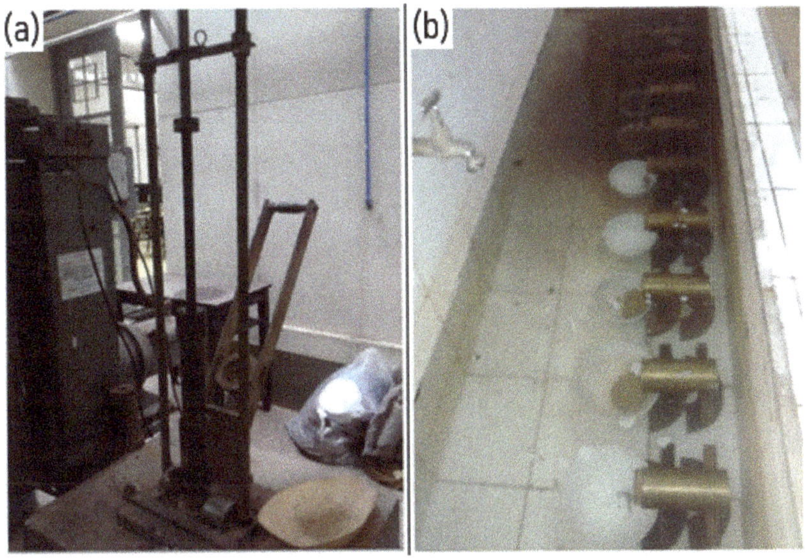

Figure 2. MCT methodology. (**a**) MCT equipment. (**b**) Immersion of the samples immediately after compaction.

2.3.2. MCT Expedito

This study also employed the quick method of tablets, as proposed by Nogami and Villibor [41]. This method consists of molding tablets using stainless steel rings with a diameter of 20 mm and a height of 5 mm, utilizing only the material passing through a 0.42 mm sieve. It allows for the determination of the MCT geotechnical classification groups, developed in a quick manner and adapted for tropical soils.

For the preparation of the tablets, approximately 100 g of soil was placed on a ground glass plate, where water was added, and the mixture was carefully spatulated until a homogeneous paste was formed. The tablets were air-dried for 24 h, although the original method suggests drying in an oven at 60 °C for 12 h. After drying, the diameter of the tablets was measured with a caliper, and then they were subjected to saturation in water for 2 h. Subsequently, penetration measurement was performed using a standard needle with a diameter of 1.3 mm and a total weight of 10 g, through a penetrometer. To ensure the representativeness of the results, three tablets were molded for each type of soil and sample under study. The test results allowed for the determination of two classification parameters: contraction (Ct) and penetration (consistency). The contraction value is correlated with the coefficient c′, used in the MCT classification, providing a clear indication of the behavior of the analyzed soils. The interpretation of the results was conducted based on the classification graph of the tablet method, presented in Figure 3.

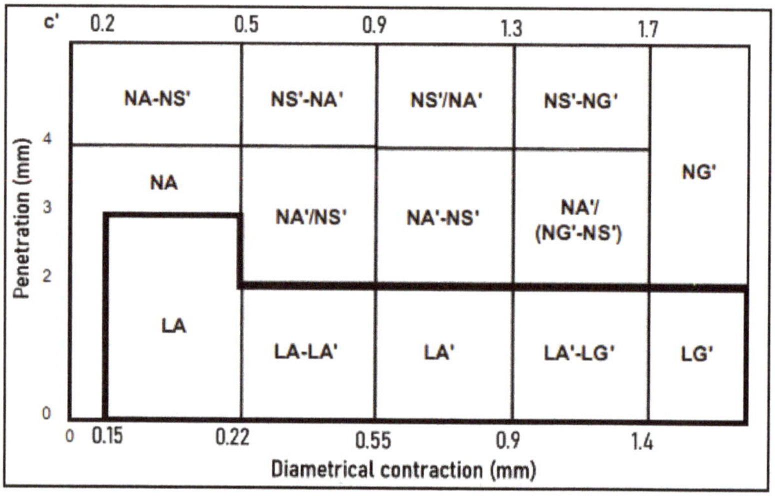

Figure 3. Grain size distribution curves of the aggregates.

2.3.3. Scanning Electron Microscopy (SEM) and Energy-Dispersive X-Ray Spectroscopy (EDX)

The samples were morphologically analyzed using a QUANTA FEG 250 microscope manufactured by FEI (Brno, Czech Republic). They were coated with gold and by utilizing a Leica ACE600 high-vacuum coating chamber, both of which belong to the electron microscopy laboratory of IME, Rio de Janeiro, Brazil. SEM analysis was performed with the following parameters: an electron beam power of 20 kV, a working distance ranging between 10.5 and 13 mm, and a spot size of 5. The magnifications used for specific samples were as follows: PRS had an image magnification of 2500×, LCS0 had 500×, and the IMW was at 300×, utilizing the secondary electron detector. For EDX analysis, a Bruker detector was employed, coupled to the microscope column.

2.3.4. X-Ray Diffraction (XRD)

The X-ray diffractograms of the samples, obtained using the powder method, were analyzed using a Bruker D4 Endeavor equipment available at COAM/CETEM, under the following operating conditions: Co$K\alpha$ radiation (40 kV/40 mA); goniometer speed of 0.02° 2θ per step with a counting time of 0.5 s per step, collected from 4 to 80° 2θ, with a position-sensitive detector (LynxEye). Qualitative interpretations of the spectrum were performed by comparing it with standards from the PDF02-ICDD database using Bruker DiffracPlus software.

2.3.5. Environmental Analysis

For the environmental characterization, leaching and solubility tests were conducted according to the guidelines of NBR 10004 [42]. The analyses were carried out at the laboratory of the Brazilian Association of Portland Cement (ABCP) in São Paulo, Brazil.

2.4. *Mechanical Characterization*

Resilience Modulus Test

The triaxial test for determining the resilient modulus (RM) aimed to evaluate the behavior of the soils and mixtures containing IMW regarding their elastic behavior. Alternatively, the California bearing ratio (CBR) test is a repeated load triaxial test that assesses the performance of compacted test specimens subjected to cyclic loading, more accurately reproducing the stresses that occur in a road pavement. The RM test was performed in duplicate; that is, two cylindrical specimens (10 × 20 cm) were molded in a split mold with the assistance of a mechanical compactor for each mixture and soil studied, under optimal moisture conditions and bulk specific mass, utilizing intermediate Proctor compaction energy, to verify the repeatability of the results.

The test was conducted following the standard testing method outlined in the DNIT 134 standard [43], using a Brazilian-made machine (SIGEO, Rio de Janeiro, Brazil) for dynamic triaxial tests, available in the soil laboratory of IME. During the resilient modulus test, eighteen pairs of confining stresses (σ_3) and deviator stresses (σ_d) were applied after the conditioning phase of the specimen. The loading cycle lasted 1s, with 0.1 s of load application and a frequency of 1 Hz (60 cycles per min). In the conditioning phase, the specimens were exposed to three sets of stresses and subjected to 500 loading cycles for each set. Subsequently, they were subjected to 18 additional sets of stresses, with 100 loading cycles for each set, totaling 3300 cycles per test. The values of the applied stresses are presented in Table 1.

Table 1. Stress pairs for the resilient modulus test.

	Conditioning Phase		
Par	**σ_3(KPa)**	**σ_d(KPa)**	**σ_3/σ_1**
1	70	70	2
2	70	210	4
3	105	315	4
	Loading Phase		
Par	σ_3(KPa)	σ_d(KPa)	σ_3/σ_1
1		20	2
2	20	40	3
3		60	4
4		35	2
5	35	70	3
6		105	4
7		50	3
8	50	100	2
9		150	3
10		70	2
11	70	140	3
12		210	4
13		105	2
14	105	210	3
15		315	4
16		140	2
17	140	280	3
18		420	4

Figure 4 shows the test specimen molded within the split mold and in the dynamic triaxial equipment.

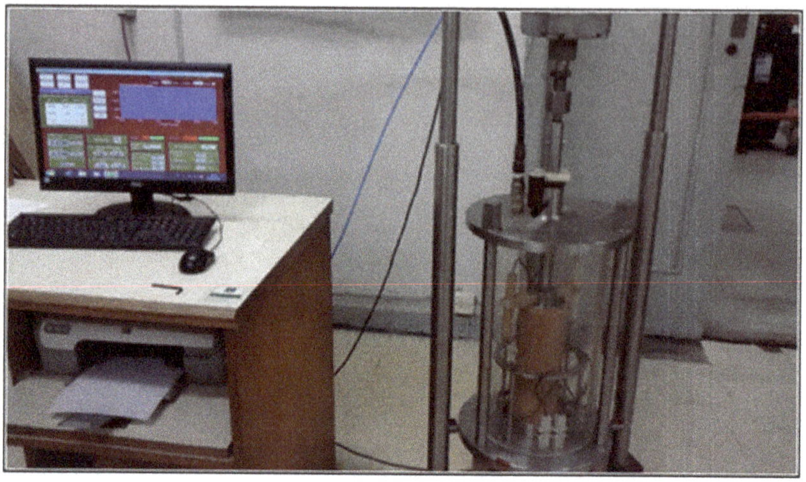

Figure 4. RM test specimen molded inside the tripartite mold dynamic triaxial equipment.

RM is obtained from the results of the repeated load triaxial tests. It is defined as the ratio between the deviant stress (σ_d) and the resilient axial strain (Δr), as presented in Equation (1).

$$RM = \frac{\sigma_d}{\Delta r} \tag{1}$$

where

- Δr is the ratio of Δh to h_0. Δh is the maximum vertical displacement, and h_0 is the initial reference length of the cylindrical specimen.

3. Results and Discussion

3.1. Physical Characterization

Figure 5 presents the grain size curves of the IMW, PRS, and the mixtures with 20%, 40%, and 50% of IMW. Figure 6 shows the grain size curves of the LCS and its mixtures. The analysis of these curves reveals that the addition of the sand IMW resulted in a coarser grain size distribution in the mixtures compared to the pure soils. This change is attributed to the higher proportion of fine sand present in the composition of the IMW.

The results shown in Table 2 indicate that the sand IMW (IMW) exhibited non-plastic and non-liquid behavior, consistent with its granular nature and the absence of plastic characteristics during testing. Although it contains 32% fine particles (passing through sieve No. 200), these particles, derived from iron ore processing, do not share the properties of natural soils, suggesting they will not exhibit the expected behavior of naturally formed clay or silt soils. The PRS displayed a liquid limit of 40.4%, sufficient to indicate cohesion, although it was classified as non-plastic. This characteristic aligns with the lamellar shape of the predominant particles, which complicates the determination of plasticity, as noted by Nogami and Villibor [40]. The addition of IMW reduced the liquid and plastic limits of the mixtures, confirming its granular behavior. All mixtures containing the IMW were classified as non-plastic and non-liquid. Furthermore, the density (D) increased with the amount of IMW due to the higher mineral density of this material, showing the contribution of the IMW to the formation of denser and more stable mixtures.

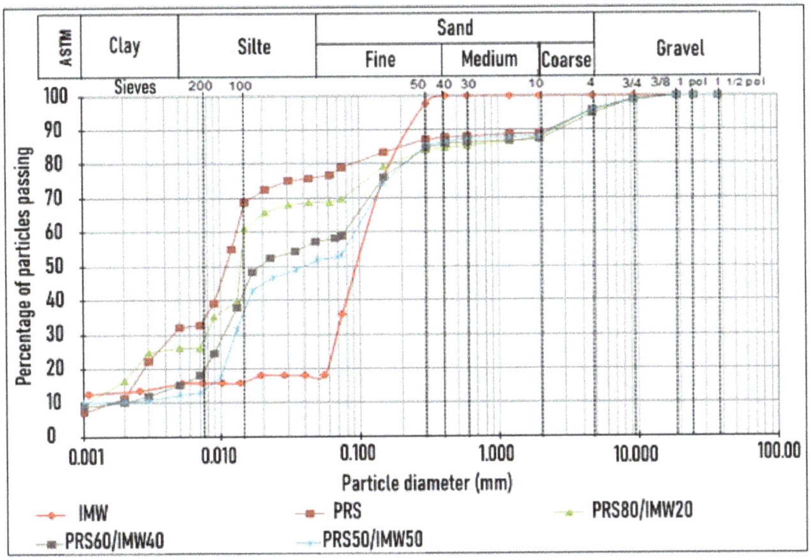

Figure 5. Grain size distribution curves of the PRS and its mixtures: PRS80/IMW20, PRS60/IMW40, and PRS50/IMW50.

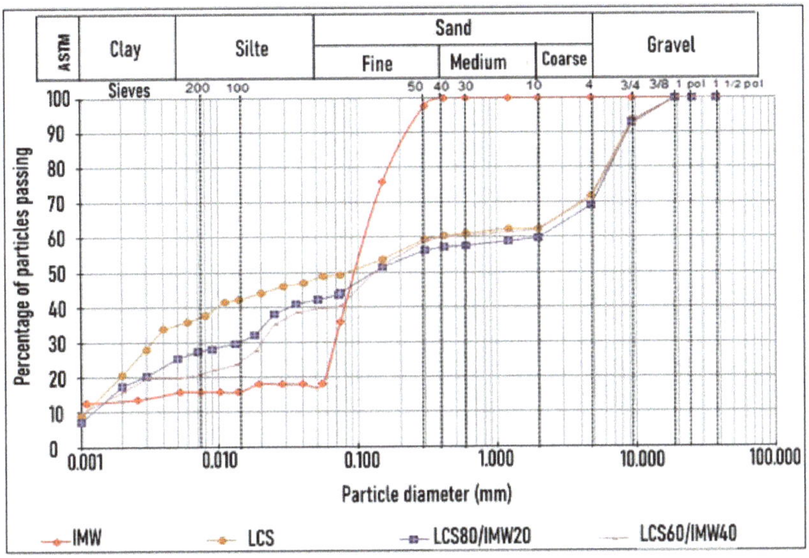

Figure 6. Grain size distribution curves of the LCS and its mixtures: LCS80/IMW20 and LCS60/IMW40.

Table 3 shows that the PRS sample had a significant expansion of 6.1%. However, adding 50% IMW to the mix resulted in a substantial reduction in this expansion, which fell to just 1%. This value is especially relevant, as it meets the maximum expansion limit established by NBR 9895 brazilian standard [35] for applications in granulometrically stabilized sub-base and subgrade reinforcement, as well as minimizing the risks associated with volumetric variation, which can compromise the durability and performance of the pavements.

The addition of IMW to the LCS was effective in keeping expandability at 0% in all the mixtures tested. This positive performance can be attributed to the mineralogical characteristics of laterite soils, which, according to Guimarães et al. [2], favor partially irreversible cementation after drying, resulting in greater structural stability.

Table 2. Consistency limits and density of samples.

Samples	LL (%)	LP (%)	IP (%)	D (g/cm³)
IMW	NL	NP	NP	2.959
PRS	40.4	NP	NP	2.556
PRS80/IMW20	NL	NP	NP	2.625
PRS60/IMW40	NL	NP	NP	2.703
PRS50/IMW50	NL	NP	NP	2.741
LCS	43.2	29.2	14	2.653
LCS80/IMW20	35.3	25.8	9.5	2.718
LCS60/IMW40	26.5	17.2	9.3	2.761

NL (No Liquid); NP (Non-Plastic).

Additionally, the inclusion of IMW modifies the physical properties of the soil mixtures, especially due to the higher density of the waste material in comparison to natural soils. As shown in Table 3, the density increases with the addition of IMW, which has a density of 2.959 g/cm^3, while PRS and LCS have densities of 2.556 g/cm^3 and 2.653 g/cm^3, respectively. This increase in density is a direct result of the higher density of IMW, leading to a denser mixture structure. Subsequently, the influence of this alteration on the mechanical behavior of the mixtures will be further analyzed in relation to the RM, allowing a more detailed analysis of the response of the materials under conditions of repeated loading.

Table 3. Properties of materials or mixtures.

Material/Mixture	Optimal Moisture (%)	Density (g/cm^3)	Expansion (%)
PRS	20.5	1.693	6.1
PRS80/IMW20	19.2	1.768	3.9
PRS60/IMW40	14.7	1.875	2.4
PRS50/IMW50	14.2	1.930	1.0
LCS	16.1	1.963	0.0
LCS80/IMW20	13.4	2.010	0.0
LCS60/IMW40	11.7	2.093	0.0

3.2. Geotechnical Characterization

The results shown in Table 4 indicate that the PRS was classified as clay, and the addition of IMW to the mixtures did not significantly alter its behavior, although it changed its classification to silt. These mixtures belong to the category of fine-grained soils with low compressibility, according to the USCS. The IMW and some mixtures incorporating it were classified as silty sands in the USCS and as A-4 in the TRB system, indicating low suitability for direct use in pavement construction. However, this performance is more favorable than that of pure soils, which are classified as A-7-6, a category with lower performance in the TRB system.

It is important to note that the USCS and TRB systems, although widely used, have limitations, especially for tropical soils. The attempt to categorize complex soils based on simple tests does not always reflect their actual geotechnical behavior [40,44]. This has motivated the development of the MCT classification system, which is deemed more appropriate for tropical soils, taking into account the influence of clay minerals and the formation of aggregates.

The soil classifications, both by the traditional method and the expedient MCT method, showed compatibility in the results obtained. The PRS and its mixtures, PRS80/IMW20,

PRS60/IMW40, and PRS50/IMW50, were identified as NA-NS', while the LCS and its mixtures were classified as LA'-LG'.

Table 4. Soil and mixture classification.

Samples	SUCS	TRB	MCT	Expedited MCT
IMW	Silty Sand	Silty (A-4)	NA-NS'	NA-NS'
PRS	Clay	Clayey (A-7-6)		
PRS80/IMW20	Silt	Silty (A-4)		
PRS60/IMW40	Silt	Silty (A-4)		
PRS50/IMW50	Silt	Silty (A-4)		
LCS0	Silty Sand	Clayey (A-7-6)	LA'-LG'	LA'-LG'
LCS80/IMW20	Silty Sand	Silty (A-4)		
LCS60/IMW40	Silty Sand	Silty (A-4)		

3.3. Mineralogical, Chemical and Environmental Analysis of Materials

3.3.1. Scanning Electron Microscopy (SEM) and Energy-Dispersive X-Ray Spectroscopy (EDX)

Figure 7 illustrates the SEM images of the IMW, PRS, and LCS samples, with magnifications of 300, 2500, and 500×, respectively.

Figure 7a reveals the presence of predominantly rounded grains. The grains are not individualized but interconnected by a matrix that appears to be amorphous. This characteristic suggests cohesion among the grains and indicates the presence of internal voids. In Figure 7b, a more complex morphology is observed. The grain structure is less uniform, with variations in texture and size. The grains exhibit more angular surfaces, suggesting greater morphological diversity in this fraction. This morphological complexity may be attributed to the formation and geological history of the material. In Figure 7c, the texture is clearly visible, with grains exhibiting a "popcorn" or "sponge" appearance. The rounded contours of the grains and the presence of internal voids are evident, suggesting a structure that retains water and possibly nutrients. This morphology is typical of lateritic soils, as mentioned by Villibor et al. [45], mirroring the cementation that can be attributed to the presence of iron and aluminum oxides.

Figure 7. SEM images of the sample fractions at different magnifications: (**a**) IMW at 300×, (**b**) PRS at 2500×, and (**c**) LCS at 500×.

The EDS analyses of the three samples provided relevant information about their composition and characteristics.

Figure 8 presents the quantitative chemical analysis of the IMW, revealing a significant composition of oxygen (39.80%), silicon (28.83%), and iron (22.27%), suggesting the presence of iron oxides and silicates. These elements are characteristic of mining waste, which retains residual minerals after processing. Traces of aluminum (0.65%) and nitrogen (8.45%) indicate the presence of impurities and complexity in the material matrix. However, the waste can be characterized as non-expansive, as no elements indicative of expansiveness, such as sodium or magnesium, which are associated with minerals that expand upon contact with water, were identified.

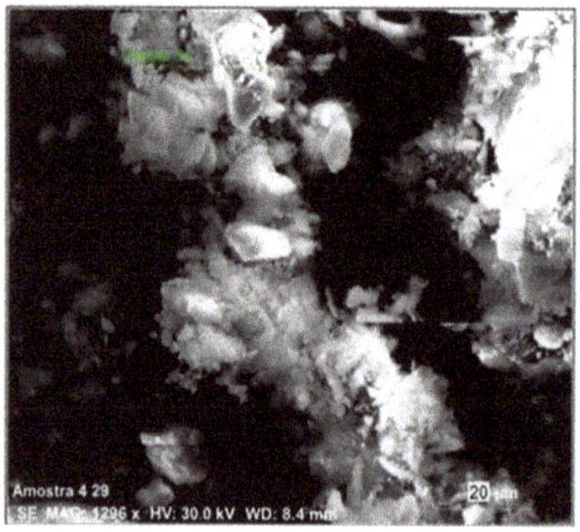

Element	wt.%
Oxygen (O)	39.80
Silicon (Si)	28.83
Iron (Fe)	22.27
Nitrogen (N)	8.45
Aluminium (Al)	0.65
Carbon (C)	0.00

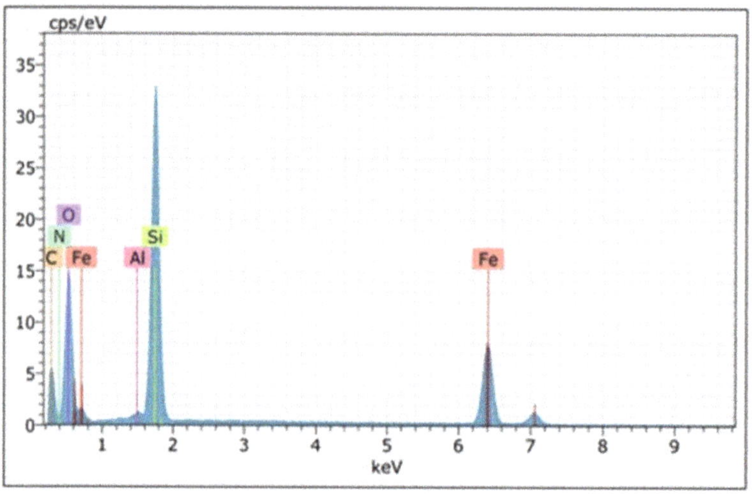

Figure 8. EDX map of IMW.

Figure 9 presents the quantitative chemical analysis obtained from the EDS test of the PRS, showing a chemical composition rich in oxygen (40.90%) and carbon (29.02%), along with a significant amount of iron (22.27%). The presence of aluminum (9.01%) and silicon (8.38%) reflects the formation of silicate minerals and oxides. Furthermore, elements in smaller proportions, such as potassium (2.35%) and copper (0.52%), suggest the presence of possible traces of minerals that compose the soil formation or contaminating materials. The presence of potassium may indicate the existence of minerals that, depending on environmental conditions, could exhibit expansive behavior.

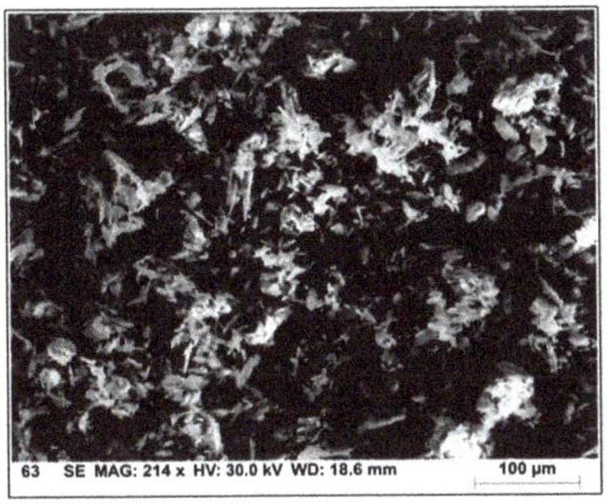

Element	**wt.%**
Oxygen (O)	40.90
Carbon (C)	29.02
Iron (Fe)	22.27
Aluminium (Al)	9.01
Silicon (Si)	8.38
Potassium (K)	2.35
Copper (Cu)	0.52
Sodium (Na)	0.00

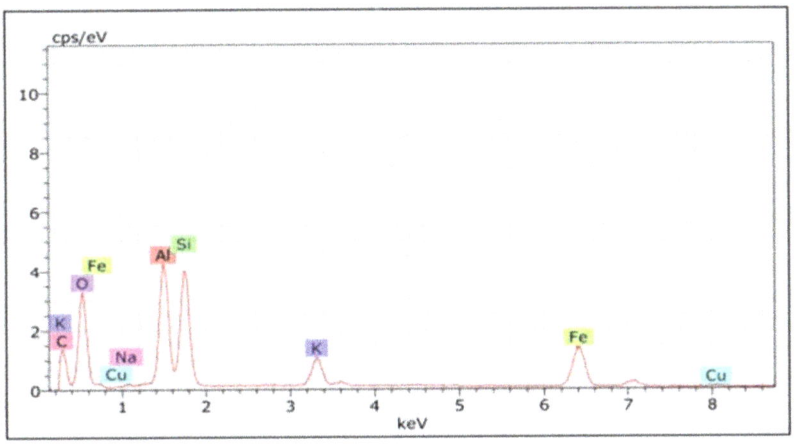

Figure 9. EDX map of PRS.

The chemical compositions obtained from the EDS test for the LCS are presented in Figure 10. The chemical profile is dominated by oxygen (47.77%) and aluminum (17.09%), suggesting the predominance of aluminosilicate oxides and clay minerals. The significant presence of silicon (16.68%) and iron (12.57%) is characteristic of lateritic soils, which are generally enriched with metallic oxides. The detection of titanium (1.09%) shows the mineralogical complexity of the soil.

This soil exhibits a high concentration of iron and aluminum oxides and hydroxides that coat the clay mineral kaolinite. These oxides not only reduce the water adsorption capacity of the clay minerals but also act as natural cementing agents, promoting the formation of stable aggregates in the presence of water. Therefore, even before the application of the methodology proposed by Nogami and Villibor [40] to definitively assess the lateritic behavior of the soil under study, the results obtained from the chemical analyses already suggest such characteristics.

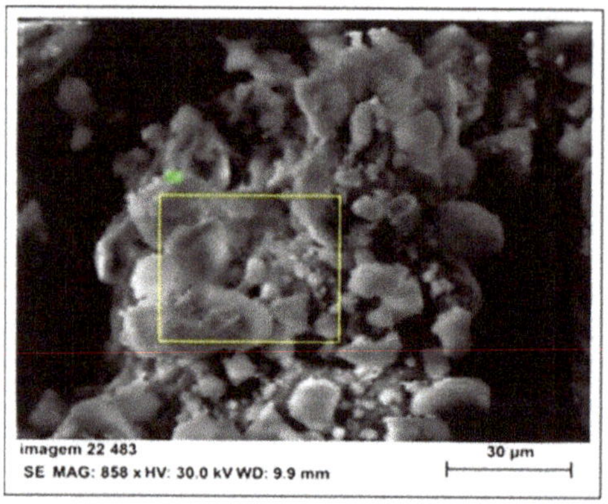

Element	wt.%
Oxygen (O)	47.77
Aluminium (Al)	17.09
Silicon (Si)	16.68
Iron (Fe)	12.57
Carbon (C)	4.77
Titanium(Ti)	1.09
Sulfur (S)	0.01
Phosphorus (P)	0.00

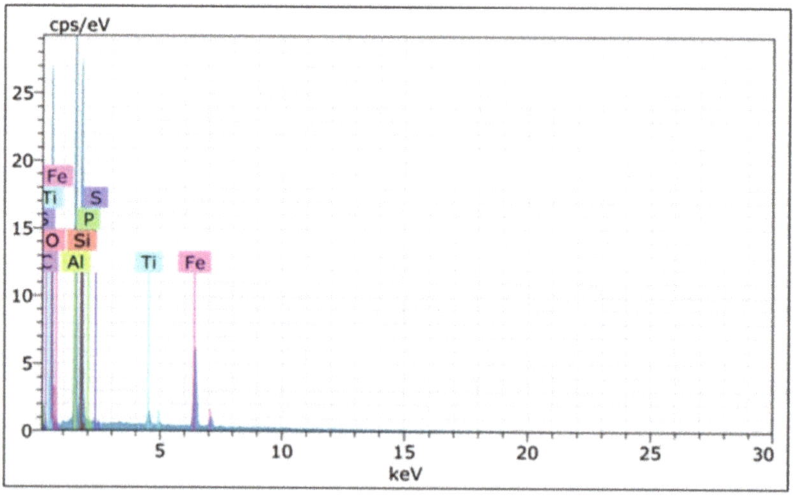

Figure 10. EDX map of LCS.

3.3.2. X-Ray Diffraction (XRD)

The XRD curves of the IMW, PRS and LCS samples, shown in Figures 11–13, respectively, highlight the peaks corresponding to the minerals identified in each sample. The mineralogical composition of these materials, including the approximate chemistry of the main minerals, is summarized in Table 5. In the IMW, quartz (SiO_2) is the predominant mineral, while hematite (Fe_2O_3) and goethite FeO(OH)) appear as subordinate minerals. The presence of these iron oxides is significant in justifying the high density of the material. The density of hematite ranges between 5.17 and 5.18 g/cm^3, and that of goethite ranges between 5.158 and 5.180 g/cm^3, while quartz has a lower specific weight of 2.65 g/cm^3. This composition explains both the observed real density values in the IMW and its non-expansive behavior, which is unlike other types of IMW, such as steel slag, which can exhibit expansion under certain moisture conditions [16]. Moreover, the studied IMW is characterized by a predominance of rigid minerals that are not susceptible to volumetric changes in the presence of water.

The PRS is classified as non-lateritic according to the MCT methodology, but it presents kaolinite as the predominant mineral—an argillomineral often associated with lateritic soils. Additionally, the presence of 10Å halloysite was identified, which has the same chemical composition as kaolinite but with an additional layer of water, facilitating its dehydration [46]. According to Santos [47], non-lateritic soils tend to contain more active argillominerals, such as smectite and illite, although they may be predominantly composed of kaolinite. The presence of anhydrite $CaSO_4$) in the phyllite soil may be the main cause of the expansive nature identified in the expansion tests. In a broader context, Barbosa et al. [48] discuss the importance of accessible methods for identifying expansive soils, especially in regions where infrastructure is limited. Their study proposes a simplified approach that correlates the plasticity index and the silt/clay content, offering useful tools to mitigate the risks associated with the volumetric variation of soils, which can be relevant for understanding the behavior of the PRS and its implications in paving projects.

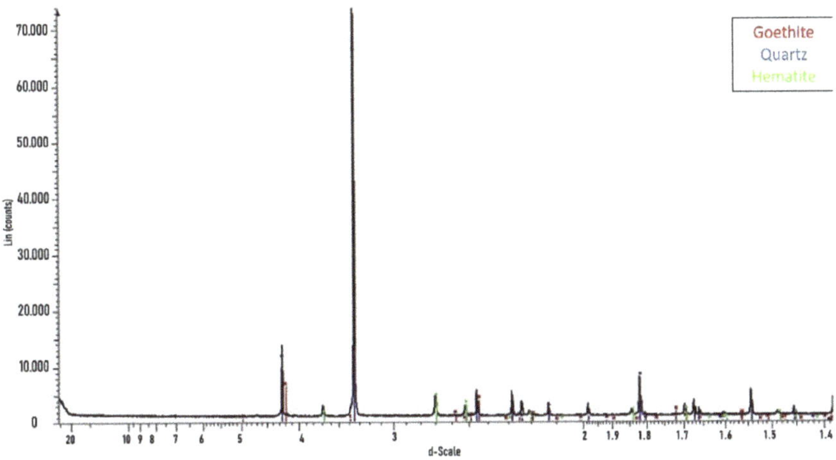

Figure 11. XRD for the IMW sample, identifying peaks corresponding to goethite, quartz, and hematite.

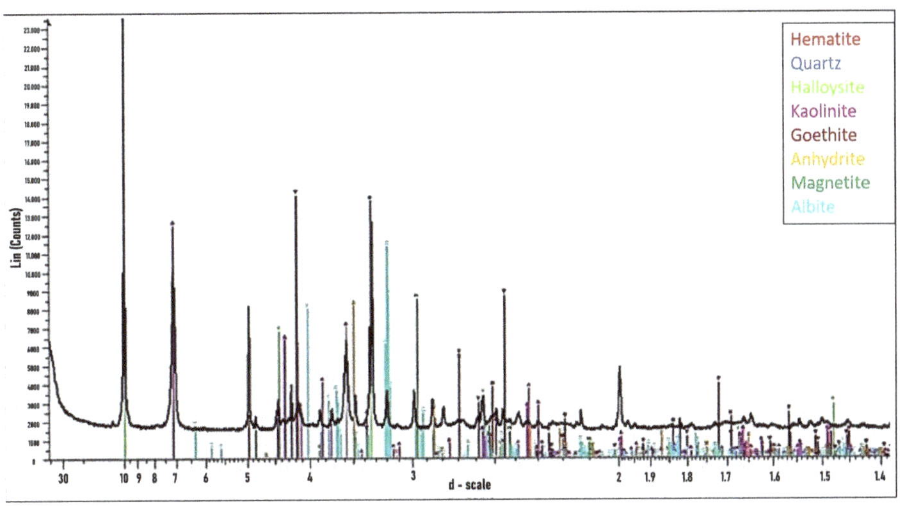

Figure 12. XRD for the PRS sample, highlighting the predominant minerals.

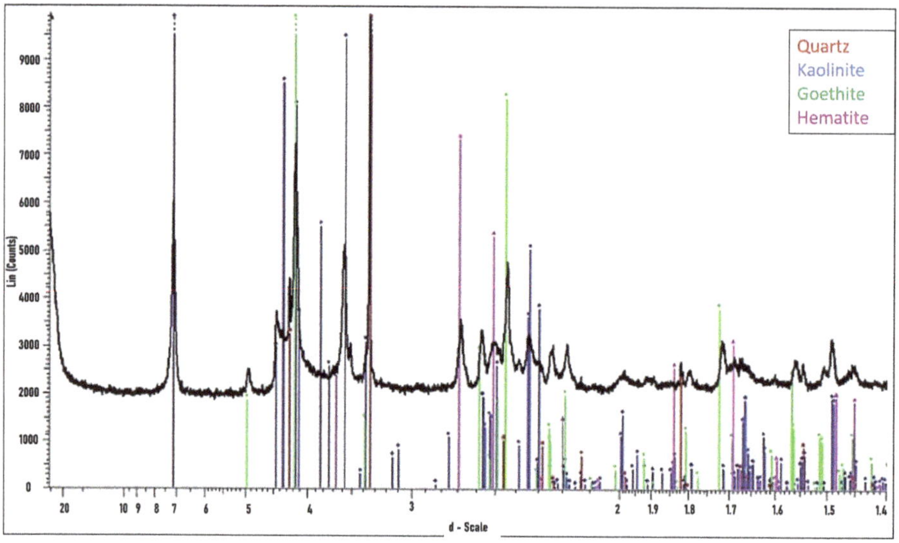

Figure 13. XRD for the LCS sample, indicating the presence of quartz, kaolinite, goethite, and hematite.

Table 5. Mineralogical composition of the soil samples (IMW, PRS, and LCS), indicating the main minerals and their approximate chemistry.

Sample	Minerals	Approximate Chemistry
IMW	Goethite Quartz Hematite	$FeO(OH)$ SiO_2 Fe_2O_3
PRS	Hematite Quartz Halloysite Kaolinite Goethite Anhydrite Magnetite Albite	Fe_2O_3 SiO_2 10Å $Al_2(Si_2O_5)(OH)_4$ $Al_2(Si_2O_5)(OH)_4$ $FeO(OH)$ $CaSO_4$ Fe_3O_4 $Na(AlSi_3O_8)$
LCS	Quartz Kaolinite Goethite Hematite	SiO_2 $Al_2(Si_2O_5)(OH)_4$ $FeO(OH)$ Fe_2O_3

On the other hand, the LCS exhibited a mineralogy typical of lateritic soils. The main minerals identified were quartz, kaolinite, goethite, and hematite. The predominant argillomineral is kaolinite, whose 1:1 structure is known for being non-expansive in the presence of water, providing stability to the material even under wet conditions. The joint presence of iron minerals, such as goethite and hematite, is characteristic of lateritic soils and contributes to their high density, mechanical resistance, and cementing properties [45]. As discussed by Benatti and Miguel [49], the cementation observed in lateritic soils results from the migration of particles and soluble chemical compounds in the unsaturated zone, a process that culminates in the formation of metastable structures with high porosity. Additionally, Guimarães et al. [50] emphasize that the mineralogical characteristics of these tropical soils provide superior performance when used as a pavement layer, particularly due to the partially irreversible cementation that occurs after drying. Laterization, in turn, can be quantitatively assessed by the ratio of silica SiO_2 remaining in the soil in comparison to the amounts of accumulated Fe_2O_3 and Al_2O_3, as proposed by Morin and Toder [51]. This integrated understanding of the morphology and mineralogical characteristics of the sample fractions is crucial for predicting their behavior.

3.3.3. Environmental Analysis

The leachate extracts were obtained according to NRB 10004 [42]. Table 6 shows the results of the inorganic constituents leached from the IMW sample, while Table 7 displays the results of the organic constituents leached from the IMW sample.

Table 6. Concentration of leached inorganic constituents.

Constituent		Unit	Results	Limits (mg/L)	
				Quantification Limits	NBR 10004 (max)
Silver	(Ag)	(mg/L)	N.D.	0.025	5.0
Arsenic	(As)	(mg/L)	N.D.	0.025	1.0
Barium	(Ba)	(mg/L)	N.D.	0.010	70.0
Cadmium	(Cd)	(mg/L)	N.D.	0.003	0.5
Chromium	(Cr)	(mg/L)	N.D.	0.010	5.0
Lead	(Pb)	(mg/L)	N.D.	0.010	1.0
Selenium	(Se)	(mg/L)	N.D.	0.025	1.0
Mercury	(Hg)	(mg/L)	N.D.	0.0002	0.1
Fluoride	(F^-)	(mg/L)	N.D.	0.02	150.0
pH	-	-	4.92	-	

N.D. means not detected.

Table 7. Concentration of leached organic constituents.

Constituent	Unit	Results	Limits (mg/L)	
			Quantification Limits	NBR 10004 (max)
Aldrin + Dieldrin	(mg/L)	N.D.	0.00006	0.003
Chlordane (isomers)	(mg/L)	N.D.	0.00006	0.02
DDT (isomers)	(mg/L)	N.D.	0.00009	0.2
2,4-D	(mg/L)	N.D.	0.0015	0.5
Endrin	(mg/L)	N.D.	0.0003	0.06
Heptachlor and Heptachlor Epoxide	(mg/L)	N.D.	0.00006	0.003
Lindane (γ-BHC)	(mg/L)	N.D.	0.0003	0.2
Methoxychlor	(mg/L)	N.D.	0.00003	2.0
Pentachlorophenol	(mg/L)	N.D.	0.0015	0.9
Toxaphene	(mg/L)	N.D.	0.000375	0.5
2,4,5-T	(mg/L)	N.D.	0.0015	0.2
2,4,5-TP	(mg/L)	N.D.	0.0015	1.0
Benzene	(mg/L)	N.D.	0.0030	0.5
Benzo(a)pyrene	(mg/L)	N.D.	0.0015	0.07
Vinyl Chloride	(mg/L)	N.D.	0.0030	0.5
Chlorobenzene	(mg/L)	N.D.	0.0030	100
Chloroform	(mg/L)	N.D.	0.0030	6.0
o-Cresol	(mg/L)	N.D.	0.0015	200.0
m,p-Cresol	(mg/L)	N.D.	0.0015	200.0
1,4-Dichlorobenzene	(mg/L)	N.D.	0.0015	7.5
1,2-Dichloroethane	(mg/L)	N.D.	0.0030	1.0

Table 7. Cont.

Constituent	Unit	Results	Limits (mg/L)	
			Quantification Limits	NBR 10004 (max)
1,1-Dichloroethylene	(mg/L)	N.D.	0.0030	3.0
2,4-Dinitrotoluene	(mg/L)	N.D.	0.0015	0.13
Hexachlorobenzene	(mg/L)	N.D.	0.0015	0.1
Hexachlorobutadiene	(mg/L)	N.D.	0.0015	0.5
Hexachloroethane	(mg/L)	N.D.	0.0015	3.0
Methyl Ethyl Ketone	(mg/L)	N.D.	0.009	200.0
Nitrobenzene	(mg/L)	N.D.	0.0015	2.0
Pyridine	(mg/L)	1.13	0.0015	5.0
Carbon Tetrachloride	(mg/L)	N.D.	0.0030	0.2
Tetrachloroethylene	(mg/L)	N.D.	0.0030	4.0
Trichloroethylene	(mg/L)	N.D.	0.0030	7.0
2,4,5-Trichlorophenol	(mg/L)	N.D.	0.0015	400.0
2,4,6-Trichlorophenol	(mg/L)	N.D.	0.0015	20.0

N.D. means not detected.

The solubilized extracts were obtained according to the NBR 10006 standard [52]. Table 8 shows the results of the tests for determining the solubilized inorganic constituents of the IMW sample. Table 8 presents the results of the tests for determining the solubilized organic constituents of the IMW sample.

Table 8. Content of solubilized inorganic constituents.

Constituent		Unit	Results	Limits (mg/L)	
				Quantification Limits	NBR 10004 (max)
Silver	(Ag)	(mg/L)	N.D.	0.010	0.05
Aluminum	(Al)	(mg/L)	N.D.	0.010	0.2
Arsenic	(As)	(mg/L)	N.D.	0.008	0.01
Barium	(Ba)	(mg/L)	N.D.	0.010	0.7
Cadmium	(Cd)	(mg/L)	N.D.	0.003	0.005
Chromium	(Cr)	(mg/L)	N.D.	0.010	0.05
Copper	(Cu)	(mg/L)	N.D.	0.010	2.0
Iron	(Fe)	(mg/L)	0.027	0.010	0.3
Manganese	(Mn)	(mg/L)	0.008	0.005	0.1
Lead	(Pb)	(mg/L)	N.D.	0.007	0.01
Selenium	(Se)	(mg/L)	N.D.	0.008	0.01
Zinc	(Zn)	(mg/L)	N.D.	0.003	5.0
Mercury	(Hg)	(mg/L)	N.D.	0.0002	0.001
Sodium	(Na)	(mg/L)	1.85	0.060	200.0
Sulfates	(SO_4^{2-})	(mg/L)	1.85	0.5	250.0
Nitrates	(N)	(mg/L)	0.89	0.5	10.0
Chlorides	(Cl^-)	(mg/L)	0.24	0.05	250.0
Fluorides	(F^-)	(mg/L)	N.D.	0.02	1.5
Cyanides	(CN^-)	(mg/L)	N.D.	0.05	0.07
Phenols	-	(mg/L)	N.D.	0.01	0.01
pH	-	-	6.28	-	-

N.D. means not detected.

Based on the results presented in Tables 6 and 8, it can be concluded that the leaching of inorganic metals and the solubilization tests showed values below the maximum limits established by the NBR 10004 standard [42]. Similar trends were observed in the leaching and solubilization tests of organic compounds (Tables 7 and 9), which exhibited null results, i.e., below the detection limits of the technique used and also below the normative maximum limits. Therefore, the IMW sample derived from iron ore processing was classified as Class II B—non-hazardous and inert—according to NBR 10004 [42]. This classification indicates that the metals are in a stable form, not susceptible to leaching or solubilization, thus eliminating environmental contamination risks in its intended application.

Table 9. Content of solubilized organic compounds.

Constituent	Unit	Result	Limits (mg/L)	
			Quantification Limits	NBR 10004 (max)
Aldrin + Dieldrin	(mg/L)	N.D.	0.00006	0.00003
Chlordane (isomers)	(mg/L)	N.D.	0.00006	0.0002
DDT (isomers)	(mg/L)	N.D.	0.00009	0.002
2,4-D	(mg/L)	N.D.	0.0009	0.03
Endrin	(mg/L)	N.D.	0.00003	0.0006
Heptachlor + Heptachlor Epoxide	(mg/L)	N.D.	0.00006	0.00003
Lindane (γ-BHC)	(mg/L)	N.D.	0.00003	0.002
Methoxychlor	(mg/L)	N.D.	0.00003	0.02
Toxaphene	(mg/L)	N.D.	0.000375	0.005
2,4,5-T	(mg/L)	N.D.	0.0009	0.002
2,4,5-TP	(mg/L)	N.D.	0.0009	0.030
Hexachlorobenzene	(mg/L)	N.D.	0.0009	0.001
Surfactants	(mg/L)	0.132	0.1	5.0

N.D. means not detected.

3.4. Mechanical Characterization

The resilient behavior of the materials and mixtures analyzed is presented in Table 10, which contains the results of the RM, obtained through the composite model that considers the deviator stress (σ_d) and the confining stress (σ_3).

Table 10. Regression values and determination coefficients of the composite model for each material or mixture in this study.

Samples	Composite Model $RM = k_1 \cdot \sigma_3^{k_2} \cdot \sigma_d^{k_3}$			
	k_1	k_2	k_3	R^2
PRS - CP1	23.78	0.188	−0.768	0.973
PRS - CP2	25.60	0.182	−0.758	0.967
PRS80/IMW20-CP1	56.13	0.243	−0.682	0.925
PRS80/IMW20-CP2	62.54	0.368	−0.750	0.953
PRS60/IMW40-CP1	62.70	0.190	−0.568	0.761
PRS60/IMW40-CP2	65.20	0.238	−0.566	0.750
PRS50/IMW50-CP1	96.87	0.246	−0.574	0.724
PRS50/IMW50-CP2	91.85	0.272	−0.540	0.743
LCS-CP1	240.18	−0.041	−0.409	0.923
LCS-CP2	234.82	0.029	−0.447	0.909
LCS80/IMW20-CP1	388.32	0.094	−0.316	0.923
LCS80/IMW20-CP2	292.39	0.113	−0.407	0.874
LCS60/IMW40-CP1	307.41	0.254	−0.409	0.768
LCS60/IMW40-CP2	354.27	0.283	−0.401	0.749

The regression parameters are presented in Table 10 along with the values of the coefficient of determination (R^2). The composite model was selected for this study due to its adequate coefficient of determination and simplicity. Moreover, this model aligns with the Brazilian mechanistic-empirical pavement design method (MeDiNa), which is used to characterize the stiffness of subgrade soils and granular materials [44,53–55].

In analyzing the results, the coefficient k_2/k_3, related to the confining stress, has a greater impact on the RM compared to the coefficient k_3, associated with the deviator stress. All values of k_3 were negative, indicating that for the materials analyzed, the deviator stress did not contribute to an increase in the RM. While this positive influence can be observed in sandier soils, such behavior was not identified in the mixtures of this study.

Furthermore, the analysis of the ratio k_2/k_3 highlights that with the increase in the addition of IMW, there was a reduction in the influence of deviator stress and a greater relevance of confining stress. For instance, in the LCS80/IMW20 mixture, k_2 increased to 0.094, while k_3 was −0.316, compared to LCS, which exhibited k_2 =−0.041 and k_3 =−0.409. This increase in the ratio k_2/k_3 reflects a higher sensitivity to confining stress in the mixtures with IMW, indicating greater stiffness and cohesion due to the addition of IMW.

The RM values presented in Table 11 are the averages obtained based on the 18 stress pairs established by the DNIT 134 brazilian standard [43]. These stress pairs were organized into three categories: low deviator stress (20 KPa), medium deviator stress (140 KPa), and high deviator stress (420 KPa). The categorization was defined to facilitate the comparative analysis of the behavior of the mixtures under different loading conditions and to reflect the practical variation found in PRS and LCS and their mixtures, allowing for the identification of material performance in various loading scenarios and contributing to optimization recommendations for the composition of the mixtures.

It is observed that RM varies substantially among the stress categories. For example, LCS exhibited the highest RM value under low-stress conditions (1270 MPa), indicating high stiffness in this category. Under medium and high stress, however, RM reduced to 464 MPa and 351 MPa, respectively. This behavior indicates that LCS has particularly high performance under light load conditions, making it suitable for upper pavement layers where loads are less concentrated.

The addition of 20% IMW to LCS (LCS80/IMW20 mixture) presented an RM of 922 MPa under low stress, with a slight reduction to 455 MPa and 382 MPa under medium and high stresses, respectively. These results suggest that incorporating 20% IMW into LCS maintains high stiffness, especially under low and medium stress conditions. This performance indicates that the controlled addition of IMW at this proportion still allows for suitable behavior for application in upper pavement layers, where stiffness is required to support loads without compromising pavement integrity.

Table 11. RM results.

Sample		**MR (MPa)**	
	(σ_d = Lower)	**(σ_d = Medium)**	**(σ_d = High)**
PRS	230	50	46 *
PRS80/IMW20	290	82	61
PRS60/IMW40	250	88	69
PRS50/IMW50	298	115	92
LCS	1270	464	351
LCS80/IMW20	922	455	382
LCS60/IMW40	561	311	278

* MR (σ_d = 275 kPa).

However, increasing the IMW proportion to 40% (LCS60/IMW40 mixture) shows a more pronounced reduction in RM values, which varied from 561 MPa under low stress to 311 MPa and 278 MPa under medium and high stresses. This decline may be attributed to the impact of IMW on the soil structure, reducing cohesion between LCS particles and lowering the stiffness of the mixture.

For PRS, RM was more modest, with values of 230 MPa, 50 MPa, and 46 MPa at low, medium, and high stresses, respectively. Nevertheless, PRS mixtures with IMW showed considerable improvements in RM values. For instance, the PRS50/IMW50 mixture achieved an RM of 298 MPa at low stress and 115 MPa and 92 MPa at medium and high stresses, respectively. This progressive increase in RM in PRS mixtures with IMW, even under higher stresses, suggests that IMW contributes to increasing the stiffness of these mixtures, favoring their use in low-traffic pavement layers.

Comparing these values with literature data reinforces the suitability of these mixtures for paving applications. For example, Lima et al. [56] recorded RM values between 129 and 366 MPa for subgrade soils on the BR-116/RJ, while Ramos and Motta [57] recorded values between 312 and 387 MPa for unbound granular materials. Similarly, the values obtained for LCS mixtures approximate the typical behavior of lateritic soils found in Brazil. Guimarães et al. [50] recorded MR values between 566 and 585 MPa for lateritic gravel in Acre, while Veloso [58] reported MR values ranging from 325 to 836 MPa for a laterite sample in Pará. These results confirm that the controlled addition of IMW to soils is suitable for applications in various pavement layers, from subgrades to granular bases.

4. Conclusions

This study explored the potential use of sandy waste from the iron ore beneficiation process at Samarco Mineração S.A., located in Mariana (MG), in mixtures with natural soils for road pavement applications. The research focused on two types of soils: phyllite residual soil (PRS) and lateritic clay soil (LCS). The investigation aimed to assess the physicochemical, geotechnical, and mechanical behaviors of these soil–waste mixtures, including an environmental analysis of the waste. The main conclusions obtained by this research are listed as follows:

- A significant reduction in plasticity indices was observed, particularly in mixtures with a higher content of waste, due to the low plasticity of this material, which is predominantly composed of quartz. This effect correlated with an increase in the specific weight of the mixtures, attributed to the higher specific mass of the waste compared to the pure soils.
- The compaction tests demonstrated that increasing the proportion of waste in the mixtures led to a decrease in optimum moisture content and an increase in maximum dry density, conditions that favor the use of the mixtures in situations where compaction and density are critical factors.
- The PRS alone is unsuitable for pavement applications due to its low resilient modulus (RM). However, adding 50% waste significantly improved its properties, increasing the RM values by approximately 29.6% under low stress, 130% under medium stress, and 100% under high stress, while also reducing expansibility. This improvement makes the phyllite–waste mixture a viable and sustainable option for reinforcing sub-base and subgrade layers in pavement structures.
- The RM results indicated that the addition of waste had a moderate impact on the LCS compared to the significant improvements observed in the PRS. Specifically, the LCS mixture with 20% waste showed a slight RM increase of approximately 8.8% under high-stress conditions, although there was a reduction of 27.4% and 1.9% under low and medium stress, respectively. The LCS mixture with 40% waste, however, displayed considerable reductions in RM across all stress levels, suggesting that higher waste content may adversely affect the stiffness of lateritic clay soils. These findings support the use of 20% waste in LCS for certain applications, particularly in base or sub-base layers, while highlighting the need for further optimization to balance stiffness and sustainability in pavement layers.

Thus, it can be concluded that iron mining waste from Samarco Mineração S.A. shows significant potential for reuse in road pavement layers, especially in stabilized subgrades and sub-bases. Its favorable physical and chemical properties, combined with its classification as inert and non-hazardous, support its use without environmental risks. However, further research is essential to optimize waste content in various regional soil types and to identify the ideal proportions for each application. Future studies should also focus on evaluating permanent deformation behavior through repeated load triaxial tests, providing a deeper understanding of these mixtures under cyclic loading conditions and their suitability for broader pavement applications.

Author Contributions: Conceptualization, D.C.G., M.M.N. and L.M.C.; methodology, D.C.G., M.M.N., C.A.M., L.M.C. and A.C.R.G.; software, M.M.N. and L.M.C.; formal analysis, S.N.M., L.M.C. and A.C.R.G.; investigation, D.C.G. and L.M.C.; data curation, D.C.G. and L.M.C.; writing—original draft preparation, D.C.G., C.A.M., M.M.N. and L.M.C.; writing—review and editing, L.M.C., A.C.R.G. and S.N.M.; supervision, A.C.R.G.; project administration, A.C.R.G.; funding acquisition, S.N.M. All authors have read and agreed to the published version of the manuscript.

Funding: This study was financed in part by the Coordination for the Improvement of Higher Education Personnel-Brazil(CAPES)—Finance Code 001.

Institutional Review Board Statement: Not applicable.

Informed Consent Statement: Not applicable.

Data Availability Statement: The original contributions presented in the study are included in this article; further inquiries can be directed to the corresponding authors.

Acknowledgments: The authors would like to thank the Samarco mining S.A. company for their partnership.

Conflicts of Interest: The authors declare no conflicts of interest.

Abbreviations

AASHTO	American Association of State Highway and Transportation Official
ABCP	Brazilian Association of Portland Cement
ABNT	Brazilian Association of Technical Standards
CBR	California bearing ratio
COAM/CETEM	Laboratory of the Mineral Analysis Coordination at the Mineral Technology Center
EDX	Energy-dispersive X-ray
IME	Military Institute of Engineering
IMW	Iron mining waste
LCS	100% Lateritic clay soil
LCS80/IMW20	80% Lateritic clay soil and 20% iron mining waste
MCT	Miniature, Compacted, Tropical
PRS	100% phyllite residual soil
PRS50/IMW50	50% phyllite residual soil and 50% iron mining waste
PRS60/IMW40	60% phyllite residual soil and 40% iron mining waste
PRS80/IMW20	80% phyllite residual soil and 20% iron mining waste
R2	Coefficient of determination
RAP	Reclaimed asphalt pavement
RM	Resilient modulus
SEM	Scanning electron microscopy
SUCS	Soil Use and Classification System
TRB	Transportation Research Board
XRD	X-ray diffraction

References

1. CNT—Confederação Nacional de Transporte. *Boletim Unificado-Agosto 2024*; CNT—Confederação Nacional de Transporte: Brasília, DF, Brazil, 2024. Available online: https://cnt.org.br/boletins (accessed on 27 October 2024).
2. Guimarães, A.; Motta, L.D. *A Study on Permanent Deformation of Lateritic Soils Including the Shakedown Concept*, 1st ed.; CRC Press: Boca Raton, FL, USA, 2008; p. 8.

3. Yaghoubi, E.; Yaghoubi, M.; Guerrieri, M.; Sudarsanan, N. Melhorando subleitos de argila expansiva usando vidro reciclado: Características de módulo resiliente e desempenho do pavimento. *Construção E Mater. De Construção* **2021**, *302*, 124384. [CrossRef]
4. Maichin, P.; Jitsangiam, P.; Nongnuang, T.; Boonserm, K.; Nusit, K.; Pra-ai, S.; Binaree, T.; Aryupong, C. Stabilized High Clay Content Lateritic Soil Using Cement-FGD Gypsum Mixtures for Road Subbase Applications. *Materials* **2021**, *14*, 1858. [CrossRef] [PubMed]
5. Hasan, M.M.; Islam, M.R.; Tarefder, R.A. Characterization of subgrade soil mixed with recycled asphalt pavement. *J. Traffic Transp. Eng. (English Ed.)* **2018**, *5*, 207–214. [CrossRef]
6. de Almeida Silva, P.; Sergio de Souza Almeida, M.; Ítalo Miller Machado Pereira.; Rucas Andrade Achy, A.; Gustavo Silva Costa, W.; Dias Amancio de Lima, C.; Almeida Brito, J. Mechanistic-empirical evaluation of soil-RAP mixtures for use as unbound granular layers of asphalt pavements. *Transp. Geotech.* **2024**, *47*, 101290. [CrossRef]
7. Boaventura, A.C.C.; Almeida, M.S.d.S.; Costa, W.G.S.; Pereira, I.M.M.; São Mateus, M.d.S.C.; Brito, J.A.; Fiuza-Jr, R.A. The influence of residual asphalt material in the mechanical behaviour of soil-RAP mixtures for use in paving. *Int. J. Pavement Eng.* **2024**, *25*, 2407901. [CrossRef]
8. Coelho, L.; Guimarães, A.; de Azevedo, A.; Monteiro, S. Sustainable Reclaimed Asphalt Emulsified Granular Mixture for Pavement Base Stabilization: Prediction of Mechanical Behavior Based on Repeated Load Triaxial Tests. *Sustainability* **2024**, *16*, 5335. [CrossRef]
9. Coelho, L.; Kox, R.; Guimarães, A.; Travincas, R.; Monteiro, S. Influence of Curing Time on the Mechanical Behavior of Cold Recycled Bituminous Mix in Flexible Pavement Base Layer. *Appl. Sci.* **2024**, *14*, 7612. [CrossRef]
10. Rodrigues, K.H.d.P.; Silva, T.O.d.; Pitanga, H.N.; Almeida, M.S.d.S.; Pedroti, L.G.; Rodrigues, M.H.R.; Lopes, E.C. Effect of carbonation of soil-slag mixtures on the resilient behaviour and structural response of an asphalt pavement. *Road Mater. Pavement Des.* **2024**, *25*, 2101–2120. [CrossRef]
11. Yao, Y.; Li, J.; Ni, J.; Liang, C.; Zhang, A. Effects of gravel content and shape on shear behaviour of soil-rock mixture: Experiment and DEM modelling. *Comput. Geotech.* **2022**, *141*, 104476. [CrossRef]
12. Dantas Neto, S.A.; Pereira, C.G.F.; Abreu, A.A. Stabilization of sandy soil with high content of asphalt emulsion. *REM-Int. Eng. J.* **2020**, *73*, 163–169. [CrossRef]
13. Purohit, S.; Panda, M.; Kumar Das, A. Performance of waste polyethylene modified bituminous paving mixes containing reclaimed asphalt pavement and recycled concrete aggregate. *Constr. Build. Mater.* **2022**, *348*, 128677. [CrossRef]
14. Navarrete, C.; Guimarães, A.; Marques, M.; Castro, C.; Toulkeridis, T. Resistance to Fatigue in Asphalts Used in Military Airports of the Brazilian Amazon through the Use of Nickel-Holding Ash. *Appl. Sci.* **2022**, *12*, 9134. [CrossRef]
15. Coelho, L.; Guimarães, A.; Alves Moreira, C.; dos Santos, G.; Monteiro, S.; da Silveira, P. Feasibility of Using Ferronickel Slag as a Sustainable Alternative Aggregate in Hot Mix Asphalt. *Sustainability* **2024**, *16*, 8642. [CrossRef]
16. Guimarães, A.C.R.; Costa, K.A.; Reis, M.d.M.; Santana, C.S.A.; Castro, C.D. Study of controlled leaching process of steel slag in soxhlet extractor aiming employment in pavements. *Transp. Geotech.* **2021**, *27*, 100485. [CrossRef]
17. Gomes, M.B.B.; Guimarães, A.C.R.; Nascimento, F.A.C.D.; Santos, J.T.A.D. Ballast with siderurgic aggregates: Variation analysis of the shape parameters of particles submitted to triaxial tests through 3D scanner. *Rev. Eng.* **2023**, *46*, e2023006122. [CrossRef]
18. Guimarães, A.C.R.; Silva Filho, J.C.; Castro, C.D. Contribution to the use of alternative material in heavy haul railway sub-ballast layer. *Transp. Geotech.* **2021**, *30*, 100524. [CrossRef]
19. Demchuk, Y.; Gunka, V.; Sidun, I.; Solodkyy, S. Comparison of Bitumen Modified by Phenol Formaldehyde Resins Synthesized from Different Raw Materials. In *Proceedings of EcoComfort 2020*; Blikharskyy, Z., Ed.; Lecture Notes in Civil Engineering; Springer: Cham, Switzerland, 2021; Volume 100. [CrossRef]
20. Xu, H.; Ji, W.; Jiang, W.; Cui, H.; Tan, Y. Preparation of inorganic binder-stabilised material with iron tailings: Strength formation and gradation optimisation. *Road Mater. Pavement Des.* **2023**, *25*, 618–636. [CrossRef]
21. Divandari, H.; Heidari, M.; Zalnezhad, M.; Zalnezhad, A. The study of the feasibility and manufacture of slurry seal surface treatment with steel slag industry wastes: Use of basic oxygen furnace (BOF) slag filler as raw material replacement. *Road Mater. Pavement Des.* **2023**, *25*, 967–987. [CrossRef]
22. Gunka, V.; Hidei, V.; Sidun, I.; Demchuk, Y.; Stadnik, V.; Shapoval, P.; Sobol, K.; Vytrykush, N.; Bratychak, M. Wastepaper Sludge Ash and Acid Tar as Activated Filler Aggregates for Stone Mastic Asphalt. *Coatings* **2023**, *13*, 1183. [CrossRef]
23. Usanga, I.N.; Ikeagwuani, C.C.; Okafor, F.O.; Ambrose, E.E. Exploration of calcined clays and nanoclays as a potential construction material for sustainable development: A comprehensive and state-of-the-art review. *Road Mater. Pavement Des.* **2024**, 1–34. [CrossRef]
24. Friber, M.; Guimarães, A.; Martins, C.; Soares, J. Study of the Mining Waste in the Production of Calcined Aggregate for Use in Pavement. *Minerals* **2023**, *13*, 1543. [CrossRef]
25. Apaza Apaza, F.R.; Rodrigues Guimarães, A.C.; Marcos Vivoni, A.; Schroder, R. Evaluation of the performance of iron ore waste as potential recycled aggregate for micro-surfacing type cold asphalt mixtures. *Constr. Build. Mater.* **2021**, *266*, 121020. [CrossRef]
26. Silveira, V.; Guimarães, A.; Coelho, L.; dos Santos, W.; da Silveira, P.; Monteiro, S. Recycling Iron Ore Waste through Low-Cost Paving Techniques. *Sustainability* **2024**, *16*, 5570. [CrossRef]
27. Guimarães, A.; Arêdes, M.; Castro, C.; Coelho, L.; Monteiro, S. Evaluation of the Mechanical Behavior of Asphaltic Mixtures Utilizing Waste of the Processing of Iron Ore. *Mining* **2024**, *4*, 889–903. [CrossRef]

28. Institute for Applied Economic Research (IPEA). *Research Report: Diagnosis of Urban Solid Waste*; Research report; IPEA: Brasília, Brazil, 2012.
29. ABNT–Associação Brasileira de Normas Técnicas. *NBR 6457:2024. Amostras de sol -Preparação para Ensaios de Compactação e Ensaios de Caracterização*; ABNT–Associação Brasileira de Normas Técnicas: Rio de Janeiro, Brazil, 2024.
30. ABNT–Associação Brasileira de Normas Técnicas. *NBR 6458:2017. Grãos de Pedregulho Retidos na Peneira de 4,8 mm -Determinação da Massa Específica, da Massa Específica Aparente e da Absorção de áGua*; ABNT–Associação Brasileira de Normas Técnicas: Rio de Janeiro, Brazil, 2017.
31. ABNT–Associação Brasileira de Normas Técnicas. *NBR 6459:2016. Solo-Determinação do Limite de Liquidez*; ABNT–Associação Brasileira de Normas Técnicas: Rio de Janeiro, Brazil, 2016.
32. ABNT—Associação Brasileira de Normas Técnicas. *NBR 7180:2016. Solo-Determinação do Limite de Plasticidade*. ABNT—Associação Brasileira de Normas Técnicas: Rio de Janeiro, Brazil, 2016; NBR 7180.
33. ABNT—Associação Brasileira de Normas Técnicas. *NBR 7181:2016. Solo-Análise granulométrica*; ABNT—Associação Brasileira de Normas Técnicas: Rio de Janeiro, Brazil, 2016.
34. ABNT—Associação Brasileira de Normas Técnicas. *NBR 7182:2020. Solo-Ensaio de compactação*; ABNT—Associação Brasileira de Normas Técnicas: Rio de Janeiro, Brazil, 2020.
35. ABNT—Associação Brasileira de Normas Técnicas. *NBR 9895:2017. Solo-Índice de Suporte Califórnia (ISC)- Método de Ensaio*; ABNT—Associação Brasileira de Normas Técnicas: Rio de Janeiro, Brazil, 2017.
36. *AASHTO M145-91*; Classification of Soils and Soil-Aggregate Mixtures for Highway Construction Purposes. American Association of State Highway and Transportation Officials: Washington, DC, USA, 2012;
37. *DNIT 259/2023*; CLA: Soils—Classification of Fine Tropical Soils for Road Purposes Using Compacted Specimens in Miniature Equipment—Classification. National Department of Transport Infrastructure (DNIT): Brasilia, Brazil, 2023; Brazilian Standard.
38. *DNIT Standard 258/2023*; ME: Soils—Compaction Using Miniature Equipment— Mini-MCV Tests and Mass Loss by Immersion—Test Method. National Department of Transport Infrastructure, Brasília, Brazil, 2023.
39. *DNER-ME 256; Solos Compactados com Equipamento Miniatura —Determinação da Perda de Massa por Imersão*. Departamento Nacional de Estradas de Rodagem: Rio de Janeiro, Brazil, 1994.
40. Nogami, J.; Villibor, D. *Pavimentação de Baixo Custo com Solos LaterfTicos*; Ed. Villibor: São Paulo, SP, Brazil, 1995; p. 213.
41. Nogami, J.; Villibor, D. Identificação expedita dos grupos da classificação MCT para solos tropicais. In Proceedings of the Anais do X Congresso Brasileiro de Mecânica dos Solos e Engenharia de Fundações, Foz do Iguaçu, PR, Brazil, October 1994; Volume 4, pp. 1293–1300.
42. ABNT–Associação Brasileira de Normas Técnicas. *Resíduos Sólidos-Classificação*; ABNT–Associação Brasileira de Normas Técnicas: Rio de Janeiro, Brazil, 2004; NBR 10004.
43. DNIT–Departamento Nacional de Infraestrutura de Transportes. *134/2018-ME. Módulo de Resiliência-Método de Ensaio*. DNIT–Departamento Nacional de Infraestrutura de Transportes: Rio de Janeiro, Brazil, 2018.
44. Pascoal, P.T.; Sagrilo, A.V.; Baroni, M.; Specht, L.P.; da Silva Pereira, D. Lateritic soil deformability regarding the variation of compaction energy in the construction of pavement subgrade. *Soils Rocks* **2023**, *46*, e26755475. [CrossRef]
45. Villibor, D.F.; Nogami, J.S.; Cincerre, J.R.; Serra, P.R.M.; Neto, A.Z. *Pavimentos de Baixo Custo Para Vias Urbanas*; Arte e Ciência: Campina Grande, Brazil, 2009; p. 196.
46. Dana, J.D.; Hurlbut, C.S. *Manual de Mineralogia*; Livros Técnicos e Científicos, Editora S.A.: Rio de Janeiro, Brazail, 1974; Volume 2, p. 624.
47. Santos, P.S. *Ciência e Tecnologia das Argilas*, 2nd, ed.; Editora Edgard Blucher Ltda: Sao Paulo, Brazil, 1992; Volume 3, p. 1089.
48. Barbosa, V.H.; Marques, M.E.S.; Guimarães, A.C.R. Predicting Soil Swelling Potential Using Soil Classification Properties. *Geotech. Geol. Eng.* **2023**, *41*, 1–13. [CrossRef]
49. Benatti, J.C.B.; Miguel, M.G. A proposal of structural models for colluvial and lateritic soil profile from southwestern Brazil on the basis of their collapsible behavior. *Eng. Geol.* **2013**, *153*, 1–11. [CrossRef]
50. Guimarães, A.C.R.; Lima, C.D.A.d.; Aragão, F.T.S.; Motta, L.M.G.d.; Santos, J.T.A.d. Mechanical characterization of an alternative laterite gravel used as pavement material. *Soils Rocks* **2024**, *47*, 1–13. [CrossRef]
51. Morin, W.; Toder, P. *Laterite and Lateritic Soils, and Other Problem Soils of Africa*; Lyon Associates: Kumisasi, Ghana, 1971; p. 290.
52. ABNT–Associação Brasileira de Normas Técnicas. *NBR 10006:Solubilização de Resíduos -Procedimento*. ABNT–Associação Brasileira de Normas Técnicas: Rio de Janeiro, Brazil, 2004.
53. Guimarães, A.; Motta, L. Mechanical Behavior of Basaltic Rocks from Serra Geral Formation Used as Road Material in Santa Catarina State, Brazil. *Soils Rocks* **2016**, *39*, 203–210. [CrossRef]
54. Pascoal, P.; Sagrilo, A.; Baroni, M.; Specht, L.; Pereira, D. Evaluation of the Influence of Compaction Energy on the Resilient Behavior of Lateritic Soil in the Field and Laboratory. *Soils Rocks* **2021**, *44*, e2021071321. [CrossRef]
55. Sagrilo, A.V.; Pascoal, P.T.; Baroni, M.; Back, A.H.; Pinheiro, R.J.B.; Specht, L.P.; Guimarães, A.C.R. Contribution to Resilient and Permanent Deformation Investigation of Unbound Granular Materials with Different Geological Origins from Rio Grande do Sul, Brazil. *Soils Rocks* **2023**, *46*, e2023009822. [CrossRef]
56. de Lima, C.D.A.; de Oliveira Ubaldo, M.; Fritzen, M.A.; da Motta, L.M.G.; de Proença Franco, F.A.C. Análise de Módulo de Resiliência de Solos Finos Tropicais para Avaliação e Dimensionamento de Pavimentos. In Proceedings of the XX Congresso Brasileiro de Mecânica dos Solos e Engenharia Geotécnica, IX Simpósio Brasileiro de Mecânica das Rochas, IX Simpósio Brasileiro

de Engenheiros Geotécnicos Jovens, VI Conferência Sul Americana de Engenheiros Geotécnicos Jovens, ABMS, Campinas, SP, Brasil, 15–18 September 2020. [CrossRef]

57. Ramos, C.R.; Motta, L.M.G. Study of the Resilient Behavior of Various Soils and Crushed Materials from the Municipality of Rio de Janeiro. In Proceedings of the 35th Annual Pavement Meeting, Rio de Janeiro, Brazil, 11–12 November 2004; Volume 1.
58. Veloso, J.F.; Guimarães, A.C.R.; Marques, M.E.S.; Souza, T.D.D. Analysis of Tropical Soils for Paving the Road BR-163/PA. *Braz. J. Dev.* **2020**, *6*, 53471–53493. [CrossRef]

sustainability

MDPI

Article

Assessment of Safe and Sustainable Operation for Freight Transportation Company Based on Tire Set Configurations Used in Its Trucks' Fleet

Vidas Žuraulis, Robertas Pečeliūnas * and Tomas Misevičius

Department of Automobile Engineering, Faculty of Transport Engineering, Vilnius Gediminas Technical University, Plytinės Str. 25, 10105 Vilnius, Lithuania; vidas.zuraulis@vilniustech.lt (V.Ž.)
* Correspondence: robertas.peceliunas@vilniustech.lt; Tel.: +370-52744791

Abstract: This article investigates the safety potential of a freight transportation company, considering tire set selection as one of the most important aspects to ensure safe driving and a reliable transportation service. The revision of tire sets selection in large vehicle fleets is attributed to a new regulation from the United Nations to maintain non-deteriorating tire wet braking performance up to a minimum allowable wear limit, encouraging both safety and sector sustainability, as a significant part of tires are currently replaced before reaching a tread depth of 3 mm. In this research, an experimental test was conducted that involved four maneuvers with a truck using ten different sets of tires (including new and retreaded) to determine which set performs better in critical driving conditions. The results are then analyzed using the TOPSIS method where the most efficient set of tires and the best alternatives are selected. Finally, the safety of trucks on the road using the appropriate set of tires is evaluated by the estimated accident reduction potential. It should be mentioned that the optimal selection of the truck tire set is also important for sustainable transportation, as the pollution of worn tires remains a relevant environmental issue.

Keywords: tire set; safe driving; truck braking; wet grip; accident reduction

1. Introduction

Smooth and efficient freight transport on land roads depends mainly on traffic and weather conditions, and to consistently reduce the number of accidents, road infrastructure and vehicle safety is being improved, driver offenses monitored, and road safety training and education provided. Despite this, around 1.19 million people a year are still killed on the world's roads and between 20 and 50 million suffer non-fatal injuries [1]. According to American accident statistics, fatalities involving heavy-duty vehicles (HDV) account for approximately 11% of the total, while in Europe this is up to 14% [2,3].

In addition to continually improving safety technology, it is important to consider the proper maintenance and operation of the vehicle fleet when analyzing the causes of vehicle accidents. These processes are particularly relevant for freight transport companies and focus mainly on road safety [4,5]. The use of defective or worn tires can be one aspect of poor fleet maintenance, and tire-related poor grip is directly related to both the cause and the consequences of an accident [6]. Tires are an essential part of the costs associated with truck operation; therefore, the rational and safety-optimized selection of the tire set is an important task in organizing the operation of a congested vehicle fleet.

The rational use of tires is important not only from a safety point of view, but also from an environmental perspective [7]. It is estimated that with the new regulation from the

United Nations (UN R117-04) to maintain non-deteriorating tire wet braking performance up to a minimum allowable wear limit, there will be a reduction in tire production-based CO_2 emissions in Europe by 6.6 million metric tons. This step toward tidiness in the transport sector is driven by the trend that almost 50% of tires are discarded before they reach the 3 mm wear threshold.

The tires of commercial vehicles used for the regular commercial transport of goods are replaced in axles as required; that is, all tires are not replaced synchronously due to unequal wear, which would directly increase the costs for a logistic company. This often leads to the search for the most cost-effective tire replacement principles, while the complexity of predicting driving safety means that the matching of these combinations is usually not evaluated. This research aims to determine to what extent the handling characteristics of HGV vary when different tire sets are used and to what extent a different tire set can increase the safety potential of a company's managed fleet.

The paper is organized as follows. The literature review is described in Section 2, and the materials and methods used for experimental research, including a description of the tire sets, are presented in Section 3. In Section 4, the experimental research with evaluation of the collected data is presented. Section 5 discusses the road safety assessment based on the estimated data for the truck fleet. In Section 6, the results and discussion regarding processed data are presented. Finally, Section 7 concludes the paper.

2. Literature Review

Theoretical and experimental research are aimed at assessing the most appropriate choice of vehicle structure, component parts, or materials to achieve the best possible dynamic and safety performance [8]. Here, variable speed conditions for a non-stationary system's dynamics require specific studies which include experimental research with data-driven strategies [9,10]. In this case, tire interaction with the road surface is a key factor in the safe and reliable handling of the vehicle and the effective operation of the active safety systems [11,12]; therefore, experimental and theoretical tests are carried out to discover the relationship between road safety and tire properties or changes in tire state [13].

Experimental tests on the whole vehicle are the most widely used to assess braking efficiency and handling under different conditions. Stoklosa and Bartnik [14] conducted experimental brake tests on dry, wet, and snowy road surfaces using summer and winter tires, respectively. The tests examined the effect of tire inflation pressure on stopping distances. The braking tests from a speed of 50 km/h showed an increase in stopping distances of 0.49 m on dry asphalt pavement using summer tires with a pressure of 1.5 bar. Above the recommended tire pressure, stopping distances increased by 0.25 m on average. With winter tires, the braking distance increases in all cases, with the greatest increase of 0.33 m at 1 bar. Similar to this, Kreitzberg et al. [15] summarized a test carried out in winter conditions with full stop braking and stopping with obstacle avoidance. Measurements were carried out in 30 different passenger cars for a total of 200 tests: 69 straight-line braking tests from 50 km/h and 131 braking tests with an obstacle on the road. The test site was covered with a 3–5 cm thick layer of packed snow with ice cracks underneath and the temperature ranged from −7 °C to +3 °C. The hypothesis that heavier vehicles generally have a longer braking distance in slippery conditions was rejected. The main factor for the shorter stopping distances was the tread depth of the tires, with the year of manufacture of the tires remaining a secondary factor.

Malmivuo and Luoma [16] conducted a study to investigate the traction of studded and non-studded tires on snowy and icy surfaces. Nokian Hakkapeliitta 8 adapted tires were used and 150 runs were carried out on a track. The test also aimed to assess how the friction coefficient changes when different numbers of vehicles pass over the test section.

For studded tires, the effect of the test lane loading on friction was found to be relatively small, but for non-studded tires, the lane loading resulted in a 24–31% reduction in grip under braking and a 13–19% reduction under acceleration.

Heavy vehicle tire testing is much rarer, as it is more costly and requires an expensive test vehicle, which can only be offered by new truck dealerships and logistics or haulage companies [17,18]. Tuononen [19] describes the characteristics of heavy-duty vehicle tires, which are estimated from optically measured tire carcass deformations. Accurate information on tire forces is important for the development of active safety systems and the prediction of vehicle stability under critical driving conditions. For research purposes, a dedicated sensor module has been developed to estimate tire deformation and forces that act on the tire in real time based on the displacement of the tire carcass. A Volvo FH12 truck is used for this test, with sensors inside the rim. The test was carried out on a snow-covered frozen lake, with a 150 m arc turning maneuver selected. The vehicle was accelerated to a threshold speed at which the towing vehicle starts to slip. The test showed that the chosen optical sensor technology is generally valid: the lateral and longitudinal force estimates are accurate in most cases, but the vertical force estimate is not reliable under heavy braking.

In addition to this, it is accepted that tire testing under variable road conditions is particularly relevant, but its accuracy is more difficult to ensure. Following this, tire contact patch deformation on wet surfaces was investigated by Niskanen and Tuononen [20]. An integrated optical sensor in the tire provides insight into the relationship and context of partial and full aquaplaning. Researchers have developed a real-time algorithm to detect partial aquaplaning at speeds up to 40 km/h. Optical, strain, or acceleration sensing inside the tire is relevant for all types of tires, while more promising applications have been tested on large or agriculture tires [21]; however, the mass integration of these sensors in tires is not yet ready, even though the installation of accelerometers is easier [22].

In contrast to the experimental assessment of overall braking performance or vehicle handling, tire grip properties are often analyzed separately. In this way, vehicle tire test benches offer many opportunities to investigate a wide range of tire traction, wear, noise, and other parameters; however, they are cost and time consuming and, compared to a detailed mathematical model (computerized calculations), do not always provide an accurate assessment of all the desired parameters. In these conditions, it is difficult to assess the load on the tire and choose the right road surface [23]. Furthermore, on test benches, the tire is usually not in natural contact with the road surface; that is, a large diameter roller that results in an incomplete contact patch [24] or a conveyor belt [25].

The comparison between experimental data and simulation results is a common approach in current research. Lugaro et al. [26] have found that both the measurement and simulation results show similar slip performance results; however, the measured stopping distances between the different tests differ by up to 2 m in individual cases. This is mainly due to the different properties of the pavement in the calculation model and in the test area; however, surface roughness causes high nonlinearities for tire grip and wear properties; therefore, these contact features must be taken into account during tire performance testing [27]. In general, stopping distances are an important aspect in assessing the technical capability of trucks to improve safety, as they determine the probability of avoiding an accident or minimizing its consequences. For heavy-duty vehicles, this safety criterion is very important because the stopping distance is longer than for other road users due to the high mass. In addition, the high weight of trucks often leads to multi-crashes, especially on highways or mountain roads [28]. Similarly, stopping sight distance (SSD) is often evaluated in road safety projects and it consists of stopping distance, driver reaction time, and brake system deployment time. For a truck with a trailer, the safe stopping sight distance at 90 km/h is 175 m, while for an average passenger car, this distance is 145 m [29].

A potential collision situation with a slower vehicle in front was analyzed by Christ [8], who found that for each of the parameters tested (vehicle speeds, distance between vehicles, driver reaction time, brake response time, and tire-pavement grip), the relative risk level was reduced; however, it was highlighted that a 5% increase in grip would result in a 4% reduction in accident rates. The author also points out that the increased use of advanced safety systems will reduce the influence of the driver and increase the impact of the grip on the risk of a crash [30,31]. Despite the fact that the preventive measures are always a positive step towards road safety, when an accident occurs, the driver's actions and vehicle condition remain the only factors. When a vehicle is braking, the tire contact patch with specific tread design and rubber compound is the only place the vehicle comes into contact with the road surface [32].

In this context, individual efforts or trends to significantly reduce accident rates are not sufficiently effective. A freight transport fleet is an appropriate ecosystem for the assessment of a wide range of different safety measures, such as the following: driver selection and training, improving driving culture, route planning and management systems, driver workplace improvements, safety technology, and fleet maintenance [33,34]. There is no single best strategy to improve road safety, and the most effective is usually a combination of road measures that affect individual fleets differently. Automated fleet management and routing is another important tool to improve safety in freight transport companies [35]. Route planning allows you to find the shortest or fastest route, avoid congestion, and reduce driver anxiety and fatigue. Such smart systems also allow one to monitor driver's well-being and drowsiness, and avoid the urge to engage in extraneous and irritating activities. Smart vehicle monitoring systems include systems that allow the monitoring of technical conditions, brake wear, maintenance intervals, tire inflation, or even general wear.

The global automation of the freight fleet can improve road safety if all vehicles meet at least the minimum requirements of the model. In line with this, Wang et al. [36] estimated in their study that, by achieving certain vehicle and road surface parameters, using a safe distance model and vehicle communication, rear-end collisions can be completely avoided. In general, vehicle safety assessment is a broad and widely researched area, including tire testing; however, the research on trucks is more specific, especially when it comes to assessing the effects of different tires and their features. The specifics of freight transport companies lead to a compromise between rational tire selection and maximum possible driving safety; therefore, in this paper, the option of individual tire set selection as a sustainable alternative is planned after experimental tests and a truck fleet safety assessment. From the sustainability point of view, retreaded tires are included in this research as a valuable option to reduce the amount of waste [37].

3. Research Conditions and Used Tire Sets

Test sessions on braking performance, hill climb, and cornering were conducted on an enclosed proving ground to determine which tire set gives the best handling for the truck (Figure 1). Three selected maneuvers were all carried out on the same MAN TGX 18.470 truck with a tri-axle trailer and measuring equipment mounted on a cabin (Figure 2). The inertia measurement unit (IMU) together with the GPS sensor from "Race Technology" (Nottingham, UK) were used to log vehicle acceleration, yaw rate, and GPS data. The truck was driven by the same instructor, experienced in driving such vehicles.

Figure 1. Maneuvers on a proving ground: 1—braking on wet asphalt; 2—braking on artificially slippery surface; 3—driving downhill with a corner; 4—slippery hill climb.

Figure 2. Test truck with measuring equipment and prepared tire sets.

The brake test, as one of the most important in a potential emergency situation, was carried out in two phases: on a wet asphalt surface with an initial speed of 60 km/h and on an artificially slippery surface with a water film from an initial speed of 40 km/h (Figure 1). In both braking maneuvers, the driver applies the brake pedal fully to achieve the shortest braking distance. All the assisted driving and safety systems of the truck were active during braking.

The downhill slippery cornering maneuver was carried out with a steady steering input at 20 km/h (Figure 1). For safety reasons, this maneuver was carried out without a trailer. Finally, the slippery hill climb was carried out from the same starting position each time, with constant accelerator pedal depression, and with an active traction control system.

The truck was tested with seven different commonly used tire sets consisting of new and used tires from different manufacturers and with different seasonality. To examine whether cheaper retreaded tires are as safe on the road as new tires, new and retreaded tires were compared on different tire sets. The trailer braking test used retreaded universal tires (three sets) of different patterns, retreaded by different tire retreaders. This range of selected tires is usually used by freight companies in Central and North Europe where the road surface is usually wet or snowy in autumn and winter seasons. Table 1 shows a list of tires used and a short description. Each tire description also includes the tread hardness measured during the test according to Shore A class [38].

Table 1. Tires with their short description and assigned coding are used for experimental research.

Tire Tread Design	Tire Properties	Tire Tread Design	Tire Properties
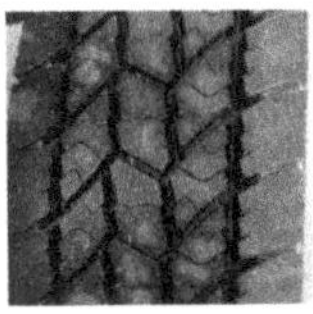	UGMAXS New, winter 'Goodyear' (tread hardness 66.7) (Colmas-Berg, Luxembourg)	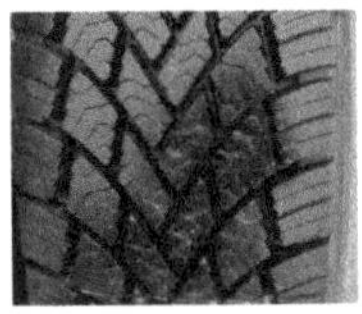	ENRUGMAXD Retreaded from 3-year-old carcass 'Goodyear KMAX D' (tread hardness 63.5) (Colmas-Berg, Luxembourg)
	NF501 New, all-season 'Kama NF501' (tread hardness 68.5) (Kruševac, Serbia)		GUMRZT220 Retreaded from a 3-year-old carcass 'Vipal VT220' (tread hardness 68.6) (Nova Prata, Brasil)
	KMAXSG2 New, all-season 'Goodyear KMAX S' (tread hardness 71.9) (Colmas-Berg, Luxembourg)		GUMM788 Retreaded from 3-year-old carcass, all-season 'Bandag' (tread hardness 57.6) (Muscatine, IA, United States)
	MOTHKPLD Retreaded from a 3-year-old carcass 'Noktop' (tread hardness 63.4) (Nokia, Finland)	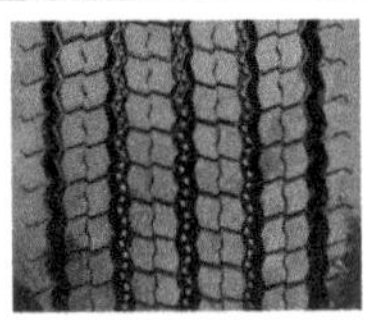	MOTHKPLF2 Retreaded from 3-year-old carcass, all-season 'Kraiburg' (tread hardness 65.6) (Geretsberg, Austria)
	UGMAXD Retreaded from 3-year-old carcass, all-season 'Nokian Hakkapeliitta Truck E2' (tread hardness 65.4) (Nokia, Finland)		ENRWSS Retreated from 3-year-old carcass, all-season 'Ringtread Blackline WSS' (tread hardness 67) (Henstedt-Ulzburg, Germany)
	MOTHKPLE New, winter 'Goodyear UltraGrip Max D' (tread hardness 66.5) (Colmas-Berg, Luxembourg)		

Although seven different tire configurations are used for the tractor and three for the trailer, most of them use the same tires on a single axle (Table 2). In this way, the first three sets of tires consisted of identical tires for the rear axle of the truck and different tires for the front axle of the truck, so that for the first three sets of tests, only the front tires were compared. By evaluating tires in this way, it is theoretically possible to select a tire set that will be even more effective in braking tests.

Table 2. Tire sets used in experimental research.

No of Tire Set	Vehicle Axis		
	Front (Steered)	Rear (Drive)	Trailer
	Code for Tire Set		
1	UGMAXS		–
2	NF501	MOTHKPLD	–
3			–
4		UGMAXD	–
5		MOTHKPLE	–
6		ENRUGMAXD	–
7	KMAXSG2		–
8			GUMM788
9		GUMRZT220	MOTHKPLF2
10			ENRWSS

4. Experimental Research

4.1. Evaluation of Brake Efficiency

The mean fully developed deceleration (MFDD) was calculated for the braking efficiency analysis that does not consider the initial and final braking stages and is calculated from raw longitudinal acceleration data [39]:

$$\text{MFDD} = \frac{v_a^2 - v_b^2}{2(S_b - S_a)}, \tag{1}$$

here, v_a—vehicle speed corresponding to 80% of the initial speed ($0.8v_o$); v_b—vehicle speed corresponding to 80% of the initial speed ($0.1v_o$); S_b—distance traveled in braking mode between v_o and v_b; S_a—distance traveled in braking mode between v_o and v_a; and v_o—initial vehicle speed before braking.

The brake data on wet and artificially slippery surfaces are summarized in Table 3.

Table 3. Data evaluated from two braking tests.

No of Tire Set	Average Deceleration, m/s^2	Maximum Deceleration Value, m/s^2	MFDD, m/s^2	Braking Distance, m	Braking Time, s
Braking on wet asphalt pavement					
Front axle under consideration					
1	6.05	7.46	6.67	28.32	3.15
2	5.66	6.79	6.05	28.59	3.37
3	6.25	7.83	6.78	27.73	3.11

Table 3. *Cont.*

No of Tire Set	Average Deceleration, m/s^2	Maximum Deceleration Value, m/s^2	MFDD, m/s^2	Braking Distance, m	Braking Time, s
		Rear axle under consideration			
4	6.59	7.90	7.17	26.06	2.95
5	6.36	7.96	6.79	24.74	2.92
6	6.49	8.07	7.11	24.90	2.89
7	6.51	8.14	7.00	24.30	2.91
		Trailer			
8	6.31	7.97	6.90	26.70	2.99
9	5.82	7.50	6.53	27.37	3.18
10	6.31	7.76	7.02	25.35	2.92
Braking on artificially slippery surface					
		Front axle under consideration			
1	2.11	4.74	2.48	46.22	6.25
2	2.55	4.65	2.89	37.79	5.39
3	2.24	5.25	2.44	39.04	5.76
		Rear axle under consideration			
4	2.18	4.89	2.37	39.67	5.89
5	1.90	4.67	2.01	39.91	6.21
6	1.82	4.67	2.05	48.12	6.94
7	1.94	4.86	2.11	42.48	6.35
		Trailer			
8	1.78	4.59	2.17	48.58	6.89
9	1.80	4.39	2.09	51.07	7.08
10	1.77	4.64	2.01	46.02	6.73

The braking performance results show certain trends between tire sets, where the most efficient tire sets also change with different pavement conditions, making it difficult to clearly identify which tire set is the most efficient and the best possible alternatives for safe driving. Here, the technique for order of preference by similarity to ideal solution (TOPSIS) methodology is used for a more accurate assessment of results. This approach is used in a wide range of areas where different possible alternatives are analyzed [40,41].

In this multi-criteria decision making (MCDM) method, it is assumed that the tire sets to be tested are defined as alternatives denoted by $A_1, A_2, \ldots, A_m$, and the criteria (MFDD, stopping distances, and other indicators) against which the alternatives are evaluated are denoted by $C_1, C_2, \ldots, C_n$. Also, x_{ij} ($i = 1, 2, \ldots, m; j = 1, 2, \ldots, n$) denotes the value assigned to criterion i of alternative j and $X = [x_{ij}]_{mn}$ is the decision matrix. The associated weight value of each criterion is denoted by $X = [w_1, w_2, \ldots, w_n]$, when $\sum_{j=1}^{n} w_j = 1 \sum w_j = 1$ [42].

The decision matrix is then normalized according to the following equation:

$$\overline{r_{ij}} = \frac{x_{ij}}{\sqrt{\sum_{k=1}^{m} x_{kj}^2}};\ i = 1, \ldots, m;\ j = 1, \ldots, n, \tag{2}$$

here, r_{ij}—the normalized value i of criterion j for alternative A_i.

A weighted normalized decision matrix is then calculated:

$$v_{ij} = w_j \cdot r_{ij};\ i = 1, \ldots, m;\ j = 1, \ldots, n, \tag{3}$$

here, w_j—the weight of criterion or attribute j.

The ideal positive and ideal negative solutions for each criterion are then determined:

$$A^{+} = \{v_1^{+}, \ldots, v_n^{+}\}, \tag{4}$$

$$A^{-} = \{v_1^{-}, \ldots, v_n^{-}\}, \tag{5}$$

here, A^{+}—specifies the ideal positive solution; and A^{-}—ideal negative solution. If criterion j is useful, then it is assumed that $v_j^{+} = max\{v_{ij}, i = 1, \ldots, m\}$ and $v_j^{-} = min\{v_{ij}, i = 1, \ldots, m\}$. If the criterion is useless, then $v_j^{+} = min\{v_{ij}, i = 1, \ldots, m\}$ and $v_j^{-} = max\{v_{ij}, i = 1, \ldots, m\}$.

The distances from each of the alternatives i to the best positive and negative solution are then calculated:

$$D_i^{+} = \sqrt{\sum_{j=1}^{n} \left(v_{ij} \cdot v_j^{+}\right)^2}, i = 1, \ldots, m, \tag{6}$$

$$D_i^{-} = \sqrt{\sum_{j=1}^{n} \left(v_{ij} \cdot v_j^{-}\right)^2}, i = 1, \ldots, m. \tag{7}$$

Finally, each alternative is evaluated in terms of its relative proximity to the ideal solution and distance from the negative solution:

$$y_i = \frac{D_i^{-}}{D_i^{+} \cdot D_i^{-}} \tag{8}$$

The performance indicator shows which of the alternatives, in this case a set of tires, is the best in terms of the selected evaluation criteria.

This analysis requires the selection of appropriate values or weights for the criteria. This is already a human factor, but for a more accurate selection of the weights, a correlation analysis of the different braking criteria is carried out using Microsoft Excel Solver software (v16.0) to discover the relationship between the different measured braking test criteria and the extent to which they are dependent on each other. The strategy chosen to weight the criterion most heavily was to use the less correlated measurements, as, because they are less correlated, they are less likely to influence the results of the analysis. The weights of the other criteria were chosen based on the practical use of the measurement results. In studies where the selection of weights is important, the TOPSIS method often uses expert interviews [43,44], but in this case, it was decided that were would be a focus on the opinion of the researchers who conducted the test. The weights for the "Race Technology" accelerometer performance criteria are shown in Table 4.

Table 4. Values of alternative criteria.

Criteria	Criteria Value, w_j
Average deceleration, m/s^2	0.1
Maximum deceleration, m/s^2	0.1
MFDD, m/s^2	0.35
Braking distance, m	0.35
Braking time, s	0.1

The analysis of the numerical results shows that the relationship between stopping distance, stopping time, and the MFDD value remains, despite the fact that the MFDD value does not take into account the initial braking pulse and the end of the braking (braking repulsion). Some of the values indicate that the full deceleration may not have been developed from the start of braking (because of tire slip, delay in ABS activation, and delay in brake force control).

After analyzing the braking maneuver test measurements using TOPSIS analysis, the specific rankings of the tire sets obtained according to the braking criteria evaluated are prepared in Table 5.

Table 5. Results of the brake test according to the TOPSIS performance index y_i.

No of Tire Set	Braking on Wet Asphalt Pavement	Braking on Artificially Slippery Surface	Evaluation Ranking Position
	Front axle under consideration		
1	0.74	0.05	3
2	0.21	0.89	2
3	1.00	0.53	1
	Rear axle under consideration		
4	0.56	0.86	1
5	0.68	0.51	3
6	0.83	0.16	4
7	0.86	0.46	2
	Trailer		
8	0.56	0.62	2
9	0.16	0.24	3
10	0.93	0.61	1

The most effective braking on wet asphalt was achieved with tire set 3, consisting of a new Goodyear KMAX S all-season tire in the front and a NOKTOP retreaded tire in the rear (Table 1), but under different braking conditions, such as on artificially slippery surfaces (imitation of ice), tire set 2, consisting of a new Kama NF501 all-season tire in the front axle of the towing vehicle and a retreaded winter tire with a Noktop tread in the rear axle, was more effective. However, the difference in test results was not sufficient to allow the overall performance of set 2 to exceed that of set 3. The most efficient tires were those with many small tread blocks with transverse grooves. Since the artificially slippery road surface corresponds to ice-covered asphalt and the geometry of the described tires is designed to cling in sipes to a low-grip road surface, the results seem to justify the predicted tire performance.

When comparing tire sets for the rear axle of the truck, tire set 4, consisting of Goodyear KMAX S all-season tires for the front axle of the truck and retreaded NOKIAN all-season tires, was the most effective in the braking test. However, this is only an optimal option, but not the best one for braking on different surfaces. In the tests, tire set 6 was the most effective on wet asphalt and tire set 4 on artificially slippery surfaces, but these tire sets were less effective on other surfaces.

Of the three different tire configurations for the trailer, the most efficient is tire set 10, with the finest tread and fine grooves angled at 45 degrees. The truck with this set of tires stopped 1.95 m shorter and had an MFDD of 0.36 m/s^2 higher than the worst performing trailer tire.

The results partially supported the hypothesis that a single set of tires would not be the most efficient for braking on both surfaces, due to differences in tread pattern properties, seasonality, and tread compound hardness. On wet asphalt, tires with a harder compound and fewer transverse grooves were more effective, but large tread blocks and fewer transverse grooves did not grip as well on artificially slippery surfaces, which had a higher water film, and tires with these geometries are more likely to suffer aquaplaning.

4.2. Hill Climb Evaluation

The hill climb allows the traction of the drive axle tires to be assessed when starting from a standstill on slippery road surfaces. Two parameters were chosen for the qualitative evaluation of the test: the average longitudinal acceleration over a distance of 35 m and the time taken to reach the distance of 35 m. The test used the same truck as in the braking test, without a trailer, with tire changes on the drive axle. The same maneuver start location was chosen for the runs, and the end location was determined after 35 m from the start of the maneuver based on GPS sensor data. The accelerator pedal was released after the end of the maneuver mark, thus avoiding braking or accelerator release before the maneuver was completed. The test results were analyzed using the TOPSIS analysis method described in the previous section and the results are presented in Table 6.

Table 6. The results of a hill climb for a truck on an artificially slippery slope.

No of Tire Set	Average Value of Longitudinal Acceleration After 35 m, m/s^2	Time Taken to Drive 35 m Climbing, s	TOPSIS Performance Indicator y_i	Evaluation Ranking Position
1	1.20	15.21	0.34	4
4	1.34	13.89	0.81	2
5	1.26	15.19	0.42	3
6	1.19	16.72	0.12	5
7	1.38	13.25	1	1

According to the data, the most effective tire set for the artificially slippery climb was tire set 7, which consisted of a new Goodyear KMAX S all-season tire, which did not affect grip in this maneuver, and a retreaded rear tire with Vipal VT220 winter tread. These tires had the hardest tread on the Shore A scale. Compared to braking tests, where the softest tires were the most effective on wet asphalt, the results are reversed in this test: the second best-performing tire was the NOKIAN HAKKAPELIITTA TRUCK E2 retreaded winter tire, while the tire set 5 (the new Goodyear UltraGrip Max D tires) was better than the Goodyear KMAX S and KMAX D retreaded tires with similar tread, but this may also be due to the tire's sipes and the hardness of the rubber compounds.

4.3. Driving Downhill with a Corner Evaluation

During the driving downhill with a corner experiment, the average lateral acceleration and yaw rate acting on the truck from its initial onset (the start of the track curve) to the full stop were analyzed. This test checked whether the truck was turning enough to maintain its trajectory and whether the unladen drive axle (without trailer) was slipping.

The measuring equipment has a GPS sensor, so it was possible to identify a single point during each test run at which cornering starts, and to sample the exact point at which cornering intensity starts to increase. From the results of the measurements (Table 7), it can be tentatively stated that the truck handles best under these conditions with tire sets 1, 3, and 6, but the values given depend on the slope and angle of the track, and the maneuver path and the driver's action (maintaining a constant speed). Therefore, the yaw rate and driving speed data in the time domain were used to evaluate this maneuver.

Table 7. Results of a driving downhill with a corner evaluation.

No of Tire Set	Mean Lateral Acceleration, m/s^2	Mean Yaw Rate, deg/s
1	3.637	19.34
2	2.424	18.66
3	2.947	20.11
4	2.703	17.72
5	2.800	18.22
6	2.912	20.39
7	2.034	16.12

These characteristics show whether the steering or drive axles were slipping, as well as the speed at which the maneuver was carried out at the apex of the turn, when the maximum yaw rate is developed. All maneuvers were started at an average speed of 26 km/h and the downhill driving was at idle speed in the engaged gear, without the accelerator or the service brake pedal, maintaining the same trajectory (driver factor). Figure 3 shows three runs of the truck with tire set 6.

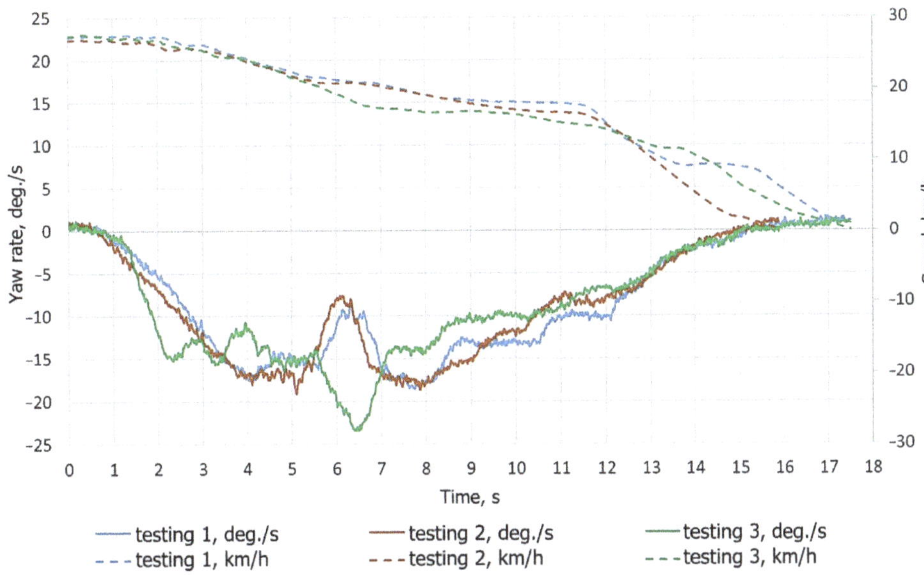

Figure 3. Testing of tire set 6; plot of yaw rate and driving speed.

The graph shows that between the sixth and seventh second, all three tests showed a significant variation in the yaw rate. This is because the rear axle of the truck was slipping when the truck was in the turning circle, resulting in a sharper rotation around the vertical axis, while the reduced intensity indicates the slipping of the steering axle (vehicle understeering). The front axle of the tractor slipped in the first and second tests and the rear (drive) axle lost traction in the third maneuver, so this set of tires did not give stable results in this test. For all tire sets, tire set 5 was the most stable one (Figure 4).

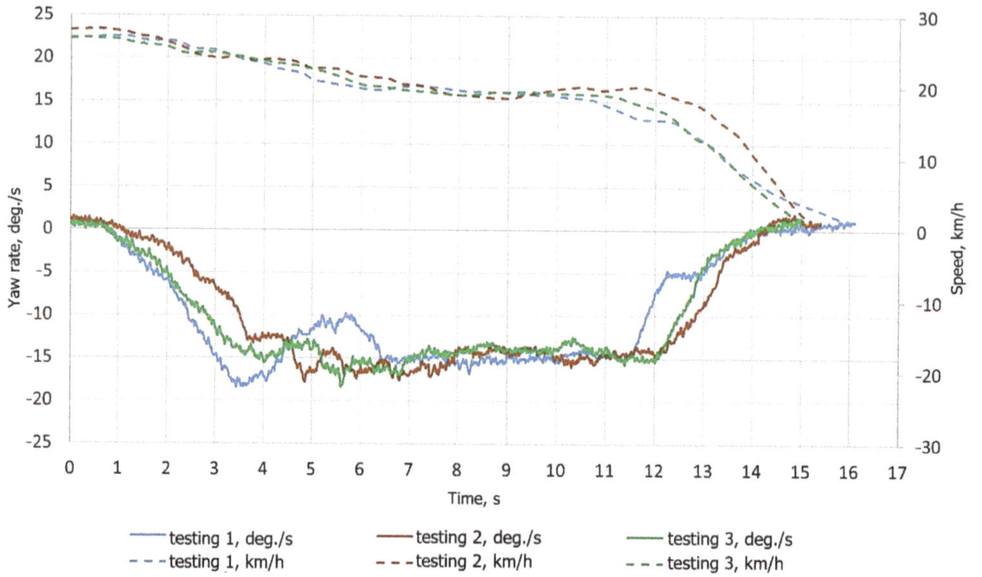

Figure 4. Testing of tire set 5; plot of yaw rate and driving speed.

It was observed that a stable yaw rate was maintained with tire set 5. The graph does not show any significant spikes in the yaw motion, so it can be concluded that there was no significant tire slippage and the handling of the truck remained smooth and predictable. The average initial speed of the three tests was 26 km/h, which was not lower than that for the other set of tires.

5. Safety Assessment of the Truck Fleet

One of the most important issues discussed in the field of road transport is the maximum permissible speed limits and their link with the frequency of accidents. Most drivers believe that they are capable of choosing a safe speed, but there is not a country that has not introduced speed limits. However, despite driver training and education, 30–50% of all drivers exceed the maximum speed limit. Speeding significantly increases the risk of accidents and injury, or even death [45]. Elvik et al. describe the emergence of the Power model and its application to road safety assessment, where the aim is to assess how a change in speed determines the risk of certain factors related to road safety. This is carried out by reviewing and summarizing a large number of studies and accident statistics that have assessed the effect of speed changes on the number and severity of accidents. It also compares whether the Power model can be accredited as suitable for use in practice, and what factors may lead to a discrepancy between the theoretical calculations.

The model is suitable for describing the severity of an accident; however, it does not specify the probability of an accident. The model has implications for the consequences of accidents involving injuries and fatalities, as it derives a degree indicator. According to the model, the number of fatal accidents, accidents with serious injuries, and all accidents with injuries recorded by the police (including fatal and serious injury accidents) varies in proportion to the change in the speed of driving, when raised by a certain degree, which is dependent on the severity of the accident or the injury under investigation. In general, the model predicts the change in the number of accidents or the number of injured or killed road users as the speed changes.

When analyzing specific aspects of road safety, the degree to which the speed differential is raised varies depending on the aspect being analyzed, as all aspects vary differently. A specific dataset is defined for the degrees and their thresholds used to assess the different severities of an accident or injury [45].

Based on this, the report of Institute of Transportation Engineers [46] used the accident rate of change method to assess the road safety risks of the experimental test. The crash modification factor (CMF) is calculated by raising the ratio of the original and modified speeds by a certain grade:

$$\mathrm{CMF} = \left(\frac{v_a}{v_b}\right)^x \tag{9}$$

here, v_a—initial speed or speed before traffic regulation changes; v_b—actual speed or speed after traffic regulation changes; and x—grade indicator by type of accident or injury.

This approach is used to select or modify the maximum speed limit on a given section of traffic, but in this case, it is applied to an experimental study on tires and how different tire performance results can determine the road safety potential.

On the basis of the braking tests carried out in this research, it was chosen to assess the difference in speeds during full braking on different surfaces. The CMF is calculated by selecting the initial speed of the most efficient tire set in terms of braking distance and the speeds of the other tire sets in the same test series in the denominator of Formula (9). As the speed is selected at the time of braking, a situation is modeled in which there is already a probability of an accident, i.e., an obstacle in the road, and therefore no rate of change in accident occurrence is calculated, only the relative rate of change in the different injuries. The speed data for the conducted braking tests are available at 100 Hz; therefore, it was chosen to assess the CMF over the entire braking maneuver range. This allows the performance of different tire configurations to be seen and the variation throughout the maneuver to be seen, as an obstacle collision can occur at any time during braking or at any speed in the braking interval. Although a CMF of 1 as a reference value is derived from the most efficient speed parameters of the tire set, it can be seen from this analysis whether there are times during braking when a more useful CMF value for road safety is calculated for a vehicle with an experimentally captured longer braking distance.

The CMF analysis of the experimental tests compared the front, rear, and semi-trailer tire configurations on wet asphalt and artificially slippery surfaces. For the CMF assessment, the grade $x = 4.1$ is selected for fatalities and $x = 2.2$ for severe injury change. Among the rear tire configurations of the truck, the shortest braking distance was found in the initial analysis with tire set 7 on wet asphalt, resulting in a CMF value of 1 for this set of tires.

Figure 5 shows that the instantaneous (residual) speed during braking varies differently for each set of tires, giving a different CMF value. Tire set 3 has a lower relative CMF than the others in the entire speed range, indicating that it is the most unsafe tire in an accident. The vehicle with the tire set 6 maintains a higher speed than the reference tire set 7 (CMF < 1) during half of the braking maneuver, but after slowing to a speed of $v_a = 24$ km/h, the CMF has a lower speed value than set 7 (starts to generate more effective braking) and thus a reduced change in fatalities in the accident. At the end of braking, as the speed drops to 11–12 km/h, there is a sharp drop in CMF of sets 4 and 6, indicating better wet grip of set 7 at low towing speeds.

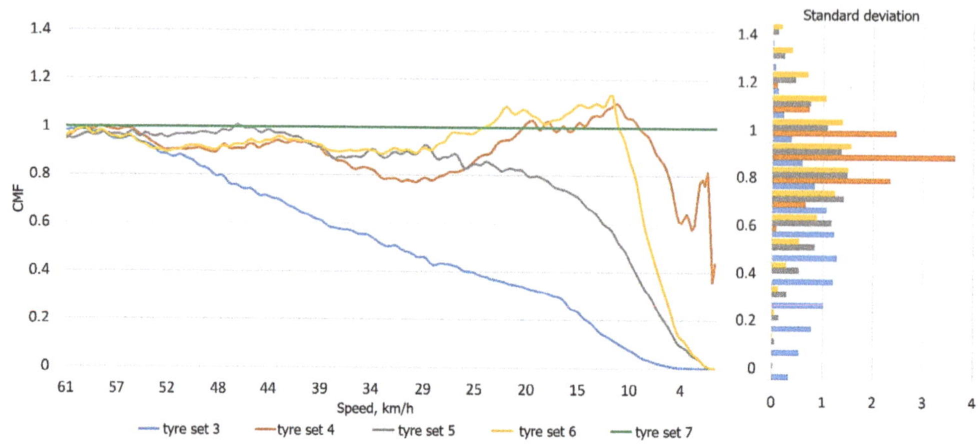

Figure 5. The change in CMF for road fatalities; rear axle tire sets on wet asphalt is evaluated.

In order to visually evaluate the most efficient tire configuration and its alternatives, a histogram of the distribution of values is prepared and presented next (vertically) to the CMF values. The width spread of the histogram indicates the mean CMF value and the height of the histogram indicates the standard deviation.

The CMF graphical comparison curves for the risk of injury to road users do not differ much from the relative value of the risk of fatality, because CMF calculation uses the same speeds for the same experimental analysis, but the grade is selected $x = 2.2$. Figure 6 shows a CMF plot of the front tire test on an artificially slippery surface, evaluating the change in road traffic injuries.

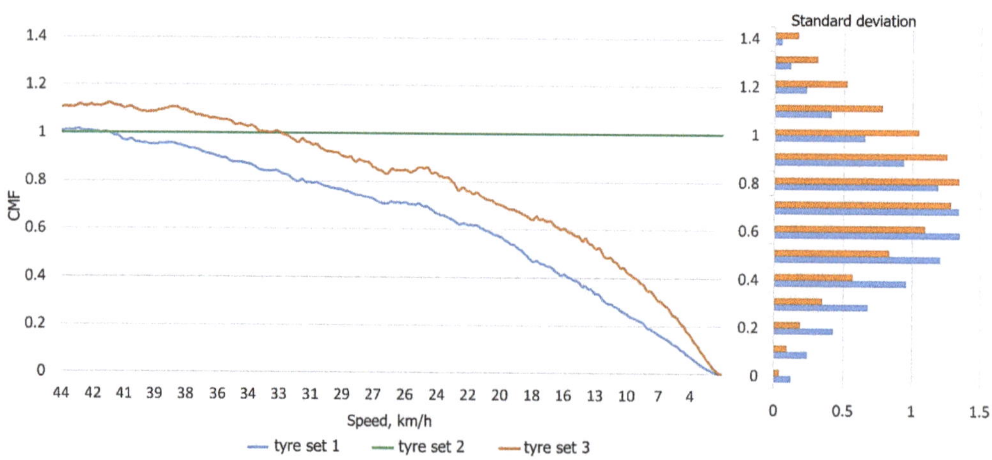

Figure 6. The change in CMF for road traffic injuries; front axle tire sets on artificially slippery surface evaluated.

The most effective in terms of braking distance in the initial analysis is tire set 2, and its CMF is used as a reference. From the graph it can be seen that the initial speed at the start of braking was not completely uniform, since the CMF curve for set 3 starts higher than for set 2. This shows that the denominator of the CMF equation had a lower speed than the numerator, so the vehicle with set 3 had a lower speed at the start of the brake.

If the curve starts below the reference CMF value, then the speed at the start of braking is higher. The CMF curves for tire sets 1 and 3 decrease uniformly, which means that the value of the speed difference increases almost equally throughout the braking period. At a reference braking speed of 20 km/h, an increase in injury change during the accident of 29% with tire set 3 and 42% with tire set 1 is calculated.

A situation can be modeled where X collisions occurred at 20 km/h when braking from 60 km/h on a snow surface, causing injuries to Y road users, with a vehicle with tire set 2 in use. The calculations presented here would then predict that the injury rate would be 29% higher under these conditions with tire set 3.

It is difficult to estimate from the CMF graph exactly how much the risk of fatality increases with the most ineffective tire set; therefore, the mean CMF value and standard deviation for each alternative tire set were calculated. A CMF value less than one indicates an increase in the change or number of injuries in the accident under investigation, based on statistical data, while values greater than one indicate a decrease in the change in the accident under investigation when compared to the reference speed. Comparisons of CMF values converted into percentages for the front tires are shown in Table 8, for the rear tires in Table 9, and for the trailer in Table 10.

Table 8. CMF comparison of front axle tire sets under braking on different surfaces.

Road Surface	No of Tire Set			No of Tire Set		
	1	2	3	1	2	3
	CMF for Road Fatalities			CMF for Road Traffic Injuries		
Wet asphalt	10% (0.27)	10% (0.38)	–	7% (0.20)	14% (0.29)	–
Artificially slippery	50% (0.33)	–	29% (0.41)	35% (0.29)	–	21% (0.30)

Table 9. CMF comparison of rear axle tire sets under braking on different surfaces.

Road Surface	No of Tire Set					No of Tire Set				
	3	4	5	6	7	3	4	5	6	7
	CMF for Road Fatalities					CMF for Road Traffic Injuries				
Wet asphalt	49% (0.31)	10% (0.11)	21% (0.27)	13% (0.25)	–	34% (0.28)	5% (0.06)	14% (0.21)	9% (0.20)	–
Artificially slippery	–	−5% (0.40)	31% (0.31)	47% (0.28)	35% (0.27)	–	−1% (0.16)	20% (0.22)	30% (0.23)	21% (0.20)

Table 10. CMF comparison of trailer tire sets under braking on different surfaces.

Road Surface	No of Tire Set			No of Tire Set		
	8	9	10	8	9	10
	CMF for Road Fatalities			CMF for Road Traffic Injuries		
Wet asphalt	28% (0.32)	9% (0.44)	–	19% (0.25)	10% (0.31)	–
Artificially slippery	11% (0.14)	16% (0.35)	–	6% (0.08)	13% (0.28)	–

According to the safety potential evaluation, the mean value of the coefficient of variation of tire set 4 for road traffic injuries when braking on wet asphalt pavement is $\overline{x} = 0.949$ and the standard deviation is $\sigma = 0.059$. Therefore, the vehicle with tire set 4 results in a 5.1% increase in the change in road fatalities (Table 9) compared to the most efficient tire set 7. The CMF values are reported in the same way in Tables 8–10. Negative values indicate that the tire set has a higher safety potential than the reference for the entire braking interval.

6. Results and Discussion

An analysis of the experimental test results identified the tire sets that were the most effective in the selected maneuvers. In the braking tests with the trailer, tire set 10 was the most effective one, with a braking distance of 1.98 m shorter than the other alternatives on wet asphalt surfaces. The analysis also identified the best tires for the maneuver on different surfaces, as well as the alternatives in order of performance, so that other tire combinations are possible if necessary to adapt to road conditions or financial resources.

According to the Shore A measurement of the hardness of the tread rubber compound, it can be seen that harder tires were more effective on wet asphalt and softer tires were more effective on artificially slippery surfaces. Tire set 3 has the softest tread (63.4 Shore A) and is the most effective at braking on artificially slippery surfaces but the least effective on wet asphalt, while tire set 7 has the hardest tread (68.6 Shore A) and is therefore the most effective on wet asphalt and the least effective on slippery surface. The same correlation was observed between the tread hardness and braking performance of the trailer tire sets, while no correlation was observed between the tire sets of the steering axle of the truck.

During the uphill maneuver, the test with the tire set 7 was the most efficient, reaching the target distance of 35 m in 13.25 s. It was 0.7 s faster than the second-best tire set (3.5 s faster than the worse tire set). The tires from set 7 had the highest tread hardness of all the tire sets compared, while the least effective tire set was the one with the lowest measured hardness.

The graphs of yaw rate and driving speed were used to assess the front and rear axle slip and the stability of its intensity during downhill driving with the cornering maneuver. The most efficient and constant tests were carried out with the tire set 5 (medium tread hardness), as no significant front or rear axle slip was observed (as it was for other tire seats) and an average lateral acceleration of 2.8 m/s^2 was achieved.

The TOPSIS analysis evaluates each alternative against a set of criteria to identify the alternative that best meets these criteria. The method is based on the concept of "closeness to the ideal solution", where the ideal solution represents the best possible result for each criterion and the worst solution represents the worst possible result. While TOPSIS offers a systematic approach to ranking alternatives according to their proximity to the ideal solution, it is important to recognize that real solutions often involve uncertainty and subjective judgment, as the weights of the criteria are chosen by the expert carrying out the analysis. In these types of analyses, it is important to ensure the robustness and reliability of the results, so a sensitivity test is carried out to see whether changing the weight of a criterion changes the position of the objects in the final ranking.

In the TOPSIS sensitivity analysis for braking tests, the weights of the criteria were changed by dividing the values equally or by giving one or the other of the criteria the majority of the weight. In the analysis of deceleration performance, giving more weight to the mean and maximum accelerations changes the ranking of the tire sets on the trailer: on wet asphalt, the tenth and eighth tire sets are ranked first and second, respectively, and on slippery surfaces, vice versa.

The best tire sets for each axle of the truck have remained the same, regardless of which feature is given more weight. This may be due to the fact that most of the criteria are quite highly correlated and interdependent. The best sets do not change with changing weights, but if alternatives are considered, such as the next best set of tires (if the first set is not available), the sensitivity test may lead to alternatives being ranked. It is therefore important to consider the importance of each of the criteria more explicitly, or to use a different method of analysis.

7. Conclusions

This paper describes the safety assessment of trucks based on experimentally tested different tire sets in four maneuvers on a vehicle test track with two types of low friction surfaces. A total of 11 tire sets were divided into three groups by axle: front axle, rear axle, and trailer. From the experimental test results, the most significant indicators of the truck's driving dynamics for initial comparison were determined: mean fully developed deceleration (MFDD), braking distance, braking time, and mean and maximum longitudinal acceleration. On the basis of that the processing of the test data was carried out by applying a multi-criteria decision analysis, in which the output criteria were evaluated by selecting different weights, and a sensitivity analysis was carried out to test whether changing the weights of the criteria changes the ranking of the tire sets.

In the braking tests on the trailer truck, the most effective tire set consisted of new all-season front tires and a retreaded all-season rear tires. The main feature that led to shorter brake distances on both surfaces is the size and layout of the tread blocks. Larger tread blocks and grooves angled at 45 degrees resulted in better braking on wet asphalt, while smaller tread blocks resulted in shorter braking distances on artificially slippery surfaces that replicate ice traction.

A slippery hill climbing maneuver compared tire sets on the drive axle of a truck and the results showed that that the tire set with retreaded rear tires with a small asymmetric pattern had the best grip. In this course, the downhill cornering maneuver was conducted to test the ability of the truck to maintain stability and handling on slippery road surfaces. The most effective tire in this maneuver was a set of new all-purpose front and rear tires designed for the winter season.

In the case of the rear axle tire assemblies, the retreaded tires were more effective in braking and hill climbing maneuvers, while the new tires performed better in the downhill cornering maneuver. The new rear tire configuration was ranked third out of five in the final braking maneuver ranking. The three sets of tires tested in the maneuvers on the front axle of the truck were new and the most effective was an all-season tire with the largest tread blocks and 45-degree grooves for water ejection. All sets of tires in the trailer maneuver had the same tread block arrangement, but the size of the blocks varied, with the most efficient tires having the smallest tread blocks.

The analysis of the accident's crash modification factor (CMF) from braking tests allowed the relative change in the safety of the vehicle fleet to be determined when the tire set with the shortest braking distance was taken as a reference. It shows that the use of less efficient tire sets than the reference tire set increases the change of the fatality rate in an accident by 19.7% on wet asphalt and by 26.6% on an ice-imitating pavement. The rate of change in injuries under the same conditions increases by 13.9% in an accident on wet asphalt and by 16.1% on artificially slippery surfaces.

The limitations of this study are due to the relatively small number of different tires (models) used, but the aim of the study was to identify the tire combinations used by one of the largest road haulage companies in Europe and to suggest best tire combinations. External factors such as wind or air temperature were not measured separately, but on the testing day, these factors did not differ and their effect was considered negligible. In addition, selected tire sets and truck type (including load and trailer configurations) do not completely represent global fleet compositions; however, the context of the research is best suited to the central and northern European regions.

The methodology described and agreed between the experiment and the CMF model can be applied in other cases, when it is desired to predict the positive or negative effect of any other factor on one of the accident rates. For example, in the case of a vehicle in bad technical condition (reduced braking performance), in the case of eco-driving mode selection (speed tolerance on downhill gradients with cruise control), or in the case of an independent study of the increase in reaction time of a fatigued driver. In our future research, additional tire models and maneuvers are planned at higher initial speeds for even more real traffic conditions typical for rural roads and highways where freight transportation is organized.

Author Contributions: Conceptualization, V.Ž. and T.M.; methodology, V.Ž.; software, T.M.; validation, V.Ž., T.M. and R.P.; formal analysis, V.Ž.; investigation, T.M.; resources, T.M.; data curation, R.P.; writing—original draft preparation, T.M.; writing—review and editing, V.Ž. and R.P.; visualization, T.M.; supervision, V.Ž.; project administration, V.Ž.; funding acquisition, R.P. All authors have read and agreed to the published version of the manuscript.

Funding: This research received no external funding.

Institutional Review Board Statement: Not applicable.

Informed Consent Statement: Not applicable.

Data Availability Statement: The original contributions presented in the study are included in this article; further inquiries can be directed to the corresponding authors.

Conflicts of Interest: The authors declare no conflict of interest.

References

1. *Global Status Report on Road Safety 2023*, 1st ed.; World Health Organization: Geneva, Switzerland, 2023; ISBN 978-92-4-008651-7.
2. Blincoe, L.; Miller, T.R.; Wang, J.-S.; Swedler, D.; Coughlin, T.; Lawrence, B.; Guo, F.; Klauer, S.; Dingus, T. *The Economic and Societal Impact of Motor Vehicle Crashes, 2019 (Revised)*; National Highway Traffic Safety Administration: Washington, DC, USA, 2023; p. 297.
3. Schindler, R.; Jänsch, M.; Johannsen, H.; Bálint, A. An Analysis of European Crash Data and Scenario Specification for Heavy Truck Safety System Development within the AEROFLEX Project. *arXiv* **2021**, arXiv:2103.05325.
4. *ISO 39001:2012*; Road Traffic Safety (RTS) Management Systems—Requirements with Guidance for Use. ISO: Geneva, Switzerland, 2012.
5. Camden, M.C.; Hickman, J.S.; Hanowski, R.J.; Walker, M. Critical Issues for Large Truck Safety. In *International Encyclopedia of Transportation*; Elsevier: Amsterdam, The Netherlands, 2021; pp. 200–209, ISBN 978-0-08-102672-4.
6. *Selected Issues in Passenger Vehicle Tire Safety*; National Transportation Safety Board: Washington, DC, USA, 2015; p. 65.
7. Hassan, M.R.; Rodrigue, D. Application of Waste Tire in Construction: A Road Towards Sustainability and Circular Economy. *Sustainability* **2024**, *16*, 3852. [CrossRef]
8. Christ, D. Simulating the Relative Influence of Tire, Vehicle and Driver Factors on Forward Collision Accident Rates. *J. Saf. Res.* **2020**, *73*, 253–262. [CrossRef]
9. Chen, Y.; Liu, X.; Rao, M.; Qin, Y.; Wang, Z.; Ji, Y. Explicit Speed-Integrated LSTM Network for Non-Stationary Gearbox Vibration Representation and Fault Detection under Varying Speed Conditions. *Reliab. Eng. Syst. Saf.* **2025**, *254*, 110596. [CrossRef]
10. Ji, Y.; Huang, Y.; Zeng, J.; Ren, L.; Chen, Y. A Physical—data-Driven Combined Strategy for Load Identification of Tire Type Rail Transit Vehicle. *Reliab. Eng. Syst. Saf.* **2025**, *253*, 110493. [CrossRef]
11. Singh, K.B.; Sivaramakrishnan, S. An Adaptive Tire Model for Enhanced Vehicle Control Systems. *SAE Int. J. Passeng. Cars—Mech. Syst.* **2015**, *8*, 128–145. [CrossRef]
12. Lorenčič, V. The Effect of Tire Age and Anti-Lock Braking System on the Coefficient of Friction and Braking Distance. *Sustainability* **2023**, *15*, 6945. [CrossRef]
13. Pečeliūnas, R.; Žuraulis, V.; Droździel, P.; Pukalskas, S. Prediction of Road Accident Risk for Vehicle Fleet Based on Statistically Processed Tire Wear Model. *Promet-Traffic Transp.* **2022**, *34*, 619–630. [CrossRef]
14. Stokłosa, J.; Bartnik, M. Influence of Tire Pressure on the Vehicle Braking Distance. *Arch. Automot. Eng. Arch. Motoryz.* **2022**, *97*, 60–73. [CrossRef]
15. Kreicbergs, J.; Zalcmanis, G.; Grīslis, A. Vehicle In-Use Tyre Characteristics Evaluation during Winter Driving Training. *Agron. Res.* **2016**, *14*, 1635–1644.
16. Malmivuo, M.; Luoma, J. Effects of Winter Tyre Type on Roughness and Polishing of Road Surfaces Covered with Ice and Compact Snow. *Eur. Transp. Res. Rev.* **2017**, *9*, 2. [CrossRef]

17. Xiong, Y.; Tuononen, A. Rolling Deformation of Truck Tires: Measurement and Analysis Using a Tire Sensing Approach. *J. Terramechanics* **2015**, *61*, 33–42. [CrossRef]
18. Becker, C.; Els, S. Motion Resistance Measurements on Large Lug Tyres. *J. Terramech.* **2020**, *88*, 17–27. [CrossRef]
19. Tuononen, A. On-Board Estimation of Dynamic Tyre Forces from Optically Measured Tyre Carcass Deflections. *Int. J. Heavy Veh. Syst.* **2009**, *16*, 362. [CrossRef]
20. Niskanen, A.J.; Tuononen, A.J. Three 3-Axis Accelerometers Fixed inside the Tyre for Studying Contact Patch Deformations in Wet Conditions. *Veh. Syst. Dyn.* **2014**, *52*, 287–298. [CrossRef]
21. Pegram, M.S.; Botha, T.R.; Els, P.S. Full-Field and Point Strain Measurement via the Inner Surface of a Rolling Large Lug Tyre. *J. Terramechanics* **2021**, *96*, 11–22. [CrossRef]
22. Singh, K.B.; Taheri, S. Accelerometer Based Method for Tire Load and Slip Angle Estimation. *Vibration* **2019**, *2*, 174–186. [CrossRef]
23. O'Neill, A.; Gruber, P.; Watts, J.F.; Prins, J. Predicting Tyre Behaviour on Different Road Surfaces. In *Advances in Dynamics of Vehicles on Roads and Tracks*; Klomp, M., Bruzelius, F., Nielsen, J., Hillemyr, A., Eds.; Lecture Notes in Mechanical Engineering; Springer International Publishing: Cham, Switzerland, 2020; pp. 1899–1908, ISBN 978-3-030-38076-2.
24. Ružinskas, A.; Giessler, M.; Gauterin, F.; Wiese, K.; Bogdevičius, M. Experimental Investigation of Tire Performance on Slush. *Eksploat. I Niezawodn. Maint. Reliab.* **2021**, *23*, 103–109. [CrossRef]
25. Žuraulis, V.; Kilikevičius, A. Quarter Car Test Rig for Extended Dynamics Research in Laboratory Conditions. In *Advances in Dynamics of Vehicles on Roads and Tracks*; Klomp, M., Bruzelius, F., Nielsen, J., Hillemyr, A., Eds.; Lecture Notes in Mechanical Engineering; Springer International Publishing: Cham, Switzerland, 2020; pp. 1425–1434, ISBN 978-3-030-38076-2.
26. Lugaro, C.; Niedermeier, F.; Wassertheurer, B.; Schüling, J. Method for Virtual Tyre and Braking Distance Simulation. *ATZ Worldw.* **2017**, *119*, 16–21. [CrossRef]
27. Becker, C.; Els, S. Effect of Surface Roughness on Tyre Characteristics. *J. Terramech.* **2022**, *102*, 27–48. [CrossRef]
28. Chen, Z.; Wen, H.; Zhu, Q.; Zhao, S. Severity Analysis of Multi-Truck Crashes on Mountain Freeways Using a Mixed Logit Model. *Sustainability* **2023**, *15*, 6499. [CrossRef]
29. Bassan, S. Review and Evaluation of Stopping Sight Distance Design—Cars vs. Trucks. *ATS Int. J.* **2012**, 5–16.
30. Taglione, L.; Bernal, M.; Corno, M.; Savaresi, S.M. The Effect of Different Tyres on the Performance of an Anti-Lock Braking System for Continuous Modulation Actuators. *IFAC-Pap.* **2024**, *58*, 726–731. [CrossRef]
31. Matara, N.N.; Suresh Kumar, P.; Karthik Krishnan, O.; Katte, Y. Real-Time Testing of AI Enabled Automatic Emergency Braking System for ADAS Vehicle Using 3D Point Cloud and Precise Depth Information. *Internet Things* **2024**, *27*, 101302. [CrossRef]
32. Fathi, H.; Ly, A.; Pathak, T.; El-Sayegh, Z. Sensitivity Analysis of Truck Tire Tread Material Properties for On-Road Applications. *Trans. Can. Soc. Mech. Eng.* **2024**, *48*, 341–354. [CrossRef]
33. Camden, M.C.; Hickman, J.S.; Hanowski, R.J. Reversing Poor Safety Records: Identifying Best Practices to Improve Fleet Safety. *Safety* **2021**, *8*, 2. [CrossRef]
34. Zaranka, J.; Pečeliūnas, R.; Žuraulis, V. A Road Safety-Based Selection Methodology for Professional Drivers: Behaviour and Accident Rate Analysis. *Int. J. Environ. Res. Public Health* **2021**, *18*, 12487. [CrossRef]
35. Alonso, M.; Mántaras, D.A.; Luque, P. Toward a Methodology to Assess Safety of a Vehicle. *Saf. Sci.* **2019**, *119*, 133–140. [CrossRef]
36. Wang, H.; Zhang, S.; Quan, W.; Liu, X. Study on Safety Distance Model of Fleet Based on Vehicle Communication. *Procedia—Soc. Behav. Sci.* **2013**, *96*, 2698–2705. [CrossRef]
37. Valentini, F.; Pegoretti, A. End-of-Life Options of Tyres. A Review. *Adv. Ind. Eng. Polym. Res.* **2022**, *5*, 203–213. [CrossRef]
38. *ASTM D2240-15*; Test Method for Rubber Property—Durometer Hardness. D11 Committee ASTM International: West Conshohocken, PA, USA, 2021. [CrossRef]
39. *ISO 11157:2005*; Road Vehicles—Brake Lining Assemblies—Inertia Dynamometer Test Method. ISO: Geneva, Switzerland, 2005. Available online: https://www.iso.org/standard/36952.html (accessed on 13 January 2025).
40. Araujo, J.V.G.A.; Moreira, M.Â.L.; Gomes, C.F.S.; Santos, M.D.; Costa, I.P.D.A.; Corriça, J.V.D.P.; Manso De Azevedo, C.; Pereira, D.A.D.M. Selection of a Vehicle for Brazilian Navy Using the Multi-Criteria Method to Support Decision-Making TOPSIS-M. *Procedia Comput. Sci.* **2023**, *221*, 261–268. [CrossRef]
41. Adhikari, D.; Gazi, K.H.; Giri, B.C.; Azizzadeh, F.; Mondal, S.P. Empowerment of Women in India as Different Perspectives Based on the AHP-TOPSIS Inspired Multi-Criterion Decision Making Method. *Results Control. Optim.* **2023**, *12*, 100271. [CrossRef]
42. Wang, P.; Zhu, Z.; Huang, S. The Use of Improved TOPSIS Method Based on Experimental Design and Chebyshev Regression in Solving MCDM Problems. *J. Intell. Manuf.* **2017**, *28*, 229–243. [CrossRef]
43. Dua, R.; Almutairi, S.; Bansal, P. Emerging Energy Economics and Policy Research Priorities for Enabling the Electric Vehicle Sector. *Energy Rep.* **2024**, *12*, 1836–1847. [CrossRef]
44. Wang, Z.; Li, P.; Cai, W.; Shi, Z.; Liu, J.; Cao, Y.; Li, W.; Wu, W.; Li, L.; Liu, J.; et al. Identifying Administrative Villages with an Urgent Demand for Rural Domestic Sewage Treatment at the County Level: Decision Making from China Wisdom. *Sustainability* **2025**, *17*, 800. [CrossRef]

45. Elvik, R.; Christensen, P.; Amundsen, A. *Speed and Road Accidents An Evaluation of the Power Model*; Institute of Transport Economics: Oslo, Norway, 2004; p. 134.
46. Forbes, G.J.; Institute of Transportation Engineers (Eds.) *Methods and Practices for Setting Speed Limits: An Informational Report*; Institute of Transportation Engineers: Washington, DC, USA, 2012; ISBN 978-1-933452-65-4.

sustainability

MDPI

Article

Modeling the Impact of Interaction Factors for Transport System Elements on Quality of Life Using Multi-Criteria Decision-Making and Applied Statistical Methods

Henrikas Sivilevičius and Vidas Žuraulis *

Transport and Logistic Competence Centre, Vilnius Gediminas Technical University (VILNIUS TECH), 10223 Vilnius, Lithuania; henrikas.sivilevicius@vilniustech.lt
* Correspondence: vidas.zuraulis@vilniustech.lt

Abstract: This paper presents an improved model of the interaction among transport system elements, which identifies six levels of interaction and their analysis. A framework of eight factors (criteria) and their interactions that improve the quality of life is developed. The Analytic Hierarchy Process, Average Rank Transformation into Weight, and rank correlation methods were used to calculate the normalized weights and compatibility indicators of the 54 expert criteria. The results show that the most significant improvements in quality of life are the development of transport infrastructure and its quality, the reduction of road traffic accidents, and the reduction of environmental pollution caused by the transport sector, all of which directly contribute to a sustainable urbanized society. The improvement in quality of life is least affected by the increase in the number of vehicles and their technical parameters, as well as by the increase in the share of the country's Gross Domestic Product generated in the transport sector. The variance of the weights of the criteria was shown to depend on the different magnitudes of the fundamental scale of the pairwise comparison chosen by the experts, the variance (stability) of the experts' personal opinions, and the net variance of the weights assigned to the criteria by the expert team. The results of this study can be used to develop a national or regional transport sector development strategy that takes into account the factors that have the greatest impact on quality of life.

Keywords: transport system; interactions of elements; quality of life; sustainable transportation; criteria importance; MCDM; expert opinion consistency

1. Introduction

The transport system is fundamental to our economy and society and plays an important role in human society [1]. Mobility is one of the most important human needs that affects quality of life, and, at the same time, it is an important social and economic element in the life of the state. Moreover, transport must be sustainable in light of the new challenges we face [2]. Nevertheless, continuous transport development and its impact on quality of life are sensitive in many aspects. In general, transport must therefore be developed in a sustainable way to meet the mobility needs of the public, as well as to reduce environmental pollution and improve safety. This requires appropriate interaction among transport system elements (TSE) while maintaining the right relationships between them. Economic regulation through environmental and social action would address contemporary sustainability issues, and multi-criteria decision-making (MCDM) approaches with integrated criteria selection and decision-making solutions can be applied to transport system development [3,4].

The aim of this paper is to present a model of the interaction among transport system elements and to determine the significance of the results of this interaction for quality of life by applying MCDM and applied statistical methods. The area under analysis is fairly extensive and complex; therefore, the use of expert evaluation methods will allow subjective opinions to be expressed in terms of the objective factors of TSE.

The remaining part of this paper has been arranged as follows: Section 2 presents the literature review of the main transport system elements that have an impact on quality of life. Section 3 details the research background and design of the model of transport system elements. The methodology of the research, including criteria selection, experts, and developed calculation methods, is described in Section 4. Then, Section 5 presents the results of the conducted research along with their interpretation. Finally, Section 6 summarizes the conclusions and limitations of this study and proposes directions for future research.

2. Literature Review

The global pandemic of the COVID-19 virus has slowed national economies, with the EU needing around EUR 1.5 trillion to mitigate its negative impact. The transport and tourism sectors will account for the largest share of this money, as mobility has been hit hardest by COVID-19. It is established that political factors (e.g., funding support) and economic externalities (e.g., agglomeration) must be implemented to achieve a positive economic impact and the general development of any region [5]. Furthermore, sustainable infrastructure projects deliver social well-being benefits [6,7]. Adding to this, John F. Kennedy has said "Our prosperity does not create infrastructure, but the transport infrastructure creates prosperity for us" [8].

In developed countries, road transport is the dominant mode of mobility since it allows for "door-to-door" transportation. However, high-speed rail transport shows better environmental friendliness compared to road transportation; nonetheless, its environmental issues related to high energy demand and associated emissions cannot be ignored [9]. Several aspects encourage renewed interest in rail systems: (1) a reduction in traffic congestion, (2) an improvement in public transport, (3) better access to the city center, (4) environmental protection, (5) a reduction in energy expenditure, (6) an increase in safety, and (7) an incentive for economic and residential growth [10].

In general, travel conditions are directly influenced by the state of the pavement, traffic signs and markings, and the location of service areas, which contribute to a certain level of safety, comfort, attractiveness, and speed [11]. Silva et al. [12] show that the satisfaction of citizens is mainly related to health and safety, well-being, recreation, public transport, walkability, social inclusion and participation, primary services, city connectivity, and the environment. The Jonek-Kowalska article focuses on housing infrastructure as a determinant of quality of life in selected Polish Smart Cities [13]. Here, the quality of life of residents is associated with the development and implementation of a city concept that includes all necessary services, with mobility as an essential component [14]. The concept of smart cities is often expressed through the implementation of sustainable activities, especially focusing on the quality of life for future generations.

Unfortunately, road transport has a major drawback—road safety issues—as shown by the high number of deaths and injuries on the roads [15]. In contrast to economic benefits, at this point, road transport results in a loss of public money. Pedestrians, as vulnerable road users, often suffer from head injuries in adult pedestrian-involved road accidents; therefore, preventing and minimizing head injuries has become an important target for improving external vehicle safety design [16]. Nevertheless, significant differences are found in the analysis of pedestrian accidents at crossings between different countries: some of the lowest rates are in Great Britain and Germany, while the highest rates are in Poland and Lithuania [17]. In this context, physical, sensory and cognitive limitations lead to a higher accident rate involving older pedestrians, which, at the same time, leads to more severe consequences due to greater physical vulnerability [18].

The progress in pedestrian safety enhancement is the result of multi-stage work, which is based on the development of vehicle active and passive safety and appropriate traffic organization. The complete separation of vehicle traffic and pedestrians would lead to zero accidents; however, it seems to be impossible to achieve this today [19]. Road safety issues were analyzed using machine learning for the MCDM method, which showed promising results in the selection of policy recommendations and potential practical solutions for economic investment [20,21]. In line with this kind of research, traffic accidents on Lithuanian roads were investigated using geographic information systems and the multi-criteria method of AHP [22]. This study describes the methodology for selecting and ranking road sections with high accident concentrations from the national significance road network.

The fight against speeding on the roads and the accidents it causes has led to the use of speed enforcement systems on the roads of various countries for many years already, which has had a positive effect on reducing the number of accidents [23]. Here, vehicle dynamics are taken into account. The potential for vehicle safety, together with the comfort level, highly depends on the road infrastructure and its quality. The interaction between the wheel and the road pavement, with respect to its design parameters, can be characterized by the type of pavement and its smoothness [24]. From the same point of view, the adhesion of the tire to the road plays a significant role in the generation of grip force and consequently influences vehicle safety [25].

After road safety improvements, the transport system will address the reduction of environmental pollution. Within the transport sector, road transport is responsible for the largest part of carbon dioxide (CO_2) emissions; however, a more rapid transition to alternative-fuel vehicles can contribute to significant reductions in CO_2 emissions [26]. Sustainable transport solutions, such as electric and hydrogen vehicles or biofuel adoption, were investigated by combining MCDM with a fuzzy theory-based approach [27,28].

Transport noise is another major issue that directly affects the quality of life of local residents and is an important part of EU policy, as growing cities lead to an increasing number of vehicles and associated public health problems [29]. The main problems are related to stress, sleep disturbances, and even heart diseases; therefore, non-motorized traffic could lead to an efficient reduction of these negative impacts, and together with noise-reducing asphalt pavement technologies, it will improve the quality of life in urbanized areas [30,31].

The Analytical Hierarchy Process (AHP) method is used in various transport areas to determine the impact of interactions among transport elements on road safety [32], to evaluate the relationship between citizens' mental satisfaction and infrastructure quality [33], to select off-road vehicle equipment for military units [34], and to select the sites of distributed photo voltaic power stations along high-speed railways [9]. Hruška et al. [35] describe a general model for the selection of transport mode using AHP, including its application to the appropriate selection of a mode of transport from three potential transport modes. Here, the brainstorming method was used to identify criteria such as transport price, average transport time, transport quality, transport reliability, ecological impact, and first and last mile.

The MCDM approach is also widely used for other transport-related concerns. Passengers who choose to travel by train instead of by bus or air transport take into account the advantages and disadvantages of these modes of public transport, which are holistically assessed using MCDM methods [36,37].

As a result, the multifaceted nature of the elements of the transport system in the life of a society is a sign of their importance, and the complexity in such cases can be addressed by applying modified methods of expert opinion survey analysis.

3. The Improved Model for the Interaction of Transport System Elements

The transport system (TS) consists of a set of transport modes. A country may use all modes of transport or only some of them, depending mainly on its geographical conditions and size. Each mode of transport is designed to transport cargo and passengers (people) (criterion H), while reducing empty mileage, environmental pollution, and the number of road accidents. The process of transporting them requires vehicles (criteria B) and transportation ways (infrastructure) (criteria C). The tri-poral system, consisting of "participants of traffic and cargo", "vehicles", and "transportation ways (infrastructure)", is stable and, in the model (Figure 1), represents a necessary condition for the existence of any transport means. The autointeraction of individual elements (I) and mutual interaction (II) must be safe, i.e., they should reduce the number of road accidents (criteria F) in the country.

The transportation process takes place in an environment (criteria A) that consists of people, objects of energy, flora, buildings, air, water, soil, and fauna. Roads and vehicles must not make the natural state of the environment worse, i.e., they must not pollute it in an unacceptable way (interaction III).

There are five main means of transport for passengers and cargo: road transport, railway transport, air transport, sea transport, and inland waterway transport. These means of transport can only be used as alternatives if all modes of transportation are available within a country or between countries. Each of them has advantages and disadvantages, which are evaluated by taking into account the specific characteristics of the route (road), distance, travel time, weather conditions, load capacity, cost-effectiveness, environmental friendliness, safety, and comfort. The best means of transport can be comprehensively selected using MCDM methods, such as the distributive and ideal modes of the AHP method. This approach addresses the main sustainable transport targets, such as efficient supply chain operation and urban freight distribution [38,39].

The interaction of different means of transport (level IV in Figure 1) must be based on the principles of cooperative, non-discriminatory competition, complementing or substituting one means for another in the pursuit of the country's common economic, environmental, and social objectives that improve the quality of life.

The results of the interaction of the TS elements are influenced by the physical direct and derivative (computational) parameters, which are shown in Figure 1, inside the oval with hollow arrows pointing toward the center of the model. According to these parameters, traffic safety or accident rates, as well as the emission and diffusion of pollutants, also have a significant impact on quality of life. Noise and vibration, excessive vehicle speed, and acceleration all contribute to the deterioration of people's quality of life.

TS serves the purpose of providing services to the country's productive and non-productive sectors. There are no sectors that do not use means of transport for essential activities. The model (Figure 1) shows, on the outside of the circle (oval), the 18 most important sectors in the country that transport cargo and passengers using private, public, and hired vehicles within public transport infrastructure. The interaction of the TS elements with the country's productive and non-productive sectors is shown at level V. The most important result of this interaction is the share of the economy (Gross Domestic Product—GDP) generated by the country's transport and logistics sector (level VI). The efficiency and profitability of transport companies contribute to this share of GDP (criteria G).

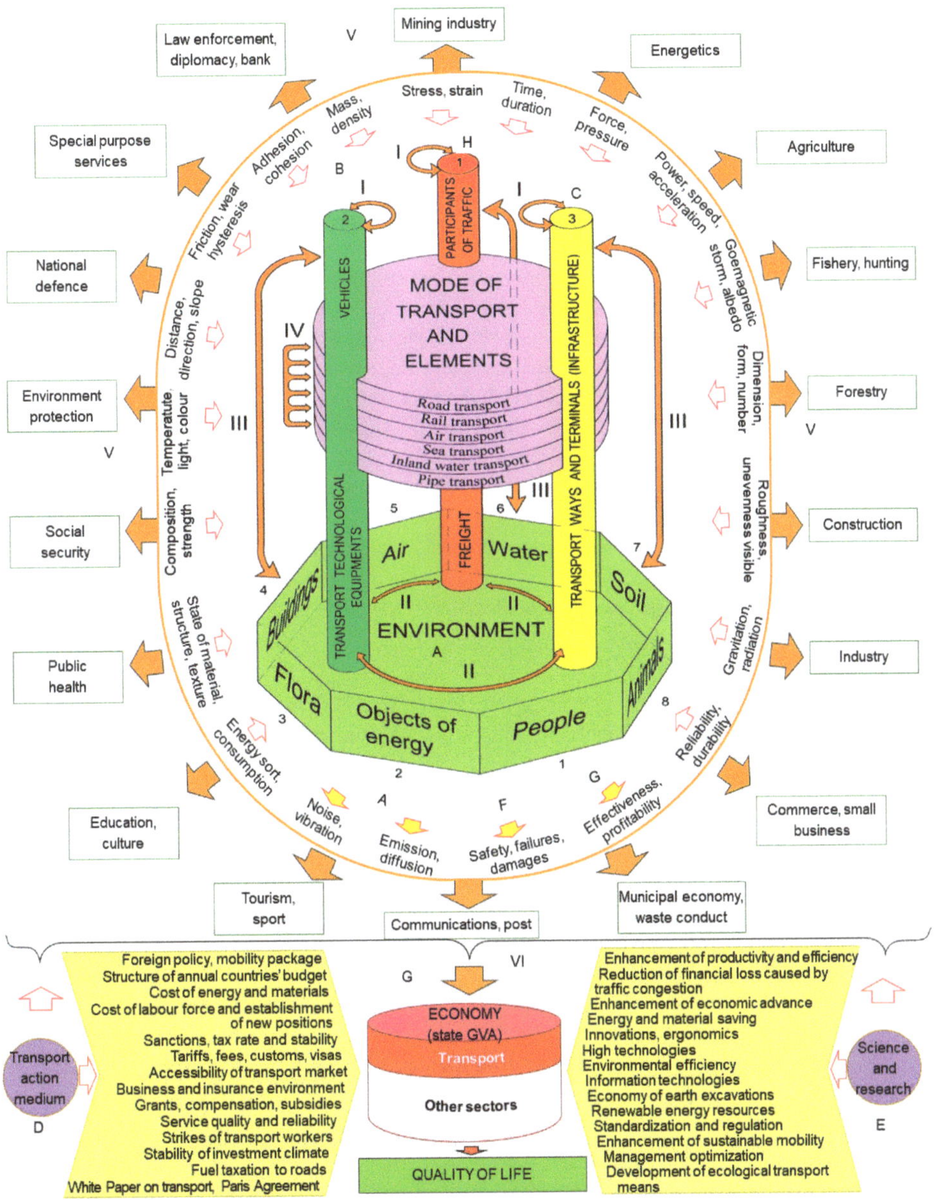

Figure 1. The Model of Transport System Element Interaction and its impact on economics and quality of life.

The success of a TS is influenced by the transport action medium, which consists of the 15 factors (criteria D) found at the bottom left of the model (Figure 1). Some factors improve, and others worsen, the environment of the transport sector, which determines the share of GDP generated by this sector.

TS is continuously improved through 15 scientific and research results (criteria E). The implementation of these results leads to improved mobility, increased operational efficiency and road safety, reduced environmental pollution, reduced empty mileage, and other objectives that contribute to a higher share of the country's GDP, thereby improving quality of life.

4. Research Methodology

4.1. Criteria

The model (Figure 1) identifies eight factors, hereafter referred to as criteria, that improve quality of life. Two options are usually used to establish the criteria (the method based on the processed literature review and the brainstorming method); however, as these approaches are not contradictory, they have been combined for this research. Using the AHP approach chosen for this study, a rational number of criteria could be seven plus or minus two, that is, between five and nine [40,41]. This method allows for the comparison of at least three criteria, as there is no random index R.I. value when comparing two criteria. This rational number of 8 criteria allowed us to define a hierarchical structure with three levels: Goal—Level 1, Main criteria—Level 2, Sub-criteria—Level 3 [42–44]. Without this hierarchical structure of criteria, the task is simpler to solve, and the expert has a better understanding of the essence of the criteria (formulation) and can compare them with each other.

The general framework diagram for the following research is presented in Figure 2.

This research assessed the factors (criteria) for the interaction of transport system elements (TSE) that improve quality of life:

A. Reduction in environmental pollution (exhaust emissions, noise, vibrations) caused by the transport sector;
B. Increase in the number of vehicles and improvement of the quality (technical parameters, condition, comfort, economy);
C. Developing the transport infrastructure network (roads, railways, airports, seaports, terminals) and improving its quality;
D. Improvement of the operating environment of the transport system (business, tax, price, investment, competition, grant, subsidies, etc.) and strikes by transport workers;
E. Developing science, research, and innovation in the transport sector and putting the results into practice;
F. Reduction in accidents, breakdowns, and damage and increase in comfort in the transport sector;
G. Increase in the share of a country's Gross Domestic Product (GDP) generated by the transport sector;
H. Improvement of the discipline, awareness, caution, courtesy, and travel behavioral culture of road users (drivers, pedestrians, passengers).

A description of the eight criteria in these formulations was provided in a questionnaire filled out by each expert. At the beginning of the questionnaire, the expert ranked all criteria, assigning the most important one a rank of 1, the second most important a rank of 2, the third most important a rank of 3, and so on, with the least important receiving a rank of 8. There were no duplicate ranks (ties) in the questionnaires.

Furthermore, the expert, either independently or with the assistance of the authors of this research, completed the pairwise comparison matrix using Saaty's [45] scale of inter-rater importance from 1/9 to 9 or a narrower scale. Some experts used an incomplete scale; that is, the highest level of importance was 8, 7, or 6. The sequence for completing this matrix, which avoids the errors associated with non-compliance with the transitivity clause, has not been described in scientific papers. In the process of filling it in, without knowing the method for checking correctness, correctness is only determined by calculating the consistency ratio (*C.R.*).

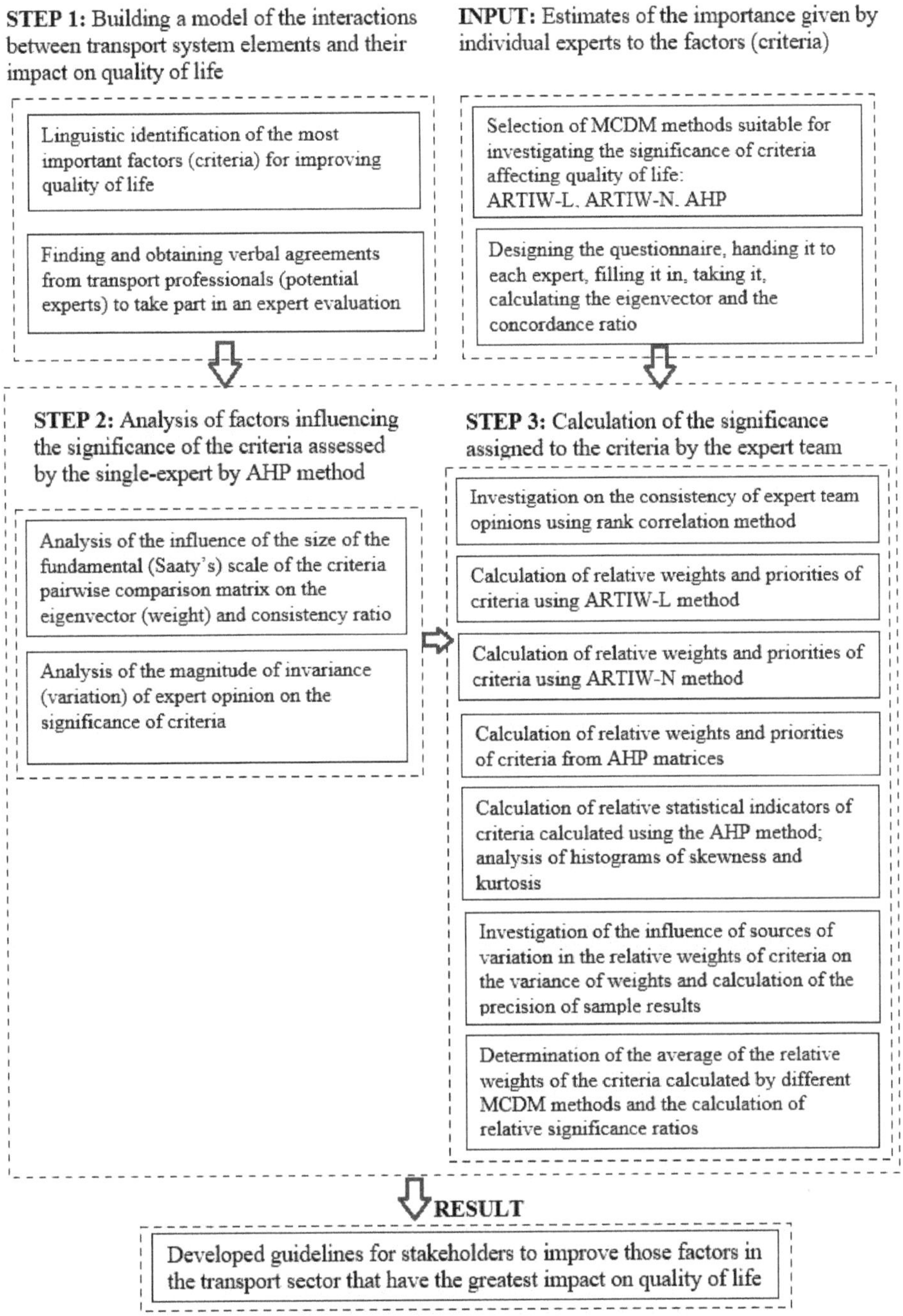

Figure 2. Research framework diagram of this study.

4.2. *Experts*

The problem under investigation requires the involvement of experts from a wide range of fields. The reliability of the team assessments depends on the knowledge of each expert and the number of experts. Assuming that the experts are sufficiently accurate assessors, as the number of experts increases, the reliability of the entire team's research also increases. The experts included 22 Ph.D. researchers in the field of land transport and 5 practitioners working in the fields of transport engineering and civil engineering. In addition, 27 students from the final (second) year of the Master's degree course who participated in the study "Interaction of Transport Elements and Assessment Methods" at the Faculty of Transport Engineering were interviewed. None of the selected experts had a particular bias toward one or another area of research; on the contrary, their broad academic knowledge of the transport sector enabled them to provide a neutral and high-quality expert assessment. All the experts were Lithuanian scientists and practitioners, so the questionnaire was written in Lithuanian. The questionnaire form translated into English is given in Appendix A.

According to the requirement that the number of experts on the team must be greater than the number of criteria [46], the number of experts who agreed to participate in this study and the 54 experts who were interviewed exceeded the number of 8 criteria by almost 7 times. This indicates that the number of experts is sufficiently high. This number is also sufficient, in line with the common practice that the minimum number of data required for a study (sample size) should be between 30 and 50. The minimum sample size of 30 is supported by the central limit theorem [47]. When the dispersion (variation) of an indicator is low, this sample size n (number of experts) is sufficient; however, when the dispersion is high, it may be too small.

The minimum number of experts can be calculated using the sample size equation [48], which takes into account the overall standard deviation, σ_ω, of the weights of the investigative subject's (quality of life) criteria, the required precision, $\Delta\omega$, and the Student's test statistic, t, which is dependent on the probability, P, chosen, as follows:

$$n_{\min} = \frac{t^2\sigma_\omega^2}{\Delta_\omega^2}. \quad (1)$$

Reducing the number of experts reduces both the cost of this study and the reliability of the results; when a single expert opinion is used, which is less reliabile than the average of the opinions of a team of experts, the cost of this study is significantly reduced. The assessments of three experts were used to select the mode of transport using the AHP method [35]. The opinions of four observers were analyzed by Kendall and Gibbons [49], explaining the rank correlation method and the concordance coefficient. The opinions of five experts or decision-makers were used in the investigations by Farhan and Fwa [50], Badi et al. [15] and Gudien et al. [51]. Saaty's [45] research work analyzed single-expert criterion significance and/or pairwise comparison matrices of alternatives. Data from the matrix collectively filled in by the expert team are rarely presented and hardly discussed [52].

The opinion of one expert can be taken as a reliable result of this study if the experts on the interviewed team share the same opinion. Without prior knowledge of the variation in the opinions of the experts in this study, it is difficult to determine the appropriate size of the team. A pilot project can be carried out for this purpose.

4.3. *Average Rank Transformation into Weight (ARTIW) Methods*

The expert gives each criterion a rank, R_{ij}, from most important (rank 1) to least important (rank 8). It is preferable that the evaluation does not contain duplicate (linked) ranks, which indicate that several criteria are of absolutely equal importance to the expert.

It is straightforward to rank (i.e., by place number) the significance of criteria. However, it is more convenient to rank a more important criterion with a higher quantitative weight (number) and a less important criterion with a lower weight. Moreover, it is rational to normalize them by calculating the weights, i.e., so that the sum of the weights of all the criteria is equal to one. This determines the relative importance of the criteria, i.e., their normalized weights.

Given a table of ranks for the criteria (i = 1, 2, ..., m) and the experts (j = 1, 2, ..., n), calculate the average of the ranks of all the j-th experts for each of the i-th criteria as follows:

$$\overline{R}_i = \frac{\sum_{j=1}^{n} R_{ij}}{n}. \tag{2}$$

where R_{ij} is the rank given by the j-th expert for the i-th criterion.

The averages of the criterion ranks, $\overline{R}_i$, calculated using Equation (1) are used to calculate the significance of the criteria in terms of the normalized weights, $\omega_i^{ARTIW-L}$, using the Average Rank Transformation into Weight-Linear (ARTIW-L) method as follows:

$$\omega_i^{ARTIW-L} = \frac{m - \overline{R}_i + 1}{\sum_{i=1}^{m} \overline{R}_i} \tag{3}$$

Similarly, the normalized weights, $\omega_i^{ARTIW-N}$, of the criteria are calculated from the rank averages, $\overline{R}_i$, using the Average Rank Transformation into Weight-Non-linear (ARTIW-N) method according to the following equation:

$$\omega_i^{ARTIW-N} = \frac{\min\limits_i \overline{R}_i}{\overline{R}_i \cdot \sum_{i=1}^{m} \frac{\min\limits_i \overline{R}_i}{\overline{R}_i}} \tag{4}$$

4.4. Consistency of the Team of Experts' Opinions on the Significance of the Criteria

The average of the significance ratings (ranks or weights) of the criteria of an expert team can only reasonably be taken as the result of a task when all experts are in agreement (not conflicting) [49]. In the absence of tied ranks, the concordance of the opinions of a team of experts is determined by calculating the variance coefficient of concordance, W, (Kendall coefficient of concordance) according to Equation (5). Only the ranks of the criteria, R_{ij}, are suitable for its calculation. If the criteria are evaluated differently by the experts (in terms of scores, percentage weights, unit fractions, a ten-point system, or the Saaty fundamental scale for pairwise comparisons), it is necessary to rank the criteria beforehand as follows:

$$W = \frac{12 \cdot \sum_{i=1}^{m} \left[\sum_{j=1}^{n} R_{ij} - \frac{n(m+1)}{2} \right]^2}{n^2 m (m^2 - 1)} \tag{5}$$

where $\frac{n(m+1)}{2} = \overline{R}$ is the overall average of the ranks.

The value of the concordance coefficient, W, was compared with its threshold minimum value, $W_{\min}$, which is calculated using Formula (6) by obtaining the number of degrees of freedom $\vartheta = m - 1$ and the usual $\alpha = 0.05$ or strict $\alpha = 0.01$ significance level as follows:

$$W_{\min} = \frac{\chi^2_{v,\alpha}}{n(m-1)} \tag{6}$$

where n—number of experts (j = 1, 2, ..., n), and m—number of criteria (i = 1, 2, ..., m).

The agreement between the experts is also indicated by the χ^2 value calculated using Equation (7), which is compared with the critical value, $\chi^2_{v,\alpha}$, as follows:

$$\chi^2 = \frac{12 \cdot \sum_{i=1}^{m} \left[\sum_{j=1}^{n} R_{ij} - \frac{n(m+1)}{2} \right]^2}{nm(m+1)} \tag{7}$$

The ratio of the concordance coefficients, W and $W_{\min}$, as well as the Pearson chi-squared statistics, χ^2 and $\chi^2_{v,\alpha}$, is known as the compatibility coefficient, k_c, which is computed using the following equation [53]:

$$k_c = \frac{W}{W_{\min}} = \frac{\chi^2}{\chi^2_{\alpha,v}} \tag{8}$$

It shows the number of times that the estimated values of the agreement between the ranks of the criteria derived from the expert study are higher than the critical values. A value of k_c that is greater than 1 indicates that the experts' opinions are not inconsistent.

4.5. Analytic Hierarchy Process Method

The Analytic Hierarchy Process (AHP) is a decision analysis method combining qualitative and quantitative approaches to solving complex multiobjective problems [54]. AHP is widely used in different decision-making problems, especially in solving problems with criteria weighting factors [34].

This very popular MCDM method allows initially ranked criteria to be compared in pairs against each other in order of importance using Saaty's nine-point scale of five levels of importance (1-3-5-7-9). The core of the method is the pairwise comparison matrix, $\mathbf{A} = \|a_{ij}\|_{m \times m}$, $(i, j = 1, 2, \ldots, m)$. Its criterion intensity elements, a_{ij}, can be thought of as ratios of the values of the quality-of-life factors, C_i and C_j, of the interaction between the elements of the transport system and, after the normalization of these values, as the relative weights of the factors. When the elements, a_{ij}, are ratios of unknown weights, the matrix, $\mathbf{A}$, is inversely symmetric ("reciprocal"), i.e., $a_{ij} = 1/a_{ji}$, actually $a_{ij} = \frac{\omega_i}{\omega_j}$ and $a_{ji} = \frac{\omega_j}{\omega_i}$.

The most accurate of the approximate algorithms for calculating the weights of the criteria is based on calculating the geometric means of the products of the elements a_{ij} and $1/a_{ji}$ of the pairwise comparison matrix. Pairwise comparisons are made between the criteria on the left-hand side of matrix A and the criteria on the top of the matrix.

For the expert-completed matrix with the transitivity condition ensured, the eigenvector of each criterion is taken as a normalized weight, which is calculated according to the following geometric means formula [42,55]:

$$\omega_i^{AHP} = \frac{\sqrt[m]{\prod_{j=1}^{m} a_{ij}}}{\sum_{i=1}^{m} \sqrt[m]{\prod_{j=1}^{m} a_{ij}}} \tag{9}$$

The correctness of the completed matrix is checked by calculating its consistency ratio, *C.R.*, which is equal to the ratio of the consistency index, *C.I.*, and the random index *R.I.* (*C.R.* = *C.I.*/*R.I.*).

The consistency index, *C.I.*, is calculated using the following equation:

$$C.I. = \frac{\lambda_{\max} - m}{m - 1} \tag{10}$$

where m—number of criteria (i = 1, 2, . . ., m), and λ_{max}—the largest eigenvalue of the pairwise comparison matrix, which is calculated using the following equation:

$$\lambda_{max} = \frac{1}{m}\sum_{i=1}^{m} \frac{\sum_{j=1}^{m} a_{ij}.\omega_i^{AHP}}{\omega_i^{AHP}} \quad (11)$$

The random index, R.I., which depends only on the number of criteria, m, in the pairwise comparison matrix, is found in table [45] or calculated using Equation (12) when there are 15 or more criteria as follows [53]:

$$R.I. = 1.98\frac{m-2}{m} \quad (12)$$

If $C.R. < 0.1$, the consistency of the assessment matrix is acceptable, and the eigenvector is the weight vector of the evaluation criteria [9].

The average of the weights of the criteria calculated using the three different MCDM methods is more reliable than the weights calculated using any one method. Since none of the three MCDM methods has a theoretical advantage over the other methods, the average of the weights, $\overline{\overline{\omega}}_i$, for each of the i-th criteria is calculated using the following equation:

$$\overline{\overline{\omega}}_i = \frac{\omega_i^{ARTIW-L} + \omega_i^{ARTIW-N} + \overline{\omega}_i^{AHP}}{3} \quad (13)$$

where $\omega_i^{ARTIW-L}$ is the weight calculated using the ARTIW-L method for the i-th criterion; $\omega_i^{ARTIW-N}$ is the weight calculated using the ARTIW-N method for the i-th criterion; and $\overline{\omega}_i^{AHP}$ is the average of the weights of the team of experts for the i-th criterion calculated using the AHP method.

The variation in the numerical estimates of the significance of the i-th criterion, as assessed by the expert team, can be determined by calculating the standard deviation of the relative weights, $\sigma_{\omega i}$. For this purpose, only the weights, ω_i^{AHP}, calculated using the AHP method are used, since the ARTIW-L and ARTIW-N methods do not assess the opinions of each expert. A low value of $\sigma_{\omega i}^{AHP}$ indicates that the opinions of the experts assessing the significance of the criterion are close.

4.6. The Model of the Criteria Weight Variance

Variation in expert opinion is inevitable and scarcely researched [56]. Its magnitude determines the minimum number of experts in a team. This variation and its components (source of variation) can be estimated with a new sum-of-variance model.

The total variance, $\sigma_{\omega i}^2$, of the relative weights, ω_i, given to the i-th criterion by the team of experts is calculated using a new mathematical additive model showing the sources of variation as follows:

$$\sigma_{\omega i}^2 = \sigma_{\omega i,\ ss}^2 + \sigma_{\omega i,\ oe}^2 + \sigma_{\omega i,\ tnv}^2 \quad (14)$$

where $\sigma_{\omega i,\ ss}^2$—variance due to the scale size difference of the Saaty scale chosen by the experts, i.e., due to unequal $\max a_{ij}$ values in the matrix; $\sigma_{\omega i,\ oe}^2$—variance due to the variability of one expert's criterion ratings, i.e., the variability of his opinion; $\sigma_{\omega i,\ tnv}^2$—the variance of the team net variation of the criteria weights of a team of n number of experts, i.e., the variance if all the experts in the team were to use a Saaty scale of the same width for the evaluation of the object of study ($\sigma_{\omega i,\ ss}^2 = 0$) and if the significance of the weights of each expert's criterion were constant over time ($\sigma_{\omega i,\ oe}^2 = 0$).

The subject of this study (quality of life) consists of m criteria ($i = 1, 2, \ldots, m$), for which the average of the standard deviations of the weights assigned by a team of experts to an equal number n is calculated using the following equation:

$$\sigma_\omega = \sqrt{\frac{\sum_{i=1}^{m} \sigma_{\omega i}^2}{m}} \tag{15}$$

where $\sigma_{\omega i}^2$—sample variance of the weights for the i-th criteria calculated according to the assessment of the team of experts.

In order to be able to calculate the pooled standard deviation of the weights, σ_ω, of all the criteria in Formula (15), it is necessary to make sure that the variances of the weights of the individual criteria are statistically equal and that the weights of the criteria are normally distributed. Here, the hypothesis of equality of several variances, called the null hypothesis, is tested.

When the population is normally distributed, samples of size n (weighted by a team of n experts) taken from the population are compared with each other by testing the homogeneity of the variances of their weights, $\sigma_{\omega i}^2$, according to Cochran's test. If the weight of each criterion corresponding to a normal distribution is derived from a different number of n expert assessments, then Bartlett's test would be appropriate for determining the homogeneity of the variance of the weights, $\sigma_{\omega i}^2$ [57]. This test is very sensitive to the normality assumption of the deviation of the sample elements from a normal distribution, so inferences about the homogeneity of variances should be treated with caution. Consequently, when the validity of this assumption is doubtful, Bartlett's test should not be used.

Only the mean of the variances of statistically equal weights can be used to estimate the total variance of the criteria for the whole subject.

The equality of variance of the weights, $\sigma_{\omega i}^2$, of the m criteria is tested by testing the null hypothesis using Cochran's test statistic, $\hat{G}_{\max}$, as follows [58]:

$$\hat{G}_{\max} = \frac{\max_i \sigma_{\omega i}^2}{\sum_{i=1}^{m} \sigma_{\omega i}^2} \tag{16}$$

where $\max \sigma_{\omega i}^2$—the variance of the weights of the criterion with the highest variation, and m is the number of criteria of the study object.

The calculated value of the sample statistic, $\hat{G}_{\max}$, is compared with the critical $G_C(\alpha,\ k,\ m)$ value, which depends on the significance level, α (usually taken as $\alpha = 0.05$), the number of degrees of freedom, $k = n - 1$ (in this study, $k = 54 - 1 = 53$), and the number of criteria, m (in this study, $m = 8$). The critical values, $G_C(\alpha,\ k,\ m)$, are given in the table [58]. The variances of the weights are statistically the same when $\hat{G}_{\max} < G_c(\alpha, k, m)$.

It is convenient and simple to check the compliance of the weights of the criteria with a normal distribution by calculating skewness and kurtosis. The empirical skewness and kurtosis coefficients of the weights for the i-th criterion from the expert test must be smaller than their standard deviations, s_{sk} and s_{ku}, multiplied by 3 and 5, respectively. When $|sk| < 3s_{sk}$ and $|ku| < 5s_{ku}$, it is reasonable to assume that the opinions of a team of experts of size n in assessing the significance of the i-th criterion comply with a normal distribution. The standard deviations of skewness and kurtosis depend only on the size of the team of experts, n, and are calculated using the following equation:

$$s_{sk} = \sqrt{\frac{6n(n-1)}{(n-2)(n+1)(n+3)}} \tag{17}$$

$$s_{ku} = \sqrt{\frac{24n(n-1)^2}{(n-3)(n-2)(n+3)(n+5)}}. \quad (18)$$

Other tests can be used also to check the normality of the distribution, e.g., Pearsson χ^2, Kolmogorov λ.

5. Results and Discussion

5.1. *Effect of Saaty's Scale Size on Criteria Significance*

The expert chooses the size of the Saaty scale by comparing the most important criteria with the least important criteria. In this study, the expert filled in four pairwise comparison matrices according to eight equally ranked criteria, where the largest matrix element is as follows: $\max a_{ij} = 3$, $\max a_{ij} = 5$, $\max a_{ij} = 7$, $\max a_{ij} = 9$ (Figure 3). The eigenvectors, ω_i, of the criteria compared in these matrices using different Saaty scales are different, but the trend of priority is the same; the most important criterion is criterion C, and the least important is criterion D (Figure 4). The difference between the highest and lowest weights of each criterion, Δ, tends to decrease when the importance of the criterion is decreasing.

$\max a_{ij} = 3$

(a)

Quality of life		Factor (criterion) $j = 1, 2, \ldots, m$								Eigenvector (weight) ω_i
		A	B	C	D	E	F	G	H	
Factor (criterion) $i = 1, 2, \ldots, m$	A	1	1	1/2	3	2	1	1	2	0.1433
	B	1	1	1	3	2	1	2	3	0.1793
	C	2	1	1	3	2	1	2	3	0.1955
	D	1/3	1/3	1/3	1	1/2	1/3	1/2	1	0.0556
	E	1/2	1/2	1/2	2	1	1/2	1	1	0.0883
	F	1	1	1	3	2	1	1	2	0.1563
	G	1	1/2	1/2	2	1	1	1	2	0.1146
	H	1/2	1/3	1/3	1	1	1/2	1/2	1	0.0671

$C \succ B \succ F \succ A \succ G \succ E \succ H \succ D$ *C.R.* = 0.0120

$\max a_{ij} = 5$

(b)

Quality of life		Factor (criterion) $j = 1, 2, \ldots, m$								Eigenvector (weight) ω_i
		A	B	C	D	E	F	G	H	
Factor (criterion) $i = 1, 2, \ldots, m$	A	1	1/2	1/3	3	2	1	1	3	0.1196
	B	2	1	1	4	3	1	3	4	0.2115
	C	3	1	1	5	4	2	3	4	0.2586
	D	1/3	1/4	1/5	1	1/2	1/4	1/3	1/2	0.0385
	E	1/2	1/3	1/4	2	1	1/2	1	1	0.0700
	F	1	1	1/2	4	2	1	2	3	0.1550
	G	1	1/3	1/3	3	1	1/2	1	2	0.0908
	H	1/3	1/4	1/4	2	1	1/3	1/2	1	0.0560

$C \succ B \succ F \succ A \succ G \succ E \succ H \succ D$ *C.R.* = 0.0158

$\max a_{ij} = 7$

(c)

Quality of life		Factor (criterion) j = 1, 2, ..., m								Eigenvector (weight) ω_i
		A	B	C	D	E	F	G	H	
Factor (criterion) i = 1, 2, ..., m	A	1	1/2	1/4	4	2	1	1	3	0.1085
	B	2	1	1/2	6	4	1	3	5	0.1974
	C	4	2	1	7	5	3	5	6	0.3358
	D	1/4	1/6	1/7	1	1/2	1/5	1/3	1	0.0326
	E	1/2	1/4	1/5	2	1	1/3	1	1	0.0567
	F	1	1	1/3	5	3	1	2	4	0.1500
	G	1	1/3	1/5	3	1	1/2	1	2	0.0774
	H	1/3	1/5	1/6	1	1	1/4	1/2	1	0.0416

$C \succ B \succ F \succ A \succ G \succ E \succ H \succ D$ *C.R.* = 0.0159

$\max a_{ij} = 9$

(d)

Quality of life		Factor (criterion) $j = 1, 2, \ldots, m$								Eigenvector (weight) ω_i
		A	B	C	D	E	F	G	H	
Factor (criterion) $i = 1, 2, \ldots, m$	A	1	1/3	1/4	6	3	1/2	2	4	0.1073
	B	3	1	1/2	8	5	2	4	6	0.2321
	C	4	2	1	9	6	3	5	7	0.3277
	D	1/6	1/8	1/9	1	1/4	1/7	1/5	1/3	0.0189
	E	1/3	1/5	1/6	4	1	1/4	1/2	2	0.0489
	F	2	1/2	1/3	7	4	1	3	5	0.1591
	G	1/2	1/4	1/5	5	2	1/3	1	3	0.0721
	H	1/4	1/6	1/7	3	1/2	1/5	1/3	1	0.0339

$C \succ B \succ F \succ A \succ G \succ E \succ H \succ D$ *C.R.* = 0.0350

Figure 3. Pairwise comparison matrices of the impact on quality of life of factors interacting with elements of the transport system using a mutual intensity scale: (**a**) $\max a_{ij} = 3$; (**b**) $\max a_{ij} = 5$; (**c**) $\max a_{ij} = 7$; (**d**) $\max a_{ij} = 9$.

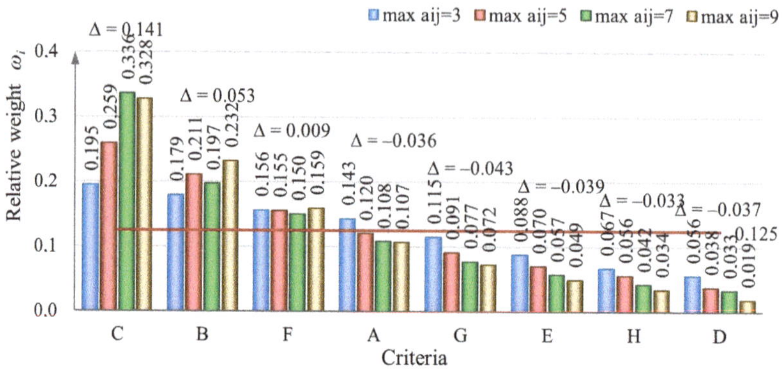

Figure 4. Normalized weights of the criteria calculated using the AHP method, where the ranks of the criteria given by the expert are the same but a different scale (3, 5, 7 and 9) is used for the intensity of the relative importance [41].

Sufficient consistency of the matrices is indicated by consistency ratio (*C.R.*) values less than 0.1. The *C.R.* of these four matrices increases with the value of the scale used, $\max a_{ij}$. This dependency is not theoretically proven, as the subjectivity of the expert's decisions when filling in the pairwise comparison matrix may be significant for the variation of the results.

When a team of n experts assesses the significance of the m criteria that comprise an object of study, not all experts choose the same scale size. The effect of scale size on the total variance, $\sigma^2_{\omega,\,ss}$, of the eight criterion weights is investigated in this paper by simulating a sample of $n = 10$, consisting of 10% ($\max a_{ij} = 3$), 10% ($\max a_{ij} = 5$), 30% ($\max a_{ij} = 7$), and 50% ($\max a_{ij} = 9$) of different scale series (Table 1). This percentage proportion is close to the magnitude of the Saaty scales used in the pilot project. Thus, Table 1 shows the weights of the criteria calculated using the AHP method when ten experts choose to use different levels of the scale.

Table 1. Dependence of the variance, $\sigma^2_{\omega i,ss}$, of the ω_i modeled weights, ω_i, from the survey data on the size of the chosen Saaty scale, $\max a_{ij}$, for the interaction factors of the transport system elements that improve quality of life.

Saaty's Scale Size $\max a_{ij}$	Factor (Criteria) $i = 1, 2, \ldots, m$							
	A	B	C	D	E	F	G	H
3	0.1433	0.1793	0.1955	0.0556	0.0883	0.1563	0.1146	0.0671
5	0.1195	0.2115	0.2586	0.0385	0.0701	0.1550	0.0908	0.0560
7	0.1085	0.1974	0.3358	0.0326	0.0567	0.1500	0.0774	0.0416
7	0.1085	0.1974	0.3358	0.0326	0.0567	0.1500	0.0774	0.0416
7	0.1085	0.1974	0.3358	0.0326	0.0567	0.1500	0.0774	0.0416
9	0.1073	0.2321	0.3277	0.0189	0.0489	0.1591	0.0721	0.0339
9	0.1073	0.2321	0.3277	0.0189	0.0489	0.1591	0.0721	0.0339
9	0.1073	0.2321	0.3277	0.0189	0.0489	0.1591	0.0721	0.0339
9	0.1073	0.2321	0.3277	0.0189	0.0489	0.1591	0.0721	0.0339
9	0.1073	0.2321	0.3277	0.0189	0.0489	0.1591	0.0721	0.0339
Mean	0.1125	0.2144	0.3100	0.0286	0.0573	0.1557	0.0798	0.0417
St. Deviat.	0.0115	0.0203	0.0463	0.0122	0.0128	0.0042	0.0135	0.0113
Skewness	2.6557	−0.4984	−2.1759	1.2249	1.9075	−0.6725	2.3085	1.6073
Kurtosis	7.1860	−1.3941	4.2220	1.3590	3.5456	−1.5779	5.4334	1.9527

When the sample size is $n = 10$, the critical triple value of skewness standard deviation is $3s_{sk} = 2.07$, and the kurtosis quintuple is $5s_{ku} = 6.67$. Therefore, the normal distribution is not satisfied by skewness criteria A, C, and G, with Sk = 2.66, Sk = −2.18, and Sk = 2.31, respectively, or by kurtosis criteria A, with Ku = 7.19. The distributions of the other criteria follow a normal law, which allows us to compare the homogeneity of the variances of the weights of the criteria, $\sigma^2_{\omega i,\ ss}$, by calculating the Cochran's test statistic, $\hat{G}_{\max}$, according to Formula (16) and to determine the mean of the variances of the weights of the criteria, $\sigma^2_{\omega,\ ss}$. After discarding the largest standard deviations, $\sigma_{\omega C} = 0.046$ and $\sigma_{\omega B} = 0.0203$, the variances of the weights of the remaining six criteria, $\sigma^2_{\omega i,\ ss}$, are statistically equal because $\hat{G}_{\max} = 0.236$ is less than the critical value of $G_C(0.05,\ 6,\ 9) = 0.368$.

The total variance of the weights of all the quality-of-life criteria, depending on the size of the scale, is $\sigma^2_{\omega,\ ss} = 0.000416$, and for statistically identical criteria (excluding C and B), $\sigma^2_{\omega,\ ss} = 0.000129$. Their values were used to calculate the effect of the magnitude of the scale on the percentage of variation in the weights of the criteria. If all experts on the team used the same Saaty scale size to determine the significance of the criteria, the proportion of the total variance of the criteria weights that depends on the scale size would be zero.

5.2. Uniformity of Opinion of One Expert

The matrix of the eight criteria compared in pairs using the AHP method by the same first E1 expert was filled in again 10 times at intervals of several weeks. The aim was to make the expert forget the ranks and the intensities of mutual importance he had previously assigned to the criteria. This study allowed us to determine the extent to which the expert's opinion is constant (unchanging) over a long period of time when assessing the importance of the same criteria, i.e., what the margin of error is in the expert's judgments.

Each expert's assessments have an element of subjectivity, which suggests that they may have slightly different opinions over time. The researchers organizing this study cannot tell the expert what size scale to use; therefore, the expert starts by comparing the most important criterion with the least important one, which results in the size of the scale. It is likely that one expert's opinion, expressed as ranks or weights, on the significance of the criteria is more stable (with less dispersion) than the opinions of the team of experts consisting of individuals regarding the significance of these criteria. Also, the value of the concordance coefficient, which indicates the degree of agreement between the opinions of a single expert, W_{one}, is significantly higher than that of a team of experts consisting of individuals, W_{team}. The results of this study show that the ranks assigned by an expert for the same criterion do not differ significantly. The concordance coefficient was calculated, $W_{one} = 0.89$, as well as its minimum value, $W_{\min} = 0.201$, at the level of significance, $\alpha = 0.05$. The compatibility coefficient, $k_c = 4.43$, is approximately 89% of its maximum value, $k_{c\max} = 4.98$.

The AHP method calculates the highest weights, ω_i, for criterion C and the lowest weights for criterion D (Table 2). The average weights for these and other criteria, which are calculated using the 10 pairwise comparison matrices completed by the same expert, show little variation. The standard deviations, $\sigma_{\omega i}$, of the weights, σ_i, vary from 0.0101 (criteria D) to 0.0752 (B).

The skewness and kurtosis coefficients of the estimated weights for all the criteria have moduli $|Sk|$ and $|Ku|$ that are below their critical values, $3s_{sk} = 2.06$ and $5s_{ku} = 6.67$, suggesting that the weights assigned to the criteria by one expert, ω_i, follow a normal distribution. Their normal distribution allowed us to use Cochran' test to compare the sample variances of the weights of the criteria, $\sigma^2_{\omega i,\ oe}$, determining their equality. Dividing the maximum sample variance of the weights of criterion B, $\max\sigma^2_{\omega B,\ oe} = 0.00565$, by the sum of all the variances, $\sum_{i=1}^{8} \sigma^2_{\omega i,\ oe} = 0.01326$, produced a Cochran' test statistic,

$\hat{G}_{\max} = 0.426$, greater than $G_C(0.05, 8, 9) = 0.293$, indicating that the sample variances of the weights are not equal. After discarding the maximum variance of 0.00565, the second maximum variance of criteria C, $\sigma^2_{\omega C,\ oe} = 0.00245$, which is the ratio of the sum of the variances of all the weights, $\sum_{i=1}^{7} \sigma^2_{\omega i,\ oe} = 0.00760$, is equal to $\sum_{i=1}^{7} \sigma^2_{\omega i,\ oe} = 0.00760$ and less than the critical value of the statistic, $G_C(0.05, 7, 9) = 0.326$.

Table 2. Subjective weights were given to one expert's criteria for the interaction of transport system elements improving quality of life using the AHP method, when criteria were evaluated 10 times with a 4- to 6-week time step.

Evaluation Time No.	**Factor (Criterion) *i* = 1, 2, . . ., *m***								*C.R.*
	A	**B**	**C**	**D**	**E**	**F**	**G**	**H**	
E1-1	0.1106	0.3682	0.2237	0.0262	0.0191	0.1423	0.0672	0.0427	0.0626
E1-2	0.1763	0.2249	0.3254	0.0332	0.0229	0.1005	0.0424	0.0744	0.0492
E1-3	0.2190	0.1058	0.3350	0.0486	0.0186	0.1711	0.0712	0.0307	0.0386
E1-4	0.1184	0.2119	0.3749	0.0428	0.0309	0.1348	0.0636	0.0227	0.0446
E1-5	0.2138	0.1560	0.3567	0.0211	0.0445	0.1005	0.0775	0.0299	0.0546
E1-6	0.0759	0.2871	0.2560	0.0238	0.0471	0.1750	0.1025	0.0326	0.0475
E1-7	0.1170	0.1701	0.3351	0.0344	0.0465	0.2069	0.0658	0.0242	0.0524
E1-8	0.1475	0.2127	0.3762	0.0223	0.0310	0.0917	0.0733	0.0453	0.0366
E1-9	0.1264	0.1450	0.3272	0.0220	0.0343	0.2314	0.0676	0.0461	0.0399
E1-10	0.1514	0.2177	0.3525	0.0186	0.0334	0.1058	0.0748	0.0458	0.0482
Mean $\overline{\omega}_i$	0.1456	0.2099	0.3263	0.0293	0.0328	0.1460	0.0706	0.0394	–
Priority	4	2	1	8	7	3	5	6	–
St. Deviat.	0.0460	0.0752	0.0495	0.0101	0.0107	0.0487	0.0148	0.0152	–
Skewness	0.4400	0.9168	−1.3013	0.9593	0.0674	0.5711	0.4261	1.3228	–
Kurtosis	−0.5324	1.1922	1.0323	−0.2284	−1.2949	−0.9180	3.2225	2.4013	–

The mean variance, $\sigma^2_{\omega,\ oe} = 0.001086$, and the mean standard deviation, $\sigma^2_{\omega,\ oe} = 0.033$, are calculated from statistically equal weights of the seven criteria. The variance of all eight statistically unequal weights is equal, $\sigma^2_{\omega,\ oe} = 0.001658$, and the standard deviation is equal, $\sigma_{\omega,oe} = 0.407$. They allow us to calculate the proportion of the variation in the weights assigned to the criteria by the team of experts, which depends on the instability (variability) of each expert's opinion.

5.3. *Assessment of the Significance of the Expert Team Criteria*

The significance of the eight criteria for the interaction of the elements of the transport system that affect quality of life has been evaluated by 54 experts, who ranked the criteria (Table 3) and compared them with each other according to the Saaty fundamental scale (Table 4). Here, Table 3 shows the calculation of the weights of the criteria from the rank means using the ARTIW-L and ARTIW-N methods.

The average of the ranks, $\overline{R}_i$, of each quality of life criterion, R_{ij}, calculated according to Formula (2), can reasonably be taken as the result of solving the problem because its concordance coefficient, $W = 0.203$, (Equation (5)) is 5.5 times higher than its minimum value, $W_{\min} = 0.037$ (Equation (6)), at a significance level of $\alpha = 0.05$.

Table 3. Ranking significance of interactions between transport system elements.

Statistical Indicator Rank	Factor (Criterion), $i = 1, 2, \ldots, m$								Total
	A	B	C	D	E	F	G	H	
Sum	237	288	135	281	274	170	298	261	1944
Mean $\overline{R}_i$	4.3889	5.3333	2.5000	5.2037	5.0741	3.1481	5.5185	4.8333	36.0000
St. Deviat.	2.11	2.36	1.42	2.07	1.90	1.97	2.19	2.34	0.0000
$\sum R_{ij} - \overline{R}$	−6	45	−108	38	31	−73	55	18	
$\left(\sum R_{ij} - \overline{R}\right)^2$	36	2025	11,664	1444	961	5329	3025	324	24,808
Skewness	0.0239	−0.4869	0.8750	−0.4429	−0.2296	0.4818	−0.5201	−0.0859	
Kurtosis	−1.0962	−1.1525	0.0190	−0.7680	−0.6878	−0.9925	−0.8599	−1.2475	
				ARTIW-L method					
$\omega_i^{ARTIW-L}$	0.1281	0.1019	0.1806	0.1054	0.1090	0.1626	0.0967	0.1157	1.0000
Priority	3	7	1	6	5	2	8	4	–
				ARTIW-N method					
u_i	0.5696	0.4688	1	0.4804	0.4927	0.7942	0.4530	0.5173	4.7760
$\omega_i^{ARTIW-N}$	0.1193	0.0982	0.2094	0.1006	0.1031	0.1663	0.0948	0.1083	1.0000
Priority	3	7	1	6	5	2	8	4	–

Table 4. Significance of the AHP method: relative weights of interaction factors between transport system elements that improve quality of life.

Expert $j = 1, 2, \ldots, n$	Factor (Criterion) $i = 1, 2, \ldots, m$								C.R.
	A	B	C	D	E	F	G	H	
E1	0.1106	0.3682	0.2237	0.0262	0.0191	0.1423	0.0672	0.0427	0.0626
E2	0.1096	0.0307	0.1315	0.0569	0.0986	0.3531	0.2028	0.0168	0.0670
E3	0.3043	0.0361	0.0747	0.1117	0.0511	0.1635	0.2312	0.0274	0.0142
E4	0.0773	0.1974	0.3277	0.0390	0.1071	0.0633	0.1492	0.0390	0.0135
E5	0.0468	0.1570	0.2586	0.3006	0.0908	0.0420	0.0764	0.0278	0.0439
E6	0.0763	0.2091	0.3916	0.0347	0.0482	0.1319	0.0273	0.0809	0.0219
E7	0.0728	0.1513	0.2338	0.3350	0.0287	0.1011	0.0461	0.0312	0.0586
E8	0.2758	0.1056	0.2780	0.1388	0.0487	0.0381	0.0286	0.0864	0.0504
E9	0.2735	0.0679	0.1686	0.0467	0.0317	0.1503	0.0927	0.1686	0.0120
E10	0.0424	0.1207	0.2724	0.0671	0.0271	0.1510	0.2970	0.0223	0.0267
E11	0.0816	0.2260	0.2882	0.0352	0.0557	0.1657	0.0217	0.1259	0.0732
E12	0.0299	0.0243	0.0740	0.0482	0.1476	0.3212	0.0847	0.2701	0.0293
E13	0.0510	0.1384	0.0677	0.1015	0.2408	0.0263	0.0176	0.3567	0.0798
E14	0.0503	0.0237	0.1627	0.2379	0.3027	0.0752	0.1133	0.0342	0.0298
E15	0.0446	0.0282	0.3172	0.0808	0.1192	0.1988	0.1533	0.0579	0.0246
E16	0.0575	0.2651	0.1413	0.1198	0.0415	0.0924	0.0250	0.2574	0.0258
E17	0.1955	0.0361	0.1563	0.0388	0.1162	0.3452	0.0326	0.0793	0.0659
E18	0.1792	0.0241	0.3245	0.0400	0.0171	0.2398	0.0628	0.1125	0.0701
E19	0.1596	0.0298	0.2526	0.0199	0.0465	0.3133	0.0692	0.1089	0.0950
E20	0.0238	0.0572	0.2491	0.1652	0.1227	0.0796	0.2671	0.0353	0.0480

Table 4. *Cont.*

Expert $j = 1, 2, \ldots, n$	Factor (Criterion) $i = 1, 2, \ldots, m$								*C.R.*
	A	B	C	D	E	F	G	H	
E21	0.1072	0.2603	0.2919	0.0454	0.0286	0.1688	0.0185	0.0793	0.0582
E22	0.2207	0.0740	0.1435	0.0515	0.1131	0.3333	0.0258	0.0381	0.0180
E23	0.2246	0.0968	0.0666	0.0432	0.0310	0.3436	0.0225	0.1717	0.0298
E24	0.3072	0.0214	0.1650	0.0300	0.1034	0.2604	0.0434	0.0692	0.0239
E25	0.1248	0.2009	0.3555	0.0167	0.0296	0.0780	0.0396	0.1549	0.0743
E26	0.1242	0.2550	0.2549	0.1929	0.0480	0.0287	0.0739	0.0224	0.0584
E27	0.2988	0.1051	0.1621	0.0391	0.0533	0.0706	0.1621	0.1089	0.0240
E28	0.0254	0.1177	0.2682	0.1485	0.2682	0.1048	0.0473	0.0199	0.0513
E29	0.0426	0.0262	0.0762	0.0192	0.1620	0.3287	0.1107	0.2344	0.0450
E30	0.0788	0.1715	0.2644	0.0494	0.0196	0.2551	0.0310	0.1303	0.0514
E31	0.1172	0.049	0.3414	0.0708	0.0264	0.1454	0.0193	0.2305	0.0438
E32	0.0316	0.0721	0.2422	0.3176	0.0209	0.1113	0.0452	0.1591	0.0275
E33	0.2564	0.0194	0.1635	0.0554	0.1041	0.3096	0.0283	0.0633	0.0638
E34	0.0421	0.1147	0.3275	0.0174	0.1604	0.2328	0.0344	0.0707	0.0564
E35	0.1689	0.0289	0.0839	0.0534	0.1031	0.2223	0.0372	0.3023	0.0464
E36	0.0442	0.0232	0.2121	0.1901	0.1063	0.0601	0.0226	0.3414	0.0537
E37	0.1591	0.0339	0.3277	0.0488	0.1073	0.0721	0.0189	0.2322	0.0347
E38	0.1065	0.0327	0.0479	0.0713	0.2319	0.3281	0.1585	0.0231	0.0291
E39	0.2319	0.0327	0.1585	0.0713	0.0479	0.3280	0.1066	0.0231	0.0291
E40	0.1331	0.2006	0.1749	0.0226	0.0458	0.3303	0.0602	0.0325	0.0435
E41	0.2284	0.0157	0.1602	0.0506	0.1142	0.3224	0.0310	0.0774	0.0675
E42	0.0800	0.3013	0.1732	0.0409	0.0555	0.2131	0.0322	0.1038	0.0123
E43	0.1632	0.0138	0.3196	0.0782	0.0476	0.1164	0.2323	0.0289	0.0830
E44	0.3371	0.0181	0.1469	0.0273	0.0753	0.2438	0.0421	0.1095	0.0593
E45	0.0426	0.0261	0.3071	0.2583	0.0716	0.1126	0.0167	0.1650	0.0940
E46	0.1570	0.0230	0.0800	0.0610	0.0310	0.3280	0.0460	0.2750	0.0046
E47	0.2287	0.0178	0.3376	0.0243	0.1206	0.0726	0.1581	0.0403	0.0658
E48	0.0170	0.0243	0.1558	0.0653	0.1065	0.2438	0.0377	0.3499	0.1044
E49	0.0446	0.0308	0.2340	0.1839	0.1196	0.0721	0.2945	0.0205	0.0307
E50	0.0721	0.1033	0.1687	0.0154	0.2235	0.0389	0.0232	0.3549	0.0906
E51	0.0208	0.0700	0.1588	0.2318	0.0471	0.3327	0.1072	0.0316	0.0234
E52	0.0230	0.033	0.0480	0.1070	0.2320	0.3280	0.1580	0.0710	0.1092
E53	0.1620	0.0367	0.2664	0.1104	0.0636	0.3168	0.0257	0.0184	0.0317
E54	0.0299	0.0455	0.3154	0.2311	0.0736	0.1117	0.0234	0.1694	0.0313
Weight Sum $\overline{\omega}_i^{AHP}$	6.7141	4.9924	11.3938	5.0841	4.9537	10.159	4.3999	6.3042	
Mean $\overline{\omega}_i^{AHP}$	0.1243	0.0925	0.2110	0.0942	0.0917	0.1881	0.0815	0.1167	
Priority	3	6	1	5	7	2	8	4	–
St. Deviat.	0.0907	0.0879	0.0927	0.0841	0.0689	0.1099	0.0759	0.1021	

The chi-squared (χ^2) statistic, calculated according to Equation (7), is equal to 76.57, and its critical value is $\chi^2_{7,\,00.5} = 14.07$. Therefore, their ratio, which is called compatibility coefficient k_c, is equal to 5.5 in Equation (8).

The estimated skewness and kurtosis moduli of the ranks of all criteria are below their critical values, $3s_{sk} = 0.974$ and $5s_{ku} = 3.194$, i.e., the ranks follow a normal distribution. Cochran's test of Formula (16) shows that the estimated statistics, $\hat{G}_{\max} = 0.1637$, are less than their critical values, $G_C(0.05,\ 8,\ 53) = 0.1956$, which indicates that the variances of the ranks of all criteria, σ^2_{Ri}, are statistically equal. The normality of the criteria and the homogeneity of the variances of all their ranks leads to the calculation of the average variance of the ranks, σ^2_R, which is 4.265, and the average standard deviation, $\sigma_R = 2.065$. They can be used to calculate the minimum size of a team of experts according to Formula (1) by taking the rank variance, σ^2_R, instead of the variance of the weights, σ^2_ω.

From the average ranks of the criteria, $\overline{R}_i$, the ARTIW-L and ARTIW-N methods are used to calculate the normalized relative weights of these criteria, $\omega_i^{ARTIW-L}$ and $\omega_i^{ARTIW-N}$, according to Formulas (3) and (4). Table 3 shows the sequence of these criteria in order from most to least important: C ≻ F ≻ A ≻ H ≻ E ≻ D ≻ B ≻ G (Figure 5).

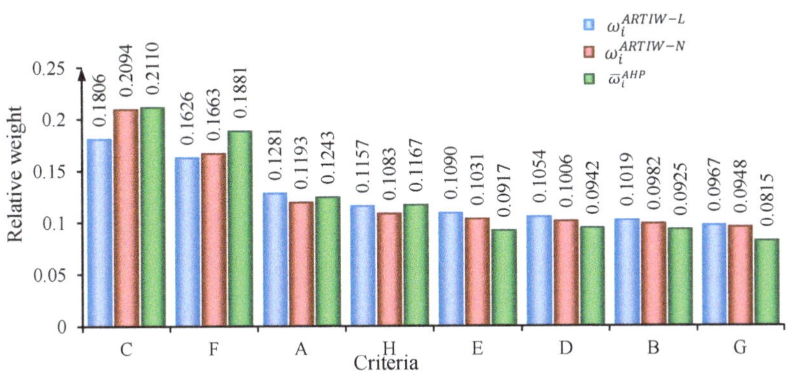

Figure 5. Significance of the interaction effects of elements of the transport system that improve quality of life identified using the different MCDM methods.

By applying the AHP method to test the significance of the criteria, each expert filled in a pairwise comparison matrix, the correctness of which is indicated by a consistency ratio (*C.R.*) of 0.1. Only two experts (E48 and E52) exhibited matrix *C.R.*s that were marginally higher than 0.1. It was calculated by dividing the consistency index (*C.I.*) (Equation (10)) by the random index (*R.I.* = 1.41) (Equation (12)). The eigenvectors of the criteria, calculated using Equation (9), are taken as their normalized relative weights (Table 4). The average of the weights of each criterion shows the following sequence of priority of the quality of life criteria, as determined by a team of 54 experts: C ≻ F ≻ A ≻ H ≻ D ≻ B ≻ E ≻ G. This sequence is the same for the four most important criteria, as found using the ARTIW-L and ARTIW-N methods. The priorities of the less important criteria, D ≻ B ≻ E, determined using the AHP method are different from those determined using the $\omega_i^{ARTIW-L}$ and $\omega_i^{ARTIW-N}$ values, i.e., the priorities are reversed (E ≻ D ≻ B) (Figure 5).

The weights and averages of the interaction factors (criteria) of the elements of the transport system that improve the quality of life calculated using the three MCDM methods, according to Formula (13), are presented in Table 5.

Table 5. Average values of relative weights of criteria calculated using different MCDM methods.

Importance Indicator	Factor (Criterion), $i = 1, 2, \ldots, m$							
	C	F	A	H	E	D	B	G
Mean weight	0.2003	0.1723	0.1239	0.1136	0.1013	0.1001	0.0975	0.0910

The number of times that one criterion is more important for quality of life than another criterion is indicated by the ratios of the pairwise comparisons (Table 6) of the relative weights (Table 5). This shows that the most important criterion, C, is 2.2 times more important than the least important criterion, G. The weights of the other criteria are smaller.

Table 6. Comparative relationships between the relative weights of the importance of factors that contribute to quality of life.

Quality of Life		Factor (Criterion), $j = 1, 2, \ldots, m$							
		C	F	A	H	E	D	B	G
Factor (criterion), $i = 1, 2, \ldots, m$	C	1	1.16	1.62	1.76	1.98	2.00	2.05	2.20
	F		1	1.39	1.52	1.70	1.72	1.77	1.89
	A			1	1.09	1.22	1.24	1.27	1.36
	H				1	1.12	1.13	1.17	1.25
	E					1	1.01	1.04	1.11
	D						1	1.03	1.10
	B							1	1.07
	G								1

5.4. Magnitude of Sources of Variation and Sampling Precision of the Weights of the Criteria Determined Using the AHP Method

The dispersion of the weights of the criteria estimated using the ARTIW-L and ARTIW-N methods, expressed as standard deviations, is not determined; therefore, the sources of variance of these weights are not known. The weights of the criteria calculated using the AHP method are reported for each of the 54 experts (Table 4), which allows the standard deviations of these weights to be determined.

The histograms of the weights of the criteria, $\overline{\omega}_i^{AHP}$ (Figure 6), and the skewness values show that only the distributions of the most important criteria, C, F, and A, follow the normal law. The histograms of the other five less important criteria have a positive asymmetry, and their asymmetry coefficients (skewness), $\overline{\omega}_i^{AHP}$, are larger than the critical value, $3s_{sk} = 0.974$ (Figure 7). The curvilinear modulus, $|Ku|$, of all the criteria is less than their critical values, $5s_{ku} = 3.194$, indicating that the weights of all the criteria follow a normal distribution.

Testing the homogeneity of the variances, $\sigma_{\omega i}^2$, of the criterion weights, $\overline{\omega}_i^{AHP}$, calculated using the AHP method (null hypothesis) with Cochran's test showed that the statistic, $\hat{G}_{\max} = 0.1851$, calculated using Equation (16) is below its critical value, $G_C(0.05,\ 8,\ 53) = 0.1956$. The statistically uniform variances of the criterion weights allowed for the calculation of the mean, σ_{ω}^2, of the variances of the eight criterion weights, which is 0.008157, and the pooled standard deviation, $\sigma_{\omega}^{AHP} = 0.0903$.

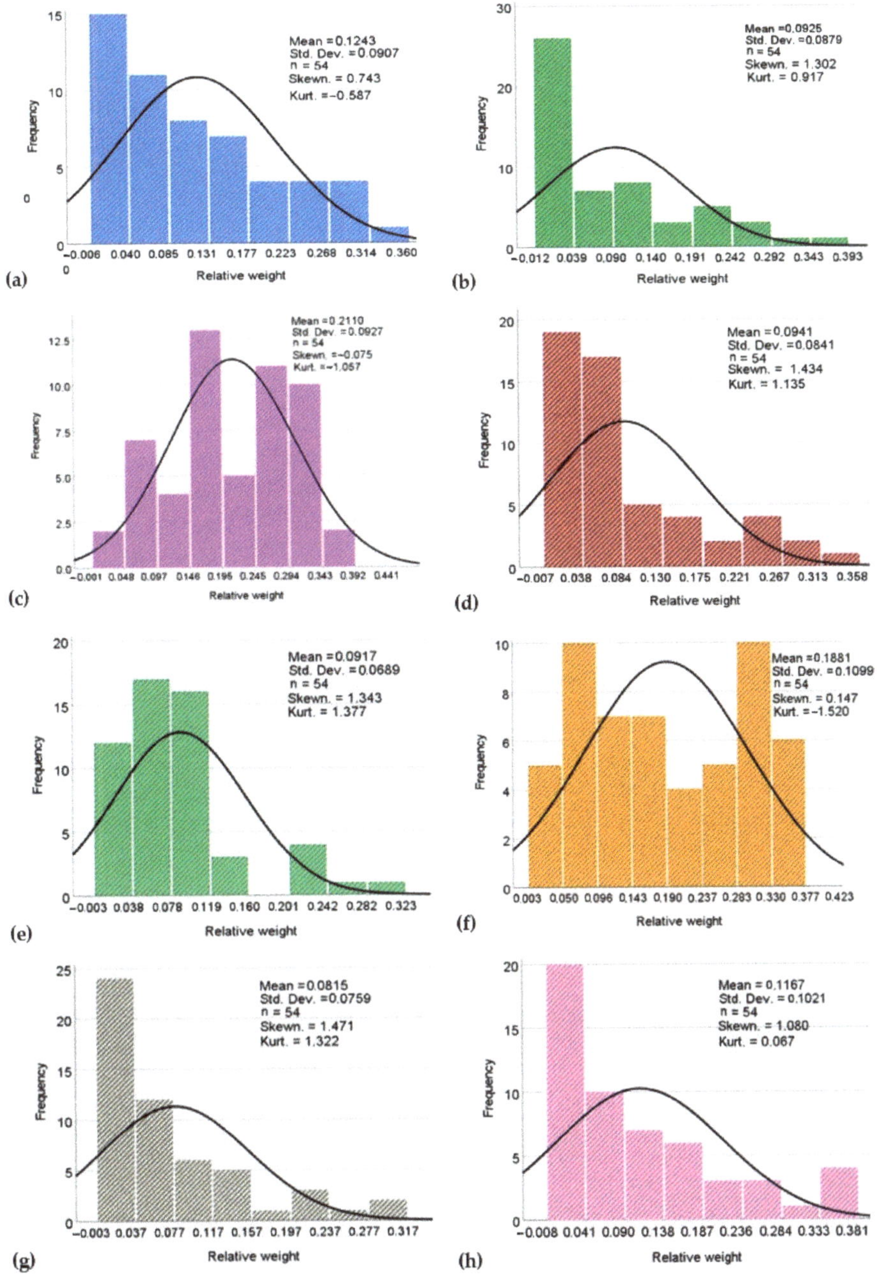

Figure 6. Histograms of the weights, theoretical curves of the normal distribution, and statistical indices of the weights obtained by applying AHP method to the interaction criteria of the transport elements that improve quality of life identified by the team of experts: (**a**)—A, (**b**)—B, (**c**)—C, (**d**)—D, (**e**)—E, (**f**)—F, (**g**)—G, (**h**)—H.

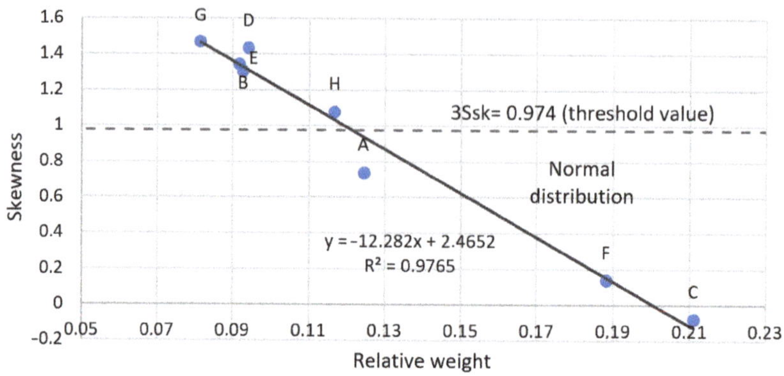

Figure 7. Influence of criterion weights determined using the AHP method on skewness values.

Using the mathematical model in (14), it is possible to calculate the proportion of the total variance of the weights of the quality of life criteria, σ^2_ω, that represents the team net variation as follows:

$$\sigma^2_{\omega,tnv} = \sigma^2_\omega - \sigma^2_{\omega,ss} - \sigma^2_{\omega,oe} \quad (19)$$

The net variance of the equal variances of the criterion weights remaining after discarding the significantly different variances of 2 (criteria B and C) or 1 (criterion B), $\sigma^2_{\omega B,ss}$, $\sigma^2_{\omega C,ss}$, and $\sigma^2_{\omega B,oe}$, is as follows:

$$\sigma^2_{\omega,tnv} = 0.008157 - 0.000129 - 0.001086 = 0.006942$$

It accounts for 85.1% of the total variance, σ^2_ω (Table 7). The variance due to differences in the magnitude of the Saaty's scale used in the expert study accounts for 1.6% of the total variance, σ^2_ω. The variance due to the variability in the weighting of the criteria by one expert accounts for 13.3% of the total variance, σ^2_ω.

Table 7. Sources of variation in the weights of the quality-of-life criteria calculated using the AHP method and their percentage impact on the magnitude of the variances.

Level of Homogeneity of Variances of Criteria Weights	Statistical Indicator	Total (Complete) Variation σ^2_ω	Sources of Variation		
			Different Saaty Scale Size	Variability in the Opinion of One Expert	Divergence in Opinions of the Team of Experts
Equal variances 6–7	Variance of weights	0.008157	0.000129	0.001086	0.006942
	Percentage	100.0	1.6	13.3	85.1
Variances for all 8 criteria	Variance of weights	0.008157	0.000416	0.001658	0.006083
	Percentage	100.0	5.1	20.3	74.6

The net variance of the weights of all eight quality-of-life criteria according to Formula (19) is as follows:

$$\sigma^2_{\omega,tnv} = 0.008157 - 0.000416 - 0.001658 = 0.006083$$

It accounts for 74.6% of the total variance, σ^2_ω (Table 7). The variance, $\sigma^2_{\omega,ss}$, accounts for 5.1%, and the variance, $\sigma^2_{\omega,oe}$, accounts for 20.3% of the total variance, σ^2_ω.

The variation in the weights of the criteria increases the number of experts, $n_{\min}$, to be interviewed, which is calculated using Equation (1). From this, the accuracy Δ can be calculated for a sample of 54 experts, taking the 95% probability (t = 1.96) and the pooled standard deviation of the criterion weights derived from this study, $\sigma_\omega = 0.0903$, as follows:

$$\Delta_\omega = \sqrt{\frac{t^2 \cdot \sigma_\omega^2}{n}} = \sqrt{\frac{1.96^2 \cdot 0.0903^2}{54}} = 0.0241$$

With 95% probability, we can say that within the estimated precision Δ range $\pm$ 0.0241 is the weight of the criterion population, ω. As the required precision and the variation of the criterion weights decrease, the minimum number of experts decreases, $n_{\min}$. For a precision of Δ = 0.03, 35 experts need to be interviewed, and for a precision of = 0.025, the number of experts on the team has to be 50. Justification of the optimum required precision is linked not only to the reliability of the results of this study but also to the time and monetary costs involved.

6. Conclusions

Factors affecting the performance of the transport sector have a significant impact on quality of life. A new model of the interaction between the elements of the transport system has been developed, showing the relationship with other productive and non-productive activities, the national economy, and quality of life (Figure 1). It identifies eight factors (criteria) whose impact on quality of life has been investigated using multi-criteria decision-making (MCDM) methods. Rank correlation, Analytic Hierarchy Process (AHP), and Average Rank Transformation into Weight-Linear (ARTIW-L) and Nonlinear (ARTIW-N) methods were used to determine the significance of the criteria. Mathematical statistical methods were also used to analyze the data from the expert survey. The normality of the distributions of the criterion weights was checked by calculating skewness and kurtosis, and the homogeneity of their variances was checked by applying the Cochran test.

The relative weights of the criteria were determined by a team of 54 experts. The weights, $\omega_i^{ARTIW-L}$ and $\omega_i^{ARTIW-N}$, and $\overline{\omega}_i^{AHP}$ averages, $\overline{\overline{\omega}}_i$, of the three MCDM methods show that the development of the transport infrastructure network and the improvement of its quality, $\overline{\overline{\omega}}_C = 0.2003$, the reduction of accidents, breakdowns, and damage, the increase in comfort in the transport sector $\left(\overline{\overline{\omega}}_F = 0.1723\right)$, and the reduction of environmental pollution caused by the transport sector have the greatest impact on quality of life ($\overline{\overline{\omega}}_A = 0.1239$). The results of this study stand out from others in that the presented model of the interaction between transport system elements shows that the impact of infrastructure is not coincidentally determined to be so important, as the quality of the roads has an impact on travel safety and comfort, as well as the speed and cost of transport-related services. In addition, improvements in road users' discipline, awareness, caution, courtesy, and culture of travel ($\overline{\overline{\omega}}_H = 0.1136$); the development of science, research, and innovation in the transport sector and the application of the results in practice ($\overline{\overline{\omega}}_H = 0.1136$); and the improvement of the operating environment of the transport system and strikes by transport workers ($\overline{\overline{\omega}}_D = 0.1001$) are of medium importance for quality of life. Experts consider that the least important factors for quality of life are an increase in the number and quality of vehicles ($\overline{\overline{\omega}}_C = 0.0975$) and an increase in the share of the country's Gross Domestic Product generated by the transport sector ($\overline{\overline{\omega}}_G = 0.0910$).

The weights of the most important criteria determined using the AHP method are closest to a normal distribution in terms of skewness. As the average weight of the criteria decreases, the skewness modulus increases, i.e., the distribution "moves away" from the normal distribution and exhibits positive asymmetry. According to the estimated values of the coefficient of excess (kurtosis), the weights of all eight criteria follow a normal distribution. The normal distribution of the weights of the criteria allowed us to compare their variances using the Cochran test and to calculate the mean of the equal variances of all the quality-of-life criteria.

The "narrower" (lower maxa_{ij}) the Saaty scale used to assess the importance of the criteria, the smaller the difference between the weights of the most and least important criteria and the more harmonized (lower consistency ratio $C.R.$) the pairwise comparison matrix. The width of the scale is theoretically not related to the number of criteria to be compared pairwise (matrix rank). However, in practice, the tendency is that the larger the matrix row, m, the wider the scale of peer intensity (larger maxa_{ij}) used to assess the significance of the criteria.

The variation in the test vector (relative weights) of the ranks of the criteria and the pairwise comparison matrices filled in 10 times by a single expert over large intervals of time is smaller than the variation in the significance estimates of these criteria for the individual experts in the team. The concordance coefficient of the ranks of the eight criteria, $W = 0.89$, is close to unity (compatibility coefficient $k_c = 4.43$). This indicates that within the level of significance, $\alpha = 0.05$, the expert's opinion is not absolutely uniform over time. The standard deviations of the weights of the criteria, $\sigma_{\omega i}$, calculated using the AHP method for this expert ranged from 0.0101 (criteria D) to 0.0752 (criteria B). The average standard deviation of the weights of all normally distributed and Cochran's test statistically identical criteria for the subject (quality of life) is as follows: $\sigma_{\omega} = 0.033$.

The new model of criterion weight variances shows that 1.6–5.1% of the total variance of the weights calculated from the weights of the pairwise comparison matrices completed by the team of experts using the AHP method is accounted for by the variance of the Saaty fundamental scale chosen by the team of experts with different magnitudes; 13.3–20.3% is accounted for by the variance of the personal opinion of each of the experts; and 74.6–85.1% is accounted for by the net variance of the weights assigned to the criteria by the expert team.

This research managed to avoid the typical errors associated with the AHP method, which often arise in other studies when completing the criterion significance and pairwise comparison matrices. Also, in this study, all 54 matrices were matched according to a strict transitivity condition. This study stands out from the rest because the AHP method was also applied to an exceptionally large team of experts for this type of analysis, which required considerable effort from both the researchers and the experts interviewed.

The limitations of this study are related to the limited number of criteria used to describe the quality of life through transport system elements; however, there is a reason for this limited number of criteria: when using the AHP method with a larger number of criteria, it becomes increasingly difficult to fill in the pairwise comparison matrix, or the results become unreliable and inconsistent.

Future research will be carried out using the AHP and the ARTIW-L and ARTIW-N methods described in this paper to analyze related problems in the transport sector, such as the importance of road infrastructure elements (e.g., asphalt pavement properties or maintenance) on road quality and the development of micro-mobility measures in cities.

Author Contributions: Conceptualization, H.S. and V.Ž.; methodology, H.S.; software, V.Ž.; validation, H.S. and V.Ž.; formal analysis, V.Ž.; investigation, H.S.; resources, H.S.; data curation, H.S.; writing—original draft preparation, H.S.; writing—review and editing, V.Ž.; visualization, V.Ž.; supervision, H.S. All authors have read and agreed to the published version of the manuscript.

Funding: This research received no external funding.

Institutional Review Board Statement: Not applicable.

Informed Consent Statement: Informed consent was obtained from all subjects involved in this study.

Data Availability Statement: The original contributions presented in this study are included in this article; further inquiries can be directed to the corresponding authors.

Conflicts of Interest: The authors declare no conflicts of interest.

Appendix A

The questionnaire presented to the experts is as follows:

Dear Expert,

I kindly ask you to use your personal expertise, experience, and qualifications to carefully read the description of the factors of interaction between the elements of the transport system that improve people's quality of life and to give them ranks (places) and fill in the AHP method pairwise comparison matrix, where the principle of positivity applies.

1. Factor inventory—factors that interact between elements of the transport system that improve people's quality of life (people live better because...):

Rank	Factor
	A. Reduction in environmental pollution (exhaust emissions, noise, vibrations) caused by the transport sector.
	B. Increase in the number of vehicles and improvement of the quality (technical parameters, condition, comfort, economy).
	C. Developing the transport infrastructure network (roads, railways, airports, seaports, terminals) and improving its quality.
	D. Improvement of the operating environment of the transport system (business, tax, price, investment, competition, grant, subsidies, etc.) and strikes by transport workers.
	E. Developing science, research, and innovation in the transport sector and putting the results into practice.
	F. Reduction in accidents, breakdowns, and damage and increase in comfort in the transport sector.
	G. Increase in the share of a country's Gross Domestic Product (GDP) generated by the transport sector.
	H. Improvement of the discipline, awareness, caution, courtesy, and travel behavioral culture of road users (drivers, pedestrians, passengers).

2. Pairwise comparison matrix $\mathbf{A} = \|a_{ij}\|; i, j = 1, 2, \ldots, m; m = 8$.

Quality of Life		Factor (Criterion) $j = 1, 2, \ldots, m$								Eigenvector (Weight) ω_i
		A	B	C	D	E	F	G	H	
Factor (criterion) $i = 1, 2, \ldots, m$	A	1								
	B		1							
	C			1						
	D				1					
	E					1				
	F						1			
	G							1		
	H								1	

Sequence: from the most important to the least important criterion

3. Scale of relative importance (fundamental scale).

Intensity of Importance	Definition	Interpretation
1	Equal importance	Factors are equally important
3	Moderate importance	One factor is slightly more important than another
5	Strong importance	One factor is more important than another
7	Very strong importance	One factor is much more important than the other
9	Extreme importance	One factor is incomparably more important than another
2, 4, 6, 8	Intermediate values	

References

1. Zhou, W.; Xu, Z.; Skačkauskas, P. Mapping knowledge domain of "transport": A bibliometric study of its status quo and emerging trends. *Transport* **2019**, *34*, 741–753. [CrossRef]
2. WHITE PAPER Roadmap to a Single European Transport Area. Towards a Competitive and Resource Efficient Transport System. 2019. Available online: http://www.trt.it/wp/wp-content/uploads/2021/06/Final-report-Evaluation-of-2011-Transport-White-Paper.pdf (accessed on 1 April 2021).
3. Dadashzadeh, N.; Sucu, S.; Pangbourne, K.; Ouelhadj, D. Socially Sustainable Mobility as a Service (MaaS): A Practical MCDM Framework to Evaluate Accessibility and Inclusivity with Application. *Cities* **2024**, *154*, 105360. [CrossRef]
4. Šikšnelytė-Butkienė, I.; Štreimikienė, D.; Baležentis, T.; Agnusdei, L. Multi-Criteria Decision-Making for Solving Transport Sustainability Issues: An Overview. *Transport* **2024**, *39*, 264–286. [CrossRef]
5. Ortega Hortelano, A.; Vassallo, J.M.; Pérez, J.I. Optimal welfare price for a road corridor with heterogeneous users. *Transport* **2019**, *34*, 318–329. [CrossRef]
6. Tarhan, Y.; Kabakuş, N. Enhancing Sustainable Road Construction: Evaluation of the Mechanical and Durability Properties of Stabilized Earth-Based Pavement Materials. *Sustainability* **2024**, *16*, 10784. [CrossRef]
7. Mostafa, M.A.; El-Gohary, N.M. Stakeholder-Sensitive Social Welfare–Oriented Benefit Analysis for Sustainable Infrastructure Project Development. *J. Constr. Eng. Manag.* **2014**, *140*, 04014038. [CrossRef]
8. Puodžiukas, V. Saugios Infrastruktūros Politikos Gairės Ir Greičio Valdymas. *Liet. Keliai* **2019**, *2*.
9. Li, C.; Xu, C.; Li, X. A Multi-Criteria Decision-Making Framework for Site Selection of Distributed PV Power Stations along High-Speed Railway. *J. Clean. Prod.* **2020**, *277*, 124086. [CrossRef]
10. Šimunović, L.; Hirnig, S.; Abramović, B. Evaluating the success factors of integrating the railway into the public urban–suburban transport. *Transport* **2023**, *38*, 105–115. [CrossRef]
11. Petkevičius, K.; Maskeliūnaitė, L.; Sivilevičius, H. Determining travel conditions on motorways for automobile trasport based on the case study for lithuanian highways. *Transport* **2019**, *34*, 89–102. [CrossRef]
12. Silva, M.B.C.D.; Bebber, S.; Fachinelli, A.C.; Moschen, S.D.A.; Perini, R.D.L. City Life Satisfaction: A Measurement for Smart and Sustainable Cities from the Citizens' Perspective. *Int. J. Knowl.-Based Dev.* **2019**, *10*, 338. [CrossRef]
13. Jonek-Kowalska, I. Housing Infrastructure as a Determinant of Quality of Life in Selected Polish Smart Cities. *Smart Cities* **2022**, *5*, 924–946. [CrossRef]
14. Faria, P.A.M.; Ferreira, F.A.F.; Jalali, M.S.; Bento, P.; António, N.J.S. Combining Cognitive Mapping and MCDA for Improving Quality of Life in Urban Areas. *Cities* **2018**, *78*, 116–127. [CrossRef]
15. Badi, I.; Stević, Ž.; Radović, D.; Ristić, B.; Cakić, A.; Sremac, S. A new methodology for treating problems in the field of traffic safety: Case study of libyan cities. *Transport* **2023**, *38*, 190–203. [CrossRef]
16. Li, F.; Li, H.; Mo, F.; Xiao, S.; Xiao, Z. Investigation on risk prediction of pedestrian head injury by real-world accidents. *Transport* **2019**, *34*, 394–403. [CrossRef]
17. Uhlmann, J. Comparison of Road Traffic Death and Injury Rates at Pedestrian Crossings in Europe. *Balt. J. Road Bridge Eng.* **2021**, *16*, 16–29. [CrossRef]
18. Jasiūnienė, V.; Čygas, D. Analysis of Older Pedestrian Accidents: A Case Study of Lithuania. *Balt. J. Road Bridge Eng.* **2020**, *15*, 147–160. [CrossRef]
19. Ptak, M. Pedestrian safety: A new method to assess pedestrian kinematics. *Transport* **2019**, *34*, 41–51. [CrossRef]

20. Guo, Z.; Liu, J.; Liu, X.; Meng, Z.; Pu, M.; Wu, H.; Yan, X.; Yang, G.; Zhang, X.; Chen, C.; et al. An Integrated MCDM Model with Enhanced Decision Support in Transport Safety Using Machine Learning Optimization. *Knowl.-Based Syst.* **2024**, *301*, 112286. [CrossRef]
21. Zhou, W.; Feng, H.; Guo, Z.; Jia, H.; Li, Y.; Luo, X.; Ran, S.; Zhang, H.; Zhou, Z.; Yuan, J.; et al. Machine Learning Embedded Hybrid MCDM Model to Mitigate Decision Uncertainty in Transport Safety Planning for OAS Countries. *Socio-Econ. Plan. Sci.* **2024**, *96*, 102082. [CrossRef]
22. Jakimavičius, M. Analysis and Assessment of Lithuanian Road Accidents by AHP Method. *Balt. J. Road Bridge Eng.* **2018**, *13*, 238–260. [CrossRef]
23. Gavėnienė, L.; Jateikienė, L.; Čygas, D.; Kasperavičienė, A. Impact of Average Speed Enforcement Systems on Traffic Safety: Evidence from the Roads of Lithuania. *Balt. J. Road Bridge Eng.* **2020**, *15*, 1–18. [CrossRef]
24. Žuraulis, V.; Surblys, V.; Šabanovič, E. Technological measures of forefront road identification for vehicle comfort and safety improvement. *Transport* **2019**, *34*, 363–372. [CrossRef]
25. Pokorski, J.; Sar, H.; Reński, A. Influence of exploitation conditions on anti-skid properties of tyres. *Transport* **2019**, *34*, 415–424. [CrossRef]
26. Ivković, I.; Kaplanović, S.; Sekulić, D. Analysis of external costs of CO_2 emissions for CNG buses in intercity bus service. *Transport* **2019**, *34*, 529–538. [CrossRef]
27. Morshed, S.M.; Shakur, M.S.; Rahman, R.; Mahin, M.M.I.; Debnath, B.; Apu, A.H.; Nafiz, F.A.; Bari, A.B.M.M. An Interval-Valued Type 2 Intuitionistic Fuzzy Theory-Based Approach to Assess the Biofuel Production and Adoption Drivers in Emerging Economies: Implications for Sustainability. *Energy Nexus* **2025**, *17*, 100369. [CrossRef]
28. Mao, Q.; Fan, J.; Gao, Y. An Investment Framework for Hydro-Wind-Photovoltaic-Hydrogen Hybrid Power System Based on the Improved Picture Fuzzy Regret-PROMETHEE Model. *Int. J. Hydrogen Energy* **2025**, *106*, 565–585. [CrossRef]
29. Ugnenko, E.; Gavrish, V.; Viselga, G.; Garbinčius, G.; Turla, V.; Nagurnas, S. Experimental study of carriageway operational condition influence on acoustic roadside area pollution. *Transport* **2019**, *34*, 591–599. [CrossRef]
30. Vaitkus, A.; Šernas, O.; Vorobjovas, V.; Gražulytė, J. Selection of Constituent Materials for Asphalt Mixtures of Noise-Reducing Asphalt Pavements. *Balt. J. Road Bridge Eng.* **2019**, *14*, 178–207. [CrossRef]
31. Zagorskas, J.; Turskis, Z. Location Preferences of New Pedestrian Bridges Based on Multi-Criteria Decision-Making and GIS-Based Estimation. *Balt. J. Road Bridge Eng.* **2020**, *15*, 158–181. [CrossRef]
32. Podvezko, V.; Sivilevičius, H. The Use of AHP and Rank Correlation Methods for Determining the Significance of the Interaction between the Elements of a Transport System Having a Strong Influence on Traffic Safety. *Transport* **2013**, *28*, 389–403. [CrossRef]
33. Shafabakhsh, G.; Mirzanamadi, R.; Mohammadi, M. Pedestrians' Mental Satisfaction's Relationship with Physical Characteristics on Sidewalks Using Analytical Hierarchy Process: Case Study of Tehran, Iran. *Transp. Lett.* **2015**, *7*, 121–132. [CrossRef]
34. Starčević, S.; Bojović, N.; Junevičius, R.; Skrickij, V. Analytical hierarchy process method and data envelopment analysis application in terrain vehicle selection. *Transport* **2019**, *34*, 600–616. [CrossRef]
35. Hruška, R.; Kmetík, M.; Chocholáč, J. Selection of the Transport Mode Using the AHP Method Within Distribution Logistics of Motor Fuels. *Promet* **2021**, *33*, 905–917. [CrossRef]
36. Sivilevičius, H.; Maskeliūnaitė, L. Multiple Criteria Evaluation and the Inverse Hierarchy Model for Justifying the Choice of Rail Transport Mode. *Promet* **2018**, *30*, 57–69. [CrossRef]
37. Maskeliūnaitė, L.; Sivilevičius, H. Identifying the Importance of Criteria for Passenger Choice of Sustainable Travel by Train Using ARTIW and IHAMCI Methods. *Appl. Sci.* **2021**, *11*, 11503. [CrossRef]
38. Kechagias, E.P.; Gayialis, S.P.; Konstantakopoulos, G.D.; Papadopoulos, G.A. An Application of a Multi-Criteria Approach for the Development of a Process Reference Model for Supply Chain Operations. *Sustainability* **2020**, *12*, 5791. [CrossRef]
39. He, Z. The Challenges in Sustainability of Urban Freight Network Design and Distribution Innovations: A Systematic Literature Review. *Int. J. Phys. Distrib. Logist. Manag.* **2020**, *50*, 601–640. [CrossRef]
40. Miller, G.A. The Magical Number Seven, plus or Minus Two: Some Limits on Our Capacity for Processing Information. *Psychol. Rev.* **1956**, *63*, 81–97. [CrossRef] [PubMed]
41. Saaty, T.L. Analytic Hierarchy Process. In *Encyclopedia of Biostatistics*; Armitage, P., Colton, T., Eds.; Wiley: Hoboken, NJ, USA, 2005; ISBN 978-0-470-84907-1.
42. Brunelli, M. *Introduction to the Analytic Hierarchy Process*; SpringerBriefs in Operations Research; Springer International Publishing: Cham, Switzerland, 2015; ISBN 978-3-319-12501-5.
43. Kou, G.; Ergu, D.; Peng, Y.; Shi, Y. *Data Processing for the AHP/ANP*; Quantitative Management; Springer: Berlin/Heidelberg, Germany, 2013; Volume 1, ISBN 978-3-642-29212-5.
44. Tzeng, G.-H.; Huang, J.-J. *Multiple Attribute Decision Making*, 1st ed.; Chapman and Hall/CRC: Boca Raton, FL, USA, 2011; ISBN 978-1-4398-6158-5.
45. Saaty, T.L. *The Analytic Hierarchy Process: Planning, Priority Setting, Resource Allocation*; McGraw-Hill International Book Co.: New York, NY, USA, 1980; ISBN 978-0-07-054371-3.

46. Podvezko, V. Sudėtingų Dydžių Kompleksinis Vertinimas. *Bus. Theory Pract.* **2008**, *9*, 160–168. [CrossRef]
47. Shin, H.-S.; Yu, J.W.; Dadvar, S.; Lee, Y.-J. Statistical Evaluation of Different Sample Sizes for Local Calibration Process in the Highway Safety Manual. *Transp. Res. Rec. J. Transp. Res. Board* **2015**, *2515*, 94–103. [CrossRef]
48. Montgomery, D.C.; Runger, G.C.; Hubele, N.F. *Engineering Statistics*, 4th ed.; Jon Wiley & Sons, Inc.: Hoboken, NJ, USA; Arozon State University: Tempe, AZ, USA, 2007.
49. Kendall, M.; Gibbons, J.D. *Rank Correlation Methods*, 5th ed.; Oxford University Press: New York, NY, USA, 1990.
50. Farhan, J.; Fwa, T.F. Pavement Maintenance Prioritization Using Analytic Hierarchy Process. *Transp. Res. Rec. J. Transp. Res. Board* **2009**, *2093*, 12–24. [CrossRef]
51. Gudienė, N.; Banaitis, A.; Podvezko, V.; Banaitienė, N. Identification and evaluation of the critical success factors for construction projects in lithuania: Ahp approach. *J. Civ. Eng. Manag.* **2014**, *20*, 350–359. [CrossRef]
52. Šakalys, R.; Sivilevičius, H.; Miliauskaitė, L.; Šakalys, A. Investigation and Evaluation of Main Indicators Impacting Synchromodality Using ARTIW and AHP Methods. *Transport* **2019**, *34*, 300–311. [CrossRef]
53. Sivilevičius, H.; Martišius, M. The Significance of the Factors Increasing the Asphalt Pavement Recycling Rate in the Country, Determined Using Multiple-Criteria Decision-Making Methods. *Appl. Sci.* **2023**, *13*, 12226. [CrossRef]
54. Zhang, T.-Y.; Yang, Y.; Zhu, Y.-T.; Yao, E.-J.; Wu, K.-Q. Deploying Public Charging Stations for Battery Electric Vehicles on the Expressway Network Based on Dynamic Charging Demand. *IEEE Trans. Transp. Electrific.* **2022**, *8*, 2531–2548. [CrossRef]
55. Liao, M.-S.; Ding, J.-F.; Liang, G.-S.; Lee, K.-L. Key Criteria for Evaluating the Green Performance of Ports. *J. Test. Eval.* **2016**, *44*, 20140354. [CrossRef]
56. Ferraro, A.; Siciliano, A.; Spampinato, M.; Morello, R.; Trancone, G.; Race, M.; Guida, M.; Fabbricino, M.; Spasiano, D.; Fratino, U. A Multi-Disciplinary Approach Based on Chemical Characterization of Foreshore Sediments, Ecotoxicity Assessment and Statistical Analyses for Environmental Monitoring of Marine-Coastal Areas. *Mar. Environ. Res.* **2024**, *202*, 106780. [CrossRef] [PubMed]
57. Montgomery, D.C. *Design and Analysis of Experiments*, 8th ed.; John Wiley & Sons, Inc.: Singapore, 2013.
58. Sachs, L. *Statistische Auswertungsmethoden*; Springer: Berlin/Heidelberg, Germany, 1972; ISBN 978-3-540-05520-4.

sustainability

MDPI

Review

Prospects of Solar Energy in the Context of Greening Maritime Transport

Olga Petrychenko [1,*], Maksym Levinskyi [1], Sergey Goolak [2] and Vaidas Lukoševičius [3,*]

1 Theory and Structure of a Ship Department, National University "Odessa Maritime Academy", Didrikhson Str. 8, 65052 Odessa, Ukraine; maxlevinskyi@gmail.com

2 Department of Electromechanics and Rolling Stock of Railways, State University of Infrastructure and Technologies, Kyrylivska Str. 9, 04071 Kyiv, Ukraine; gulak_so@gsuite.duit.edu.ua

3 Department of Transport Engineering, Faculty of Mechanical Engineering and Design, Kaunas University of Technology, Studentų Str. 56, 44249 Kaunas, Lithuania

* Correspondence: o.petrychenko@onma.edu.ua (O.P.); vaidas.lukosevicius@ktu.lt (V.L.)

Abstract: The aim of this article is to examine existing technologies for the use of electrical energy and to develop proposals for their improvement on maritime vessels. As a criterion for evaluating the effectiveness of alternative energy sources on ships, factors such as greenhouse gas emissions levels, production and transportation characteristics, onboard storage conditions, and technoeconomic indicators have been proposed. The analysis of fuel types reveals that hydrogen has zero greenhouse gas emissions. However, transportation and storage issues, along with the high investment required for implementation, pose barriers to the widespread use of hydrogen as fuel for maritime vessels. This article demonstrates that solar energy can serve as an alternative to gases and liquid fuels in maritime transport. The technologies and challenges in utilizing solar energy for shipping are analyzed, trends in solar energy for maritime transport are discussed, and future research directions for the use of solar energy in the maritime sector are proposed. The most significant findings include the identification of future research directions in the application of solar energy in the maritime sector, including the adaptation of concentrated solar power (CSP) systems for maritime applications; the development of materials and designs for solar panels specifically tailored to marine conditions; the development of methods for assessing the long-term economic benefits of using solar energy on vessels; and the creation of regulatory frameworks and international standards for the use of solar energy on ships. Furthermore, for hybrid photovoltaic and diesel power systems, promising research directions could include efforts to implement direct torque control systems instead of field-orientated control systems, as well as working on compensating higher harmonics in the phase current spectra of asynchronous motors.

Keywords: greenhouse gases; alternative energy sources; maritime transport; solar energy; concentrated solar power systems; hybrid PV–diesel power supply

1. Introduction

Maritime transport dominates international trade with large volumes of cargo and has the greatest impact on supporting economic growth and globalization [1]. For many years, maritime trade has been the most important type of international transportation. It makes up more than 80% of the world's trade. The volume of maritime trade has increased by 3% each year in the last 40 years [2].

Maritime transportation and related infrastructure directly or indirectly affect a wide range of industries with a large overall impact on the economy [3,4]. Maritime transport is considered to be the backbone of global trade [3]. Many resources transported by sea are shipped to manufacturing centers. As a result, many other sectors of the economy are highly dependent on maritime transportation. Logistics hubs and port activities are components of maritime transportation. They have a major impact on the development of maritime sectors and trade. This fact contributes to economic growth and job creation.

In the European Union (EU), maritime transport accounts for more than three-quarters of external trade and one-third of internal trade [3]. Twenty-two countries are involved in maritime transport between EU members. As a result, European shipping accounts for a large share of the world fleet (more than 40%) and forms some of the largest maritime clusters in the world. Maritime transport affects not only economic growth and development but also the environment. The Green Pact identifies the problems and sets out the pollution reduction policies of the EU member states, according to which measures should be taken to reduce pollution from the maritime sector [3]. The maritime industry is ready to embrace innovative solutions that not only reduce greenhouse gas (GHG) emissions but also improve energy efficiency in the sector [5–7].

Carbon dioxide (CO_2) emissions have detrimental effects on natural ecosystems. First and foremost, CO_2 is the primary greenhouse gas significantly contributing to global warming and climate change [8–10]. This phenomenon leads to serious consequences, such as escalating Earth's surface temperature, melting glaciers, and rising sea levels [11–13]. Oceans, by absorbing excess CO_2, become more acidic, creating threats to marine ecosystems, including coral reefs and marine fauna. Changes in precipitation patterns and air temperature also affect plant and animal life, leading to decreased biodiversity. Reducing these emissions is imperative for maintaining the stability of natural systems and the planet's biological diversity [14]. Data presented in ref. [14] indicate a trend of increasing CO_2 emissions since 1990. This observation is significant in the context of combating climate change as CO_2 is one of the main greenhouse gases contributing to the greenhouse effect and global warming.

The trend of increasing CO_2 emissions poses a significant threat to our planet and emphasizes the importance of adopting long-term and effective measures to reduce them. The International Maritime Organization (IMO) requires the shipping industry to reduce annual GHG emissions by 2050 by at least 50% of 2008 levels. The fourth IMO study on greenhouse gases used 2008 as the baseline year. According to the study, greenhouse gas emissions from shipping will increase by 90–130% by 2050 without significant decarbonization. To achieve zero emissions, the shipping industry will need to take comprehensive action from a number of perspectives to counter the adverse effects of global warming [15–17].

Some approaches to this problem include a shift to renewable energy sources and improvements in energy efficiency. The last includes not only energy savings in homes and businesses but also the optimization of energy production processes, resulting in higher energy yields with reduced CO_2 emissions. Furthermore, reducing waste and improving industrial production efficiency, for example, through the implementation of cleaner technologies and optimization of resource use, also contribute to reducing overall carbon dioxide emissions [18]. Achieving sustainable development in the maritime sector requires several key actions: the development of marine technologies, the implementation of renewable energy sources, and a reduction in harmful emissions.

2. Literature Review

A promising solution to reduce GHG emissions and combat climate change in the offshore industry is the use of renewable energy sources. Such sources include solar energy, wind energy, and hybrid systems. The application of such energy sources in the shipping industry can play a crucial role in protecting the environment and combating global warming [19–21].

The focus on improving marine technology and the use of renewable energy is crucial. These measures directly contribute to reducing CO_2 emissions and addressing climate change. Furthermore, raising public awareness and creating legal and economic incentives are vital to support these efforts and ensure sustainable and environmentally responsible use of marine resources for future generations. In modern shipping, alternative fuels are being actively researched to reduce environmental impact [22].

The main options include natural gas, hydrogen, ammonia, methanol, biodiesel, and electricity liquefied natural gas (LNG), which is already widely used because of its low GHG emissions and high energy efficiency. Hydrogen is attractive for its zero emissions during combustion but faces storage and production challenges. Although special precautions must be taken, ammonia looks promising as a hydrogen carrier. Methanol is distinguished by its ease of handling and potential for production from renewable sources. Biodiesel is compatible with existing engines and provides reduced greenhouse gas emissions [23–25].

The marine industry faces an urgent environmental challenge. Researchers, industry stakeholders, policymakers, and experts in the field are looking for innovative solutions to reduce GHG emissions [26–28]. Among these, the integration of solar energy and Carbon Capture, Utilization, and Storage (CCUS) technologies has emerged as a promising pathway.

The integration of solar energy into maritime operations marks a monumental shift in how ships navigate the seas. Ships, previously synonymous with emissions and environmental harm, have undergone an extraordinary transformation. Now, through the pervasive power of the sun, they embrace environmental sustainability like never before. Modern vessels proudly display solar panels, strategically positioned on their decks and structures, silently converting sunlight into clean renewable energy. This implementation of solar technology signifies a decisive departure from traditional fuel sources, promising not only cost savings but also significant environmental impact reduction [29].

When implementing solar energy as an alternative fuel on marine vessels, technical challenges were identified, such as effective panel placement and performance under adverse weather conditions. The economic viability of solar integration on large cargo vessels remains a subject of debate [30,31].

Improving the efficiency and reliability of solar panels, expanding their realization, and exploring new materials to improve performance are the focus of much research and development. These benefits not only improve vessel operational efficiency but also make a significant contribution to the global pursuit of more environmentally friendly and responsible maritime practices [32,33].

Despite the valuable insights gained from existing research, there are notable gaps that require attention. These gaps encompass technical challenges, economic viability assessment, understanding of regulatory frameworks, knowledge dissemination, and conducting comparative studies. Addressing these gaps is crucial for further progress in the maritime industry.

Future prospects depend on technological improvements aimed at overcoming existing challenges, increasing overall efficiency, and aligning the maritime sector with global environmental goals, as noted in various studies [34]. These areas of focus promise to shape the industry's trajectory toward a cleaner and more efficient future.

The aim of the article is to review existing electrical power technologies and develop proposals for their improvement on marine vessels. The research contributions of the article are as follows: (1) an analysis of alternative fuels used on marine vessels was performed, using criteria such as greenhouse gas emissions levels, production and transportation characteristics, and onboard storage specifics; (2) technical and economic indicators were analyzed for the introduction of alternative fuels on marine vessels; (3) technologies and challenges of solar energy use in shipping were analyzed; (4) trends in solar energy for maritime transport were analyzed; and (5) future research directions for solar energy use in the maritime sector were proposed.

The article has the following structure:

- Section 3 describes research methodologies and data.
- Section 4 presents an analysis of alternative fuels used on marine vessels.
- Section 5 analyzes technologies that use solar energy in shipping and their technical and economic performance.
- Section 6 analyzes the challenges and trends in the use of solar energy for maritime transport.
- Section 7 discusses the obtained results and proposes future research directions for solar energy use in the maritime sector.
- Section 8 presents conclusions.

3. Description of Research Methodology and Data

The research methodology was chosen on the basis of the following considerations. As noted above, the use of alternative fuels on marine vessels contributes to the reduction in GHG emissions into the environment. Furthermore, the use of alternative fuels will increase the energy efficiency of both the main and auxiliary drives of a marine vessel. At the same time, each type of alternative fuel has its own specifics in production and the conditions of storage on the ship.

Energy efficiency, the cost of production, delivery to the ship, and the storage of the alternative fuel on board also affect the economics of using this type of fuel compared to conventional fuel. Therefore, the following criteria were selected to compare the performance of alternative fuels:

1. The level of GHG emissions to the environment.
2. Energy efficiency.
3. Features of production and transportation.
4. Features of storage on board ship.
5. Economic efficiency.

The following sources were used to address these research objectives: Refs. [35–39] to compare the efficiency of LNG application; Refs. [35,39–42] to compare the efficiency of methanol application; Refs. [43–47] to compare the efficiency of hydrogen application; Refs. [48–51] to compare the efficiency of ammonia application; and Refs. [52,53] to compare the efficiency of biofuel application.

This research approach allowed us to demonstrate the effectiveness of solar energy as an alternative energy source for marine vessels. The analysis of the efficiency of technologies using solar energy as an alternative power source was carried out by comparing photovoltaic (PV) panel technologies with concentrated solar power (CSP) systems. The analysis was based on the data presented in the study [54]. As a result of the analysis, the challenges related to the use of solar energy on ships were identified, and possible solutions were proposed. Since the highest energy consumption and GHG emissions are attributed to large-displacement cargo ships, the study utilized data specifically for this type of vessel.

4. Analysis of Alternative Fuels Used on Marine Vessels

Addressing climate change and reducing GHG emissions are critically important challenges. Maritime vessels, which contribute significantly to global emissions, face increasing pressure to adopt cleaner and more sustainable energy sources. Data from the Fourth IMO GHG Study 2020 indicate that the share of global anthropogenic GHG emissions attributed to the shipping industry has increased from 2.76% to 2.89% [55]. This growing impact emphasizes the urgency of aligning maritime operations with environmental objectives.

The number of vessels using various types of fuel, both in operation and under construction, continues to increase (Figure 1). When constructing Figure 1, the following number of vessels was considered, both in operation and in construction: hydrogen-powered—4485, LNG-powered—811, battery-powered—627, LNG-ready—829, LPG-powered—104, methanol-powered—56, and scrubber-equipped—8 [56].

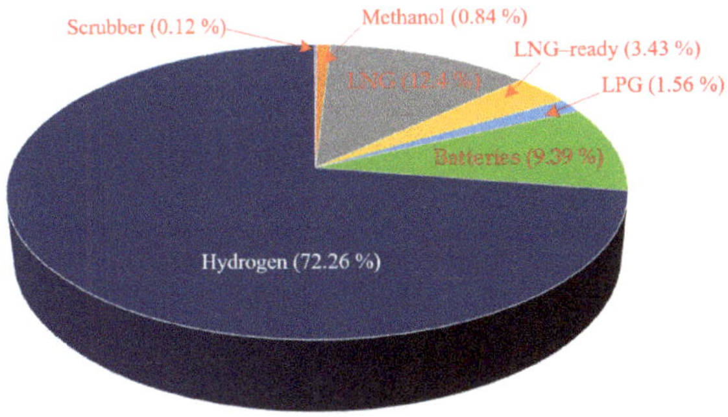

Figure 1. Number of vessels by fuel type [56].

The maritime sector is striving to reduce its carbon footprint. To supplement or replace traditional fuels, various alternative energy sources are being explored, including LNG, electric batteries, and hydrogen. Solar energy, while still at an experimental stage for larger vessels, shows potential due to its renewable nature and compatibility with other energy systems.

The development of technologies and the use of alternative energy sources contribute to reducing GHG emissions in the sector and addressing the issue of climate change [57]. Furthermore, raising public awareness and creating legal and economic incentives are vital to support these efforts and ensure the sustainable and environmentally responsible use of marine resources for future generations.

4.1. Liquefied Natural Gas

In the maritime industry, liquefied natural gas (LNG) is considered one of the most promising alternative fuels. Compared to traditional oil-based fuels, the use of LNG as a marine fuel offers several advantages. These benefits include a significant reduction in sulfur oxide (SOx) and particulate emissions, as well as a decrease in nitrogen oxide (NOx) and carbon dioxide (CO_2) emissions [58–60]. However, the implementation of LNG requires substantial investments in infrastructure and vessel modifications, which is one of the main challenges for the widespread adoption of this technology.

4.1.1. Environmental Impact of LNG-Powered Vessels

The consumers of LNG in maritime vessels are the main and auxiliary engines. During operation, these engines emit greenhouse gases into the atmosphere [61]. The primary greenhouse gas consists of methane (CH_4). At a temperature of $-160\ ^\circ C$ and atmospheric pressure, this gas becomes liquid. Compared to heavy fuel oil (HFO), which is used as the main fuel on maritime vessels, LNG is a cleaner fuel. It is used to reduce SOx, NOx, and particulate emissions [35]. The IMO's Fourth Study showed that, due to the growing number of vessels, methane emissions from the combustion of LNG in engines increased by 150% from 2012 to 2018. This led to a rise in GHG emissions [62].

In ref. [36], a comparison of fuel consumption and emissions was made for the operation of main engines (Table 1) and auxiliary engines (Table 2) in maritime vessels running on LNG and HFO.

Table 1. Data for annual GHG emissions assessment T_{tW} for main engines using LNG [36].

Fuel Type	LNG		HFO
Main engine type	Low-speed high-pressure dual-fuel engines (LS-HPDF)	Low-speed low-pressure dual-fuel engines (LS-LPDF)	Low-speed diesel cycle engines (LSD)
Engine model:	6G50ME-GI	6X-52DF	6G50ME
Average main engine load at sea and operating days	72.5%/250 days		
Engine thermal efficiency (%) at 72.5% MCR	55.30	50.20	53.9
SFC (g/kWh) at 72.5% MCR (Primary/Pilot Fuel)	126.85/3.9	141.75/1.95	147.24/ -
Lower heating value (LHV), MJ/kg	49.2	49.2	40.2
Emission factor (GHG t/t of Fuel)			
CO_2	2.75	2.75	3.144
CH_4	0.001449	0.017086	0.00006
N_2O	0.000217	0.000137	0.00016

Table 2. Emission factors for annual T_{tW} GHG emissions assessment from auxiliary engine [36].

Fuel Type	LNG	HFO
Type of auxiliary engine	Medium-speed high-pressure dual-fuel engines (MS-LPDF)	Medium-speed diesel cycle engines (MSD)
SFOC (g/kWh)	152	195
Emission factor (GHG t/t of Fuel)		
CO_2	2.75	3.144
CH_4	0.036	0.00006
N_2O	0.000131	0.00025

The nominal power of the main engine, load factor, type of fuel, engine type, and year of manufacture are factors that influence GHG emissions. Changes in the operational parameters of the vessel over time lead to variations in the power output and load factor of the main engine [37]. The output power required to propel the vessel is a function of the operational speed and is expressed as the "load", representing a share of the total installed maximum continuous rating (MCR) [37]. In this study, abstract ship log (LOG) data were used. The average main engine load at sea was assumed to be 72.5% of MCR. This load was converted into specific fuel consumption (SFC) for both HFO and LNG. For dual-fuel engines operating on LNG and HFO, the amount of fuel used is a function of the engine load.

Figure 2 shows a comparison of annual T_{tW} emission estimates from vessels using HFO and LNG in main and auxiliary engines [36]. The highest level of greenhouse gas emissions is demonstrated by a vessel operating on HFO, while vessels with LS-HPDF using LNG show the lowest indicators. The emission volumes for these engines are approximately

17,689 t for HFO, 12,463 t for LS-HPDF, and 15,806 t for LS-LPDF. Overall, emissions from auxiliary engines account for only 5.2–7.3% of total annual T_{tW} greenhouse gas emissions compared to main engines.

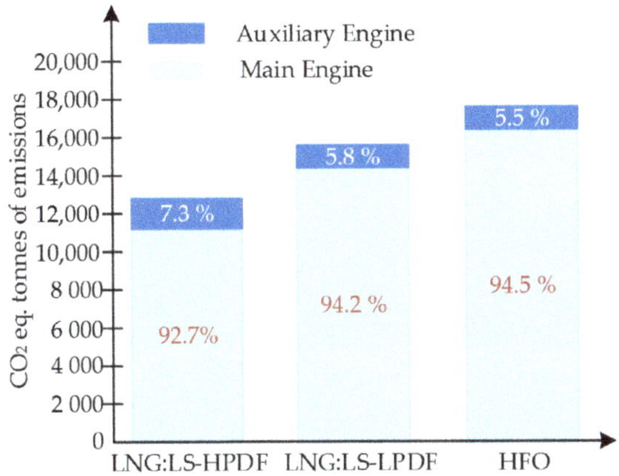

Figure 2. Annual T_{tW} GHG emissions for HFO and LNG [36].

Figure 3 shows the average greenhouse gas emission intensity for energy used onboard vessels for the given year, where all vessel emissions are considered only from main and auxiliary engines [36]. The trend in greenhouse gas emission intensity between fuel types follows a similar pattern to the annual greenhouse gas emissions shown in Figure 2. Based on the data presented in [36], Table 3 shows the specific fuel consumption (g/kWh) for HFO and LNG depending on the engine load.

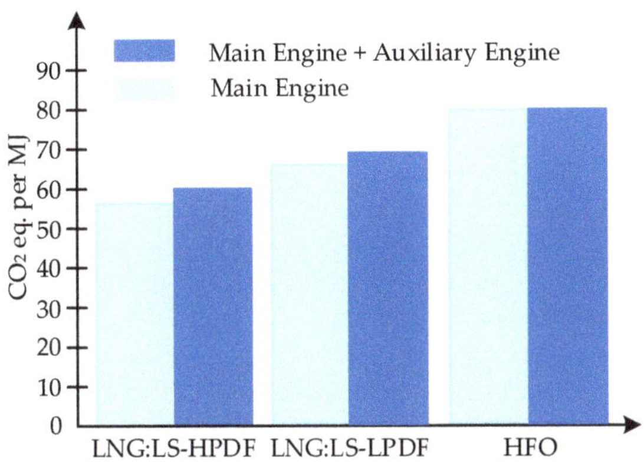

Figure 3. Average energy intensity of emissions during the operation of engines on board the vessel, running on HFO and GHG [36].

The analysis of the results presented in Table 3 indicates a dependence of LNG consumption on the fuel combustion characteristics and the combustion cycles of the engines. From the data presented in Table 3, it can be concluded that engines operating on LNG have lower fuel consumption when operating under high load conditions. In [36], GHG emissions were calculated based on specific fuel consumption data for HFO and LNG, depending on the engine load. The results of these calculations are presented in Figure 4.

Table 3. Specific fuel consumption (g/kWh) for HFO and LNG depending on engine load [36].

Load (%)	LS-HPDF		LS-LPDF	
	SFOS	**SPOS**	**SFOS**	**SPOS**
10	141.4	12.5	156.8	4.7
20	137.1	10.1	152.6	4.1
30	133.8	8.0	149.1	3.5
40	131.4	6.3	146.3	2.9
50	129.8	5.0	144.1	2.5
60	129.2	4.1	142.6	2.1
70	129.4	3.5	141.7	1.8
80	130.6	3.3	141.5	1.6
90	132.7	3.5	142.0	1.5
100	135.6	4.1	143.1	1.4

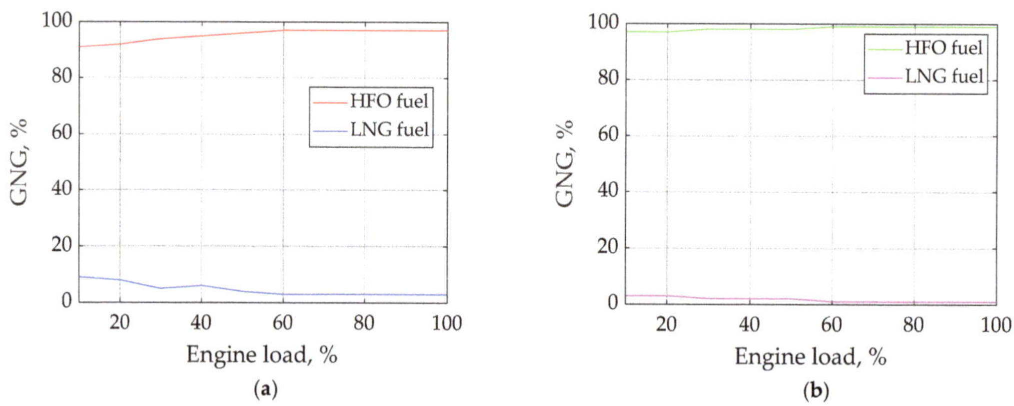

Figure 4. GHG emission ratios for HFO and LNG: LS-HPDF (**a**) and HFO and LNG: LS-LPDF (**b**) as a function of main engine load.

Figure 4 shows that when LNG is used as fuel, the GHG emissions are high at low engine loads. Furthermore, it can be concluded from Figure 4 that the GHG emissions of the LNG-operated engines under low load conditions depend on the type of engine.

4.1.2. Features of LNG Production and Transportation

The production of LNG is based on refrigeration cycles, such as steam compression and gas expansion. The main difference between these cycles is the presence of a phase change of the refrigerant in the steam compression cycle, while no such phase change occurs in the gas expansion cycle.

Both cycles include four main stages: (1) the compression of the refrigerant into a high-pressure hot stream (the compressor); (2) heat released from the compressed refrigerant (the condenser or cooler and heat exchanger); (3) the expansion of the compressed refrigerant into a low-pressure cold stream (the valve or expander); (4) heat absorbed by the cold refrigerant (the heat exchanger). In the final stage, cooling is provided for the natural gas. By repeating these four steps, natural gas can be continuously cooled [63].

LNG production can be carried out using one of the following three main technologies: Cascade, Mixed Refrigerant (MR), and Expander-based (EXP) [64]. Cascade technology uses three cooling cycles, each with different cooling levels. Each cycle has different levels of cooling. Pure propane, ethylene, and methane are used as refrigerants, respectively.

In MR technology, there is only one cooling cycle that uses a mixture of light hydrocarbons as the refrigerant. In the EXP process, pure nitrogen or methane is used as a refrigerant, which can reach low temperatures for liquefying natural gas in a single loop. Due to high energy consumption, EXP is less efficient compared to Cascade and MR technologies. Replacing the throttle valve with an expander helps recover part of the compressor work, thus reducing energy consumption during the EXP process. Table 4 demonstrates the advantages and disadvantages of the three LNG production technologies [63].

Table 4. Advantages and disadvantages of the three LNG production technologies [63].

Indicator	Cascade	MR	EXP
Application	**LS Onshore**	**LS Onshore, SS, and Offshore**	**SS Onshore and Offshore**
Energy efficiency	H	M to H	L
Equipment quantity	H	L to M	L
Heat transfer surface area	M	H	L
Operational simplicity	L	L to M	H
Ease of startup and adjustment	M	L	H
Feed gas composition adaptability	H	M	H
Vessel motion sensitivity	H	M to H	L
Space requirements	H	M	L
Hydrocarbon refrigerant storage	H	M to H	None
Capital expenditure	H	L to M	L

Abbreviations are used: H—high; M—medium; L—low; LS—large-scale; SS—small-scale.

The assessment of the criteria for the three LNG technologies is based on a relative comparison. Because the Cascade process uses multiple refrigerants, it operates at several levels of cooling temperatures. This allows for a small temperature difference between the hot and cold sides in heat exchangers [65].

Using a refrigerant composed of a carefully selected hydrocarbon mixture, the MR process can replicate the cooling curve of natural gas. In contrast to the Cascade process, MR achieves a smaller temperature difference but requires a larger heat exchange area [64].

Throughout the entire EXP process, the pure refrigerant remains in the gaseous state, thereby ensuring a constant specific heat value for the cooling curve. EXP achieves a greater temperature difference between the natural gas and the refrigerant. This, in turn, results in high energy consumption [64]. Although a larger temperature difference reduces the required heat exchange area, it leads to a lower nitrogen heat transfer coefficient compared to hydrocarbons [64].

In the liquefaction process, the most expensive equipment is the heat exchanger, drive, and compressor. Therefore, the comparison of liquefaction processes is based on the type of refrigerant and the equipment listed. Based on the analysis of data presented in various studies, the following conclusions can be drawn:

- In liquefaction processes, either a mixed or a pure refrigerant is employed. In MR technology, a specially selected blend of hydrocarbons is used to replicate the cooling curve of natural gas. In Cascade technology, several distinct pure refrigerants with cascading boiling points throughout the cycles are utilized. In EXP technology, nitrogen or methane is employed. Owing to the extremely low boiling points of these gases, the liquefaction of natural gas is accomplished in a single cycle. The ranking of the processes by the temperature difference between the refrigerant and natural gas is as follows: MR less Cascade; Cascade less EXP. A reduction in energy consumption is achieved by maintaining a smaller temperature difference. Such a smaller temperature difference can be attained by increasing the heat exchange surface area, which in turn requires greater capital investments. From the above, it follows that the optimization of the gas liquefaction process can be accomplished by balancing the refrigerant zones and the heat exchanger design [65].

- Two primary types of heat exchangers are employed in LNG production: (1) the plate-fin or brazed aluminum type (PFHE) and (2) the spiral-wound heat exchanger type (SWHE). The advantages of PFHE include low pressure drops, versatility in accommodating fluctuations in low-temperature differentials, and a greater number of competitive suppliers. In contrast, the drawbacks of PFHE encompass the necessity for meticulous design and a pronounced susceptibility to physical damage, both of which are inherent to the aluminum construction of PFHE [63]. Among the advantages of SWHE are its higher reliability and ease of operation. However, its disadvantages include a higher cost, limited flexibility with respect to the feed gas composition, and less favorable mass and dimensional characteristics. Moreover, SWHE units are produced by only a few companies. Additionally, SWHE offers a higher maximum capacity (a single PFHE is rated at 1.5 million tons per year, whereas one SWHE can reach up to 4 million tons per year). The benefits and drawbacks of these heat exchanger types ultimately determine their application in various installations. For instance, Cascade and EXP installations employ PFHE, while SWHE is used in large-scale MR installations [65,66].
- Five types of drives and two types of compressors should be considered when analyzing the equipment used in liquefaction processes. The drive and compressor are adapted to the specific technology of the liquefaction process [64].

The transportation of LNG requires intermodal transportation, which combines road and rail transport modes. At railway stations, distribution centers should be organized to facilitate the distribution of LNG from suppliers to consumers [67]. Moreover, LNG logistics require transfer points between road and rail transportation modes and storage zones for holding LNG while awaiting transfer from one mode of transportation to another [38].

4.1.3. Features of LNG Storage on Marine Vessels

LNG has been transported by gas carriers since the 1960s. Compared to other industries, this form of transport has an excellent safety record [68]. Incidents that have occurred during LNG transportation have been minor [69]. According to [68], a total of 182 incidents have been recorded involving LNG carriers, none of which were severe accidents caused by LNG-related issues. The storage and supply systems for LNG on vessels are designed and operate similarly to those used on land-based LNG systems. LNG must be stored in pressure vessels at temperatures below −140 °C.

The risks associated with the use of LNG are related to its low temperature, the consequences of potential leaks, and failures of fuel system components [70]. Ensuring the safe design and operation of a vessel is achieved through a thorough examination of hazardous scenarios during the design phase [68].

The international code for the safety of ships using gases or other low-flashpoint fuels (IGF Code) has established fundamental rules for vessels using gases or other fuels with a low flashpoint. These rules are only partially mandatory to control hazardous scenarios. Amendments to these rules help improve the overall safety of the vessel's systems. Since technology is relatively new, research on the general analysis of the safety and reliability of LNG supply systems remains limited.

4.2. Methanol

Methanol, primarily derived from renewable sources and produced through the hydrogenation of carbon dioxide, is an effective solution for reducing GHG emissions. Global scalability and compatibility with existing infrastructure (particularly LNG infrastructure) make methanol a cost-effective alternative with minimal disruption. Methanol is both efficient and economically viable, outperforming LNG, especially on new ships. Its use as a marine fuel provides flexibility without requiring special pressurized tanks.

Global trends in marine fuels emphasize fuel characteristics, availability, and environmental considerations, including factors such as policies, emissions, bunkering, and engine adaptability during the transition [71].

4.2.1. Environmental Impact of Methanol-Powered Marine Engines

In [36], comparative data are provided on fuel consumption and emissions during the operation of main (Table 5) and auxiliary (Table 6) engines of marine vessels powered by methanol and heavy fuel oil.

Table 5. The results of the comparison of the annual greenhouse gas emissions (T_{tW}) for main engines running on methanol and HFO [36].

Fuel Type	**Methanol**	**HFO**
Main Engine Type	LSD	LSD
Engine manufacturer model	6G50ME-LGIM	6G50ME
Thermal efficiency (%) at 72.5% MCR	53.9	53.9
SFC (g/kWh) at 72.5% of MCR (methanol/HFO)	307.5/13.1	147.24/-
Lower heating value (LHV), MJ/kg	19.9	40.2
Emission Factor (GNG t/ton of fuel)		
CO_2	1.375	3.144
CH_4	0.000006	0.00006
N_2O	0.00016	0.00016

Table 6. Results of the comparison of the annual greenhouse gas emissions (T_{tW}) for auxiliary engines running on LNG and HFO [36].

Fuel Type	**Methanol**	**HFO**
Auxiliary engine type	MSD	MSD
SFOC (g/kWh)	370	195
Emission factor (GHG t/t fuel)		
CO_2	1.375	3.144
CH_4	0.00006	0.00006
N_2O	0.00025	0.00025

Considering the use of abstract data (LOG) for the vessel, as in the case of LNG, the average load of the main engine at sea was assumed to be 72.5% of the MCR. The load of the main engine, using specific fuel consumption (SFC) for HFO and methanol, was converted into fuel consumption.

The comparison of annual emissions (T_{tW}) from the main and auxiliary engines of vessels using HFO and methanol is shown in Figure 5 [36]. The vessel operating on HFO produced higher greenhouse gas emissions than the vessel operating on methanol. The emissions of vessels that use these engines are approximately 17,689 t and 13,981.95 t, respectively. It can be seen in Figure 5 that the auxiliary engine emissions account for 5.5% to 7.3% of the total GHG emissions from T_{tW}.

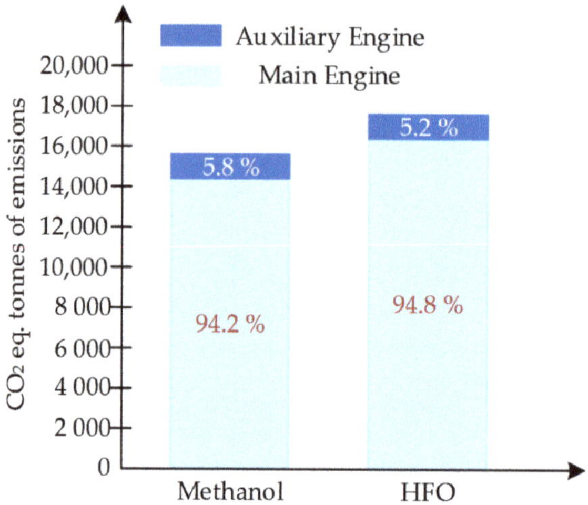

Figure 5. Annual T_{tW} GHG emissions for HFO and LNG [36].

The average energy intensity of GHG emissions throughout the year, when all the emissions of the vessel are considered only from the main and auxiliary engines, is shown in Figure 6 [36]. The trend in greenhouse gas emission intensity between fuel types is like the trend of annual greenhouse gas emissions shown in Figure 5.

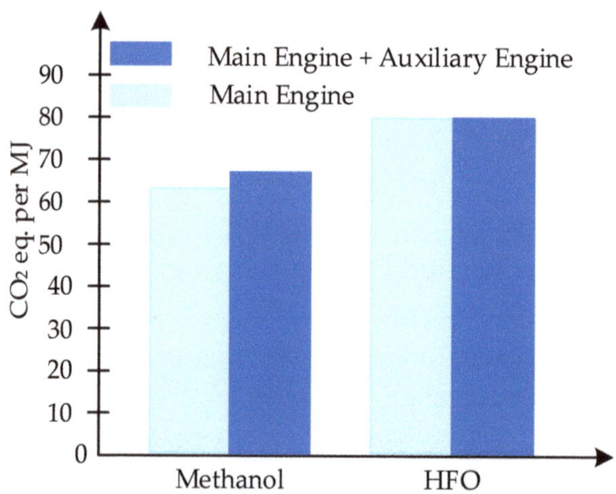

Figure 6. Average energy intensity of emissions from engines operating aboard the vessel on HFO and methanol [36].

Based on the data provided in [36], Table 7 presents the specific fuel consumption (g/kWh) for HFO and methanol depending on the engine load. The data presented (Table 7) indicate that under the main load range (40–60%), engines running on methanol will have the lowest fuel consumption. Based on the data on specific fuel consumption for HFO and methanol depending on engine load in work ref. [36], their GHG emissions were calculated. Figure 7 shows the results of these calculations.

Table 7. Specific fuel consumption (g/kWh) for HFO and LNG depending on the engine load [36].

Engine Load (%)	HFO	Methanol
10	141.4	45.9
20	137.1	33.3
30	133.8	25.1
40	131.4	20.8
50	129.8	19.7
60	129.2	21.3
70	129.4	24.8
80	130.6	29.9
90	132.7	35.8
100	135.6	41.9

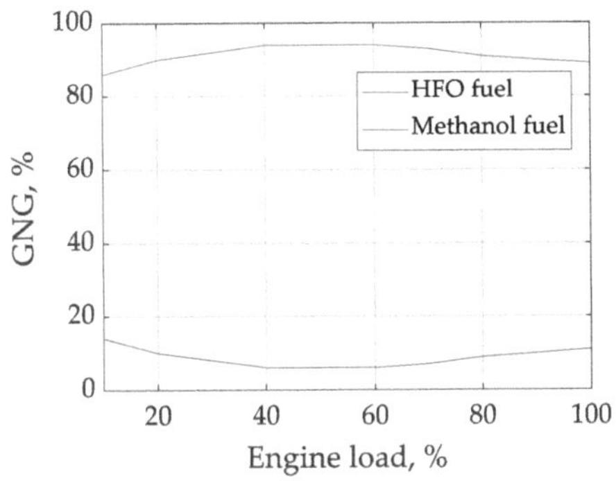

Figure 7. Ratio of GHG emissions from HFO and methanol as a function of main engine load [36].

From the analysis of the results presented in Figure 7, it can be concluded that GHG emissions when using methanol are high at low (10–20%) and high (90–100%) engine loads, while they are lowest at medium loads (40–60%).

4.2.2. Methanol Production and Transportation Features

Coal and natural gas are the primary raw materials for methanol production. Alternative sources for methanol production can include wood, household and agricultural waste, renewable resources, and even CO_2 [71]. CO_2-based methanol synthesis mitigates the greenhouse effect and provides a wide range of chemical products and clean fuel [72].

Primary production includes the synthesis of natural gas or synthesis gas. Coal serves as the raw material for the production of synthesis gas, which contains hydrogen (H2) and carbon monoxide (CO) [73]. According to [74], methanol is primarily produced from synthesis gas using heterogeneous catalysts. These catalysts require elevated temperatures (200–300 °C) and pressures (50–100 bar). In the process of methanol production from coal or coke gas, wastewater and exhaust gases are generated. This leads to significant environmental problems and pollution. Additionally, methanol production releases a significant amount of CO_2, which contributes to the greenhouse effect.

A modern trend in methanol production is the concept of carbon neutrality. As a result, there is increasing interest in the technical approach based on CO_2 hydrogenation to produce methanol [74]. CO_2 is the primary raw material to produce methanol [73].

The hydrogenation of CO_2 to methanol is an exothermic process. Hydrogen facilitates this process. Without a catalyst, the process becomes more complex, leading to the formation of undesirable by-products [40]. To maximize the output of CH_3OH and prevent the formation of by-products, the process must be optimized.

The conditions for optimizing the process are as follows: a suitable catalyst, a temperature below 150 °C, and a pressure in the range of 5–10 MPa. High temperatures increase the consumption of H_2, triggering the water–gas conversion reaction, which reduces the amount of CH_3OH produced. Additionally, temperatures above 240 °C activate CO_2.

Typically, the hydrogenation process occurs in temperature ranges of 250–300 °C and pressures of 50–100 bar, with catalysts based on Cu, Zn, or Al being used [39,41]. In this process, the most efficient use of resources is ensured precisely under these conditions. Simplifying the CO_2-to-methanol conversion process, reducing production costs, and achieving zero CO_2 emissions are the main goals of optimizing methanol production, regardless of the chosen method [71].

Compared to LNG production, methanol production is characterized by higher specific fuel consumption. This is due to the lower calorific value of the methanol production process. Despite this, methanol remains more cost-effective, even when crude oil prices exceed those of LNG [75].

The price of methanol is 38.6% cheaper per metric ton than HFO. However, an increase in the annual fuel cost of 28.16% requires a reduction in the ship's speed of 28% to maintain constant fuel expenses [76]. The methanol engine system on new ships is more economically efficient with lower investment costs compared to LNG. It is expected that as experience with methanol usage grows, these costs will decrease [71].

The advantage of using methanol as fuel on marine vessels is that it is more manageable than LNG. This is because methanol remains in a liquid form at standard temperature and pressure [77]. Methanol can be stored in plastic containers, making it convenient for transportation, refueling, storage, and use [78].

4.2.3. Methanol Storage Features on Marine Vessels

Due to its liquid form, methanol is easier to store on board a vessel compared to LNG. However, its integration into ship fuel systems requires modification of existing installations and infrastructure upgrades to facilitate regular bunkering [79]. Additionally, due to its non-static nature, methanol is easily soluble in water and can be extinguished with water in case of fire [71].

Creating an extensive network of bunkering stations worldwide is a crucial task for the widespread adoption of methanol as marine fuel among commercial vessels. Methanol requires the adaptation of onboard fuel and safety systems [42]. Methanol remains in a liquid state at ambient temperatures, so its need for modifications is less compared to LNG.

The technological readiness of using methanol as fuel for marine vessels highlights the use of existing technological components with innovations lying in the careful integration of these components. The liquid state of methanol at ambient temperature on board the vessel provides flexibility in storage. As a result, specialized pressurized tanks are not required.

According to IMO classification, methanol is not a substance that contaminates the marine environment. Therefore, according to IMO regulations for vessels carrying dangerous chemicals, methanol tanks can be placed near the hull. This requires modifying existing tanks with coatings that are compatible with methanol [80].

The use of double-hull tanks is required when storing conventional oil on board a vessel. For storing gaseous fuels on board, pressurized tanks are necessary, while for LNG (liquefied natural gas), cryogenic liquid tanks are required, which are separate pressurized tanks [81]. For vessels with critical volumes, these factors can be problematic.

Barriers to the adoption of methanol as a fuel include the lack of bunkering infrastructure and uncertainty regarding the long-term availability of the fuel [71]. As discussed above, methanol is produced from fossil fuels, particularly natural gas. This makes methanol widely available around the world. This factor should be considered when evaluating methanol as a potential fuel for the maritime sector [71].

Currently, the availability of renewable methanol is limited. This is due to its production occurring only in a few locations and in small quantities. At this stage, methanol can only serve as a temporary or transitional fuel. This provides a practical solution until renewable methanol production and availability can be significantly increased [82].

In Europe, in ports such as Antwerp and Rotterdam, as well as many ports around the world, there is already infrastructure for the storage and distribution of methanol in the chemical industry [68]. Unlike the changes required for the implementation of LNG infrastructure, converting existing infrastructure to supply methanol as marine fuel requires only minor modifications [71].

Methanol storage tanks originally designed for gasoline can be easily and quickly adapted for methanol storage [71]. The infrastructure changes required for adopting methanol as a marine fuel are smaller compared to other alternative fuels [81].

4.3. Hydrogen

A promising option for achieving the desired reduction in GHG emissions is zero-emission CO_2 fuel produced from renewable sources (hydropower, wind, or solar). For maritime transport, this electro-fuel is produced in two forms: electro hydrogen (or electro-ammonia), or carbon-based electro-fuels in the form of electro diesel or electro-liquefied natural gas.

The use of electro hydrogen and electro-ammonia requires new vessels and supply infrastructure or the modernization of existing ones. Electro diesel fuel or electro-liquefied natural gas fully mixes with their fossil counterparts, such as Marine Gas Oil (MGO) and LNG. For these blends, no modifications or the development of new infrastructure are required on modern vessels [43].

4.3.1. Parameters of Hydrogen as Fuel for Marine Vessel Engines

Hydrogen produces only water vapor when burned, so it is potentially the best fuel for marine vessels. The lower heating value (LHV) of hydrogen is about three times higher than that of conventional marine fuels. Compared to other fuels, hydrogen combustion provides the greatest amount of heat energy per 1 kg of mass.

Hydrogen can be obtained from a wide range of sources, including water electrolysis [44]. The low density of hydrogen in both the liquid and gaseous states is a major problem. The properties of hydrogen and methane are compared in Table 8. It can be seen that the condition for hydrogen liquefaction is a very low critical temperature (−240 °C, about 33 K) [44]. Long-term hydrogen storage creates several more serious problems than in the case of other gaseous fuels.

Due to the high maximum combustion speed (see Table 8), there is no adequate technology for using hydrogen as marine fuel. A comparison of energy densities shows that hydrogen's density is lower than that of ammonia. For example, liquefied hydrogen at −253 °C has an energy density of 8.5 MJ/dm^3, while compressed hydrogen at 30 MPa has a density of 2.46 MJ/dm^3, and liquefied ammonia at −33 °C has an energy density of 12.69 MJ/dm^3.

Many advantages and challenges of using gaseous hydrogen as a fuel have led to attempts to create gas mixtures containing hydrogen additives up to 30%. A reforming unit for converting liquefied gas and steam into methane, mixed with CO_2 and some hydrogen, is an alternative option for using liquefied gas.

Table 8. Comparison of hydrogen and methane properties [44].

Parameter	Units	Hydrogen	Methane
Gas density at standard conditions (ISO)	kg/m^3	0.0905	0.716
Relative gas density (air = 1)	relative units	0.07	0.554
Relative density, liquid (water = 1)	relative units	0.071	0.44–0.48
Lower heating value	MJ/kg	119.96	50
Critical temperature	°C	−240	−82.45
Boiling point	°C	−253	−161.48
Melting point	°C	−259	−187.6
Combustibility range	% (v/v)	4–77	5–14
Maximum combustion velocity	m/s	2.91	0.37
Autoignition temperature	°C	585	537
UN number		1954	1971
Additional information		Burns with an invisible flame, extremely flammable gas	Asphyxiating

The reformed gas is sufficient for use in conventional gas or dual-fuel engines without the necessity to reduce engine ratings [43]. The primary goal is to use hydrogen in fuel cells, with its use in internal combustion engines being an intermediate solution.

4.3.2. Features of Hydrogen Production and Transportation

Gaseous hydrogen is not a rich resource; it needs to be produced from other products. Gaseous hydrogen is a carrier, not a source of energy. There are several methods for hydrogen production, but not all of them are environmentally clean. Currently, about 96% of hydrogen is produced through fossil fuel reforming processes, such as natural gas (NG), heavy oil, petroleum products, or coal. Since these processes produce large volumes of CO_2, alternative methods of hydrogen production are being explored. In these methods, water and biomass are used as renewable raw materials.

In [83], various hydrogen production methods are considered and analyzed. These methods are based on different types of water electrolysis, such as: Alkaline, Polymer Electrolyte Membrane (PEM), Solid Oxide Fuel Cell (SOFC), and membraneless electrolysis. The characteristics of the listed types of water electrolysis are given in Table 9.

Table 9. Characteristics of water electrolysis technologies [83].

Technology	Energy Consumption (kWh/kg(H2))	Operating Temperature (°C)	Hydrogen Pressure (MPa)	Stack Lifetime (hours)	Deployment Scale (kW)
Alkaline	53.4	80	0.7	90,000	>100,000
PEM	54.6	80	0.5	50,000	>100,000
SOFC	36.14	>800	0.1	23,000	7.5
Membraneless	54.34	<50	0/1	-	~10

These methods have been proven technologies. The exception is membraneless electrolysis, which is a new technology but closely related to the marine industry as it was initially developed as a ballast water treatment system.

A large number of gas pipelines for NG transportation have been built and planned worldwide. The cost of building hydrogen transport systems is more than 10% higher than the cost of natural gas (NG) transport systems [84]. An active area of research among scientists around the world is the mixing of hydrogen with natural gas (NG) for transportation through existing pipelines [45].

From the analysis of studies on the problems of mixing hydrogen with NG, it is clear that, at this stage, this method is the best for efficient hydrogen transportation. Figure 8 shows the process of obtaining compressed hydrogen from NG. Gas storage facilities, main pipelines, compressor stations, reduction stations, urban distribution gas pipelines, and customer terminals constitute the key components of the existing natural gas transportation and distribution network.

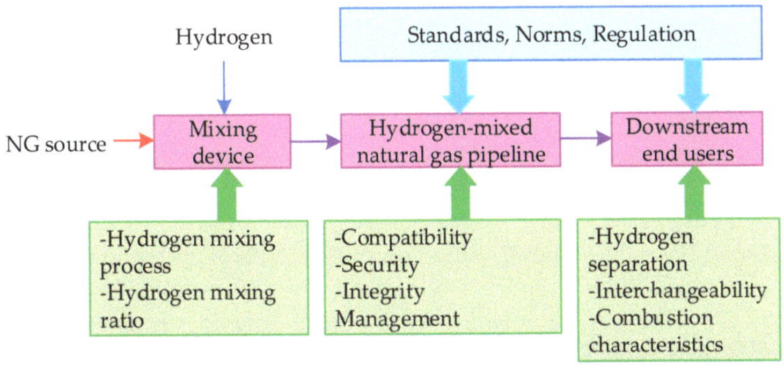

Figure 8. Schematic diagram of the compressed hydrogen natural gas process [73].

The issue of peak consumption of city gas can be effectively addressed using short-term storage methods, such as spherical tanks for natural gas (NG), high-pressure batch storage, storage at the end of main pipelines, and urban storage in high-pressure pipelines [85].

A systematic issue is determining the optimal blending ratio of NG with hydrogen. This includes factors such as the composition of NG, pipeline material, hydraulic conditions, the lifespan of the facility, environmental impact conditions, and differences in the final consumer equipment [85].

Research from the European Union's NATURALHY project suggests that safety will not be significantly impacted at a hydrogen blending ratio of 20%, and even hydrogen blending ratios up to 50% are possible but need to be evaluated according to specific situations [85].

4.3.3. Features of Hydrogen Storage on Marine Vessels

Hydrogen Usage Onboard

Hydrogen is often associated with fuel cells in energy systems, but it can also be combusted in diesel or gas engines. It can serve as the sole fuel (single-fuel engines) in these engines or be used in a dual-fuel system [86]. Before hydrogen is released and used as fuel for the engine, some storage methods require specialized reactors.

Regardless of whether hydrogen is burned or used in fuel cells, the end product is always water. The primary difference between fuel cell engines and internal combustion engines, in terms of emissions, lies in the NOx emissions of the latter [46]. Although fuel cell engines are significantly more expensive and have a shorter lifespan than internal combustion engines, they are more energy-efficient [47].

Reactor Development

The development of a reactor system capable of operating onboard is crucial for hydrogen storage systems that require a chemical conversion stage. A significant challenge, however, lies in the fact that existing chemical reactors are designed for stationary conditions and are unsuitable for the dynamic environment of a ship at sea.

This represents a major issue in the development of new reactor systems [85]. To function onboard, the reactor must be designed to remain unaffected by the ship's movement and be capable of efficiently removing released hydrogen to accelerate the reaction toward dehydrogenation [85].

Hazards of Using Hydrogen Onboard

First and foremost, it should be noted that hydrogen is generally a hazardous substance. Hydrogen has a low ignition energy (0.017 mJ) and a wide flammability range (4–75 vol. % in air), which makes it highly flammable. Its high diffusion coefficient allows it to easily seep through joints and cracks in any pipeline or storage systems.

Being a very light gas, hydrogen disperses quickly when released into the open air [85]. Although hydrogen is a non-toxic gas, it can cause asphyxiation by displacing oxygen from the air at high concentrations as it is an invisible, odorless gas.

To avoid the risk of explosion, it is essential to control the release of hydrogen in a compressed state. Since the designed tanks can withstand bullet impacts, storing compressed hydrogen inside the tank itself is considered safe [87]. When storing hydrogen in liquid form at −253 °C, appropriate materials that can withstand low temperatures without becoming brittle must be used.

Spills of liquid hydrogen can damage the ship's hull. Spills onboard ships can be particularly dangerous. They can cause cold-induced fracturing of the steel used to construct the vessel, potentially leading to hull damage [85]. Following liquid hydrogen spills, vapor clouds form, which remain at very low temperatures. These vapor clouds can pose a serious risk to onboard personnel due to their low temperatures.

Unlike gaseous fuel, liquid hydrogen spill clouds do not disperse as they contain a significant amount of water vapor. This makes them heavier than air. This delayed effect increases the hazards of asphyxiation and explosion.

4.4. *Ammonia*

Alternative fuels have the potential to drive global reductions in CO_2 emissions. Pure hydrogen is one such type of fuel. The issues related to storage and the energy requirements for compression/liquefaction make it difficult to definitively assess the prospects of hydrogen becoming a primary energy source in the transportation sector.

An alternative fuel is ammonia, which serves as an excellent hydrogen carrier and is favored in the maritime sector [48]. The International Maritime Organization (IMO) has set ambitious goals for reducing greenhouse gas emissions. By 2050, total GHG emissions should be reduced by at least 50% compared to the 2008 levels, and CO_2 emissions should be reduced by at least 70% [48].

4.4.1. Environmental Impact of Ammonia-Powered Marine Engines

Table 10 presents the CO_2 equivalent emission factors for main engines (MEs) and auxiliary engines (AEs) [49]. Marine engines running on ammonia are expected to become commercially available after 2024 [49], thus there are no published data on N_2O emissions. As a result, preliminary test data from laboratory experiments have been used.

Based on the results in Table 10, Figure 9a presents a comparative chart of greenhouse gas emissions for different types of fuel (CO_2-eq (total)) from main engines, while Figure 9b shows the emissions from auxiliary engines.

Table 10. CO_2 emission coefficients (ton/ton of fuel) for engines [49].

Engine Type	Fuel	CO_2	CH_4	CO_2-eq (CH4)	N_2O	CO_2-eq (N_2O)	CO_2-eq (Total)
Two-stroke slow-speed main engine (ME) with diesel cycle	MGO	3.21	6.53×10^{-5}	1.83×10^{-3}	2.22×10^{-4}	5.88×10^{-2}	3.27
	LNG	2.75	1.64×10^{-3}	4.58×10^{-2}	2.45×10^{-4}	6.51×10^{-2}	2.86
	Methanol	1.38	3.34×10^{-5}	9.35×10^{-4}	1.00×10^{-5}	2.65×10^{-3}	1.38
	Ammonia	0	0	0	5.42×10^{-4}	1.44×10^{-1}	0.14
Four-stroke medium-speed auxiliary engine with Otto cycle	MGO	3.21	5.35×10^{-5}	1.50×10^{-3}	1.60×10^{-4}	4.25×10^{-2}	3.25
	LNG	2.75	3.54×10^{-2}	9.92×10^{-1}	1.29×10^{-4}	3.41×10^{-2}	3.78
	Methanol	1.38	2.53×10^{-5}	7.09×10^{-4}	7.59×10^{-6}	2.01×10^{-3}	1.38
	Ammonia	0	0	0	5.02×10^{-3}	1.33	1.33

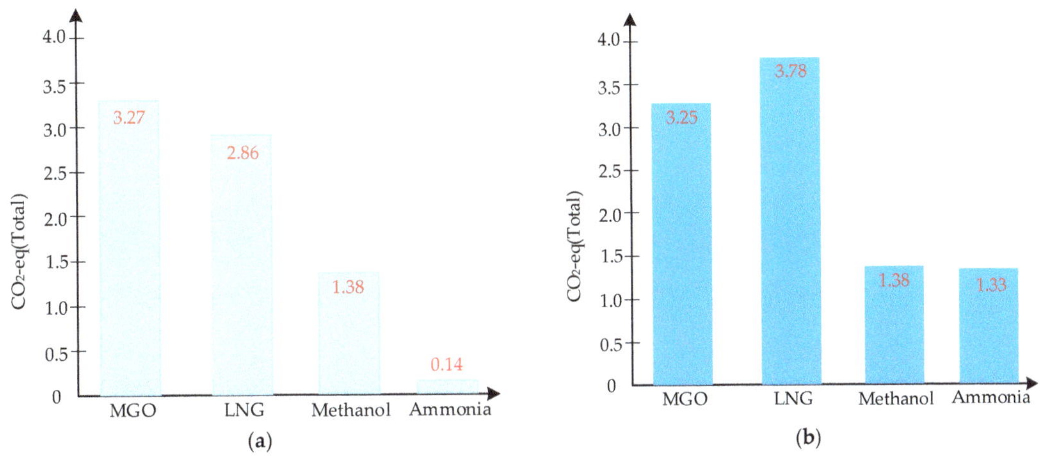

Figure 9. Comparative diagram of greenhouse gas emissions for different fuels (CO_2-eq (total)) from main (**a**) and auxiliary (**b**) engines [49].

As seen from Figure 9, when ammonia is used as fuel, it results in the lowest share of GHG emissions into the atmosphere from both main and auxiliary engines. The comparative characteristics of ammonia and diesel fuel for marine engines are presented in Table 11 [50].

Table 11. Comparative characteristics of two fuels: ammonia and diesel [50].

Fuel Property	Ammonia	Diesel
Liquid density, kg/m^3	602.8	832
Heat of combustion, MJ/kg	18.8	42.7
Octane number	110	30
Cetane number	5–7	40–55
Ignition temperature, °C	651	254–285
Laminar flame speed, m/s	0.07–0.14	0.87
Stoichiometric air-to-fuel ratio	6.06	14.5
Heat of vaporization, kJ/kg	1370	232

The unfavorable physical properties of ammonia require solutions when using it as fuel for marine engines. The combustion characteristics of ammonia differ significantly from those of oil and most renewable fuels (Table 11). Its low number of cetane (5–7 units) and high autoignition temperature (650 °C) mean that ammonia can only ignite at a compression ratio of 35:1 or higher. Considering the geometric parameters of marine engines [50], this is not practical.

The compression ratio in diesel engines reaches up to 20:1. To ensure the combustion of ammonia in such engines, fuel with good autoignition properties (such as diesel or biodiesel) is required.

4.4.2. Features of Ammonia Production and Transportation

The production of ammonia for commercial purposes uses the Haber–Bosch process. In this process, hydrogen and nitrogen are combined at high temperatures using a catalyst [88]. Solar, wind, or hydro energy can be used in the production of green ammonia. This is an advantage of green ammonia production over HFO production [88]. One of the leading global ammonia producers, Yara, plans to build a demonstration facility for green ammonia synthesis using solar energy by 2025 [88]. This will enhance the competitiveness of green ammonia production.

Ammonia, like hydrogen, can be used as a supplementary fuel in diesel engines and gas turbines and as the primary fuel in fuel cells. This makes it attractive and competitive [88]. For land transportation, ammonia is transported via pipelines, while for water transportation, tankers are used [51].

4.4.3. Features of Ammonia Storage on Marine Vessels

Innovation and safety are crucial to mitigating the risks associated with using ammonia as marine fuel. The substance's toxicity necessitates strict adherence to safety and occupational health regulations. Limited space and the absence of external assistance are challenges when installing and operating equipment on maritime vessels. A thorough risk assessment to identify potential leakage points and failure scenarios is necessary to ensure safety. Impact thresholds and alarm systems should be set regarding toxicity and flammability concentrations. To prevent prolonged and immediate exposure, the placement of gas detectors in hazardous areas is crucial. Safety measures are vital for operational security [52].

A significant issue in ammonia storage is corrosion, which creates both financial and safety risks. Liquid ammonia is stored at ambient temperatures around −33 °C under atmospheric pressure. Storage tanks for anhydrous ammonia are designed according to specific codes such as API 620 R or BS 7777 [52]. Standard materials for atmospheric ammonia tanks include certified low-temperature carbon manganese steel tested at temperatures close to −40 °C.

The susceptibility to crack formation due to corrosion increases under stress. The compatibility of the weld seam and the base material is crucial for resistance to crack formation under the influence of ammonia [52].

4.5. *Biofuel*

Biofuels are produced from organic raw materials. In the near future, they are considered one of the most competitive options for the rapid decarbonization of maritime transport. Biofuels, when blended, are compatible with conventional marine fuels. They can be used on existing vessels with minimal modifications or without any at all. Furthermore, existing bunkering infrastructure can be utilized [53].

4.5.1. Environmental Impact of Marine Engines Running on Biofuel

Depending on the raw materials used in production, biofuels are divided into four main generations. The first generation (1G) is produced from food crops and is widely used in the automotive sector. The competition for land with food production impacts the sustainability of 1G. Considering the life-cycle emissions from land-use changes (LUC), the environmental efficiency of 1G biofuels is questionable. Furthermore, to meet the demand for biofuels in shipping, vast areas of land are required.

Second-generation biofuel (2G) is produced from non-food biomass (lignocellulosic materials, waste, and residues). This fuel has significant potential for transportation because there are no food-versus-fuel disputes. However, the production of 2G biofuel still faces economic and technical challenges. As manufacturing technologies evolve, it is expected that by 2030, this type of fuel will penetrate the market significantly.

Third-generation (3G) biofuels are produced from algae. Algae are a rich resource. Compared to energy crops, this resource can theoretically yield oil 10 to 100 times more. At this stage, the low energy efficiency of the 3G system is a technological barrier to its viability. Genetically modified microorganisms and crops are used as feedstock in the production of fourth-generation (4G) biofuels. Currently, this biofuel is in the research phase [53]. In ref. [53], emission measurements were conducted accounting for voyage conditions and engine operating parameters. Considerable effort was made to perform emission measurements at engine loads closely aligned with those specified in ref [89].

The positions of the engine telegraph in the forward direction will determine five typical operating modes of the main engine: Mode 1—very slow, Mode 2—slow, Mode 3—half ahead, Mode 4—full ahead, and Mode 5—navigational full ahead. Two auxiliary diesel generators operated in all five modes of the main engines, while the third remained in reserve.

In all five engine operating modes, emissions were measured in increasing order for each fuel tested. Considering the weight coefficients recommended by ref. [89] for marine applications of type E2, the weight coefficients were established for each engine operating mode. The operation of the engine at loads exceeding 50% was assumed to have a weight coefficient of 85%.

In each mode, navigational data, engine parameters (engine power, speed, fuel consumption, etc.), and environmental conditions were monitored using ship's instruments. Table 12 shows the corresponding weight coefficients for each of the five engine operating modes. Additionally, the registered engine operating conditions for various test fuels are presented. In the case of using a biofuel mixture, for each engine condition operating in Modes 2 and 4, two values are given [53]. These values, collected under similar engine operating conditions during the use of biofuel mixture, represent different emission samples.

Table 12. Corresponding engine operating conditions for the tested biofuels [53].

Operating Mode	Mode 1	Mode 2	Mode 3	Mode 4	Mode 5
Weighting Factor	0.05	0.05	0.25	0.50	0.15
Engine Conditions Using Biofuel Mixture					
Engine Load (%)	20	32 and 37	41	61 and 63	90
Engine Speed, rpm	37	55 and 66	63	79	90
Power (kWh)	1986	3178 and 3675	4072	6058 and 6257	8939
Fuel Consumption (kg·h^{-1})	280	561 and 825	736	1179 and 1215	1780
Specific Fuel Consumption (g·kWh^{-1}·h^{-1})	141	177 and 224	181	195 and 194	199
Engine Conditions Using LSMGO					
Engine Load (%)	23	37	44	67	90
Engine Speed, rpm	36	55	63	79	87
Power (kWh)	2284	3675	4370	6654	8939
Fuel Consumption (kg·h^{-1})	304	587	762	1245	1703
Specific Fuel Consumption (g·kWh^{-1}·h^{-1})	133	160	174	187	191

Figure 10 shows the recorded data for each engine operation mode, navigation, and environmental conditions [53]. The red dashed line (Figure 10c,e) represents the NOx emission limits for Tier II. These values apply to bulk carriers operating with engine speeds below 130 rpm, specifically 14.4 g/kWh.

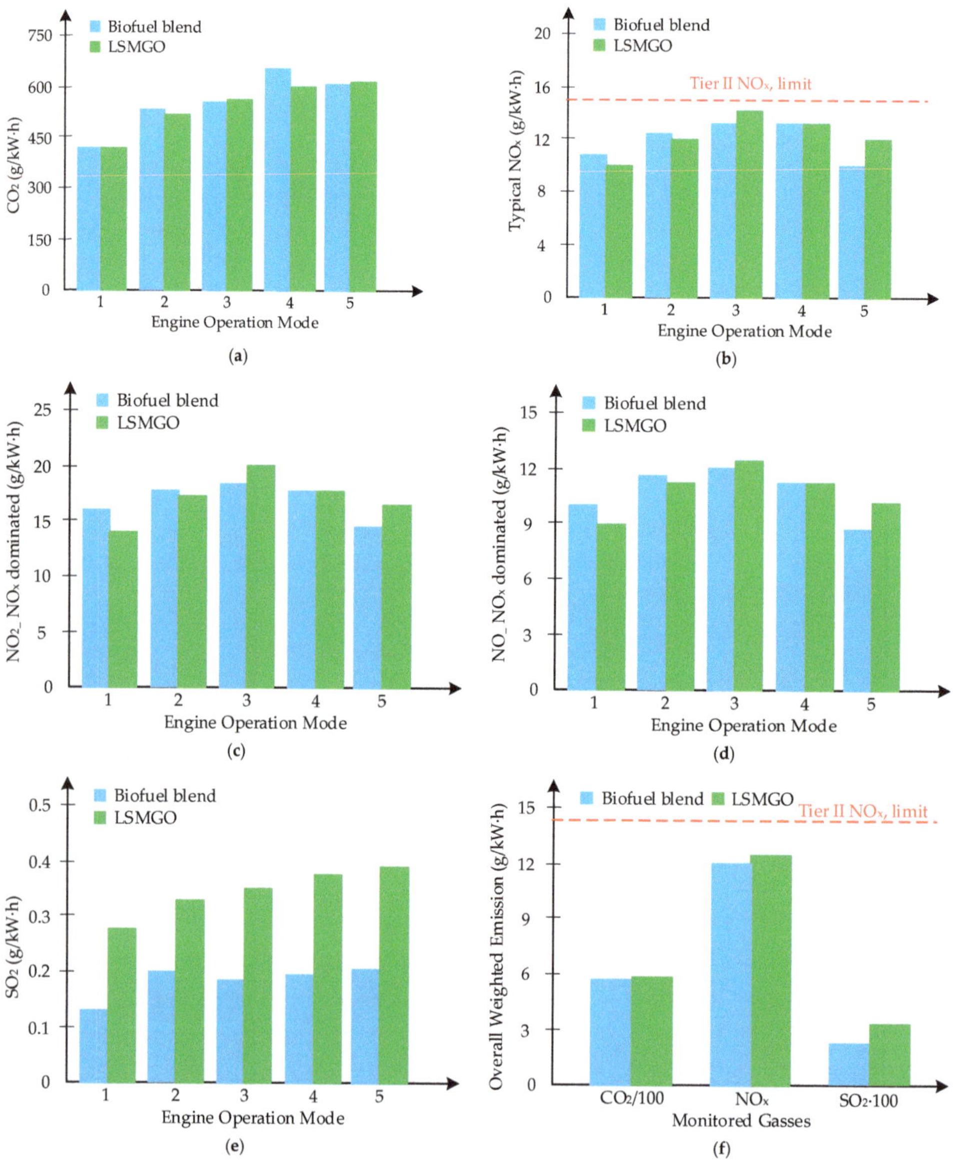

Figure 10. Emission factors for the investigated fuel types [53]: (**a**)—dependence of CO_2 emissions on engine operating modes; (**b**)—dependence of total NOx emissions on engine operating modes (data are provided for the typical contribution of NO and NO_2 to total NOx emissions, 0.85 and 0.15, respectively); (**c**)—extreme scenario of total NOx emissions where NO_2 emissions dominate per engine operating mode; (**d**)—extreme scenario of total NOx emissions where NO emissions dominate per engine operating mode; (**e**)—dependence of calculated SO_2 emissions on engine operating modes; (**f**)—total weighted emissions of CO_2, NOx, and SO_2.

4.5.2. Features of Biofuel Manufacturing and Transportation

In discussions surrounding shipping sector decarbonization, numerous alternative energy sources are under consideration. The discourse frequently centers on ammonia (NH_3), hydrogen (H_2), natural gas (primarily CH_4), and methanol (CH_3OH). For short-term biofuel applications, there are several other alternatives [90]. Table 13 presents a comprehensive overview of these candidate fuels and their respective production technologies.

Table 13. Renewable energy sources that can be used as fuel for maritime transportation and technologies for their production [90].

Renewable Carrier	Production Technology
LBG/LNG	Anaerobic digestion Gasification and cat. reaction (Sabatier process)
Ethanol	Fermentation from sugars Fermentation from lignocellulosic waste
Methanol	Anaerobic production of methane and dehydration Gasification from solid biomass
DME (dimethyl ether)	Gasification Dehydration of bio-methanol
SVO (straight vegetable oil)	Extraction from plants
FAME (fatty acid methyl esters)	Transesterification of vegetable oils
HVO (hydrotreated vegetable oil)	Hydrogenation of vegetable oils
Renewable diesel	FT, HTL, pyrolysis, HVO
Lignin diesel oil	Solvolysis

Biochemical reactions are processes facilitated by biocatalysts, such as microbial organisms, microbial communities, or enzymes. These reactions typically exhibit high feedstock specificity and occur under mild conditions, specifically at room temperature, atmospheric pressure, and neutral pH. Liquefied biogas (LBG) and ethanol have proven to be the most viable biofuel types for the maritime sector [90].

Thermochemical processes represent production pathways that rely on decomposition effects induced by heat and pressure. These processes effectively simulate the natural mechanisms of fossil oil and gas generation from biomass. Natural generation mechanisms took place over millions of years.

In contrast, thermochemical processes occur over the span of seconds or minutes. The primary biomass components—cellulose, hemicellulose, and lignin—undergo decomposition and depolymerization into oligomers. Oligomers can either break down into smaller molecules or re-polymerize into long chains of hydrocarbons or fatty acids [90].

In general, raw materials can be divided into four groups: (i) (poly)saccharides, (ii) oils and fats, (iii) lignocellulose, and (iv) lignin. These can be classified into the following biomass groups [90]:

1. Carbohydrate crops.
2. Oil crops.
3. Lignocellulosic biomass.
4. Woody biomass.
5. Algal biomass, as well as residues from their production and processing.

Marine algal biomass is particularly interesting due to its abundance and the fact that it does not occupy arable land. The processing of marine algae is complex. It faces numerous challenges in industrial applications, making algae processing an impractical solution for the fuel sector at present [43,44]. Given that this study focuses on short-term solutions for biofuels, algae were excluded from further consideration in the raw material analysis.

The waste generated from harvesting or other production processes can be used to produce biofuels. This makes a significant contribution to the circular economy and sustainable development. Thus, the European Renewable Energy Directive incentivizes the use of waste for bioenergy by introducing emission savings multipliers for biomass derived from waste [90].

One of the frequently mentioned challenges in biofuel production is the security of supply, logistics, and transportation of biomass. The supply of biomass fundamentally differs from the continuous extraction of fossil oil [90]. Harvesting is seasonal and can vary significantly month by month. Biorefineries may face downtime during the off-season, which heavily impacts their economics.

For the economic efficiency and quality of the produced biofuel, the transportation of biomass is a critical factor. For example, in the Netherlands, wood pellets are not produced in sufficient quantities, and even dried and compressed biomass degrades over time [90].

This becomes problematic when biomass must be imported from abroad, which can take 6–9 weeks or more. The study [90] also emphasizes the challenges arising from this. High-quality biomass may be purchased at a high price, but instead a partially degraded product might be received for which the bioprocess is not optimized.

4.5.3. Features of Biofuel Storage on Marine Vessels

Table 14 presents the energy characteristics of the studied types of biofuels and their fossil fuel equivalents [90]. From Table 14, it is evident that despite its moderate calorific value, HFO has the highest energy density. This is because its gravimetric density is close to that of water. All other oils are significantly lighter. This factor leads to an increased frequency of refueling.

Table 14. Energy performance of marine biofuels and their fossil counterparts [90].

Fuel	LHV (MJ/kg)	HHV (MJ/kg)	Energy Density (MJ/L)
HFO	39	42	39
ULS MDO (Ultra-Low Sufhur Marine Diesel Oil)	43	46	36
Petrol	43	47	32
LBG	48	55	21
Biomethanol	20	23	16
Bioethanol	27	30	21
SVO	37	40	34
FAME	38	40	34
HVO	43	47	33
FT diesel (Fischer–Tropsch diesel fuel)	44	–	34
UPO (Upgraded Pyrolysis Oil)	43	47	36
Lignin oil	33	35	–
HTL biocrude (Hydrothermal Liquefaction Biocrude)	34–38	–	–

LBG/LNG has the lowest calorific value and energy density. Cryogenic storage is required for LBG/LNG, which is difficult to implement in ship hulls. This is because storing LBG/LNG requires sacrificing a significant amount of cargo space. When transporting

LBG/LNG, part of the cargo can be used as fuel. Currently, this is the most common and economically efficient use of LNG in maritime shipping.

As seen in Table 14, LBG, biomethanol, and bioethanol have the lowest energy density (21, 16, and 27 MJ/L, respectively). The energy density of other biofuels ranges from 33 MJ/L (HVO) to 36 MJ/L (UPO). This implies an increase in refueling frequency by 15–8% compared to HFO.

These percentages assume that the tank volume is a limiting factor. Slow and slow steam and arrival at port on time are measures to improve efficiency. They can result in significant fuel savings, which may help balance fuel requirements.

4.6. The Techno-Economic Analysis of Alternative Fuels for Marine Vessels

4.6.1. Analysis of Technologies

Single-fuel and dual-fuel internal combustion engines (ICEs), as well as fuel cells (FC), are the main installations for fuel use in the maritime sector. Fuel cells are electrochemical devices that convert chemical energy from gaseous fuels into electrical energy. The high energy conversion efficiency and low emissions are unique advantages of this technology compared to combustion technologies [91].

The classification of fuel cells (FC) is based on the selection of the electrolyte and fuel. Currently, the following types of fuel cells are available:

1. PEMFCs—Proton Exchange Membrane Fuel Cells.
2. AFCs—Alkaline Fuel Cells.
3. PAFCs—Phosphoric Acid Fuel Cells.
4. MCFCs—Molten Carbonate Fuel Cells.
5. SOFCs—Solid Oxide Fuel Cells.
6. DMFCs—Direct Methanol Fuel Cells [92,93].

From the perspective of availability, efficiency, and technological maturity for onboard power generation, PEMFC and MCFC are the most promising types of fuel cells. Additionally, for medium- and long-range vessels, SOFCs are considered the most promising due to their potential.

Table A1 (Appendix A) presents the characteristics of alternative fuels used in maritime vessels. The data in Table A1 are sourced from [91].

4.6.2. Economic Performance Analysis

The commercial viability of alternative fuels plays a crucial role in determining the level of their deployment. Table A2 presents a cost analysis of selected alternative fuels based on different production pathways. The average price of Marine Gas Oil (MGO) at 20 global ports from 2018 to 2021 was used as a benchmark for comparison. Relative costs are compared based on the metric of USD/MJ of energy content. The data in Table A3 are sourced from [91].

Table A3 combines the costs of building and acquiring engine installations with the costs of onboard fuel storage systems. Based on the listed costs, the total capital expenditures for the ship's engine installation systems and fuel storage systems using alternative fuels are determined.

From the analysis of Tables A1–A3, it follows that hydrogen, when used as a fuel for Otto cycle engines, can be consumed in spark ignition engines and in various types of dual-fuel engines. Hydrogen-powered marine engine systems are not yet widely adopted. The cost of such systems is comparable to that of LNG/LBG-based systems.

Diesel engine systems using methanol are well established, with low modernization costs. Conversion from conventional systems to methanol/dual-fuel systems achieves similar or even higher efficiency [94].

Ammonia-based diesel engines are a relatively new concept in the maritime sector. Large ammonia engines used in marine vessels have not yet been commercialized.

The main barriers to the adoption of fuel cell (FC) technology in the maritime sector are high capital investments and relatively short, expected service life. The cost analysis of FC systems shows that, with increased production volumes, the costs of such systems tend to decrease [95].

One of the main issues for the broader use of hydrogen in the maritime sector is its storage. Releasing hydrogen from chemical storage requires higher energy or higher temperature. As a result, the general trend in hydrogen storage technologies is that hydrogen is less accessible in chemical storage than in physical storage [96].

LNG and LBG have low volumetric density. Therefore, LNG and LBG are stored on board in specially designed energy storage systems to keep them in a liquid state. In addition to membrane tanks for marine use, the IMO has defined three main types of tanks (Types A, B, and C).

For the storage of large volumes of LNG and LBG, Type A, B, and membrane tanks, which are not pressurized, are used. In smaller vessels, Type C pressurized tanks are commonly used because of their smaller capacity and the absence of the need for a secondary barrier and boil-off gas processing system.

5. Innovative Approaches and Challenges of Using Solar Energy in Shipping

Another alternative power source for ship engines is electricity. Unlike the alternative energy sources discussed in Section 4, electricity eliminates several issues associated with transportation when used to power ship engines. Additionally, the use of electricity results in no greenhouse gas emissions into the atmosphere. Solar energy, wind energy, and wave energy can be primary sources of electrical power [97].

Since the use of wave energy as a source of electricity is still poorly studied, this primary source of energy will not be considered in further analysis. Wind energy will also not be considered as a primary source of electricity due to the following factors:

- Integrating wind turbines on large cargo ships requires massive turbines and significant space for installation and energy storage [98].
- Wind gusts cause fluctuations in the engine power supply, which necessitates additional measures to stabilize the voltage, leading to increased costs [99].

5.1. Components of Solar-Powered Marine Vessels

The inclusion of fuel cells in hybrid engines is an innovative approach to improving the reliability of solar-powered ships. The chemical reaction between oxygen and hydrogen is the basis for generating electricity through fuel cells. Fuel cells, in combination with solar panels and/or wind turbines, can be used to power ship engines.

This combination of energy sources can offer increased efficiency, reduced emissions, and greater reliability compared to a system that relies solely on solar or wind energy [100–102]. Figure 11 shows a structural diagram of such a system.

The photovoltaic battery, the DC–DC converter, Power Point Tracking (PPT), the electric motor, and the battery management system constitute the structural components of the hybrid photovoltaic–diesel system for a marine vessel (Figure 11). This system incorporates elements such as a photovoltaic module, a DC–DC converter, Maximum Power Point Tracking (MPPT), a battery along with its charge-discharge controllers, a diesel generator, an inverter, and a connector.

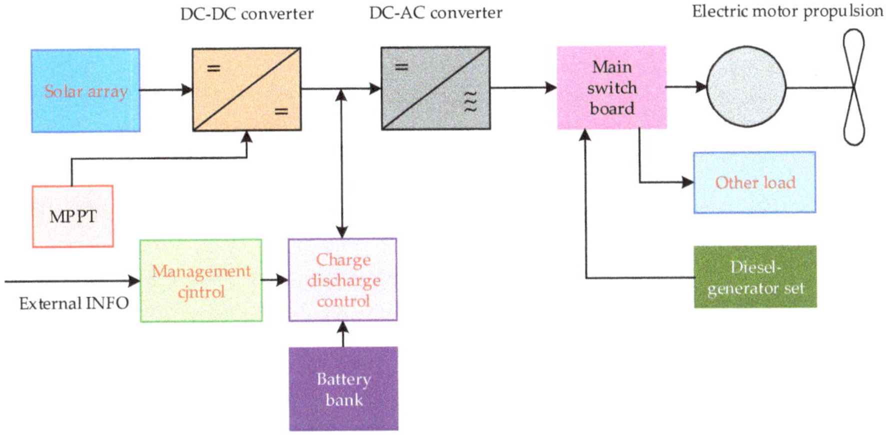

Figure 11. Topology of a hybrid photovoltaic–diesel system [93].

Another type of vessel that operates on solar power is a vessel entirely dependent on solar photovoltaic energy as the sole source for meeting the onboard load demand. This design is intended for small vessels. The power supply in such vessels is required for operating the electric motor for propulsion and for minor loads, such as lighting.

Such a watercraft simply requires power to operate the electric motor for propulsion and small loads, such as lighting. In these types of boats, PV is the only sufficient energy source to serve all purposes on a small scale. The only difference between the two topologies is the presence of a diesel engine in the hybrid system. All other components for both types of boats are the same, with slight differences in their component ratings, as this vessel caters to low loads. A 30-foot vessel typically requires 300–350 W of power on average, primarily depending on energy consumption and boat size [100].

5.1.1. PV Array

The limited space available for installing photovoltaic (PV) cells on a vessel is a major challenge when using solar energy as the primary energy source. The placement of PV modules should be optimized to absorb the maximum amount of sunlight. A photovoltaic battery, or solar battery, is an assembly of various solar modules. The output energy generated by a photovoltaic battery depends on the intensity of sunlight absorbed by its elements [103].

Utilizing equipment to convert solar energy into power for the marine vessel's engines poses stability challenges due to the load on the vessel. Another issue is partial shading, which affects the PV system's performance under different weather conditions [103].

The absence of energy production costs, the universality of the installation, energy generation aligning with peak demand, and cost savings are significant technological advantages of solar PV systems [104].

The cost of solar-powered ships will be higher than that of diesel-powered vessels, but their operating expenses will be lower. Additionally, solar-powered ships use environmentally friendly fuel, making them a promising long-term option [105].

5.1.2. DC–DC Converter as a Power Stabilizer

A power stabilizer is a crucial component that enhances system efficiency. It regulates the power output of the photovoltaic (PV) system and increases its yield when connected to the Maximum Power Point Tracking (MPPT) system.

Power stabilizers used in PV systems are DC–DC converters. Step-down (buck), step-up (boost), and step-down/step-up (buck–boost) converters are commonly used in PV systems. The boost converter is primarily used due to its high-power capability. Unlike a step-up converter, in a step-down converter, the output voltage can be lower than the input voltage. In a step-up converter, the output voltage is higher than the input voltage. In a buck–boost converter, the output voltage can be either higher or lower than the input voltage.

By means of sequential conversion of the maximum available solar energy, even under harsh weather conditions, a high efficiency of the photovoltaic generator (PVG) can be achieved [106]. A step-down converter is used for this purpose. It extracts the maximum energy from the photovoltaic source to meet the load requirements [107].

5.1.3. PV MPPT

MPPT (Maximum Power Point Tracking) is designed to increase the output power of the photovoltaic (PV) battery regardless of weather conditions. The application of MPPT allows achieving the maximum power of photovoltaic panels at different levels of solar radiation, irrespective of weather conditions. Various MPPT algorithms are used to regulate the operating cycle of the DC–DC converter [108]. MPPT methods play a crucial role in enhancing electricity production using solar cells [105].

5.1.4. Electric Motor

In [109], it is shown that the mass and size characteristics, along with lower cost, are the main advantages of the alternating current (AC) motor compared to the direct current (DC) motor. Therefore, the use of an AC motor is preferred as the main engine.

5.1.5. Battery Management System

The battery is located onboard the vessel. Solar radiation depends on the time of day and weather conditions. On sunny days, the output energy from the solar panels is higher. Therefore, solar energy cannot always meet the load requirements. The onboard battery is designed to support the continuous supply of power to the load. If the electric motor stops due to insufficient sunlight for the solar PV panels to generate energy, the battery serves as a backup. To enhance battery efficiency and lifespan, a good battery management system (BMS) is required. Lithium-ion batteries are currently the most preferred due to their long lifespan, high energy density, and environmental friendliness [110]. Internal operating parameters, such as current, voltage, and temperature, are monitored and managed by the battery management system (BMS) during charging and discharging. To enhance safety and performance, the BMS evaluates the State of Charge (SOC) and State of Health (SOH) [111]. To prevent overcharging and sudden explosions, the BMS maintains the charge level within specified limits.

5.2. The Application of Concentrated Solar Energy Systems on Marine Vessels

A promising direction for scientific research is the use of concentrated solar energy systems (CSPs) on ships. In this technology, mirrors or lenses are used to concentrate sunlight on a small focal point. This creates high temperatures, which can be used to produce steam and subsequently generate electricity.

Although CSP systems are not yet widely used on ships, their efficiency has been demonstrated in land-based applications. This technology holds significant potential for future deployment on large marine vessels. The adaptability of CSP technologies, combined with solar energy, can lead to innovative and sustainable energy solutions for marine applications, overcoming some limitations associated with traditional photovoltaic systems [109].

CSP systems include key types of technologies such as parabolic trough collectors (PTCs), linear Fresnel reflectors (LFRs), solar dishes (SDs), and solar towers (STs) (Figure 12) [102]. Figure 13 shows the main categories of CSP configurations, such as linear-focusing and point-focusing [100].

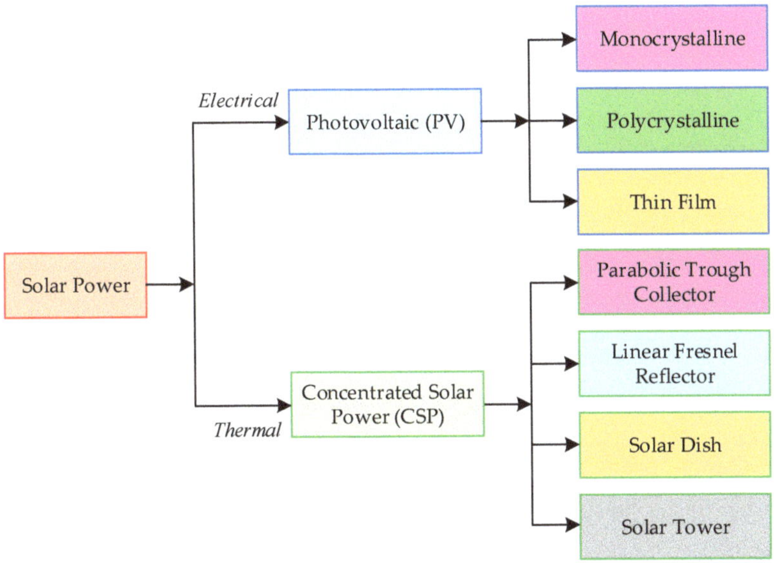

Figure 12. Leading technologies for solar energy integration [109].

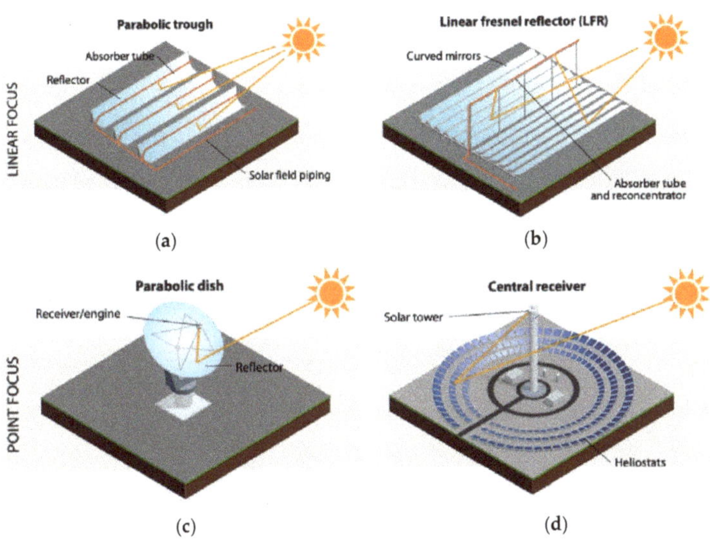

Figure 13. The four main types of CSP technologies: (**a**)—parabolic trough collector; (**b**)—linear Fresnel reflector; (**c**)—solar dish; (**d**)—solar tower [107].

All CSP technologies share the same fundamental concept but have different implementations and processes. The primary components of CSP systems are reflectors, receivers, Heat Transfer Fluid (HTF), and the associated cooling system [54].

Solar energy is concentrated on a receiver. The receiver then converts this energy into heat with the help of heliostats or controlled mirrors. Lenses or mirrors are used as reflectors to focus sunlight into a narrow beam and direct it onto the receiver. To generate electricity, mechanical energy is converted into turbine energy. This is performed using steam, which is produced by heating the Heat Transfer Fluid (HTF) and circulating it through the receiver.

The Thermal Energy Storage (TES) in CSP systems is required for storing thermal energy before it is converted into electricity. Due to its storage capability, CSP serves as a versatile renewable energy source.

To ensure continuous and reliable power supply over extended periods, TES must be integrated into the solar thermal power plant. Currently, TES is used as storage for both sensible and latent heat, as well as thermochemical storage with minimal volume and the highest energy density [112].

Figure 14 illustrates the operational diagram of an advanced solar power plant, which utilizes CSP for electricity generation [112]. This system is based on the use of a dense suspension of HTF particles. The application of highly efficient electricity generation cycles in this system uses the dense suspension of HTF particles, allowing the receiver to operate at high temperatures (above 650 °C).

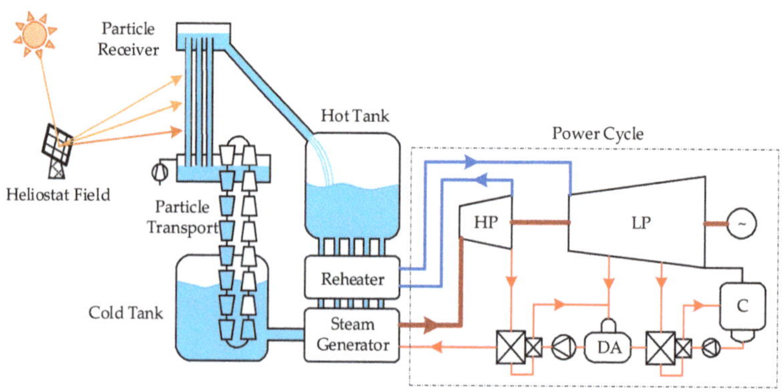

Figure 14. Schematic diagram of an advanced CSP technology for electricity generation from heat [111].

The efficiency of thermal conversion is improved by this supercritical Rankine cycle power block, increasing from 39.9% to 45.4%, which leads to a reduction in the heliostat field area by 9.6%. The study [111] notes that increasing the operating temperature range by 24.5% results in a 12.5% increase in storage density and a 22.5% reduction in overall storage capacity. The lack of heating also leads to a decrease in parasitic energy consumption. Overall, a reduction in electricity costs of approximately 10.8% is expected due to improved cycle efficiency, increased storage density, and the overall reduction in parasitic energy consumption [111].

5.3. Contrasting PV and CSP

CSP and PV panels are leading technologies in solar energy. Monocrystalline PV panels have high power and efficiency, which is why they are widely used. Due to their availability in higher power modules (over 300 W), they generate more energy compared to other types of panels [113]. Unlike CSP, PV instantly converts light into electricity. Solar cells absorb light, generating direct current, which is then converted into alternating current by inverters and supplied to the electrical grid. An effective method for utilizing solar energy through PV technology is proposed in ref. [113], involving the conversion of direct or diffuse sunlight into electricity. In ref. [113], the results of a comparison of the overall performance of monocrystalline and polycrystalline solar modules over the course of a year are presented. The comparison showed that the monocrystalline module has higher maximum efficiency and total specific output energy.

The type of technology used the light spectrum, solar radiation, ambient temperature, humidity, and wind were considered when evaluating PV modules [114]. In [115], it was assumed that, in addition to the mentioned factors, climatic and environmental factors also influence the wear and aging of solar modules. Table A4 presents the results of a comparison of other characteristics of PV and CSP.

5.4. Techno-Economic Analysis of the Use of Battery Storage Systems for Powering Marine Vessel Engines

Vessel propulsion systems powered by battery storage are more energy-efficient than internal combustion engines (ICEs) and fuel cells (FCs) [116]. To store solar energy on board, battery types such as lithium iron phosphate (LPF) or lithium nickel manganese cobalt (NMC) batteries, which have higher capacity, can be used. Due to the low volumetric density of battery systems, this technology is not suitable for long-distance vessels. Table 15 presents the costs of constructing and acquiring propulsion systems, along with the costs for on-board energy storage systems.

Table 15. Costs of construction, acquisition of propulsion systems, and costs of on-board energy storage systems [91].

Component	Capital Investments for the Construction of New Ships (Four-Stroke Engine/Two-Stroke Engine), USD/kW	Life Cycle
	Propulsion system	
Electric motor	400	30 years
	Fuel storage system	
Battery, NMC, USD/kWh	400–1000 and expected to fall to 124 in 2030	8–11 years
Battery, LPF, USD/kWh	210–1000 and expected to fall to 70 in 2030	7–13 years

Battery systems have economic drawbacks related to high capital costs. They are expensive components. When applied on marine vessels, their high cost is explained by the stricter requirements for insulation, cooling of equipment, and additional fire safety measures.

Currently, battery systems are only available for small coastal vessels. On long-range ships, battery systems can be applied as part of a hybrid power system. In such systems, the electric motor receives power partially from a diesel generator running on liquid fuel and partially from the battery.

6. Emerging Trends in Solar Energy for Maritime Transport

6.1. Analysis of the Limitations for the Use of Solar Energy on Marine Vessels

Despite the benefits of solar energy, several challenges and limitations remain. The low conversion efficiency of PV cells is a key problem in their application in maritime transportation. The low efficiency can result in insufficient energy production for the engine system, particularly under unfavorable weather conditions or periods of reduced sunlight. Improving the efficiency of PV cells is considered critical to addressing this limitation [34].

Another disadvantage is the significant deck or roof area required for the installation of a sufficient number of solar panels. To generate substantial electrical power, ample space must be allocated for a large-scale deployment of solar arrays. The availability of adequate surface area is a practical limitation, and shipbuilders must carefully plan the integration of solar panels into the vessel's design.

Cost is also a critical factor. While solar energy itself is free, the production of high-quality solar cells remains expensive. The cost of a PV system constitutes a significant portion of the initial capital investment for shipbuilders. Achieving a balance between long-term economic benefits and the upfront costs of installing solar panels presents a challenging decision for ship owners [34,117].

The harsh conditions of the marine environment pose additional challenges for the durability of photovoltaic cells. Factors such as vibration, temperature fluctuations, hail, and corrosion caused by salt or humidity can affect the performance of solar panels. Although many solar panels aboard vessels are equipped with tempered glass and atmospheric and salt-resistant protective coatings, they remain susceptible to contamination, discoloration, corrosion, and wear due to prolonged exposure to temperature variations, high salinity, and humidity. Reliable design practices and maintenance procedures are crucial to mitigating these issues and ensuring the long-term efficiency of solar energy systems on ships.

6.2. Trends in the Development of Solar Energy for Marine Transportation

The emerging trends in the utilization of solar energy for maritime transport focus on integrating advanced technologies to enhance resilience and efficiency within the sector. Key directions include the development of hybrid systems that combine solar energy with sources such as wind and hydrogen fuel cells. This combination aims to reduce greenhouse gas emissions and dependence on fossil fuels. The maritime industry stands on the brink of revolutionary changes in embracing solar energy.

A critical success factor will be improving the efficiency of PV cells. The current limitations in the efficiency of solar panels are one of the main obstacles to their widespread use on ships. Future research should concentrate on the development of new materials and designs capable of significantly increasing energy production per unit of surface area. This is particularly important given the limited space available on ships [118].

The development of hybrid systems represents another promising direction. Combining solar energy with other energy sources, such as fuel cells, wind turbines, and traditional diesel engines, will create more reliable and efficient energy systems. In the future, we can expect the emergence of more advanced hybrid systems capable of optimally utilizing different energy sources depending on the operating conditions of the vessel.

The maritime industry is also exploring the use of solar energy in conjunction with energy storage systems, such as batteries. This enables vessels to store excess energy produced during the day for use at night or during periods of low sunlight. Furthermore, advances in solar technologies, such as more efficient solar cells and lightweight, flexible panels, are making the use of solar energy on maritime vessels more feasible [77,119].

The integration of Carbon Capture, Utilization, and Storage (CCUS) technologies with solar systems opens new avenues in the fight against greenhouse gas emissions. Further development of these technologies could lead to the creation of ships with nearly zero carbon footprints. Particularly promising is the approach of converting captured CO_2 into synthetic fuel or other valuable products, which could enhance the economic efficiency and autonomy of vessels.

The adaptation of concentrated solar power (CSP) systems for marine applications represents a promising avenue for future research, with new practical applications in the maritime sector. Although CSP technology is not yet widely used on ships, its ability to generate high-temperature heat addresses specific needs in maritime operations. For example, CSP systems can be effectively utilized in desalination processes on ships, particularly in Multi-Effect Distillation (MED) systems, which require temperatures of 70–80 °C. This capability is critical for long voyages where freshwater production is essential and often energy intensive [120].

Moreover, the high-temperature heat generated by CSP systems can be used for absorption cooling systems in cargo holds and living spaces, potentially reducing the load on electrical systems on ships. On large vessels with adequate space for solar collectors, CSP technology could assist in electricity generation via Organic Rankine Cycle (ORC) systems, although this application requires further development for marine use.

These CSP applications in the maritime context align with real operational needs, offering potential improvements in energy efficiency and reducing reliance on conventional fuel sources. However, the implementation of CSP on ships faces challenges, including spatial constraints and the necessity to stabilize the system in dynamic marine environments. These challenges are active areas of current research and development in the field of marine solar energy applications.

An important area of focus will be the development of materials and designs for solar panels specifically adapted to marine conditions. Durability against corrosion, vibration, extreme temperatures, and mechanical damage is crucial for the long-term effectiveness of solar systems on vessels. Innovative solutions in this area will significantly expand the use of solar technologies in the maritime industry [121].

However, issues remain, including the necessity for substantial initial investments and the development of reliable infrastructure to support these technologies [122]. Reducing the costs of manufacturing and installing solar systems, as well as developing more efficient methods of integrating them into ship designs, will make this technology more attractive to ship owners. Additionally, it is important to develop methods for assessing the long-term economic benefits of using solar energy on ships [123,124].

The design of new types of vessels specifically developed for the maximum efficient use of solar energy could become a breakthrough direction. This may include innovative hull shapes optimized for the placement of solar panels or new concepts for the layout of decks and superstructures.

Finally, the development of a regulatory framework and international standards for the use of solar energy on ships will play a critical role in stimulating innovation and ensuring safety. Establishing clear guidelines and incentives for the adoption of solar technologies can significantly accelerate their acceptance within the maritime industry [125–127].

The future of solar energy in shipping appears promising. A combination of technological innovations, economic optimization, and legal frameworks could lead to the widespread adoption of solar technologies across various types of vessels. This would not only reduce the carbon footprint of the maritime sector but also provide new opportunities to enhance vessel efficiency and autonomy [128].

7. Discussion

This study compares alternative energy sources for marine vessels based on criteria such as greenhouse gas emissions, production, transportation, and storage challenges. The comparison was conducted for various alternative fuels: LNG, methanol, hydrogen, ammonia, and biofuels. The analysis revealed that using hydrogen as a fuel results in no greenhouse gas emissions. However, issues arise in hydrogen transportation, such as determining the mixing ratio of hydrogen, which is a systematic challenge due to various factors: natural gas composition, pipeline material, hydraulic conditions, equipment service life, environmental impact, and variations in client-specific end equipment [85]. Furthermore, when storing hydrogen on board vessels, factors such as the following should be considered:

- Hydrogen is a highly flammable fuel.
- Risks of hydrogen leaks through joints and cracks in any pipelines or storage system.
- Hydrogen disperses easily in the air.
- In high concentrations, hydrogen can displace oxygen in the air, leading to asphyxiation.
- The necessity to avoid explosion risks [85].
- The necessity of appropriate materials capable of withstanding low temperatures to prevent brittleness.
- The necessity to prevent liquid hydrogen spills to avoid asphyxiation and explosion.

Another potential energy source for powering marine vessel engines is solar energy. When the engines are powered by electricity generated through the conversion of solar energy, as in the case of hydrogen fuel, there are no greenhouse gas emissions. However, providing the necessary electrical energy to the engines is only feasible for smaller vessels [100]. Current solar energy conversion technologies do not allow for the complete powering of main engines with electrical energy [100]. For larger vessels, either a hybrid PV–diesel power system or concentrated solar power (CSP) systems are employed, with the structural configurations depicted in Figures 11 and 14, respectively. The benefits and drawbacks of PV and CSP systems are summarized in Table 15.

The main drawback of PV systems is their intermittency, which makes energy storage challenging as batteries are economically unfeasible for large-scale installations. Hybrid power systems are increasingly used in other forms of transportation, particularly in rail transport [129,130]. The issues encountered in hybrid power systems for rail transport are also relevant to marine applications. Specifically, these involve the optimization of energy flow algorithms from the PV system and the diesel engine to the main engine.

In [90], it is noted that field-oriented control (FOC) is used to manage the main engines of marine vessels with hybrid PV–diesel power. Studies in [131,132] show that both FOC and direct torque control (DTC) can be applied for managing the induction motors used as main engines on marine vessels. The comparison of the advantages and disadvantages of FOC and DTC is presented in Table A5.

The analysis of the advantages and disadvantages of the two control systems, presented in Table A5, predetermines their effectiveness for use on ships of different types. For vessels where frequent changes in movement parameters are characteristic, DTC is more efficient. This is since when the vessel stops in an FOC system, power is supplied to the main engine, but the engine does not perform useful work. This operating mode is necessary to prevent the main engine from exiting saturation mode. As a result, the energy efficiency of the propulsion system with FOC is reduced. DTC does not have this disadvantage.

Moreover, the use of DTC is more effective on large cargo vessels. This is because, for such vessels, the quality of torque control is a more important criterion. In contrast to cargo vessels, passenger ships operate at higher speeds. Additionally, the engine shaft speed may exceed the nominal value. For such vessels, FOC is more efficient because, in these modes, it demonstrates better speed control performance.

Another significant issue in the traction systems of marine vessels with hybrid PV–diesel power is the presence of an inverter (Figure 11). Technically, the inverter can be implemented using electronic switches, typically IGBTs (Insulated Gate Bipolar Transistors) and, less frequently, GTOs (Gate Turn-Off Thyristors). Pulse-width modulation (PWM) is used as the control law for the electronic switches. A key parameter of PWM is the sampling frequency. Since the main engines of marine vessels are high-power asynchronous motors, the values of the phase stator currents and voltages are substantial.

In other words, the electronic switches of the inverter must operate with high voltages and currents. This results in the fact that the high-power electronic switches have a limited sampling frequency, typically in the range of hundreds of Hertz. The consequence of this low sampling frequency is the presence of significant higher harmonics in the stator currents. This leads to increased losses in the electrical circuits of the traction system and, consequently, higher energy consumption [131,132]. Furthermore, the presence of higher harmonic components results in torque pulsations on the motor shaft—the higher the level of harmonics, the greater the torque ripple on the motor shaft.

To eliminate higher harmonic components in the spectra of phase currents of the induction motor stator, compensation devices should be used [133–135]. When compensating for higher harmonics, both frequency-domain and time-domain filtering can be used [134]. A comparison of the advantages and disadvantages of harmonic compensation systems with filtering in the frequency and time domains is presented in Table A6.

The use of harmonic compensation devices will complicate the propulsion system of the ship but will enhance its energy efficiency [134]. Moreover, while the hybrid PV–diesel system reduces greenhouse gas emissions to some extent, the issue of greenhouse gases is not fully resolved since the diesel is powered by fossil fuels.

CSP can supply electricity even at night, making it dispatchable and increasing its integration into the energy sectors. Moreover, CSP generates alternating current, thus eliminating the need for an inverter (Figure 13). As a result, this technology avoids the issues associated with inverters, as discussed earlier. Additionally, since CSP fully supplies the main engine with electrical energy, it resolves the problem of greenhouse gas emissions.

However, CSP has several drawbacks, the main one being that CSP relies solely on Direct Normal Irradiance (DNI), which fluctuates between 65% and 85% of the total global horizontal irradiance (GHI). Furthermore, the use of a steam turbine and energy systems (Figure 14) increases the mass and size of the system, which is critical for marine applications.

The adaptation of concentrated solar power (CSP) systems for marine applications represents one of the promising directions for future research, with new practical applications in the maritime sector.

Another important research direction is the development of materials and designs for solar panels specifically tailored to marine conditions []. The increase in the number of ships using solar energy requires advanced batteries. An analysis of the advantages of battery systems that may be used in maritime transport in the future is provided in Table A7.

The development of methods for evaluating the long-term economic benefits of using solar energy on ships [120,121] and the establishment of regulatory frameworks and international standards for solar energy use on vessels [129,131] are also crucial areas for future studies.

Additionally, for hybrid PV–diesel power systems, prospective research directions could include work on implementing DTC (direct torque control) instead of vector control systems, as well as efforts to mitigate higher harmonics in the phase current spectra of induction motors.

8. Conclusions

This study provides an overview of the existing types of alternative energy sources used in marine vessels and offers recommendations to improve the solar energy technologies in ships. The analysis of alternative fuels used on marine vessels is based on the following criteria: greenhouse gas emissions, production and transportation characteristics, onboard storage requirements, and technoeconomic indicators.

It has been established that when hydrogen is used as a fuel, no greenhouse gas emissions are released into the atmosphere. However, there are challenges associated with the transportation and storage of hydrogen onboard ships. These challenges are related to its destructive impact on the materials of supply and storage systems, as well as its flammability and explosiveness. Furthermore, the use of hydrogen as fuel in marine vessels requires the highest investment.

The lowest investment is required when LNG, produced from natural gas, is used as fuel, but this fuel is characterized by a relatively high level of greenhouse gas emissions.

As a result of the analysis conducted, it was found that the use of solar energy would eliminate the problems related to transportation. Two technologies were considered: hybrid photovoltaic–diesel power systems and concentrated solar power (CSP) systems. It was discovered that the PV–diesel hybrid system utilizes global horizontal irradiance (GHI) and can operate under diffuse light. However, this system generates electricity instantaneously, making energy storage challenging, and batteries are economically unfeasible for large-scale installations.

Intermittency remains the primary disadvantage of photovoltaic. Moreover, in this system, direct current (DC) must be converted into alternating current (AC) using inverters, which also requires a reduction in higher harmonics in the main engine's stator currents. This system only partially addresses the issue of greenhouse gas emissions.

The use of CSP allows for a continuous supply of electricity even at night, making it dispatchable. Moreover, CSP generates alternating current (AC), eliminating the need for inverters and, consequently, avoiding the associated issues. Additionally, CSP fully powers the main engine, effectively addressing the problem of greenhouse gas emissions. However, the use of steam turbines and energy systems leads to an increase in the mass and size of the system, which creates challenges for marine applications.

Trends in solar energy for maritime transport have been analyzed, such as the following:

1. The adaptation of concentrated solar power (CSP) systems for marine applications.
2. The development of materials and designs for solar panels is specifically intended for the marine environment.
3. The development of methods for evaluating the long-term economic benefits of using solar energy on ships.
4. The improvement of regulatory frameworks and international standards for the application of solar energy on ships.

Future research directions for the use of solar energy in the maritime sector have been proposed. Promising research areas for hybrid photovoltaic and diesel power systems include studies on the implementation of direct torque control (DTC) instead of field-oriented control (FOC), as well as efforts aimed at compensating higher harmonics in the stator current spectra of asynchronous motors. This will help improve the efficiency of the propulsion system.

Additionally, promising research directions may include studies focused on exploring the potential use of energy generated by the electric motor during the deceleration of the vessel. For ships with hybrid propulsion systems, research related to the synchronization of energy usage from various sources is of particular interest.

Author Contributions: Conceptualization, O.P., M.L. and S.G.; methodology, O.P., M.L. and S.G.; software, V.L.; validation, O.P., M.L. and S.G.; formal analysis, O.P., M.L. and S.G.; investigation, O.P., M.L. and S.G.; resources, V.L.; data curation, O.P., M.L. and S.G.; writing—original draft preparation,

O.P., M.L. and S.G.; writing—review and editing, O.P., M.L., S.G. and V.L.; visualization, V.L. All authors have read and agreed to the published version of the manuscript.

Funding: This research received no external funding.

Institutional Review Board Statement: Not applicable.

Informed Consent Statement: Not applicable.

Data Availability Statement: Not applicable.

Conflicts of Interest: The authors declare no conflicts of interest.

Appendix A

Table A1. Overview of the characteristics of alternative marine fuels in use [91].

Type of Fuel	Calorific Value (MJ/kg)	Onboard Storage Requirements	Utilization Efficiency FC (Electrical Efficiency)	ICE (Internal Combustion Engine)	Byproduct of Utilization	Availability of Commercial Utilization Technologies
Liquefied Natural Gas (LNG)	50–55	163° (under pressure or without pressure)	45–60% (SOFC) 35–45% (PEMFC)	35–40%	CO_2, H_2O, NO_x (ICE)	ICE: Yes, widely available for commercial use; FC: Underdeveloped, no commercial maritime use
Hydrogen	120–140	350–700 bar (gaseous), −252.8 °C (liquefied), material-based storage	50–60% (PEFMC/SOFC)	30–35%	H_2O, NO_x (ICE)	ICE: Yes, available for commercial use; FC: Yes, available for commercial use
Ammonia	22.5	21 °C below 8.8 bar or 33 °C at atmospheric pressure	30–60% (SOFC) 15–30% (AFC)	35–40%	H_2O, N_2, NO_x (ICE)	ICE: Underdeveloped; FC: Underdeveloped, no commercial maritime use
Methanol	22.7	Liquid at ambient temperature	30–50% (SOFC) 20–30% (DFMC)	40%	CO_2, H_2O, CO (ICE), NO_x (ICE)	ICE: Yes, widely available for commercial use; FC: Underdeveloped, no commercial maritime use
Biodiesel	38–46	Liquid at ambient temperature	10–30% (AFC)	10–30% (AFC), Varies	CO_2, H_2O, CO (ICE), NO_x (ICE)	ICE: Yes, widely available for commercial use; FC: Underdeveloped, no commercial maritime use
Electricity	-	High energy density battery systems such as NMC, LFP, NCA, and solid-state batteries	-	-	-	Currently available only for small shore-based vessels

Table A2. Cost analysis of selected alternative fuels by different production pathways [91].

Fuel	Material for Fuel Production	Fuel Production Costs/Import Price, USD/MJ	Fossil Fuel Price, USD/MJ	Total Price
LNG	Natural gas	0.0022–0.0092	0.014	0.16–0.66
LBG	Biomass	0.0081–0.032		0.57–2.29
Biodiesel	Soybean, palm oil, rapeseed	0.013–0.027		0.92–1.93
Methanol	Biomass	0.021–0.037		1.50–2.64
Methanol	Natural gas	0.006–0.022		0.43–1.57
Hydrogen SMR	Wind power, water	0.0063–0.035		0.45–2.5
Hydrogen electrolysis	Wind power, water	0.025–0.08		1.78–5.71
Hydrogen electrolysis	Nuclear energy, water	0.019–0.045		1.36–3.21
Ammonia	Natural gas	0.0093–0.036		0.66–2.57
Ammonia	Wind power	0.021–0.037		1.50–2.64

Table A3. Construction costs, propulsion system acquisition costs, and on-board fuel storage system costs [91].

Component	Capital Investments for Vessel Modernization, USD/kW	Capital Expenditures for Construction of New Vessels (Four-Stroke Engine/Two-Stroke Engine), USD/kW	Life Cycle
Propulsion system			
ICE diesel, biodiesel	-	240/460	30 years
ICE, methanol	328	265/505	30 years
ICE, ammonia	-	370/600	30 years
ICE, LNG, LBG, hydrogen	900	387/850	30 years
Fuel cell, SOFC	-	4000–9000	7–12 years
Fuel cell, PEMFC	-	730–2860	7–12 years
Electrical and generator, LNG, LBG, hydrogen	-	400	30 years
Electric motor	-	250	30 years
Fuel storage system			
Gas supply system + tank, LNG, LBG, USD/kg	270–420	270–420	30 years
Gas supply system + high-pressure tank (700 bar), Hydrogen, USD/kg	-	576–868	30 years

Table A4. Comparison of PV and CSP technologies [114].

Criteria	PV	CSP
Storage and intermittency	It generates electricity immediately, making storage difficult, and batteries are economically unfeasible for large installations. Intermittency is the main drawback of PV.	Overcoming interruptions and maintaining supply at night, TES (Thermal Energy Storage) is an attractive feature of CSP, making it dispatchable and increasing CSP's penetration in energy sectors.
Efficiency	Maximum efficiency 10–28%, with productivity decreasing over time.	Average efficiency 41%, efficiency increases with temperature.
Capacity factor	10–35%	With TES 29–33%, without TES 28–29%.

Table A4. *Cont.*

Criteria	PV	CSP
Equipment	Direct current must be converted to alternating current using inverters.	Output is alternating current combined with feeding cycles; therefore, an inverter is not needed.
Solar Irradiance	Uses Global Horizontal Irradiance (GHI) and can operate in diffuse light.	CSP uses only Direct Normal Irradiance (DNI), which varies from 65 to 85% of total GHI.

Table A5. Comparison of advantages and disadvantages of FOC and DTC [131].

Control System	Advantages	Disadvantages
FOC	1. Lower torque pulsation coefficient on the motor shaft. 2. Better speed control quality at speeds higher than the rated value.	1. Less energy efficiency. 2. Worse torque control quality.
DTC	1. Better energy efficiency. 2. Better torque control quality.	1. Higher torque pulsation coefficient on the motor shaft. 2. Lower speed control quality at speeds greater than the rated value.

Table A6. Comparison of advantages and disadvantages of FOC and DTC [134].

Filtration Type	Advantages	Disadvantages
Frequency domain	1. Higher filtering accuracy	1. More time of unstable operation of the compensator. 2. More complex algorithm for obtaining the desired signals of phase currents of the stator phase currents of the induction motor.
Time domain	1. Less time of unstable operation of compensator. 2. Simpler algorithm to obtain the desired signals of induction motor stator phase currents.	1. Less accurate filtering.

Table A7. Advantages of new battery technologies for maritime transportation.

Battery Type	Advantages
Gel battery	1. No maintenance required. 2. No leaks. 3. Can be submerged in water. 4. Spill-proof. 5. Eliminated gas emission. 6. Can be safely used near sensitive electronics and in public areas.
Pure lead thin film	1. Can withstand higher charge power. 2. Charges quickly. 3. 400 cycles of charge and discharge.
Lithium-manganese	1. Can withstand higher charge power. 2. Charges quickly. 3. 400 cycles of charge and discharge.
With absorbed glass mat	1. Good starting power. 2. Deep cycle capability.

References

1. Li, B.; Elmi, Z.; Manske, A.; Jacobs, E.; Lau, Y.Y.; Chen, Q.; Dulebenets, M.A. Berth allocation and scheduling at marine container terminals: A state-of-the-art review of solution approaches and relevant scheduling attributes. *J. Comput. Des. Eng.* **2023**, *10*, 1707–1735. [CrossRef]
2. UNCTAD. *Review of Maritime Transport 2022*; United Nations Conference on Trade and Development: Bridgetown, Barbados, 2022. Available online: https://unctad.org/rmt2022 (accessed on 15 July 2024).
3. Fratila, A.; Gavril, I.A.; Nita, S.C.; Hrebenciuc, A. The importance of maritime transport for economic growth in the European union: A panel data analysis. *Sustainability* **2021**, *13*, 7961. [CrossRef]
4. Elmi, Z.; Li, B.; Liang, B.; Lau, Y.Y.; Borowska-Stefańska, M.; Wiśniewski, S.; Dulebenets, M.A. An epsilon-constraint-based exact multi-objective optimization approach for the ship schedule recovery problem in liner shipping. *Comput. Ind. Eng.* **2023**, *183*, 109472. [CrossRef]
5. Issa, M.; Ilinca, A.; Martini, F. Ship energy efficiency and maritime sector initiatives to reduce carbon emissions. *Energies* **2022**, *15*, 7910. [CrossRef]
6. Hoang, A.T.; Foley, A.M.; Nižetić, S.; Huang, Z.; Ong, H.C.; Ölçer, A.I.; Nguyen, X.P. Energy-related approach for reduction of CO_2 emissions: A critical strategy on the port-to-ship pathway. *J. Clean. Prod.* **2022**, *355*, 131772. [CrossRef]
7. Xing, H.; Spence, S.; Chen, H. A comprehensive review on countermeasures for CO_2 emissions from ships. *Renew. Sustain. Energy Rev.* **2020**, *134*, 110222. [CrossRef]
8. Jones, M.W.; Peters, G.P.; Gasser, T.; Andrew, R.M.; Schwingshackl, C.; Gütschow, J.; Houghton, R.A.; Friedlingstein, P.; Pongratz, J.; Le Quéré, C. National contributions to climate change due to historical emissions of carbon dioxide, methane, and nitrous oxide since 1850. *Sci. Data* **2023**, *10*, 155. [CrossRef]
9. Nunes, L.J. The rising threat of atmospheric CO_2: A review on the causes, impacts, and mitigation strategies. *Environments* **2023**, *10*, 66. [CrossRef]
10. Chataut, G.; Bhatta, B.; Joshi, D.; Subedi, K.; Kafle, K. Greenhouse gases emission from agricultural soil: A review. *J. Agric. Food Res.* **2023**, *11*, 100533. [CrossRef]
11. Dervash, M.A.; Yousuf, A.; Ozturk, M.; Bhat, R.A. Global warming: Impacts of temperature escalation. In *Phytosequestration: Strategies for Mitigation of Aerial Carbon Dioxide and Aquatic Nutrient Pollution*; Springer International Publishing: Cham, Switzerland, 2023; pp. 27–36. [CrossRef]
12. Mariappan, S.; David Raj, A.; Kumar, S.; Chatterjee, U. Global warming impacts on the environment in the last century. In *Ecol. Footpr. Clim. Change: Adapt. Approaches Sustain*; Springer International Publishing: Cham, Switzerland, 2023; pp. 63–93. [CrossRef]
13. Singh, E.A.; Shindikar, M.R. A comprehensive review on climate change and its effects. *Int. J. Environ. Clim. Change* **2023**, *13*, 924–931. [CrossRef]
14. Jain, A.K. Global Carbon Budget 2022. *Earth Syst. Sci. Data* **2022**, *14*, 4811–4900. [CrossRef]
15. Issa Zadeh, S.B.; López Gutiérrez, J.S.; Esteban, M.D.; Fernández-Sánchez, G.; Garay-Rondero, C.L. Scope of the literature on efforts to reduce the carbon footprint of seaports. *Sustainability* **2023**, *15*, 8558. [CrossRef]
16. Cardoso, M.; Santos, T.; Tessarolo, L.G.A.; Aprigliano, V.; Rodrigues da Silva, A.N.; da Silva, M.A.V. Exploring the Resilience of Public Transport Trips in the Face of Urban Violence from a Gender Perspective. *Sustainability* **2023**, *15*, 16960. [CrossRef]
17. Jiwa, M.Z.; Kim, Y.T.; Mustaffa, Z.; Kim, S.; Kim, D.K. A Simplified Approach for Predicting Bend Radius in HDPE Pipelines during Offshore Installation. *J. Mar. Sci. Eng.* **2023**, *11*, 2032. [CrossRef]
18. Ding, C.; Zhu, J.; Qin, W.; Tao, X.; Fan, F.; Li, R.; Li, C. A Hydrogen Farm Strategy for Scalable Solar Hydrogen Production with Particulate Photocatalysts. *Angew. Chem. Int. Ed.* **2020**, *59*, 9653–9658. [CrossRef]
19. Tay, Z.Y.; Konovessis, D. Sustainable energy propulsion system for sea transport to achieve United Nations sustainable development goals: A review. *Discov. Sustain.* **2023**, *4*, 20. [CrossRef]
20. Munim, Z.H.; Chowdhury, M.M.H.; Tusher, H.M.; Notteboom, T. Towards a prioritization of alternative energy sources for sustainable shipping. *Mar. Policy* **2023**, *152*, 105579. [CrossRef]
21. Parhamfar, M.; Sadeghkhani, I.; Adeli, A.M. Towards the application of renewable energy technologies in green ports: Technical and economic perspectives. *IET Renew. Power Gener.* **2023**, *17*, 3120–3132. [CrossRef]
22. Sagin, S.V.; Kuropyatnyk, O.A. Using exhaust gas bypass for achieving the environmental performance of marine diesel engines. *Austrian J. Tech. Nat. Sci.* **2021**, *7–8*, 36–43. [CrossRef]
23. Shi, J.; Zhu, Y.; Feng, Y.; Yang, J.; Xia, C. A prompt decarbonization pathway for shipping: Green hydrogen, ammonia, and methanol production and utilization in marine engines. *Atmosphere* **2023**, *14*, 584. [CrossRef]
24. Bilgili, L. A systematic review on the acceptance of alternative marine fuels. *Renew. Sustain. Energy Rev.* **2023**, *182*, 113367. [CrossRef]
25. Popek, M. Alternative Fuels–Prospects for the Shipping Industry. *TransNav Int. J. Mar. Navig. Saf. Sea Transp.* **2024**, *18*, 25–33. [CrossRef]

26. Melnyk, O.; Onishchenko, O.; Onyshchenko, S. Renewable energy concept development and application in shipping industry. *Lex Portus* **2023**, *9*, 15. [CrossRef]
27. McLaughlin, H.; Littlefield, A.A.; Menefee, M.; Kinzer, A.; Hull, T.; Sovacool, B.K.; Bazilian, M.D.; Kim, J.; Griffiths, S. Carbon capture utilization and storage in review: Sociotechnical implications for a carbon reliant world. *Renew. Sustain. Energy Rev.* **2023**, *177*, 113215. [CrossRef]
28. Memon, S.U.R.; Manzoor, R.; Fatima, A.; Javed, F.; Zainab, A.; Ali, L.; Ullah, U.; Saleem, A.; Ullah, Q. A Comprehensive Review of Carbon Capture, Utilization, and Storage (CCUS): Technological Advances, Environmental Impact, and Economic Feasibility. *Sch. Acad. J. Biosci.* **2024**, *7*, 184–204. [CrossRef]
29. Petrychenko, O.; Kulakov, M.; Prytula, D. The evolution of ship's saving energy policy in the world. In Proceedings of the International Conference on Science and Technology «Navigation, Shipping and Technology», Odesa, Ukraine, 18–19 November 2021; pp. 39–46. Available online: http://onma.edu.ua/wp-content/uploads/2021/12/ISCT-NST-novyjSformuloj.pdf (accessed on 20 September 2024).
30. Sagin, S.; Kuropyatnyk, O.; Sagin, A.; Tkachenko, I.; Fomin, O.; Píštěk, V.; Kučera, P. Ensuring the Environmental Friendliness of Drillships during Their Operation in Special Ecological Regions of Northern Europe. *J. Mar. Sci. Eng.* **2022**, *10*, 1331. [CrossRef]
31. Zhu, Y.; Zhou, S.; Feng, Y.; Hu, Z.; Yuan, L. Influences of Solar Energy on the Energy Efficiency Design Index for New Building Ships. *Int. J. Hydrogen Energy* **2017**, *42*, 19389–19394. [CrossRef]
32. Budashko, V.V.; Kozminykh, M.A.; Mazur, O.M.; Onyshchenko, O.A.; Bucaros, A.Y. Energy-Efficient Cooling Systems for Special Equipment of Maritime and Land Military Complexes. In Proceedings of the XI International Scientific and Practical Conference "Modern Approaches to High-Efficiency Use of Transport Means", Izmail, Ukraine, 3–4 December 2020.
33. Karatuğ, Ç.; Durmuşoğlu, Y. Design of a Solar Photovoltaic System for a Ro-Ro Ship and Estimation of Performance Analysis: A Case Study. *Sol. Energy* **2020**, *207*, 1259–1268. [CrossRef]
34. Glykas, A.; Papaioannou, G.; Perissakis, S. Application and cost–benefit analysis of solar hybrid power installation on merchant marine vessels. *Ocean Eng.* **2010**, *37*, 592–602. [CrossRef]
35. Bezyk, Y.; Górka, M.; Sówka, I.; Nęcki, J.; Strąpoć, D. Temporal dynamics and controlling factors of CO2 and CH4 variability in the urban atmosphere of Wroclaw, Poland. *Sci. Total Environ.* **2023**, *893*, 164771. [CrossRef]
36. Ha, S.; Jeong, B.; Jang, H.; Park, C.; Ku, B. A framework for determining the life cycle GHG emissions of fossil marine fuels in countries reliant on imported energy through maritime transportation: A case study of South Korea. *Sci. Total Environ.* **2023**, *897*, 165366. [CrossRef] [PubMed]
37. Bilgili, L.; Ölçer, A.I. IMO 2023 strategy-Where are we and what's next? *Mar. Policy* **2024**, *160*, 105953. [CrossRef]
38. Animah, I.; Shafiee, M. Application of risk analysis in the liquefied natural gas (LNG) sector: An overview. *J. Loss Prev. Process Ind.* **2020**, *63*, 103980. [CrossRef]
39. Kaiser, S.; Siems, F.; Mostert, C.; Bringezu, S. Environmental and Economic Performance of CO_2-Based Methanol Production Using Long-Distance Transport for H_2 in Combination with CO_2 Point Sources: A Case Study for Germany. *Energies* **2022**, *15*, 2507. [CrossRef]
40. Biswal, T.; Shadangi, K.P.; Sarangi, P.K.; Srivastava, R.K. Conversion of Carbon Dioxide to Methanol: A Comprehensive Review. *Chemosphere* **2022**, *298*, 134299. [CrossRef]
41. Meesattham, S.; Kim-Lohsoontorn, P. Low-Temperature Alcohol-Assisted Methanol Synthesis from CO_2 and H_2: The Effect of Alcohol Type. *Int. J. Hydrogen Energy* **2022**, *47*, 22691–22703. [CrossRef]
42. Vedachalam, S.; Baquerizo, N.; Dalai, A.K. Review on impacts of low sulfur regulations on marine fuels and compliance options. *Fuel* **2022**, *310*, 122243. [CrossRef]
43. Lindstad, E.; Lagemann, B.; Rialland, A.; Gamlem, G.M.; Valland, A. Reduction of maritime GHG emissions and the potential role of E-fuels. *Transp. Res. Part D Transp. Environ.* **2021**, *101*, 103075. [CrossRef]
44. Herdzik, J. Decarbonization of Marine Fuels—The Future of Shipping. *Energies* **2021**, *14*, 4311. [CrossRef]
45. Abbas, A.J.; Hassani, H.; Burby, M.; John, I.J. An investigation into the volumetric flow rate requirement of hydrogen transportation in existing natural gas pipelines and its safety implications. *Gases* **2021**, *1*, 156–179. [CrossRef]
46. Thawko, A.; Eyal, A.; Tartakovsky, L. Experimental comparison of performance and emissions of a direct-injection engine fed with alternative gaseous fuels. *Energy Convers. Manag.* **2022**, *251*, 114988. [CrossRef]
47. Ahmadi, P.; Torabi, S.H.; Afsaneh, H.; Sadegheih, Y.; Ganjehsarabi, H.; Ashjaee, M. The effects of driving patterns and PEM fuel cell degradation on the lifecycle assessment of hydrogen fuel cell vehicles. *Int. J. Hydrogen Energy* **2020**, *45*, 3595–3608. [CrossRef]
48. Machaj, K.; Kupecki, J.; Malecha, Z.; Morawski, A.W.; Skrzypkiewicz, M.; Stanclik, M.; Chorowski, M. Ammonia as a potential marine fuel: A review. *Energy Strategy Rev.* **2022**, *44*, 100926. [CrossRef]
49. Huang, J.; Fan, H.; Xu, X.; Liu, Z. Life cycle greenhouse gas emission assessment for using alternative marine fuels: A very large crude carrier (VLCC) case study. *J. MArine Sci. Eng.* **2022**, *10*, 1969. [CrossRef]
50. Drazdauskas, M.; Lebedevas, S. Optimization of Combustion Cycle Energy Efficiency and Exhaust Gas Emissions of Marine Dual-Fuel Engine by Intensifying Ammonia Injection. *J. Mar. Sci. Eng.* **2024**, *12*, 309. [CrossRef]

51. Wang, H.; Daoutidis, P.; Zhang, Q. Ammonia-based green corridors for sustainable maritime transportation. *Digit. Chem. Eng.* **2023**, *6*, 100082. [CrossRef]
52. Chavando, A.; Silva, V.; Cardoso, J.; Eusebio, D. Advancements and challenges of ammonia as a sustainable fuel for the maritime industry. *Energies* **2024**, *17*, 3183. [CrossRef]
53. Stathatou, P.M.; Bergeron, S.; Fee, C.; Jeffrey, P.; Triantafyllou, M.; Gershenfeld, N. Towards decarbonization of shipping: Direct emissions & life cycle impacts from a biofuel trial aboard an ocean-going dry bulk vessel. *Sustain. Energy Fuels* **2022**, *6*, 1687–1697. [CrossRef]
54. Khan, M.I.; Gutiérrez-Alvarez, R.; Asfand, F.; Bicer, Y.; Sgouridis, S.; Al-Ghamdi, S.G.; Jouhara, H.; Asif, M.; Kurniawan, T.A.; Abid, M.; et al. The economics of concentrating solar power (CSP): Assessing cost competitiveness and deployment potential. *Renew. Sustain. Energy Rev.* **2024**, *200*, 114551. [CrossRef]
55. International Maritime Organization. *Fourth IMO GHG Study 2020: Executive summary*; International Maritime Organization: London, UK, 2020.
56. Peng, Y.; Li, X.; Wang, W.; Wei, Z.; Bing, X.; Song, X. A method for determining the allocation strategy of on-shore power supply from a green container terminal perspective. *Ocean. Coast. Manag.* **2019**, *167*, 158–175. [CrossRef]
57. Sagin, S.V.; Sagin, S.S.; Madey, V. Analysis of methods of managing the environmental safety of the navigation passage of ships of maritime transport. *Technol. Audit Prod. Reserves* **2023**, *4*, 33–42. [CrossRef]
58. Bilgili, L. Life cycle comparison of marine fuels for IMO 2020 Sulphur Cap. *Sci. Total Environ.* **2021**, *774*, 145719. [CrossRef] [PubMed]
59. Çelikaslan, Z.; Kılıç, A. Safety Precautions for The Use of LNG as Marine Fuel. *J. Marit. Transp. Logist.* **2023**, *4*, 11–22. [CrossRef]
60. Kuittinen, N.; Koponen, P.; Vesala, H.; Lehtoranta, K. Methane slip and other emissions from newbuild LNG engine under real-world operation of a state-of-the art cruise ship. *Atmos. Environ. X* **2024**, *23*, 100285. [CrossRef]
61. Poberezhnyi, R.V.; Sagin, S.V. Ensuring environmental performance of diesel engines of river and sea transport vessels. *Ship Power Plants: Sci. Tech. Collect.* **2020**, *41*, 5–9. [CrossRef]
62. Lehtoranta, K.; Kuittinen, N.; Vesala, H.; Koponen, P. Methane Emissions from a State-of-the-Art LNG-Powered Vessel. *Atmosphere* **2023**, *14*, 825. [CrossRef]
63. Zhang, J.; Meerman, H.; Benders, R.; Faaij, A. Comprehensive review of current natural gas liquefaction processes on technical and economic performance. *Appl. Therm. Eng.* **2020**, *166*, 114736. [CrossRef]
64. Gao, L.; Wang, J.; Binama, M.; Li, Q.; Cai, W. The design and optimization of natural gas liquefaction processes: A review. *Energies* **2022**, *15*, 7895. [CrossRef]
65. Xu, J.; Lin, W.; Chen, X.; Zhang, H. Review of unconventional natural gas liquefaction processes. *Front. Energy Res.* **2020**, *10*, 915893. [CrossRef]
66. Yuan, X.; Zhang, B.; Liang, R.; Wang, R.; Sun, Y. Environmental impact of the natural gas liquefaction process: An example from China. *Appl. Sci.* **2020**, *10*, 1701. [CrossRef]
67. Junsang, P.; Jaturanonda, C.; Wuttipornpun, T.; Watcharejyothin, M. Liquefied natural gas logistics management through optimal road-rail intermodal logistics planning considering community safety: A case study in Thailand. *Int. J. Knowl. Syst. Sci. (IJKSS)* **2023**, *14*, 1–25. [CrossRef]
68. Milioulis, K.; Bolbot, V.; Theotokatos, G. Model-based safety analysis and design enhancement of a marine LNG fuel feeding system. *J. Mar. Sci. Eng.* **2021**, *9*, 69. [CrossRef]
69. Kim, S.K.; Park, S.I.; Paik, J.K. Collision-accidental limit states-based safety studies for a LNG-fuelled containership. *Ocean Eng.* **2022**, *257*, 111571. [CrossRef]
70. Goolak, S.; Petrychenko, O.; Burmaka, I.; Vynohradova, A. The use of LNG as an alternative marine fuel: Conceptual requirements and safety measures. In Proceedings of the 28th International Scientific Conference Transport Means, Kaunas, Lithuania, 2–4 October 2024; pp. 965–970. [CrossRef]
71. Parris, D.; Spinthiropoulos, K.; Ragazou, K.; Giovou, A.; Tsanaktsidis, C. Methanol, a Plugin Marine Fuel for Green House Gas Reduction—A Review. *Energies* **2024**, *17*, 605. [CrossRef]
72. Arnaiz del Pozo, C.; Cloete, S.; Jiménez Álvaro, Á. Techno-Economic Assessment of Long-Term Methanol Production from Natural Gas and Renewables. *Energy Convers. Manag.* **2022**, *266*, 115785. [CrossRef]
73. Guil-López, R.; Mota, N.; Llorente, J.; Millán, E.; Pawelec, B.; Fierro, J.L.G.; Navarro, R.M. Methanol Synthesis from CO_2: A Review of the Latest Developments in Heterogeneous Catalysis. *Materials* **2019**, *12*, 3902. [CrossRef]
74. Simon Araya, S.; Liso, V.; Cui, X.; Li, N.; Zhu, J.; Sahlin, S.L.; Jensen, S.H.; Nielsen, M.P.; Kar, S.K. A Review of The Methanol Economy: The Fuel Cell Route. *Energy* **2020**, *13*, 596. [CrossRef]
75. Zincir, B.; Deniz, C.; Tunér, M. Investigation of Environmental, Operational and Economic Performance of Methanol Partially Premixed Combustion at Slow Speed Operation of a Marine Engine. *J. Clean. Prod.* **2019**, *235*, 1006–1019. [CrossRef]
76. Ammar, N.R. An Environmental and Economic Analysis of Methanol Fuel for a Cellular Container Ship. *Transp. Res. D Transp. Environ.* **2019**, *69*, 66–76. [CrossRef]

77. Tadros, M.; Ventura, M.; Soares, C.G. Review of Current Regulations, Available Technologies, and Future Trends in the Green Shipping Industry. *Ocean Eng.* **2023**, *280*, 114670. [CrossRef]
78. Tian, Z.; Wang, Y.; Zhen, X.; Liu, Z. The Effect of Methanol Production and Application in Internal Combustion Engines on Emissions in the Context of Carbon Neutrality: A Review. *Fuel* **2022**, *320*, 123902. [CrossRef]
79. Xing, H.; Stuart, C.; Spence, S.; Chen, H. Alternative Fuel Options for Low Carbon Maritime Transportation: Pathways to 2050. *J. Clean. Prod.* **2021**, *297*, 126651. [CrossRef]
80. Musyoka, S.K.; Khalil, A.S.; Ookawara, S.A.; Elwardany, A.E. Effect of C4 alcohol and ester as fuel additives on diesel engine operating characteristics. *Fuel* **2023**, *341*, 127656. [CrossRef]
81. Svanberg, M.; Ellis, J.; Lundgren, J.; Landalv, I. Renewable Methanol as a Fuel for the Shipping Industry. *Renew. Sustain. Energy Rev.* **2018**, *94*, 1217–1228. [CrossRef]
82. Ellis, J.; Tanneberger, K. Study on the use of ethyl and methyl alcohol as alternative fuels in shipping. *Eur. Marit. Saf. Agency* **2015**, *46*, 1–38.
83. Nikolaidis, P.; Poullikkas, A. A comparative overview of hydrogen production processes. *Renew. Sustain. Energy Rev.* **2017**, *67*, 597–611. [CrossRef]
84. Cheng, W.; Cheng, Y.F. A techno-economic study of the strategy for hydrogen transport by pipelines in Canada. *J. Pipeline Sci. Eng.* **2023**, *3*, 100112. [CrossRef]
85. Zhang, C.; Shao, Y.; Shen, W.; Li, H.; Nan, Z.; Dong, M.; Bian, J.; Cao, X. Key technologies of pure hydrogen and hydrogen-mixed natural gas pipeline transportation. *Acs Omega* **2023**, *8*, 19212–19222. [CrossRef]
86. Teoh, Y.H.; How, H.G.; Le, T.D.; Nguyen, H.T.; Loo, D.L.; Rashid, T.; Sher, F. A review on production and implementation of hydrogen as a green fuel in internal combustion engines. *Fuel* **2023**, *333*, 126525. [CrossRef]
87. Li, J.; Zhang, L.; Li, R.; Yang, X.; Zhang, T. High-pressure gaseous hydrogen storage vessels: Current status and prospects. *Energy Storage Sci. Technol.* **2021**, *10*, 1835. [CrossRef]
88. Mallouppas, G.; Yfantis, E.A. Decarbonization in shipping industry: A review of research, technology development, and innovation proposals. *J. Mar. Sci. Eng.* **2021**, *9*, 415. [CrossRef]
89. *ISO 8178-7:2015 Standard;* Reciprocating internal combustion engines — Exhaust emission measurement — Part 7: Engine family determination. International Organization for Standardization (ISO): Geneva, Switzerland, 2015.
90. Bunse, M.J. Sustainable Supply Chains for Maritme Biofuels 2021. Available online: https://platformduurzamebiobrandstoffen.nl/wp-content/uploads/2021/04/2021_PDB_Bunse_Sustainable-supply-chains-for-maritime-biofuels.pdf (accessed on 10 November 2024).
91. Wang, Y.; Wright, L.A. A comparative review of alternative fuels for the maritime sector: Economic, technology, and policy challenges for clean energy implementation. *World* **2021**, *2*, 456–481. [CrossRef]
92. Wang, Z.; Hu, H.; Zhen, L. Berth and quay cranes allocation problem with on-shore power supply assignment in container terminals. *Comput. Ind. Eng.* **2024**, *188*, 109910. [CrossRef]
93. Elkafas, A.G.; Rivarolo, M.; Gadducci, E.; Magistri, L.; Massardo, A.F. Fuel cell systems for maritime: A review of research development, commercial products, applications, and perspectives. *Processes* **2022**, *11*, 97. [CrossRef]
94. Kryshtopa, S.; Górski, K.; Longwic, R.; Smigins, R.; Kryshtopa, L. Increasing parameters of diesel engines by their transformation for methanol conversion products. *Energies* **2021**, *14*, 1710. [CrossRef]
95. Gadducci, E.; Lamberti, T.; Rivarolo, M.; Magistri, L. Experimental campaign and assessment of a complete 240-kW Proton Exchange Membrane Fuel Cell power system for maritime applications. *Int. J. Hydrogen Energy* **2022**, *47*, 22545–22558. [CrossRef]
96. Halder, P.; Babaie, M.; Salek, F.; Shah, K.; Stevanovic, S.; Bodisco, T.A.; Zare, A. Performance, emissions and economic analyses of hydrogen fuel cell vehicles. *Renew. Sustain. Energy Rev.* **2024**, *199*, 114543. [CrossRef]
97. Khurshid, H.; Mohammed, B.S.; Al-Yacoubya, A.M.; Liew, M.S.; Zawawi, N.A.W.A. Analysis of hybrid offshore renewable energy sources for power generation: A literature review of hybrid solar, wind, and waves energy systems. *Dev. Built Environ.* **2024**, *19*, 100497. [CrossRef]
98. Manneh, O.; Islam, M.S.; Danso, Y. Application of Renewable Energy Technologies in Marine Vessels; State of the Art Review. Ph.D. Dissertation, Departmental of Technical and Vocational Education (TVE), Islamic University of Technology (IUT), Board Bazar, Gazipur, Bangladesh, 2022. Available online: http://103.82.172.44:8080/xmlui/handle/123456789/1815 (accessed on 12 December 2024).
99. Adham, M.I. Analysis and Optimal Planning of Nuclear-Renewable Hybrid Energy Systems for Ships. Ph.D. Dissertation, University of Ontario Institute of Technology, Oshawa, ON, Canada, 2021. Available online: https://library-archives.canada.ca/eng/services/services-libraries/theses/Pages/item.aspx?idNumber=1344012581 (accessed on 12 December 2024).
100. Sharma, K.; Syal, P. A Review on Solar Powered Boat Design. *Int. Res. J. Adv. Sci. Hub* **2021**, *3*, 1–10. [CrossRef]
101. Wang, G.; Zhang, Z.; Lin, J. Multi-energy complementary power systems based on solar energy: A review. *Renew. Sustain. Energy Rev.* **2024**, *188*, 114464. [CrossRef]

102. Nyanya, M.N.; Vu, H.B.; Schönborn, A.; Ölçer, A.I. Wind and solar assisted ship propulsion optimisation and its application to a bulk carrier. *Sustain. Energy Technol. Assess.* **2021**, *47*, 101397. [CrossRef]
103. Bonthagorla, P.K.; Mikkili, S. Optimal PV array configuration for extracting maximum power under partial shading conditions by mitigating mismatching power losses. *CSEE J. Power Energy Syst.* **2020**, *8*, 499–510. [CrossRef]
104. Minak, G. Solar Energy-Powered Boats: State of the Art and Perspectives. *J. Mar. Sci. Eng.* **2023**, *11*, 1519. [CrossRef]
105. Tercan, Ş.H.; Eid, B.; Heidenreich, M.; Kogler, K.; Akyürek, Ö. Financial and technical analyses of solar boats as a means of sustainable transportation. *Sustain. Prod. Consum.* **2021**, *25*, 404–412. [CrossRef]
106. Murshed, N.; Tushar, M.S.K.; Chowdhury, S. Power Performance Analysis of PV Module with DC to DC Buck Converter. *Adv. J. Grad. Res.* **2020**, *8*, 27–39. [CrossRef]
107. Waseem, M.; Ahmad, M.; Parveen, A.; Suhaib, M. Battery technologies and functionality of battery management system for EVs: Current status, key challenges, and future prospectives. *J. Power Sources* **2023**, *580*, 233349. [CrossRef]
108. Lawan, M.; Aboushady, A.; Ahmed, K.H. Photovoltaic MPPT techniques comparative review. In Proceedings of the 2020 9th International Conference on Renewable Energy Research and Application (ICRERA), Glasgow, UK, 27–30 September 2020; IEEE: New York, NY, USA, 2020; pp. 344–351. [CrossRef]
109. Prastiko, D.H.; Supardi, A. Pengendali dan Monitoring Kecepatan Putar Motor Induksi 3 Fasa Berbasis Programmable Logic Controller dan Expansion dengan Human Machine Interface. *Emit. J. Tek. Elektro* **2022**, *22*, 168–176. [CrossRef]
110. Sharma, S.K.; Sharma, G.; Gaur, A.; Arya, A.; Mirsafi, F.S.; Abolhassani, R.; Rubahn, H.-G.; Yu, J.-S.; Mishra, Y.K. Progress in electrode and electrolyte materials: Path to all-solid-state Li-ion batteries. *Energy Adv.* **2022**, *1*, 457–510. [CrossRef]
111. Benlahbib, B.; Dahbi, A.; Fares, B.; Lakhdari, A.; Bouarroudj, N.; Mekhilef, S.; Abdelkrim, T. An experimental investigation design of a bidirectional DC-DC buck-boost converter for PV battery charger system. *Int. J. Power Electron. Drive Syst. (IJPEDS)* **2023**, *14*, 2362–2371. [CrossRef]
112. Andriani, D.; Bicer, Y. A review of hydrogen production from onboard ammonia decomposition: Maritime applications of concentrated solar energy and boil-off gas recovery. *Fuel* **2023**, *352*, 128900. [CrossRef]
113. Benghanem, M.; Haddad, S.; Alzahrani, A.; Mellit, A.; Almohamadi, H.; Khushaim, M.; Aida, M.S. Evaluation of the performance of polycrystalline and monocrystalline PV technologies in a hot and arid region: An experimental analysis. *Sustainability* **2023**, *15*, 14831. [CrossRef]
114. Zubeer, S.A.; Ali, O.M. Performance analysis and electrical production of photovoltaic modules using active cooling system and reflectors. *Ain Shams Eng. J.* **2021**, *12*, 2009–2016. [CrossRef]
115. Emmanuel, B.; Yuan, Y.; Gaudence, N.; Zhou, J. A review on the influence of the components on the performance of PVT modules. *Sol. Energy* **2021**, *226*, 365–388. [CrossRef]
116. Rasul, M.J.; Kim, J. Comprehensive review and comparison on battery technologies as electric-powered source in marine applications. *J. Energy Storage* **2024**, *88*, 111509. [CrossRef]
117. Dotto, A.; Satta, F.; Campora, U. Energy, environmental and economic investigations of cruise ships powered by alternative fuels. *Energy Convers. Manag.* **2023**, *285*, 117011. [CrossRef]
118. Greenp Pingkuo, L.; Jiahao, W. Study on the diffusion of CCUS technology under carbon trading mechanism: Based on the perspective of tripartite evolutionary game among thermal power enterprises, government and public. *J. Clean. Prod.* **2024**, *438*, 140730. [CrossRef]
119. Chettri, M.; Kundu, S.; Diyali, N.; Biswas, B. Using Solar Energy in Methanol Production: Efficiency, Environmental Impact and Economical Performance. *Ref. Modul. Chem. Mol. Sci. Chem. Eng.* **2024**. [CrossRef]
120. Melnyk, O.; Onyshchenko, S.; Onishchenko, O.; Lohinov, O.; Ocheretna, V. Integral Approach to Vulnerability Assessment of Ship's Critical Equipment and Systems. *Trans. Marit. Sci.* **2023**, *12*. [CrossRef]
121. Abraham, E.J.; Linke, P.; Al-Rawashdeh, M.; Rousseau, J.; Burton, G.; Al-Mohannadi, D.M. Large-scale shipping of low-carbon fuels and carbon dioxide towards decarbonized energy systems: Perspectives and challenges. *Int. J. Hydrogen Energy* **2024**, *63*, 217–230. [CrossRef]
122. Bondarenko, I.; Lukoševičius, V.; Neduzha, L. Novel 'Closed'-System Approach for Monitoring the Technical Condition of Railway Tracks. *Sustainability* **2024**, *16*, 3180. [CrossRef]
123. Harahap, F.; Nurdiawati, A.; Conti, D.; Leduc, S.; Urban, F. Renewable marine fuel production for decarbonised maritime shipping: Pathways, policy measures and transition dynamics. *J. Clean. Prod.* **2023**, *415*, 137906. [CrossRef]
124. Ampah, J.D.; Yusuf, A.A.; Afrane, S.; Jin, C.; Liu, H. Reviewing two decades of cleaner alternative marine fuels: Towards IMO's decarbonization of the maritime transport sector. *J. Clean. Prod.* **2021**, *320*, 128871. [CrossRef]
125. Daniel, H.; Trovão, J.P.F.; Williams, D. Shore power as a first step toward shipping decarbonization and related policy impact on a dry bulk cargo carrier. *eTransportation* **2022**, *11*, 100150. [CrossRef]
126. Kalikatzarakis, M.; Geertsma, R.D.; Boonen, E.J.; Visser, K.; Negenborn, R.R. Ship energy management for hybrid propulsion and power supply with shore charging. *Control Eng. Pract.* **2018**, *76*, 133–154. [CrossRef]

127. Yu, J.; Voß, S.; Tang, G. Strategy development for retrofitting ships for implementing shore side electricity. *Transp. Res. Part D Transp. Environ.* **2019**, *74*, 201–213. [CrossRef]
128. Sciberras, E.A.; Zahawi, B.; Atkinson, D.J. Reducing shipboard emissions—Assessment of the role of electrical technologies. *Transp. Res. Part D Transp. Environ.* **2017**, *51*, 227–239. [CrossRef]
129. Riabov, I.; Goolak, S.; Kondratieva, L.; Overianova, L. Increasing the energy efficiency of the multi-motor traction electric drive of an electric locomotive for railway quarry transport. *Eng. Sci. Technol. Int. J.* **2023**, *42*, 101416. [CrossRef]
130. Riabov, I.; Goolak, S.; Neduzha, L. An Estimation of the Energy Savings of a Mainline Diesel Locomotive Equipped with an Energy Storage Device. *Vehicles* **2024**, *6*, 611–631. [CrossRef]
131. Goolak, S.; Liubarskyi, B.; Riabov, I.; Lukoševičius, V.; Keršys, A.; Kilikevičius, S. Analysis of the Efficiency of Traction Drive Control Systems of Electric Locomotives with Asynchronous Traction Motors. *Energies* **2023**, *16*, 3689. [CrossRef]
132. Goolak, S.; Liubarskyi, B.; Riabov, I.; Chepurna, N.; Pohosov, O. Simulation of a direct torque control system in the presence of winding asymmetry in induction motor. *Eng. Res. Express* **2023**, *5*, 025070–025086. [CrossRef]
133. Goolak, S. Methodological Guidelines for Choosing a Method of Compensation of Higher Harmonic Current Components in Traction Drive Systems of Electric Rolling Stock. In Proceedings of the proceedings of the 27th International Scientific Conference Transport Means, Palanga, Lithuania, 4–6 October 2023; pp. 685–690. [CrossRef]
134. Goolak, S. Improving the Efficiency of Electric Rolling Stock Operation Through the Use of Adaptive Filtering Methods for High Harmonic Current Components in Traction Drive Systems. *Probl. Energeticii Reg.* **2023**, *4*, 1–13. [CrossRef]
135. Goolak, S. To the Choice of Filter Approximation in the Synthesis of the Reactive Power Compensator Control System. In Proceedings of the 28th International Scientific Conference Transport Means, Kaunas, Lithuania, 2–4 October 2024; pp. 192–197. [CrossRef]

sustainability

MDPI

Article

Sustainability of the Trans-European Transport Networks Land Infrastructure to Address Large-Scale Disasters: A Case Study in the Czech Republic

David Rehak [1,*], Martin Vlkovsky [2], Pavel Manas [3], Jiri Apeltauer [4], Tomas Apeltauer [4] and Martin Hromada [5]

1 Faculty of Safety Engineering, VSB—Technical University of Ostrava, 700 30 Ostrava, Czech Republic
2 Faculty of Military Leadership, University of Defence, 662 10 Brno, Czech Republic
3 Faculty of Military Technology, University of Defence, 662 10 Brno, Czech Republic
4 Faculty of Civil Engineering, Brno University of Technology, 602 00 Brno, Czech Republic
5 Faculty of Applied Informatics, Tomas Bata University in Zlín, 760 05 Zlín, Czech Republic
* Correspondence: david.rehak@vsb.cz; Tel.: +420-597-322-816

Abstract: In recent years, the concepts of sustainability and sustainable development of transport infrastructure have been examined mainly from an economic, ecological, and social perspective. However, an essential aspect of the sustainability of the Trans-European Transport Networks (TEN-T) land infrastructure is also the construction and technical parameters of critical infrastructure objects. For this reason, the aim of the article is to define recommendations for the sustainability and future development of the TEN-T land infrastructure. For this purpose, the article presents the procedure for defining and modelling typical transport situations using the TEN-T land infrastructure and the procedure for identifying critical infrastructure objects of the TEN-T land infrastructure. Attention is also paid to the digitalization of critical infrastructure objects, which categorizes infrastructure according to its suitability for excessive and oversized transport using map data. The main output of the article are recommendations for the sustainability and future development of the TEN-T land infrastructure. These recommendations are based on the systematic development of TEN-T in four basic layers, namely legislative, capatitive, construction-technical, and technological. These recommendations are intended primarily for the Ministry of Defence of the Czech Republic and the Ministry of Transport of the Czech Republic, but also for transport infrastructure operators, i.e., Road and Motorway Directorate and Railway Infrastructure Administration.

Keywords: TEN-T; land transport infrastructure; sustainability; large-scale disasters; Czech Republic

1. Introduction

Transport is a significant sector of European Critical Infrastructure, which is listed together with ten other sectors in the Directive (EU) on the resilience of critical entities [1]. This Directive points out the importance of predictable transportation, which can be perceived both at the construction-technical and capacity levels. The most important part of the land transport infrastructure can be considered the Trans-European Transport Networks (TEN-T). In the current situation, the essence of this transport network is to connect EU regions, support the growth of the internal market and employment, and strengthen social, economic and territorial cohesion [2]. However, TEN-T can also be used to address large-scale disasters, and it is, therefore, necessary to consider the duality of its use, as stated in the EU's Strategic Compass for Security and Defence [3].

In recent years, European territory has been increasingly exposed to threats resulting in large-scale disasters [4]. Examples include the extensive floods in the Czech Republic (2002), the earthquakes in Italy (2009) and Greece (2020), the COVID-19 pandemic (2020–2023), the tornado in the Czech Republic (2021), and the war in Ukraine (2022–present). At the same time, it should be noted that in connection with climate change, a further increase in natural disasters is predicted in the coming ten years [5].

In the fight against large-scale disasters, measures are implemented on both the prevention and response sides. While the prevention of these disasters is based primarily on risk management [6,7], crisis management information systems [8] and disaster management logistics [9] play a significant role in mitigating these disasters. Disaster logistics can be seen as a support function that ensures the timely delivery of emergency supplies and rescue services to disaster-affected areas [10].

The transport of these emergency supplies, especially in an international context, can be carried out in four basic ways: by air, by waterway, by rail, or by road. In all cases, the Trans-European Transport Networks (TEN-T) are useful [11], which include infrastructures enabling fast and efficient international transport. However, in the case of transport of excessive and oversized goods, it is advisable to use the TEN-T land infrastructure as a priority [12].

From the above, it is clear that an essential aspect of the long-term functionality of the TEN-T land infrastructure is its sustainability. Sustainable transport is one of the global goals of sustainable development, which is important not only for economic growth, but also for ensuring accessibility [13]. The sustainability of accessible transport infrastructure is an important factor not only in everyday life, e.g., for the healthcare system [14] or the economy [15], but also in the event of global crises [16].

The new Regulation (EU) of the European Parliament and of the Council emphasizes that the European transport corridors should facilitate the development of the TEN-T infrastructure, which focuses on addressing bottlenecks, improving cross-border connections and increasing efficiency and sustainability [2]. In this context, it can be stated that this sustainability should be based on the continuous development of transport infrastructure, considering the specifics of dual use [3]. The dual or multiple use of the TEN-T land infrastructure (i.e., by the civilian sector, components of the integrated rescue system and armies—both national and NATO) has a fundamental impact on the capacity utilization of the transport infrastructure in question.

In recent years, the concepts of sustainability and sustainable development of transport infrastructure have been examined mainly from an economic [17–19], environmental [20–22], and social [23–25] perspective. However, in the context of the long-term usability of the TEN-T land infrastructure, it is also necessary to pay attention to the construction-technical aspects of sustainability [26–28]. Based on these facts, the aim of the article is to define recommendations for the future development of the TEN-T land infrastructure. These recommendations will be based on the results of (1) defining and modelling the typical transport situations using the TEN-T land infrastructure and (2) defining critical infrastructure objects of the TEN-T land infrastructure.

In the context of the above-mentioned objective, the article is structured into two comprehensive sections, i.e., Materials and Methods, and Results. The first section provides a description of the TEN-T land infrastructure in the Czech Republic, a description of large-scale disasters resulting in above-standard use of the TEN-T land infrastructure, and a description of the methods used for modelling and digitizing routes for excessive and oversized transport. The second section presents the key results of the article. Within this section, typical transport situations using the TEN-T land infrastructure are defined and modelled, a procedure for identifying critical infrastructure objects of the TEN-T land infrastructure is created, a procedure for digitizing critical infrastructure objects of the TEN-T land infrastructure is presented, and recommendations for the sustainability and future development of the TEN-T land infrastructure in the context of military mobility are defined.

2. Materials and Methods

The purpose of this part of the article is to familiarize the reader with the materials and methods used. For this purpose, in relation to the thematic focus of the article, the TEN-T land infrastructure, large-scale disasters resulting in above-standard use of the TEN-T land infrastructure, and tools for transport modelling and identification of critical infrastructure objects are presented.

2.1. TEN-T Land Infrastructure in the Czech Republic

The establishment and development of Trans-European Networks was enshrined in the Maastricht Treaty in 1992 [29]. Support for this development concerned networks for (1) energy and telecommunications, and (2) transport. The Trans-European Transport Networks (TEN-T) are networks of road and rail corridors, international airports and waterways created to improve the quality and connectivity of transport infrastructure in the European Union [30]. This transport network is defined as two-layer and consists of comprehensive and core networks [31]. The comprehensive network ensures multimodal connectivity of all European regions, while the core network represents a subset of the comprehensive network and contains the most important trans-European routes.

The road and rail corridors are collectively referred to as the TEN-T land infrastructure and, according to the European Directive [1], are referred to as critical infrastructure elements. The infrastructure elements (objects) of this land infrastructure can be classified into three groups according to their topological structure [32,33]:

- Line elements—ensure the transfer, delivery, or transport of goods between two physically separate locations (e.g., road and rail corridors). They are a meaningful basic group that is in relation to all point and area elements (the length dimension is an order of magnitude greater than the width dimension).
- Point elements—represent a concentrated and closed unit located on a small area, which fulfils its function for the needs of specific line objects (e.g., bridges, tunnels), (length and width are negligible compared to line and area elements).
- Area elements—represent one unit or integral objects, where two or more point elements or two or more key technologies occur (e.g., significant transport hubs).

The Czech Republic is an important region in terms of the sustainability of European road and rail transport, as it plays the role of a transit country [34]. For this purpose, linear elements of the TEN-T land infrastructure are used, i.e., road and rail corridors. A description of these corridors is given in the following text.

The TEN-T road transport network currently includes motorways (see Figure 1) and some sections of class I roads in the Czech Republic. Other sections are currently under construction.

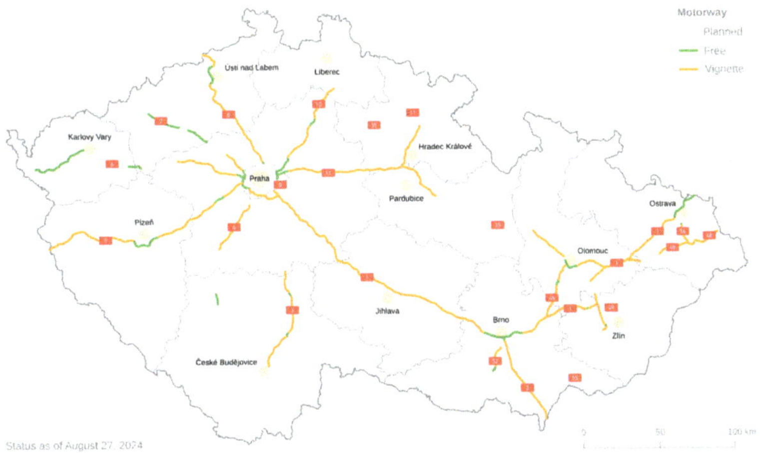

Figure 1. Motorways in the Czech Republic [35].

There are 17 motorways built in the Czech Republic with a total length of 1394 km (as of 1 July 2024). Of these, only D1, D2, D3, D5, D8, and D11 were originally built as motorways and are classified as class I motorways. The rest of the motorways were originally expressways, which were converted to class II motorways in 2016. Class I roads, which are part of the TEN-T network, total 91,028 km [36].

These TEN-T road corridors also include point and area elements. The point elements include tunnels and bridges. Specifically, there are 22 tunnels and 1871 bridges on motorways. The area elements include selected level crossings.

The TEN-T railway transport network currently includes four transit corridors in the Czech Republic (see Figure 2). A further 839 km of high-speed lines should be in operation by 2050 [37].

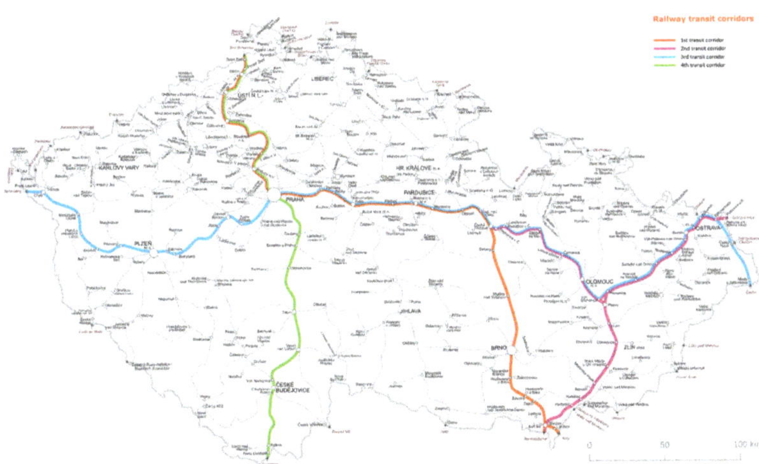

Figure 2. Railway transit corridors in the Czech Republic [38].

All four rail transit corridors have a total length of 1286 km [39]. The first corridor runs from Germany to Austria via Děčín—Prague—Česká Třebová—Brno—Břeclav. The total length of this corridor is 458 km, of which 379 km are inside the corridor and 79 km are outside the corridor. The second corridor runs from Austria to

Poland via Břeclav—Přerov—Ostrava. The Přerov—Česká Třebová section also belongs to this corridor. The total length of this corridor is 313 km, of which 7 km are within the first corridor. The third corridor runs from Germany to Slovakia via Cheb/Domažlice—Plzeň—Prague—Česká Třebová—Ostrava—Bohumín/Havířov. The total length of this corridor is 665 km, of which 370 km are inside the first and second corridors. The fourth corridor runs from Germany to Austria via Děčín—Prague—Veselí nad Lužnicí—Horní Dvořiště/České Velenice. The total length of this corridor is 365 km, of which 145 km are inside the first and third corridors.

These TEN-T rail transits also include point and area elements. Point elements include tunnels and bridges. For example, on the first transit corridor there are 14 tunnels and 697 bridges [39]. Area elements include major railway junctions and terminals.

2.2. *Large-Scale Disasters Resulting in Above-Standard Use of TEN-T Land Infrastructure*

Large-scale disasters can be defined from two perspectives. In terms of impact, they are defined as disasters that result in 100 to 1000 people being adversely affected (e.g., through displacement, injury or death) or the effects of these disasters cover 10 to 100 km^2 [40]. Large-scale disasters with a death toll of 1000 people and an affected geographical area of 100 km^2 can also be referred to as enormous or gargantuan disasters [41]. In terms of their management, they are defined as "*types of disasters affecting a society which requires national or international assistance*" [4]. It is clear from both definitions that dealing with such disasters requires extensive logistical support and, in some cases, the use of TEN-T land infrastructure.

The basic classification of large-scale disasters is made into two categories, which are natural and man-made disasters [42]. A more detailed classification of natural disasters is based on the Peril Classification [43], which includes six basic groups of hazards, namely geophysical, hydrological, meteorological, climatological, biological, and extraterrestrial. Unlike natural disasters, which are the result of natural hazards, man-made disasters have an element of negligence or human intent [44]. Such disasters include industrial accidents, terrorist acts, or armed conflicts.

The following section of the article defines large-scale disasters resulting in the above-standard use of the TEN-T land infrastructure. In this context, above-standard use is considered to be the movement of excessive or oversized transport [45] or overloading of the TEN-T land infrastructure capacities. The selection procedure for these disasters was divided into three phases (see Figure 3), which consider key criteria (i.e., excessiveness and over-dimensioning of equipment and infrastructure capacity) determining the above-standard use of TEN-T land infrastructure in managing large-scale disasters.

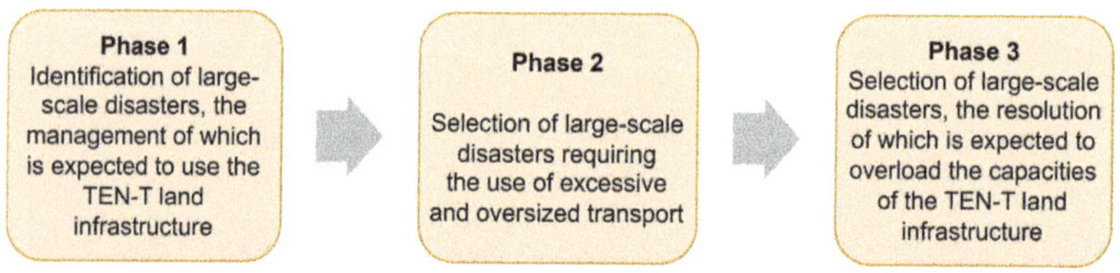

Figure 3. Selection procedure for large-scale disasters resulting in above-standard use of TEN-T land infrastructure.

The first phase of the process focused on identifying large-scale disasters, the management of which is expected to involve the use of TEN-T land infrastructure. This identification was based on the Hazard Analysis for the Czech Republic [46]. In this document, a total of 72 types of hazards were identified and divided into six groups according to their nature: abiotic, biotic, cosmic, technogenic, sociogenic, and economic.

Another important material for identifying large-scale disasters, the management of which is expected to involve the use of TEN-T land infrastructure, was the EM-DAT Classification Tree [47]. This document currently monitors 77 natural and technological disasters, including the number of their occurrences in the years 1900–2023. Based on the results of this analysis, a total of 10 large-scale disasters were identified: riverine flood, flash flood, extreme wind, landslide, earthquake, tornado, industrial accident, large-scale migration wave, large-scale lawlessness, and armed conflict.

In the second phase, a selection of large-scale disasters requiring the use of excessive or oversized transport was carried out [45]. Based on the results of this selection, 6 large-scale disasters were selected: riverine flood, earthquake, tornado, industrial accident, large-scale migration wave, and armed conflict.

In the third phase, a selection of large-scale disasters was carried out, the resolution of which is expected to overload the capacities of the TEN-T land infrastructure. On the basic of the results of this selection, only one large-scale disaster was selected, namely an armed conflict requiring extensive deployment of NATO armies. In such a case, above-standard use of the TEN-T land infrastructure would be expected.

2.3. *Description of the Methods Used for Modelling and Digitizing Routes for Excessive and Oversized Transport*

To identify critical infrastructure objects of the TEN-T land infrastructure, transport models are used—typical transport situations. The created typical transport situations on the road are designed as basic variants and, thus, represent the most demanding possible "use" of the road infrastructure objects in question. The same road trains (with the exception of low-loaders) are further used analogously for modelling within the framework of rail transport. Transport models serve to reveal bottlenecks from a construction-technical perspective, as well as from a capacity perspective, with both perspectives intertwining and influencing each other.

The road transport model considers 4 variants that copy the "most problematic" land military equipment. The technical parameters of the selected equipment exceed the parameters of civilian equipment (e.g., construction machinery), and they also demonstrate the possibilities of using the TEN-T land infrastructure not only by the Czech Army, but also by the armies of NATO member states. In the rail transport model, variants 2–4 are combined into one due to the possibility of loading the ground military equipment in question into one military train.

Transport models—typical transport situations are created under the following conditions. Firstly, it concerns the construction of separate transport flows (convoys) according to the applicable legislation and standards, where the standard size of a convoy ranging from 4 to 20 vehicles. The convoys include vehicles that represent the greatest limitation for the monitored TEN-T land infrastructure and require its above-standard use or usually combine excessive and oversized transport. The transport model is created according to the Construction Concept of the Army of the Czech Republic 2030 [48], which includes a commitment to NATO to allocate a company task force consisting of 250 people, which represents the basic building block of larger units, i.e., battalion and brigade.

The considered company task force is, among other things, expected to be separated under the auspices of the United Nations in 2023–2025. In terms of the ground military equipment transported, it is a tank company, including accompanying vehicles. Vehicle streams and complete convoys are assembled according to the requirements [49], including vehicles for protection and defence, workshop, and recovery vehicles. Rest on Movement (ROM) and Rest on Night (RON) also play an important role in the requirements for the related infrastructure. These areas must provide construction-technical and capacity-appropriate options for parking, resupply, maintenance and minor repairs, and rest. Last but not least, the models are limited to the military movements of the ground military equipment in question, not of personnel, which are (standardly) addressed separately and are not the subject of this article.

Modelling these typical transport situations requires a systematic multi-phased approach that combines the use of large datasets and specialized methodologies. The Road Databank of the Directorate of Roads and Highways [50] is a key tool for ensuring efficiency, which contains vector data with parameters such as bridge load capacity, underpass clearance, road geometry, and other relevant road infrastructure elements. This data source allows the creation of a basic throughput model that identifies critical infrastructure objects of the transport network that are insufficient for a given transport. The first step in the process is the creation of a basic throughput model and a global analysis of typical transports on a digital road network model (road database). At this stage, automated modelling of the TEN-T network throughput takes place according to the parameters of typical transports, for example, convoys of Leopard 2A7V tanks on DOLL S8E low-loaders or recovery vehicles on a P-50 N low-loader. The automation of this stage allows the rapid identification of locations such as bridges with insufficient load capacity or underpasses with limited height that do not correspond to the parameters of the transported vehicles.

If the basic GIS model identifies non-compliant elements, a detailed calculation of the passability of these critical points is carried out using the STANAG 2021 methodology [51]. This methodology was developed for military purposes and provides an accurate assessment of the load-bearing capacity of bridges by determining the so-called Military Load Classification (MLC). The calculation considers the specific characteristics of military equipment, such as weight distribution on the axles and dynamic effects during transport. It also considers the current technical condition of the bridge, or rather the element to be solved. This document [51], on the one hand, defines hypothetical military vehicles divided into 16 categories with their geometric and weight characteristics, and on the other hand, it assigns a load class to bridge structures according to the category of a hypothetical vehicle that could still cross the bridge under given conditions. In reality, each military vehicle within NATO is assigned with a certain MLC number given by a hypothetical wheeled or tracked vehicle category according to the table in Annex A of AEP-3.12.1.5 [52].

Each existing or newly designed bridge can be assigned with a certain MLC number representing the load capacity of that particular bridge. If a vehicle has an MLC number less than or equal to the MLC loading capacity of a bridge under given passing conditions, can go through. Those conditions are normal crossing (similar to normal civil transport conditions), crossing with caution and risk crossing (single vehicle on bridge span and other limitations). In some cases, according to STANAG 2021 [51], heavier vehicles can pass the bridge, but with less reliability compared to the civil load capacity given by Eurocodes or similar documents, which is significantly advantageous in a crisis situation.

Modelling of traffic flow according to STANAG 2021 includes static and dynamic analyses of bridge structures, using data on material properties and the current technical condition of the structure [53]. The results of this analysis allow either to confirm the possibility of using the element for a given transport, or to identify the need for measures such as temporary bridge reinforcement, route adaptation, or modification of transport equipment. Also, only a limited number of passages of specific type of transport can be determined.

This two-phase approach, which first uses the Road Databank data for a multi-criteria assessment and then applies the STANAG 2021 methodology to potentially non-compliant elements, ensures the effectiveness of military transport planning. It allows for the rapid identification and resolution of potential problems and the design of measures, thereby minimizing time and financial costs while increasing the safety and sustainability of military operations on the TEN-T network and other roads.

3. Results

The essence of this part of the article is to present the results of the article with the aim of supporting the sustainability of the TEN-T land infrastructure to address large-scale disasters. For this purpose, attention is first paid to the definition and modelling of typical transport situations using the TEN-T land infrastructure. Subsequently, the procedure for defining critical infrastructure objects of the TEN-T land infrastructure is presented. At the end of the article, recommendations for the future development of the TEN-T land infrastructure are defined.

3.1. Defining and Modelling Typical Transport Situations Using TEN-T Land Infrastructure

The analyses revealed that from the perspective of construction-technical parameters, the greatest challenge is military movements of both the Czech Army and NATO in the context of Host Nation Support [54]. Especially in the event of the need to deploy NATO forces and assets on the eastern flank of the alliance (e.g., in the context of the Russian–Ukrainian conflict), in connection with the New NATO Force Model, the TEN-T land infrastructure would be exposed to very above-standard use both in terms of capacity [55] and from a construction-technical point of view [56]. The aforementioned multiple use of the TEN-T land infrastructure (civilian transport, including components of the integrated rescue system and national armies, as well as NATO) also plays a crucial role, which increases the likelihood of cascading effects [57].

Typical transport situations are further divided into separate transport sets (variants) according to the type of "problematic" vehicle in the set. The most important, in terms of the impact on the TEN-T land infrastructure, is the transport set—variant 1, which contains Leopard 2A7V tanks on DOLL S8E low-loaders with a MAN SLT70 tractor. Leopard tanks, including the latest 2A8 variants to be delivered in the future, will place the greatest demands on the above-standard use of the TEN-T land infrastructure, because they represent both excessive and oversized transport [58]. Accompanying vehicles in other variants such as the KN-251 wheel loader, the SPOT-55 firefighting vehicle, and the T-815-7 VV 8 × 8 recovery vehicle on P-50 N low-loaders with a T-815 8 × 8 tractor represent a capacity problem due to the need to create complete convoys that slow down traffic.

The same ground military equipment, identified in the typical transport situations above, is loaded onto suitable railway wagons, thus, creating a railway transport model in two variants. Variant 1 is analogous to the road transport model, while variants 2–4 can be loaded onto one military train. Other vehicles/elements designated as part of the transport set are also added. The transport set in variant 1 contains Leopard 2A7V tanks on a Samms wagon (or Res54), where the related limitation is the availability of Samms railway freight wagons, which are available in limited numbers and only abroad (e.g., in Germany with Deutsche Bahn). The set also includes a mobile car workshop and a recovery tank as standard. Transport sets in variants 2–4 consist of a KN-251 wheel loader with a dismantled cabin, a SPOT-55 firefighting vehicle after dismantling the mounted streamliner, and a T-815-7 VV 8 × 8 recovery vehicle. The transport set also includes security equipment—CAS 30 T-815-7 6 × 6 medium tank trucks, T-815-7 S3 trucks, UDS 214 T-815-7 universal finishing machine, T-815-7 mobile car workshop 8 × 8, and a recovery vehicle. The above-mentioned ground military equipment also requires special railway freight cars; however, there are more options, which are also available in the Czech Republic, e.g., Res54 at ČD Cargo.

3.2. Identification of Critical Infrastructure Objects of the TEN-T Land Infrastructure

Based on the definition and modelling of typical transport situations, it is possible to identify critical objects of the TEN-T land infrastructure. This identification can be based on modelling the impacts of the critical infrastructure failure [59], using cascading impact assessment [57], or factors influencing the performance of critical land transport infrastructure elements [60].

The outputs from these tools provide relevant information, but do not reflect the criticality of infrastructure objects in the context of their use for excessive and oversized transport. For this reason, a specific procedure was created, the output of which is an assessment of the suitability of infrastructure objects for these types of transport. The proposed identification procedure includes three phases (see Figure 4), which determine activities for assessing data that are the necessary basis for comparing the technical parameters of the convoy and the construction and technical parameters of the infrastructure. Within each phase, the inputs and outputs necessary for the process and the methods that are appropriate to use within a given phase are defined.

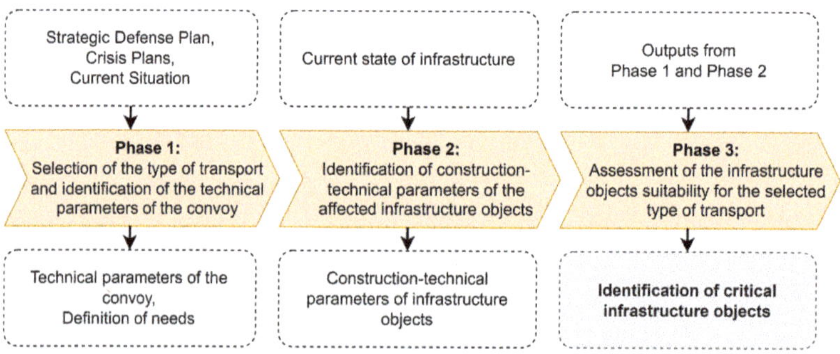

Figure 4. Procedure for identifying critical infrastructure objects of the TEN-T land infrastructure.

3.2.1. Phase 1: Selection of the Type of Transport and Identification of the Technical Parameters of the Convoy

The starting point of this phase is the definition of the assessor's needs, which result from defence and crisis plans. Based on these needs, it is first necessary to select the type of transport on the basis of which a specific convoy variant will be defined. For this convoy variant, the main technical parameters are then identified, which means the maximum passing widths and heights of all vehicles and vehicle combinations in the convoy and also the classification of all vehicles and vehicle combinations into MLC categories according to STANAG 2021 [51]. An example of the identification of technical parameters for one vehicle combination is presented in Figure 5.

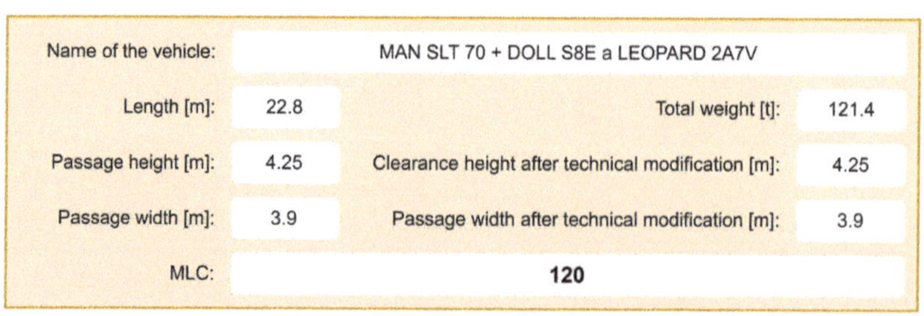

Figure 5. Example of identification of technical parameters for a vehicle combination MAN SLT 70 + DOLL S8E a LEOPARD 2A7V.

The most complex technical parameter of a vehicle or vehicle combination is the determination of the MLC category. An example of the parameters of vehicles in the MLC categories is presented in Figure 6.

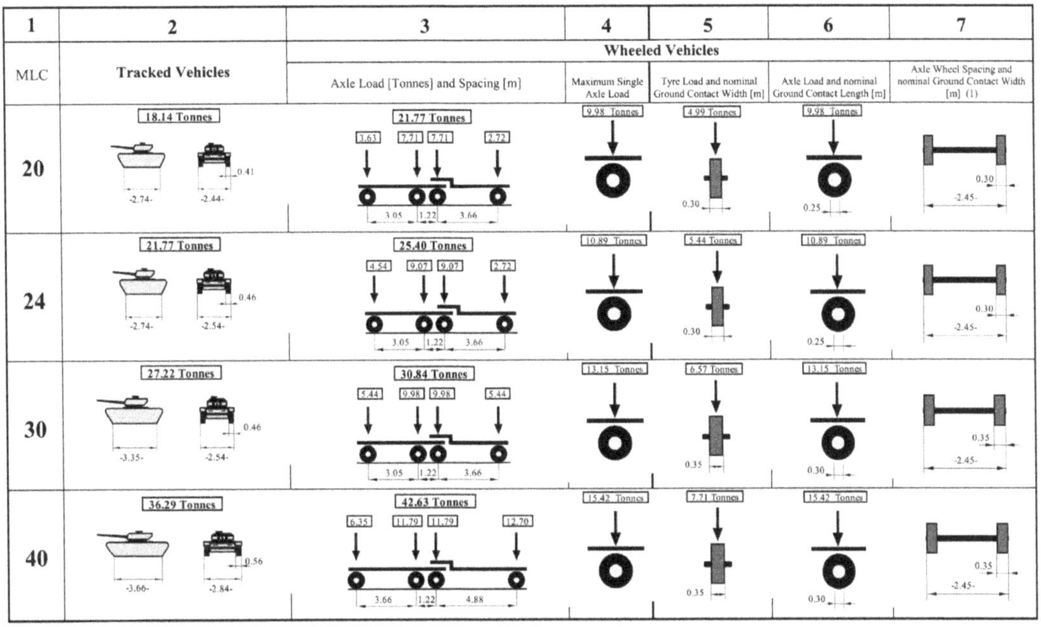

Figure 6. Vehicle parameters in load class MLC 20 to MLC 40 according to STANAG 2021 [51].

The technical parameters of the convoy determined in this way are in the vast majority of cases fixed. Under specific circumstances, the clearance height and possibly the MLC class can be manipulated to a certain extent by appropriate distribution of the load in the vehicle combinations. For military vehicles within NATO, their MLC class is known, as is the case for some typical vehicle combinations. If the civilian vehicle or vehicle combination does not have an MLC class, it can be determined by calculation according to the procedure specified in STANAG 2021 [51].

3.2.2. Phase 2: Identification of Construction-Technical Parameters of the Affected Infrastructure Objects

This phase of the process involves selecting the affected infrastructure objects and identifying their construction-technical parameters that affect the capacity of the transport network. An example of identifying construction-technical parameters for a specific bridge structure is presented in Figure 7.

These parameters enter the next stages as limiting conditions, which are either temporary or permanent in nature and at the same time, according to the definition of the term construction-technical parameter, their value is constant or variable, discrete or continuous, often in a certain interval. Construction-technical parameters and the construction-technical condition of bridge structures can be obtained from the bridge sheet, which must exist for each bridge structure and is available in the bridge structure database and is regularly updated.

In addition, regular checks of the construction-technical condition of the bridge structure allow for an up-to-date overview of these parameters, and the construction-technical parameters and condition, thus, obtained are suitable for further computer processing and enter the next stages of the process, where they can be validated, if necessary, through an extraordinary inspection of the bridge.

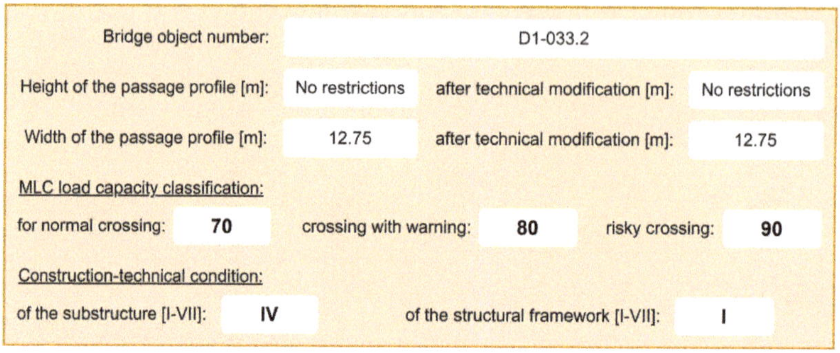

Figure 7. Example of identification of construction-technical parameters for a bridge structure.

Based on the information from the bridge sheet and regular checks, it is possible to determine the construction-technical condition of the substructure and structural framework of the bridge structure. According to the assessment results, the construction-technical condition of bridge structures is classified into seven levels (see Figure 8).

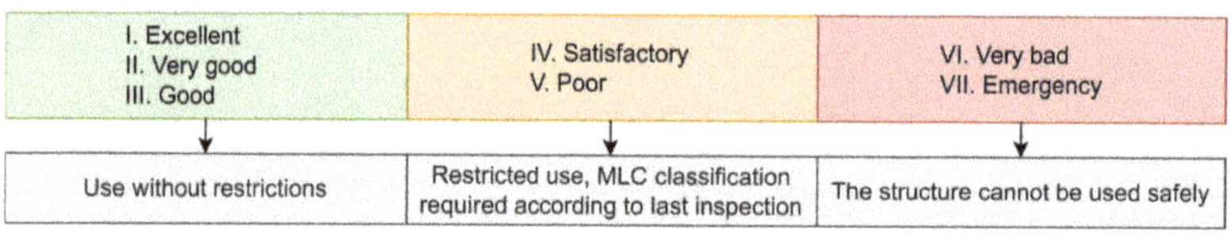

Figure 8. Condition of buildings based on the substructure and structural framework conditions.

For other infrastructure objects such as underpasses and tunnels, the main technical parameters are the height and width of the passage profile, but it is true that underpasses and tunnels limit the transport capacity of the network only in a relatively small percentage of cases compared to bridge structures. For railway infrastructure objects, their construction-technical parameters are clearly given by the category of the line, and it is practically impossible to define railway transport that would not meet the criteria for a given line.

3.2.3. Phase 3: Assessment of the Infrastructure Objects Suitability for the Selected Type of Transport

This phase assesses the infrastructure objects suitability for the selected type of transport. This assessment consists of a comparison between the technical parameters of the convoy (see Phase 1) and the construction-technical parameters of the affected infrastructure objects (see Phase 2). The comparison must always be carried out individually, i.e., one vehicle or vehicle combination against one infrastructure object. The output of this phase is the identification of critical infrastructure objects of the TEN-T land infrastructure.

The assessed parameters are the clearance height, clearance width, MLC value, and the construction-technical condition of the bridge. Based on the results of the comparison, the infrastructure object is classified into three categories according to its suitability for the selected type of transport:

- green category: the object is suitable for the given transport;
- orange category: the object is suitable for the given transport after technical modification of the object or vehicle/train (e.g., dismantling of a certain part of it) and with increased attention;

- red category: the object is not suitable for the given transport.

The classification of an object into the above categories is carried out according to the criteria shown in Figure 9.

Assessment of the passage height	Assessment of the passage width	Assessment of the MLC load capacity assessment	Assessment of the bridge structural and technical condition
The passage height of the vehicle is less than the height of the clearance profile of the bridge structure.	The passage width of the vehicle is smaller than the passing profile width of the bridge structure.	The numerical value of the vehicle MLC is less than or equal to the MLC of the bridge structure for normal crossing.	The condition of the bridge's substructure and structural framework is classified in category I-III.
The passage height of the vehicle is less than the height of the clearance profile of the bridge structure after technical modification.	The passage width of the vehicle is smaller than the width of the passage profile of the bridge structure after technical modification.	The numerical value of the vehicle MLC is greater than the MLC of the bridge structure for a normal crossing, but at the same time it is less than or equal to the MLC of the bridge structure for a crossing with a warning.	The condition of the bridge's substructure and structural framework is classified in category IV-V.
The passage height of the vehicle after technical modification is less than the height of the clearance profile of the bridge structure.	The passage width of the vehicle after technical modification is smaller than the width of the clearance profile of the bridge structure.	The numerical value of the vehicle MLC is greater than the MLC of the bridge structure for a crossing with a warning, but at the same time it is less than the MLC of the bridge structure for a risky crossing.	
The passage height of the vehicle after technical modification is greater than the height of the clearance profile of the bridge structure after technical modification.	The passage width of the vehicle after technical modification is greater than the width of the passing profile of the bridge structure after technical modification.	The numerical value of the vehicle MLC is equal to or greater than the MLC of the bridge structure for the hazardous crossing.	The condition of the bridge's substructure and structural framework is classified in category VI-VII.

Figure 9. Categorisation of the suitability of infrastructure objects for the selected type of transport.

Following this process, it is possible to validate the current state of the defined critical infrastructure objects, which creates the basis for defining construction-technical requirements.

3.3. Digitalization of Critical Infrastructure Objects of the TEN-T Land Infrastructure

An important supporting tool for identifying critical objects of the TEN-T land infrastructure is their digitalization. Digitalization allows for the precise identification of problematic points in the transport network, where there may be limitations due to the load-bearing capacity of bridges or the clearance height of underpasses and tunnels. Using the Road Databank [50] and outputs from the above presented procedure, it is possible to create detailed map documents that categorize the infrastructure according to its passability for all excessive or oversized transports.

The digitalization of critical TEN-T land infrastructure objects also plays a key role in the planning of strategic military transport. The analysis within this process focuses on the ability of bridges, tunnels, and underpasses to meet the requirements of military transport, which is essential for crisis scenarios and ensuring the smooth passage of excessive or oversized transports. The digitalization includes the evaluation of the load-bearing capacity of bridges and the clear heights of underpasses and tunnels in order to determine their passability. Bridges were divided into four transport modes: normal, warning, risk and exceptional load exceeded. Two modes were defined for tunnels and underpasses—pass and warning. Thanks to this categorisation, it was possible to create a clear model of the transport network, which provides key information for planning military movements and allows for the effective identification of critical sections that require reconstruction or other measures. An example of the digitalization of the results of the assessment of the suitability of Exit 10 of the D0 motorway (outer Prague ring road) for selected vehicles/trains is demonstrated in Figure 10.

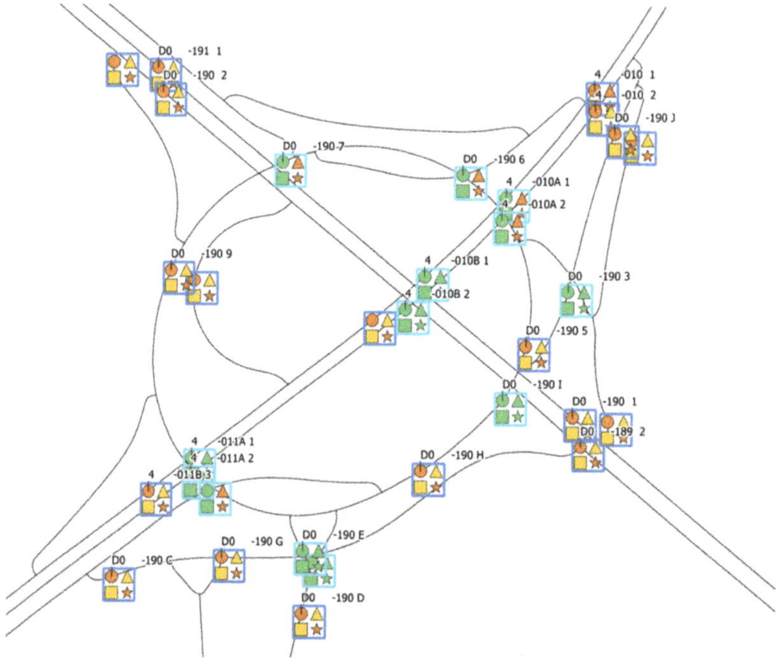

Figure 10. Results digitization example of the infrastructure objects suitability assessment for a selected type of transport.

Figure 10 was created using QGIS (v3.38), an open-source geographic information system. The process of creating the map output included two key steps. The first step was the digitization of critical land infrastructure objects, which was carried out based on detailed map data from the Roads and Highways Directorate [50], GIS layers of the road network, and the proposed Procedure for identifying critical infrastructure objects of the TEN-T land infrastructure (see Section 3.2). Using these inputs, key infrastructure objects were identified—bridges, underpasses, and tunnels, which were subsequently categorized according to their passability for typical military transport.

The second step was data processing in QGIS, where individual critical objects were imported into the software environment and assigned the corresponding attributes. Each bridge, tunnel, and underpass were classified based on their condition and impact on military transport, or possible restrictions for military equipment crossings. Subsequently, a visualization was created in QGIS, which included setting the symbology and exporting the map output as a raster image. The export was performed at high resolution so that all labels and legends were readable.

The colour and shape differentiation of infrastructure objects allows for quick recognition of restrictions and effective interpretation of results. Tunnels are marked on the map as a square with a pink outline, with green indicating full traffic and orange indicating the need for special measures. Underpasses are shown as squares with a blue outline, and their trafficability is indicated by the same colour code. Bridges are shown as squares with a dark blue outline, where green indicates full trafficability, orange and yellow warn of restrictions (yellow—warning, orange—risk), and red indicates that the object is not passable for military transport without accurate calculation. Individual symbols then indicate the following types of transport: Circle—Leopard tank, Triangle—Wheel loader, Square—Fire truck, and Star—Rescue vehicle.

The results of the digitalization confirm that there are fundamental limitations for heavy military equipment on some sections, either due to the load capacity of bridges or due to insufficient clearance of underpasses and tunnels. The identified problem areas represent a potential risk for effective military mobility, which requires the implementation of specific measures. Firstly, it is necessary to focus on the modernisation of bridges, which will increase their load capacity on strategic routes. The increased load capacity will allow for the safe passage of heavy military equipment, thereby minimizing the risk of the need to reload or slow down the convoy. Modernization should be focused primarily on bridges that showed a high proportion of cases in the alert and risk mode in the analyses.

In the case of underpasses and tunnels, the possibilities for modification are considerably limited. In the case of tunnels, any height modification is practically impossible, even in new construction, because it would lead to disproportionate investments. In the case of underpasses, the situation is somewhat more flexible, but here too the possibilities for reconstruction are limited and are more feasible in newly built roads. Given these limitations, it is necessary to focus on the means of transport themselves and their adaptation to the existing infrastructure. Alternative solutions may be modular transport systems that allow for a reduction in the height of the transported units, or greater use of rail transport where the road network does not allow for trouble-free passage. An important measure is also the introduction of monitoring and predictive systems that will enable regular monitoring of the condition of critical objects and timely planning of their modernisation. As part of the further development of digitalization, it is appropriate to regularly update digital maps to reflect changes in the infrastructure and enable effective planning of military mobility. The knowledge gained will also be used in the preparation of strategic plans and crisis scenarios, which will ensure better preparedness of transport infrastructure for both military and civilian purposes.

3.4. Defining Recommendations for the Sustainability and Future Development of TEN-T Land Infrastructure in the Context of Military Mobility

The need for a systematic definition of recommendations for future development arises from the identified gaps, especially in connection with military mobility. Military mobility in this context can be understood as the military movement and transport from places of permanent deployment (supply hubs) to the final destination (logistics base or generally to the Area of Responsibility). As part of this process, it is necessary to plan the routes and times of military convoy movements, including ROM/RON. However, there are also fundamental construction-technical and capacity aspects (constraints) in the background of this process, which are also affected by external entities. The analyses carried out show the need for systematic development of the TEN-T in four basic layers (see Figure 11).

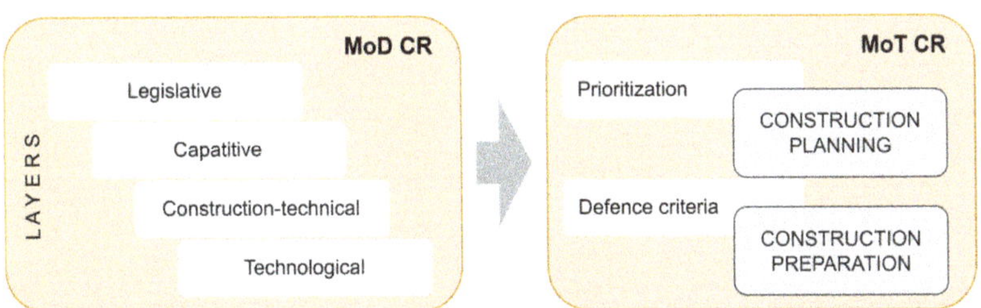

Figure 11. Recommendations for the sustainability and future development of TEN-T land infrastructure.

3.4.1. Legislative Layer: Amend the Relevant Legislation with the Possibility of Prioritizing Military Movements

Ensuring military mobility within the Czech Republic primarily has a legislative and legal dimension. A fundamental shortcoming at present is that in the event of the implementation of the worst-case scenarios that would lead to the above-standard use of TEN-T, military movements have only minimal legislative and legal support. With the exception of the new definition of a military train, which partially interferes with the prioritization of passenger transport on the railway, the key problem remains, which is the impossibility of fundamentally interfering with civilian traffic. Within the framework of the road movement, which will form the backbone of the military movement, it is only possible to ensure the temporary removal of closures or the suspension of repairs and maintenance (through agreements with the Ministry of Transport of the Czech Republic, the Road and Motorway Directorate, etc.) and to ensure management through the Military Police, or the Police of the Czech Republic. From a legislative perspective, the key problem is the time discrepancy between the requirement for military mobility and the declaration of the relevant state of crisis, when the powers of public administration entities change.

3.4.2. Capacitive Layer: FOCUS on the Specifics of Military Mobility and Conduct Relevant Military Movements Simulations

Military mobility can be understood as military movement/transportation from places of permanent deployment to the target destination. As part of this process, it is necessary to plan the routes and times of military convoy movements, including rest areas ROM/RON. The aim is to coordinate activities with regard to the capacity possibilities of the TEN-T. However, there are also construction and technical aspects (constraints) behind this process, which are also entered into by external entities. Within the Czech Republic, this is primarily the Ministry of Transport and infrastructure operators (i.e., Road and Motorway Directorate and Railway Infrastructure Administration). The entire process of moving military units of NATO member states' armies is implemented with the support of the Czech Republic (Host Nation Support). HNS is primarily intended to ensure a smooth and coordinated passage through the Czech Republic, including ensuring the necessary logistics (creating ROM/RON, refuelling, providing evacuation capacities in the event of a technical problem, etc.).

Partial capacity deficiencies are a prerequisite—especially in the worst-case scenario, which is classified (NATO Secret) and cannot be specified. In this scenario, a large number of military units will be moved from west to east across the territory of the Czech Republic. The assumption is to maintain the standard regime, i.e., a military unit, max. 48 h on our territory, incl. rest, and other necessary activities (e.g., repairs). The goal should be, in extreme situations and only temporarily, the possibility of a fundamental restriction or stoppage of civilian traffic (with the exception of the components of the Integrated Rescue System). The measure also includes the possibility of using Intelligent Transport Systems and new technologies to better manage traffic flow, especially in narrow places (i.e., connections to expressways, bridges, roundabouts, etc.), where there are dimensional or weight limits [61].

However, this depends on the ratio between the real throughput—the capacity of transport routes (TEN-T) and the number of vehicles. For these purposes, the Ministry of Defence of the Czech Republic should have at its disposal capacity simulations of selected convoys (scenarios) to ensure the smooth implementation of military movements and transports, including the resulting requirements for HNS. Simulations of selected scenarios will also allow for the detection of other expected gaps within the comprehensive provision of HNS for military movements and transports.

3.4.3. Construction-Technical Layer: Prioritize Relevant Infrastructure Objects (Primarily Bridges) in the Maintenance and Repair System in Cooperation with the Ministry of Defence of the Czech Republic/Ministry of Defence of the Czech Republic

The construction and technical aspect are very time-consuming and also require the aspect of sustainability and dynamics. Sustainability in the sense of ensuring the passage of not only the first convoy but all convoys of vehicles so that the key infrastructure elements are always in an appropriate (at least sufficient) technical condition. For example, in the case of bridges, ensuring an appropriate technical condition is a big question mark, mainly due to the above-standard use of the TEN-T, i.e., also of these specific infrastructure elements. To illustrate, the above-mentioned scenario may be a situation where from the point of view of excessive transport (higher load—total/axle load than usual), the relevant bridges will be loaded in one week of military movements more than in 1–2 "normal" years. The impact on the technical condition of bridges must, therefore, be assessed dynamically, i.e., simulate this above-standard use, moreover, in such a short period of time. This is not yet available, and it can only be assumed that during the first week of military deployment, some bridges will move from the "green" category (unrestricted use), when the bridge is in perfect (I.), very good (II.), or at least good (III.) condition, to the "yellow" category (restricted use), when the bridge is in a satisfactory (IV.) or poor (V.) condition.

The construction and technical level are in the system of maintenance, repair and modernization of the relevant bridges in the given concept. If the process is not planned with appropriate prioritization and financial resources, the condition of the given bridges may already be in the yellow category or even in the red category at the moment of the start of the implementation of the relevant scenario. For example, on the backbone road—the D1 motorway—two motorway bridges were even in the red category (in very poor condition) in 2024 and several dozen in the yellow category. The evidence for this classification can be found in bridge inspection reports in the Bridge Management System database [62].

3.4.4. Technological Layer: Use the Tools of Intelligent Transport Systems and New Technologies

Intelligent Transport Systems (ITSs) play a crucial role in the modernisation and efficient management of transport infrastructure within the TEN-T, particularly on motorways, main railway corridors and multimodal transport hubs. The integration of ITS into the TEN-T transport infrastructure ensures higher efficiency of the transport system, minimizes delays, and contributes to the achievement of the objectives of sustainable mobility and emission reduction in line with the European Transport Strategy.

The Action Plan of the Ministry of Transport of the Czech Republic for the Development of ITSs sets out strategic objectives and measures for their implementation and modernisation in the Czech Republic [63]. These objectives are based on a detailed analysis of the shortcomings of existing ITS solutions, the priorities of the Czech Republic's transport policy, related strategic documents, and legislative requirements. The plan focuses on the systematic introduction and development of ITSs across all modes of transport with an emphasis on their interconnection, interoperability, and efficient use of the existing transport infrastructure. The action plan includes a specification of specific measures, estimated financial costs for their implementation, and identification of possible sources of financing with the aim of maximizing the benefits of ITSs for the transport system and safety and sustainable mobility. The ITS implementation plan is designed as a living document with continuous updates.

ITSs are of key importance for security and rescue services, especially for the Integrated Rescue System and the Army of the Czech Republic. Connecting ITSs with security and crisis systems enables a faster and more effective response to emergencies, which leads to higher protection of the population and a reduction in the risks associated with traffic accidents or disasters. The main recommendations in the field of ITSs are:

- Strengthen the ability of ITS and professional personnel to respond to new threats, emergencies, and cooperate with the Army of the Czech Republic (e.g., movements of own and allied forces—HNS).
- Supplement ITSs according to the requirements and needs of the Army of the Czech Republic, e.g., supplement the collection of data on the condition and operation of land roads with the detection of new emergencies and transfer and sharing of these data. Use the potential of unmanned vehicles.
- Strengthen the infrastructure and build elements of active traffic management on selected sections of land roads so that it is possible to effectively manage traffic flows of civilian and military vehicles.
- Involve fleets of land road administrator vehicles/assistance service vehicles equipped with ITS and other technologies in resolving emergencies that are beyond their basic purpose.
- Increase the resilience of ITSs, existing and newly built, against cyber-attacks.
- Make static and dynamic data and information accessible and share, e.g., on the condition and operation of roadways, between civilian ITSs (including the Integrated Rescue System) and the systems of the Army of the Czech Republic. Protect sensitive data, e.g., exchange and purpose of data, data transfer, etc.

3.4.5. Summary of Recommendations

The development of transport infrastructure, especially TEN-T, is not only a key condition for the implementation of military movements and military transports of the Army of the Czech Republic but also of the armies of NATO member states. The analyses carried out show that there are relatively significant gaps within the defined layers. The priority task in this area is to resolve the issue throughout the European Union in coordination with planning at the NATO level. Nation states and the Czech Republic are no exception; they derive individual measures and their prioritization from these superior plans. The capacity shift in the context of the new NATO model, which responded to the Russian–Ukrainian conflict, means a several-fold higher burden on the transport infrastructure (and related requirements for HNS). The above requires much greater communication between interested entities (MoD CR, MoT CR, and infrastructure operators, i.e., the Road and Motorway Directorate and Railway Infrastructure Administration), including appropriate methodological documents. The methodology from the level of the Ministry of Transport of the Czech Republic is the authorship collective of the article in the certification process, as well as the Methodological Guidelines from the level of the Ministry of Defence of the Czech Republic. Both documents should clearly elaborate on the areas mentioned in the article, including the setting of communication, cooperation, and coordination between the two ministries. The first is the setting of a process model for determining the prioritization of transport structures on the TEN-T by including a new "military" criterion, which would be set by the responsible body of the Ministry of Defence of the Czech Republic (usually the military transport authorities). In the case of road structures on the TEN-T, the priorities of the Ministry of Defence of the Czech Republic can be included by excluding a specific transport structure from the process of economic evaluation of the structure by the Central Commission of the Ministry of Transport of the Czech Republic, in a similar way to the procedure for removing critical points on sections of class I roads from the point of view of removing critical points in the field of road safety.

However, the economic aspect must also be considered and the process model should be set between the Ministry of Defence of the Czech Republic and the Ministry of Transport of the Czech Republic so that specific construction and technical parameters of the structures are already defined in the preparation phase of transport structures on the TEN-T and are part of the implementation of the technical study, or documentation for a joint permit. Additional inputs and requirements in a higher stage of project development increase the cost of the actual construction and extend the project implementation period.

At the national level, it is also necessary to revise the so-called Designated Road and Railway Network, i.e., transport routes important for military mobility and, in a broader context, for the defence of the state. The revision should include bottlenecks (primarily bridges), which are already in the red/yellow category. The mentioned bottlenecks should be included as a priority in the repair/modernization plan, while adhering to the required construction and technical parameters. In cooperation with the Ministry of Transport of the Czech Republic and the Directorate of Roads and Highways, rest areas intended for the establishment of ROM/RON should be further separately addressed.

4. Conclusions

The essence of the article is to define recommendations for the sustainability and future development of the TEN-T land infrastructure to address large-scale disasters. For this purpose, a description of the TEN-T land infrastructure in the Czech Republic was first carried out, a description of large-scale disasters resulting in above-standard use of the TEN-T land infrastructure was described, and the methods used for modelling and digitizing routes for excessive and oversized transport were described.

Based on these facts, typical transport situations using the TEN-T land infrastructure were defined and modelled. Subsequently, attention was paid to the identification and digitization of critical infrastructure objects of the TEN-T land infrastructure. In this part of the article, an identification procedure consisting of three phases was proposed, namely (1) selection of the type of transport and identification of the technical parameters of the convoy, (2) identification of construction-technical parameters of the affected infrastructure objects, and (3) assessment of the infrastructure objects suitability for the selected type of transport.

The core part of the article was the definition of recommendations for the sustainability and future development of the TEN-T land infrastructure to address large-scale disasters. These recommendations were classified into five layers, which represent the requirements of the Ministry of Defence of the Czech Republic towards the Ministry of Transport of the Czech Republic. These layers consider the requirements in the areas of legislative, capacitive, construction-technical, economic-procedural, and intelligent transport systems.

The results presented in this article are primarily intended for competent employees of the Ministry of Defence of the Czech Republic and the Ministry of Transport of the Czech Republic, who are responsible for the implementation of excessive and oversized transport. However, they can also be used by lower-level entities, specifically the Road and Motorway Directorate and Railway Infrastructure Administration, in order to define specific construction-technical requirements for the TEN-T land infrastructure.

The current research included several limitations due to its large scale. Of the large number of large-scale disasters, primary attention was paid only to armed conflict, which requires extensive and above-standard use of the TEN-T land infrastructure. The research was further limited to a specific type of infrastructure object, namely bridge structures. The last and most significant limitation was the aspect of the perception of the sustainability of the transport infrastructure, which was examined only in the context of construction-technical parameters.

In conclusion, it is appropriate to state that in the current turbulent security situation, construction-technical requirements for infrastructure objects on the TEN-T need to be given due attention. Since the dual use of the transport infrastructure is assumed, follow-up research must include a comprehensive examination of the necessary level of reliability of the affected infrastructure objects when dealing with all types of crisis situations and all possible scenarios of use. It should also be emphasized that any shift in construction-technical requirements usually has a significant impact on the construction technologies used and subsequent sustainability and, thus, on the overall economy.

Author Contributions: Conceptualization, D.R. and P.M.; methodology, D.R., M.V., P.M., J.A., T.A., and M.H.; validation, P.M., M.V., and J.A.; formal analysis, D.R., M.V., P.M., and J.A.; investigation, T.A. and M.H.; resources, D.R., M.V., and P.M.; data curation, T.A. and M.H.; writing—original draft preparation, D.R., M.V., P.M., J.A., T.A., and M.H.; writing—review and editing, D.R., M.V., P.M., J.A., T.A., and M.H.; visualization, D.R.; supervision, D.R.; project administration, D.R.; funding acquisition, D.R. All authors have read and agreed to the published version of the manuscript.

Funding: This research was funded by the Technology Agency of the Czech Republic, grant number CK03000182.

Institutional Review Board Statement: Not applicable.

Informed Consent Statement: Not applicable.

Data Availability Statement: All data and information are available in this article.

Conflicts of Interest: The authors declare no conflicts of interest. The funders had no role in the design of the study; in the collection, analyses, or interpretation of data; in the writing of the manuscript; or in the decision to publish the results.

Abbreviations

The following abbreviations are used in this manuscript:

EU	European Union
HNS	Host Nation Support
ITS	Intelligent Transportation System
MoD CR	Ministry of Defence of the Czech Republic
MLC	Military Load Classification
MoT CR	Ministry of Transport of the Czech Republic
NATO	North Atlantic Treaty Organization
ROM	Rest on Movement
RON	Rest on Night
TEN-T	Trans-European Transport Networks

References

1. Directive (EU) 2022/2557 of the European Parliament and of the Council of 14 December 2022 on the Resilience of Critical Entities and Repealing Council Directive 2008/114/EC. Available online: http://data.europa.eu/eli/dir/2022/2557/oj (accessed on 10 March 2025).
2. Regulation (EU) 2024/1679 of the European Parliament and of the Council of 13 June 2024 on Union Guidelines for the Development of the Trans-European Transport Network, Amending Regulations (EU) 2021/1153 and (EU) No 913/2010 and repealing Regulation (EU) No 1315/2013. Available online: http://data.europa.eu/eli/reg/2024/1679/oj (accessed on 10 March 2025).
3. European External Action Service. *A Strategic Compass for Security and Defence*; European Union: Brussels, Belgium, 2022; Available online: https://www.eeas.europa.eu/eeas/strategic-compass-security-and-defence-1_en (accessed on 29 November 2024).
4. *Sendai Framework for Disaster Risk Reduction 2015–2030*; United Nations Office for Disaster Risk Reduction: Geneva, Switzerland, 2015.
5. Banholzer, S.; Kossin, J.; Donner, S. The Impact of Climate Change on Natural Disasters. In *Reducing Disaster: Early Warning Systems for Climate Change*; Singh, A., Zommers, Z., Eds.; Springer: Dordrecht, The Netherlands, 2014; pp. 21–49. [CrossRef]
6. Shi, P. Disaster Risk Management. In *Disaster Risk Science*; Springer: Singapore, 2019; pp. 491–539. [CrossRef]
7. Bernatik, A.; Senovsky, P.; Senovsky, M.; Rehak, D. Territorial Risk Analysis and Mapping. *Chem. Eng. Trans.* **2013**, *31*, 79–84. [CrossRef]

8. Vichova, K.; Hromada, M.; Rehak, D. The Use of Crisis Management Information Systems in Rescue Operations of Fire and rescue system in the Czech Republic. *Procedia Eng.* **2017**, *192*, 947–952. [CrossRef]
9. Gad-el-Hak, M. *Large-Scale Disasters: Prediction, Control, and Mitigation*; Cambridge University Press: Cambridge, UK, 2008. [CrossRef]
10. Jiang, Y.; Yuan, Y. Emergency Logistics in a Large-Scale Disaster Context: Achievements and Challenges. *Int. J. Environ. Res. Public Health* **2019**, *16*, 779. [CrossRef] [PubMed]
11. Directorate-General for Mobility and Transport. *Trans-European Transport Network (TEN-T)*; European Commission: Brussels, Belgium, 2024.
12. Macioszek, E. Oversize cargo transport in road transport—Problems and issues. *Sci. J. Silesian Univ. Technol. Ser. Transp.* **2020**, *108*, 133–140. [CrossRef]
13. United Nations. *Sustainable Transport*; Department of Economic and Social Affairs: New York, NY, USA, 2015; Available online: https://sdgs.un.org/topics/sustainable-transport (accessed on 4 March 2025).
14. Angelov, S.; Jimenez, C.K.; Wall, V.; O'Croínin, D. An assessment of sustainable transport infrastructure in a national healthcare system. *Surgeon* **2024**, *22*, 203–208. [CrossRef]
15. Bénichou, W. Sustainable development of transport infrastructure in the system of national economy. *J. Econ. Innov. Manag. Entrep.* **2024**, *1*. [CrossRef]
16. Sobczuk, S.; Borucka, A. Recent Advances for the Development of Sustainable Transport and Their Importance in Case of Global Crises: A Literature Review. *Appl. Sci.* **2024**, *14*, 10653. [CrossRef]
17. Bland, M.; Burke, M.I.; Bertolaccini, K. Taking steps toward healthy & sustainable transport investment: A systematic review of economic evaluations in the academic literature on large-scale active transport infrastructure. *Int. J. Sustain. Transp.* **2024**, *18*, 201–220. [CrossRef]
18. Zhang, Y.; Cheng, L. The role of transport infrastructure in economic growth: Empirical evidence in the UK. *Transp. Policy* **2023**, *133*, 223–233. [CrossRef]
19. Badassa, B.B.; Sun, B.; Qiao, L. Sustainable Transport Infrastructure and Economic Returns: A Bibliometric and Visualization Analysis. *Sustainability* **2020**, *12*, 2033. [CrossRef]
20. Fokina, O.; Mottaeva, A.; Mottaeva, A. Transport infrastructure in the system of environmental projects for sustainable development of the region. In Proceedings of the International Scientific Conference Transport Technologies in the 21st Century (TT21C-2024), Rostov-on-Don, Russia, 8–10 April 2024; Curran Associates, Inc.: Sydney, Australia, 2024; Volume 515, p. 01015. [CrossRef]
21. Fang, W.; Ma, C.; Lei, Z. Research on Sustainable Development of Transport Infrastructure Based on Corporate Culture and Low-Carbon Perspective. *J. Environ. Public Health* **2022**, *1*, 4629422. [CrossRef] [PubMed]
22. Ghent, C. Mitigating the Effects of Transport Infrastructure Development on Ecosystems. *Consilience: J. Sustain. Dev.* **2018**, *18*, 58–68. [CrossRef]
23. Bao, L.; Kusadokoro, M.; Chitose, A.; Chen, C. Development of socially sustainable transport research: A bibliometric and visualization analysis. *Travel Behav. Soc.* **2023**, *30*, 60–73. [CrossRef]
24. Winata, D.T.; Rarasati, A.D. Sustainable infrastructure transportation to improve society wellbeing in Karawang. In Proceedings of the 2nd International Tropical Renewable Energy Conference (i-TREC), Bali, Indonesia, 3–4 October 2017; IOP Publishing Ltd.: Bristol, UK, 2018; Volume 105, p. 012008. [CrossRef]
25. Oberg, M.; Nilsson, K.L.; Johansson, C. Major transport corridors: The concept of sustainability in EU documents. *Transp. Res. Procedia* **2017**, *25*, 3694–3702. [CrossRef]
26. Petru, J.; Krivda, V. The Transport of Oversized Cargoes from the Perspective of Sustainable Transport Infrastructure in Cities. *Sustainability* **2021**, *13*, 5524. [CrossRef]
27. Milani, L.; Mohr, D.; Sandri, N. *Built to Last: Making Sustainability a Priority in Transport Infrastructure*; McKinsey & Company: New York, NY, USA, 2021; Available online: https://www.mckinsey.com/industries/infrastructure/our-insights/built-to-last-making-sustainability-a-priority-in-transport-infrastructure (accessed on 4 March 2025).
28. Mead, L. *The Road to Sustainable Transport*; International Institute for Sustainable Development: Winnipeg, MB, Canada, 2021; Available online: https://www.iisd.org/articles/deep-dive/road-sustainable-transport (accessed on 4 March 2025).
29. Treaty on European Union (92/C 191/01). Available online: https://eur-lex.europa.eu/legal-content/EN/TXT/?uri=oj:JOC_1992_191_R_TOC (accessed on 10 March 2025).
30. Bottcher, B. The Trans-European Transport Network (Ten-T): History, Progress and Financing. *Commun.-Sci. Lett. Univ. Zilina* **2006**, *8*, 51–54. [CrossRef]
31. Regulation (EU) No 1315/2013 of the European Parliament and of the Council of 11 December 2013 on Union Guidelines for the Development of the Trans-European Transport Network and Repealing Decision No 661/2010/EU. Available online: http://data.europa.eu/eli/reg/2013/1315/oj (accessed on 10 March 2025).

32. Rehak, D.; Senovsky, P.; Hromada, M.; Lovecek, T. Complex Approach to Assessing Resilience of Critical Infrastructure Elements. *Int. J. Crit. Infrastruct. Prot.* **2019**, *25*, 125–138. [CrossRef]
33. Fekete, A. *Common Criteria for the Assessment of Critical Infrastructures*; Federal Office of Civil Protection and Disaster Assistance: Bonn, Germany, 2011.
34. Czech, M. PAN—European transport corridors in the policy of the European Union. *Sci. J. Silesian Univ. Technol. Ser. Transp.* **2021**, *112*, 51–62. [CrossRef]
35. Hudec, S. *Motorways in the Czech Republic*; Czech Motorways: Prague, Czech Republic, 2024; Available online: https://www.ceskedalnice.cz/wp-content/uploads/zpoplatnene-velka.png (accessed on 3 December 2024).
36. Road and Motorway Directorate. Communication Lengths and Other Data. 2024. Available online: https://web.archive.org/web/20250125192752/https://www.rsd.cz/web/guest/silnice-a-dalnice/delky-a-dalsi-data-komunikaci#zalozka-silnice-i-tridy (accessed on 3 December 2024).
37. Railway Administration. High-Speed Lines: Sections and Maps. 2024. Available online: https://vrtky.cz/useky-a-mapy (accessed on 3 December 2024).
38. Kryze, P. *Railway Transit Corridors*; Railway Administration: Prague, Czech Republic, 2024; Available online: https://www.spravazeleznic.cz/documents/50007830/51605747/tranzitn%C3%AD+koridory.pdf/d09016c4-89b8-4815-b9d9-bd68bdf28d94 (accessed on 3 December 2024).
39. K-Report. Railway Corridors of the Czech Republic: Characteristics of Corridors. 2024. Available online: https://www.k-report.net/koridory/histori3.htm (accessed on 3 December 2024).
40. Kelman, I. Addressing the root causes of large-scale disasters. In *Large-Scale Disasters: Prediction, Control, and Mitigation*; Gad-el-Hak, M., Ed.; Cambridge University Press: Cambridge, UK, 2009; pp. 94–119. [CrossRef]
41. Gad-el-Hak, M. (Ed.) The art and science of large-scale disasters. In *Large-Scale Disasters: Prediction, Control, and Mitigation*; Cambridge University Press: Cambridge, UK, 2009; pp. 5–68. [CrossRef]
42. Baxter, P.J. Catastrophes—Natural and Man-Made Disasters. In *Conflict and Catastrophe Medicine*; Ryan, J., Mahoney, P.F., Greaves, I., Bowyer, G., Eds.; Springer: London, UK, 2002; pp. 27–48. [CrossRef]
43. IRDR. *Peril Classification and Hazard Glossary (IRDR DATA Publication No. 1)*; Integrated Research on Disaster Risk: Beijing, China, 2014.
44. Stallings, R.A. Man-made Disasters. *Int. J. Mass Emergencies Disasters* **1998**, *16*, 225. [CrossRef]
45. Council Directive 96/53/EC of 25 July 1996 Laying Down for Certain Road Vehicles Circulating Within the Community the Maximum Authorized Dimensions in National and International Traffic and the Maximum Authorized Weights in International Traffic. Available online: http://data.europa.eu/eli/dir/1996/53/oj (accessed on 10 March 2025).
46. Paulus, F.; Kromer, A.; Petr, J.; Cerny, J. *Hazard Analysis for the Czech Republic*; Fire Rescue Service of the Czech Republic: Prague, Czech Republic, 2015.
47. EM-DAT. Disaster Classification System. 2023. Available online: https://doc.emdat.be/docs/data-structure-and-content/disaster-classification-system/#fn:5 (accessed on 10 December 2024).
48. Ministry of Defence. *Czech Armed Forces Development Concept 2035*; The Ministry of Defence of the Czech Republic: Prague, Czech Republic, 2024.
49. NATO. *Road Movements and Movement Control (AMovP-1)*; Nort Atlantic Treaty Organization: Brussels, Belgium, 2004.
50. Road and Motorway Directorate. *Road databank and National Traffic Information Centre*; Road and Motorway Directorate: Prague, Czech Republic, 2025; Available online: https://web.archive.org/web/20230203134022/https://www.rsd.cz/rsd/silnicni-databanka-a-ndic (accessed on 12 January 2025).
51. NATO. *Military Load Classification of Bridges, Ferries, Rafts and Vehicles (STANAG 2021)*, 8th ed.; North Atlantic Treaty Organization: Brussels, Belgium, 2017.
52. NATO. *Military Load Classification of Bridges, Ferries, Rafts and Vehicles (AEP-3.12.1.5), Edition A, Version 1*; North Atlantic Treaty Organization: Brussels, Belgium, 2017.
53. Lenner, R.; Keuser, M.; Sykora, M. Safety Concept and Partial Factors for Bridge Assessment under Military Loading. *Adv. Mil. Technol.* **2014**, *9*, 5–20.
54. Polach, M.; Vlkovsky, M. Resilience of the Railway Infrastructure in an Urban Agglomeration with Regard to the Transport of Specific Material. In Proceedings of the 2023 Smart City Symposium Prague (SCSP), Prague, Czech Republic, 25–26 May 2023; Ruzicka, J., Ed.; IEEE: Prague, Czech Republic, 2023. [CrossRef]
55. NATO. New NATO Force Model. In *NATO Summit in Madrid*; North Atlantic Treaty Organization: Madrid, Spain, 2022.
56. Jagelcak, J.; Vrabel, J.; Kikrotva, M. Using of Different Road Vehicle Combinations for Abnormal Transports Based on the Assessment of Load Rating of Bridges in Slovakia for the Most Critical Vehicle Combination. In Proceedings of the 5th International Conference on Road and Rail Infrastructure, Zadar, Croatia, 17–19 May 2018; Lakusic, S., Ed.; University of Zagreb: Zagreb, Croatia, 2018. [CrossRef]

57. Rehak, D.; Patrman, D.; Brabcova, V.; Dvorak, Z. Identifying Critical Elements of Road Infrastructure Using Cascading Impact Assessment. *Transport* **2020**, *35*, 300–314. [CrossRef]
58. Galor, W.; Galor, A. *Carriage and Securing of Oversize Cargo in Transport*; Akademia Morska: Szczencin, Poland, 2011.
59. Rehak, D.; Novotny, P. Bases for Modelling the Impacts of the Critical Infrastructure Failure. *Chem. Eng. Trans.* **2016**, *53*, 91–96. [CrossRef]
60. Patrman, D.; Splichalova, A.; Rehak, D.; Onderkova, V. Factors Influencing the Performance of Critical Transportation Infrastructure Elements. *Transp. Res. Procedia* **2019**, *40*, 1518–1524. [CrossRef]
61. Tichy, T.; Broz, J.; Ruzicka, J.; Hejciarova, E.; Belinova, Z.; Jisova, J. Information Modelling and Smart Approaches at the Interface of Road and Rail Transport. In Proceedings of the 2023 Smart City Symposium Prague (SCSP), Prague, Czech Republic, 25–26 May 2023; Institute of Electrical and Electronics Engineers: New York, NY, USA, 2023; pp. 1–8. [CrossRef]
62. Brighenti, F.; Caspani, V.F.; Costa, G.; Giordano, P.F.; Limongelli, M.P.; Zonta, D. Bridge management systems: A review on current practice in a digitizing world. *Eng. Struct.* **2024**, *321*, 118971. [CrossRef]
63. Ministry of Transport. *Implementation Plan for the Action Plan for the Development of Intelligent Transport Systems (ITS) in the Czech Republic Until 2020 (with a View to 2050)*; Ministry of Transport of the Czech Republic: Prague, Czech Republic, 2016. Available online: https://md.gov.cz/Dokumenty/Strategie/ITS/Implemetacni-plan-k-Akcnimu-planu-rozvoje-intelige?returl=/Dokumenty/Strategie/ITS (accessed on 4 March 2025). (In Czech)

sustainability

MDPI

Article

Enhancing Sustainability in Port Infrastructure Through Innovation: A Case Study of the Spanish Port System

Javier Vaca-Cabrero, Javier Domínguez Rastrojo, Nicoletta González-Cancelas * and Alberto Camarero-Orive

Department of Transport, Territorial and Urban Planning Engineering, Universidad Politécnica de Madrid, 28040 Madrid, Spain; javier.dominguez.rastrojo@alumnos.upm.es (J.D.R.)

* Correspondence: nicoleta.gcancelas@upm.es

Abstract: This research explores the role of innovation in fostering sustainability within the Spanish Port System, emphasizing its implications for transport infrastructure. It examines the intersection of innovation and sustainability, addressing key challenges such as maritime traffic growth, energy efficiency, waste management, and community integration. It identifies opportunities for technological advancements, collaborative initiatives, and circular economy strategies that contribute to the sustainable development of port infrastructure. The findings highlight the necessity of implementing innovative solutions to enhance operational efficiency, mitigate environmental impact, and strengthen stakeholder engagement. The application of advanced technologies and cooperative frameworks among port stakeholders emerges as a critical driver for achieving sustainability objectives within maritime transport systems.

Keywords: innovation in transport infrastructure; sustainable ports; port system efficiency; circular economy in ports; business observation tool

1. Introduction

In recent decades, sustainability has become a central issue on the global agenda due to the environmental, economic, and social challenges facing our planet [1]. Within this context, transportation infrastructure, particularly port systems, plays a fundamental role as a critical enabler of international trade and economic development. However, the increasing demands on port infrastructure, driven by global trade and maritime traffic growth, pose significant challenges in terms of operational efficiency, environmental impact, and resource management [2].

The Spanish Port System holds strategic importance in Europe, acting as a key connection hub between Europe, Africa, and America [3]. The rise in international trade and tourism has placed growing pressure on Spanish ports to enhance efficiency while ensuring sustainability. This challenge requires innovative solutions for port infrastructure and transportation management to reduce environmental impact, optimize resource use, and foster social responsibility.

Innovation emerges as a crucial enabler for achieving sustainability in port transportation and infrastructure. The implementation of new technologies, more efficient operational practices, and innovative business models can contribute to enhancing competitiveness, resilience, and long-term environmental sustainability in maritime transport [4].

This article analyzes the scope of innovation in promoting sustainable port transportation and infrastructure in Spain. It examines the intersection between innovation and sustainability, addressing economic, environmental, and social dimensions. Additionally,

https://doi.org/10.3390/su17062593

this study identifies key challenges such as maritime traffic congestion, energy efficiency, waste management, and integration with local communities, while exploring innovation-driven solutions. The Business Observation Tool (BOT) methodology is applied to evaluate how innovation can be strategically leveraged to enhance sustainability in the Spanish Port System [3].

This paper is structured as follows: First, a review of the existing literature on sustainability in port transportation and infrastructure highlights the role of innovation as a driver of change. Next, the main sustainability challenges for Spanish ports are presented, followed by a discussion of innovation opportunities. Case studies and best practices, both national and international, are analyzed to identify effective strategies for sustainable port development. Finally, conclusions and recommendations are provided to encourage the adoption of innovative practices in the Spanish Port System.

By integrating innovation into port sustainability strategies, this study aims to provide actionable insights for stakeholders involved in port infrastructure, maritime transport, and logistics management. The findings will contribute to the discussion on how technological advancements, collaboration, and circular economy principles can drive the transition toward sustainable and resilient transportation systems.

This article explores the impact of innovation on the sustainability of the Spanish port system, with a focus on maritime transport infrastructure. Through the application of the Business Observation Tool, opportunities are identified to improve operational efficiency, reduce environmental impact, and strengthen the integration of ports with their local environments.

Since ports are critical nodes in global shipping, their sustainability has a direct impact on logistics efficiency and reducing emissions in the sector. Innovation applied to these infrastructures is key to transforming maritime transport into a more resilient and sustainable system.

This study focuses on a selected group of Spanish ports that exhibit significant sustainability challenges and innovation opportunities. Specifically, the ports of Barcelona, Valencia, and Bilbao were analyzed due to their strategic roles in European maritime trade and their recent sustainability initiatives. These ports serve as representative cases for the broader Spanish Port System, providing insights that can be extrapolated to other port environments across the country.

2. Innovation for Sustainable Port Infrastructure and Transport

The state of the art on innovation in the sustainability of the Spanish Port System is essential to understand and address the current and future challenges in the sector. The search for innovative solutions has become a priority to ensure competitiveness, operational efficiency, and the mitigation of environmental impacts in Spanish ports. This state of knowledge aims to review and analyze the existing literature on innovation in port sustainability, with a particular focus on the Spanish context.

The integration of innovation into sustainable port infrastructure and transportation is crucial for addressing contemporary challenges in maritime logistics. The search for technological advancements and operational efficiency has become a priority to ensure the competitiveness of ports while mitigating environmental and social impacts [4]. Ports are not only logistic hubs but also complex infrastructures that require continuous adaptation to new regulatory, technological, and environmental demands. This section reviews the existing literature on innovation strategies aimed at enhancing sustainability in port transportation and infrastructure, with a particular focus on Spain.

2.1. Dimensions and Approach to Innovation in Port Sustainability

Sustainable port innovation encompasses three core dimensions: environmental, economic, and social [4].

- Environmental innovation focuses on clean energy adoption, waste management, emissions reduction, and eco-friendly port designs.
- Economic innovation improves operational efficiency, logistics optimization, and infrastructure resilience.
- Social innovation aims to integrate local communities, enhance employment, and develop corporate social responsibility strategies.

These dimensions must be supported by innovative approaches, including the following [2]:

- Technological advancements: electrification of port operations, AI-driven logistics, and digital twin applications for predictive maintenance.
- Sustainable operational management in port logistics relies on internationally recognized environmental management certifications, green supply chain strategies, and circular economy applications. These approaches aim to reduce environmental impact while enhancing operational efficiency.

Innovation in port sustainability encompasses multiple interrelated dimensions. First, the environmental dimension focuses on the adoption of practices and technologies that reduce negative impacts on the marine and terrestrial ecosystem. This includes implementing efficient waste management systems, optimizing energy efficiency, reducing greenhouse gas emissions, and conserving marine biodiversity [5].

The social dimension of innovation in port sustainability focuses on improving the quality of life of local communities and promoting a harmonious relationship between ports and their environment. This involves the inclusion of communities in decision-making, the creation of local employment, respect for human rights, and the promotion of corporate social responsibility [6].

In terms of the economic dimension, innovation in port sustainability seeks to improve the efficiency and competitiveness of ports, while promoting the creation of long-term economic value. This is achieved through the implementation of advanced technologies that allow for more efficient management of the logistics chain, the diversification of port activities, and the promotion of the circular economy [7].

There are several innovation approaches that are applied in the pursuit of sustainability in ports. One of them is technological innovation, which involves the adoption of clean and efficient technologies, such as the electrification of port machinery, the use of renewable energies, and the implementation of advanced environmental monitoring systems [8].

Another important focus is innovation in management and operational practices. This includes the implementation of environmental management strategies, such as ISO 14001 certification, the integration of quality and environmental management systems, and the adoption of cleaner production and circular economy practices [9].

Within the innovation applied to sustainability in ports, the digitalization of port infrastructure and maritime transport has played a fundamental role. Technologies such as terminal automation, the use of artificial intelligence in maritime traffic management, and the integration of Internet of Things (IoT) systems make it possible to optimize logistics flows and improve operational efficiency. These innovations not only reduce operating costs, but also contribute to reducing energy consumption and greenhouse gas emissions, facilitating the transition to a more sustainable shipping system [8].

2.2. Results of Innovation in Port Sustainability

Innovation in port sustainability has proven to be beneficial from an environmental, economic and social perspective [10]. In environmental terms, the implementation of sustainable technologies and practices has led to the reduction in greenhouse gas emissions, the improvement of air and water quality, and the conservation of marine biodiversity.

A key aspect in the sustainable transformation of ports is the development of smart port infrastructure. The concept of 'Smart Ports' has emerged as an innovative strategy to integrate advanced technologies, such as the use of smart grids, digital management platforms, and automated control systems. These solutions allow for not only a more efficient use of resources, but also a better ability to respond to fluctuations in maritime transport demand, thus contributing to the sustainability of the sector. In various studies, it has been shown that the implementation of smart ports can improve efficiency in loading and unloading, reduce waiting times, and minimize environmental impact through the use of renewable energy and emission monitoring technologies [8].

In the economic sphere, innovation in port sustainability has generated business opportunities and increased operational efficiency, which translates into greater competitiveness of ports [4]. In addition, the implementation of sustainable practices has led to the optimization of resources and the reduction in costs in the long term.

In social terms, innovation in port sustainability has fostered greater participation of local communities in decision-making, as well as local employment generation and capacity building. It has also improved relations between ports and communities, strengthening trust and promoting more equitable and sustainable development [11].

The state of knowledge on the scope of innovation in port sustainability evidences the importance and benefits of implementing innovative practices in this sector. The integration of environmental, social, and economic dimensions in port innovation contributes to a transformation towards more sustainable development. The adoption of technological, managerial, and operational approaches has demonstrated positive results in terms of reducing environmental impacts, improving economic efficiency and promoting social welfare [12]. However, a holistic approach and greater collaboration between different actors is required to ensure the effective implementation of innovation in port sustainability and its replication in global contexts [13].

2.3. Innovation in the Sustainability of the Spanish Port System

Although the Spanish Port System comprises a diverse network of ports with varying operational structures, this research focuses on key ports that have demonstrated leadership in sustainable innovation. By narrowing the analysis to Barcelona, Valencia, and Bilbao, we ensure that the findings are relevant to ports that play a major role in international logistics while facing significant environmental and operational challenges.

To understand the scope of innovation in the sustainability of the Spanish Port System, it is necessary to establish the key concepts of innovation and sustainability. Innovation refers to the introduction of new ideas, products, processes, or practices that generate value and improve efficiency and competitiveness. On the other hand, sustainability involves balancing economic, environmental, and social aspects to meet present needs without compromising future generations.

The intersection between innovation and sustainability in the port sector is presented as an opportunity to address current challenges, such as climate change, energy efficiency, waste management, and integration with local communities. Technological innovation, efficient operating practices, and collaboration between the different actors in the logistics chain are identified as key aspects to achieve sustainability in ports [14].

2.3.1. Challenges and Opportunities in the Spanish Port System

If we talk about the challenges for the Spanish Port System, we should talk about the growth of maritime traffic, energy efficiency and emission reduction, waste management and protection of the marine environment, and the relationship with local communities. [15].

The growth of maritime traffic represents one of the main challenges in terms of sustainability for the Spanish Port System. The increase in demand for port services raises the need to optimize operations and resources to reduce congestion and minimize associated environmental impacts [16].

The growth of maritime traffic in Spanish ports not only represents a challenge in terms of congestion, but also highlights the need for resilient and sustainable infrastructure. In this sense, the implementation of sustainable mobility technologies, such as the use of alternative fuels (LNG, hydrogen and biofuels) and the electrification of port operations, has been one of the main strategies to reduce the carbon footprint of maritime transport. Likewise, the digitalization of customs and logistics processes has allowed for better coordination between port infrastructures and land transport systems, improving the efficiency of intermodal transport and reducing energy consumption [16].

Energy efficiency and emission reductions are key elements in the sustainability of the port sector. Energy consumption and greenhouse gas emissions associated with port operations must be reduced through the implementation of cleaner technologies, the use of renewable energy, and the adoption of sustainable operating practices [17].

Proper waste management and the protection of the marine environment are major challenges to ensure the sustainability of Spanish ports. The implementation of recycling systems, efficient wastewater management, and pollution prevention are essential aspects of preserving marine ecosystems and promoting biodiversity [18].

The relationship with local communities is a determining factor in the sustainability of the Spanish Port System. Citizen participation, dialogue, and collaboration with communities near ports are critical to addressing social concerns, improving residents' quality of life, and promoting sustainable port development [19].

Innovation opportunities in the Spanish Port System will preferably correspond to technological innovation, collaboration and cooperation, and circular economy [20].

2.3.2. Advancements and Research Gaps in Port Sustainability

Technological innovation plays a crucial role in the search for sustainable solutions in the port sector. Implementing advanced technologies, such as process automation, equipment electrification, and the use of artificial intelligence, can improve operational efficiency, reduce emissions, and optimize resource management [21].

Collaboration and cooperation between different players in the sector, including port authorities, operators, government agencies, and private companies, offer opportunities to drive innovation and promote sustainability. The creation of strategic alliances and the sharing of good practices can generate synergies and facilitate the implementation of innovative solutions [22].

The transition to a circular economy in the Spanish port sector can be an opportunity for innovation and sustainability. Implementing reuse, recycling, and waste reduction practices can generate economic, environmental, and social benefits, while promoting efficiency and resilience in ports [23].

The state of the art on innovation in the sustainability of the Spanish Port System highlights the need to address current and future challenges through innovative solutions. The intersection of innovation and sustainability offers opportunities to improve operational efficiencies, reduce environmental impacts, and strengthen relationships with local communities. The implementation of advanced technologies, collaboration between actors in the sector, and the transition to a circular economy are key aspects to achieve sustainable port development in Spain.

Despite the increasing emphasis on sustainability in port infrastructure, there remains a significant gap in research regarding the structured evaluation of innovation-driven sustainability strategies within the Spanish Port System. While previous studies have addressed various aspects of sustainability in ports—such as emissions reduction, energy efficiency, and digitalization—few have systematically integrated these elements into a comprehensive analytical framework that considers macro-environmental, economic, socio-cultural, and political factors simultaneously.

Most existing research tends to focus on isolated case studies or sector-specific solutions without fully exploring the interplay between technological, economic, and policy-driven innovations. Additionally, there is a lack of empirical assessments that measure the actual impact of sustainability initiatives using standardized methodologies like the Business Observation Tool (BOT). This research seeks to bridge this gap by providing a holistic evaluation of sustainability innovation within the Spanish Port System, identifying key drivers, constraints, and opportunities for improvement.

Furthermore, limited studies have examined the role of institutional and regulatory mechanisms as catalysts for sustainability-oriented technological adoption in ports. The findings of this paper contribute to the literature by offering an integrated assessment that links institutional support, economic viability, and technological advancements to the broader goal of sustainable port operations. By addressing this gap, this study aims to inform policymakers, port authorities, and industry stakeholders on how to optimize sustainability strategies through evidence-based decision-making.

Despite the increasing adoption of sustainable practices in European ports, research on the integration of innovation and sustainability in Spanish ports remains limited. While studies have examined the role of digitalization and energy efficiency separately, there is a lack of comprehensive frameworks that assess their combined impact on operational resilience and environmental performance. This study aims to fill that gap by applying the Business Observation Tool (BOT) to evaluate sustainability-driven innovations in selected Spanish ports.

3. Methodology and Results

The Business Observation Tool (BOT) methodology provides a structured approach to assessing the sustainability and innovation potential of port infrastructure. In the context of sustainable transport, the BOT allows for a comprehensive analysis of port efficiency, energy consumption, environmental impact, and stakeholder engagement. By integrating key indicators related to maritime logistics, digitalization, and decarbonization strategies, this methodology helps in identifying pathways to enhance the resilience and sustainability of transport infrastructures.

This study provides an overall assessment of the Spanish Port System as a whole, rather than focusing on individual ports. While it is acknowledged that some ports, such as Barcelona, Valencia, and Bilbao, may be more advanced in sustainability efforts, the aim of this research is to evaluate the general state of the system. The data sources include aggregated sustainability reports from various Spanish ports, stakeholder interviews, and operational performance metrics collected from port authorities and national maritime organizations. By taking a holistic approach, this study aims to reflect the broader trends and challenges faced by the Spanish port sector in its journey toward sustainability.

The aim is to develop the tool (BOT). It is a tool that allows one to start and recognize minimum elements to be considered, when formulating a business idea, through observation. It is a management tool and an alternative to PESTEL (Political, Economic, Social, Technological, Ecological, Legal).

The model aims to determine the reality of the conditions of the environment in which the business is expected to develop and to establish possible scenarios for the business to evolve beyond the observed reality.

The methodology developed is shown in Figure 1.

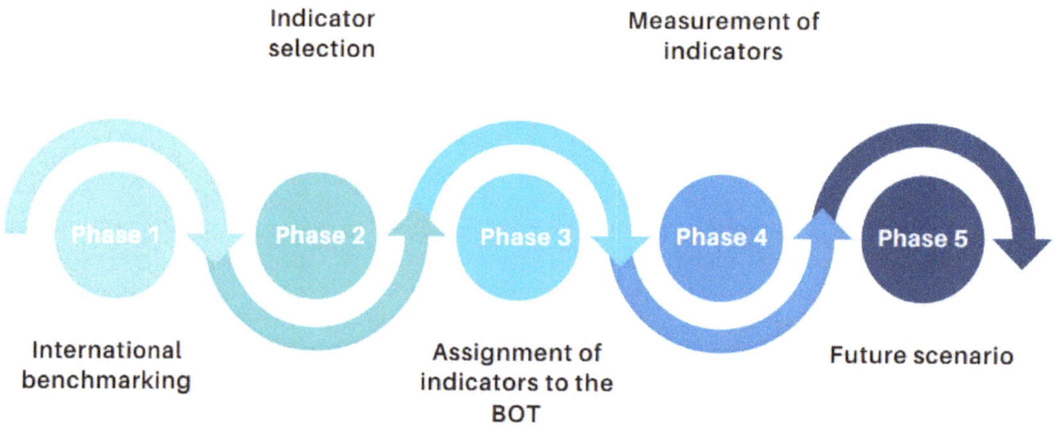

Figure 1. Methodology used, phases, and blocks. Source: own elaboration.

3.1. Generation of the Work Scenario

This study is based on the research and compilation of information on the current situation in terms of sustainability and more specifically on port innovation to achieve it, from, firstly, considering the world's ports and then focusing on those ports belonging to the Spanish Port System. All this has helped us to obtain the necessary information to define the context in which sustainability is developed in Spanish ports and their strategic objectives in relation to this matter, thus allowing us to address the objective of our study.

One of the key aspects of this methodology is its ability to analyze transport efficiency within port infrastructures. The BOT framework allows for the identification of inefficiencies in logistics chains, assessment of modal shift opportunities towards more sustainable transport modes (rail and short-sea shipping), and evaluation of energy-efficient technologies such as shore power supply, automation, and digital twins for predictive maintenance. This approach ensures that sustainability measures align with the operational needs of port logistics and intermodal transportation

The first step is to conduct an exhaustive study of the entire sustainability environment, and more specifically of innovation in relation to it, by compiling information from ports worldwide first and then focusing on Spanish ports. This analysis must be supported by the establishment of clear and defined foundations on which the final objective of this project will be based.

3.2. "Business Observation Tool" Analysis

The Business Observation Tool (BOT) is an observation and analysis tool used in the business environment to collect data and gain insights into organizational practices and processes. It is a methodology that is based on the direct observation of activities and behaviors in the business environment, which allows one to obtain valuable on operational efficiency, decision-making, interaction between employees, and other relevant aspects of the organization.

The Business Observation Tool (BOT) follows a structured five-phase methodology designed to assess the integration of innovation into sustainable port management. These phases include the following: (1) identification of relevant sustainability challenges; (2) selection of key performance indicators through international benchmarking; (3) assigning qualitative and quantitative values to measure sustainability impact; (4) evaluation of indicators based on expert input and real-world data; and (5) strategic decision-making based on insights from the BOT framework. The international benchmarking process involved an extensive literature review and case study analysis of leading European ports (e.g., Rotterdam, Hamburg, and Barcelona). The selection of indicators was carried out using a multi-criteria decision-making approach, ensuring that each indicator reflected industry best practices in sustainability and operational efficiency.

The BOT can be applied in different business contexts, such as manufacturing, service sector, and retail, among others. By directly observing and analyzing the day-to-day activities of the company, researchers or practitioners can identify opportunities for improvement, identify bottlenecks in processes, evaluate the effectiveness of policies and procedures, and make informed decisions to optimize organizational performance.

The BOT (Figure 2) allows us to identify the minimum elements to be considered when formulating a business idea; in this case, the 'business' will be that of innovation in sustainability in the Spanish Port System.

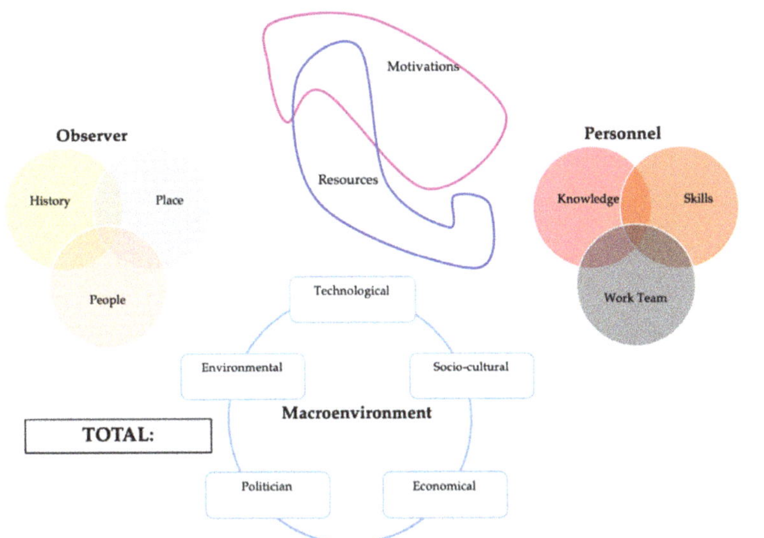

Figure 2. Business Observation Tool (BOT) analysis. Source: adapted from Álvaro Morales. https://www.lafabricatcm.cl/, accessed on 5 February 2025.

In this section, we will present the objective of choosing indicators for their subsequent evaluation, using measurement variables and scoring rules that are most appropriate for the case study.

The application of the BOT methodology in the Spanish Port System is particularly relevant for addressing sustainability challenges in maritime transport. Through a structured evaluation of port operations, energy consumption, and stakeholder interactions, the BOT allows for the development of targeted strategies to enhance sustainability. Key aspects include optimizing port traffic flow, implementing smart grid solutions, fostering collaboration between maritime and land transport operators, and integrating digital monitoring systems to track emissions and resource consumption.

This study aims to provide a comprehensive diagnostic of the Spanish Port System as a whole, rather than focusing exclusively on specific ports. The methodology evaluates sustainability trends, challenges, and opportunities across multiple Spanish ports, considering data from port sustainability reports, stakeholder interviews, and operational performance metrics. While Barcelona, Valencia, and Bilbao are highlighted due to their strategic relevance and advanced sustainability initiatives, the findings are meant to reflect broader trends applicable to the entire Spanish port network.

The BOT is divided into large blocks of work to determine the future scenario:

- Motivations and resources to move forward: Motivations and resources are often the most important scenarios. Therefore, they usually facilitate the viability of the project, enabling its execution with a high impact. We have studied the factors that can motivate us to carry out innovation in sustainability within Spanish ports, as well as the resources available to implement this innovation. To do this, a series of questions are answered:
    - ○ What is the motivation to develop the project and ensuring high impact?
    - ○ What resources are available for developing this project?
    - ○ What technological innovations can be integrated into port infrastructures to improve energy efficiency and operational sustainability?
- Formation of working groups: In this group, the composition of the labor system is analyzed. This takes into account the knowledge and skills that the agents involved (port operators and agents related to port operations) have or we are expected to have within the Spanish system. It is also be important to understand the shortcomings they may have in order to carry out innovation in sustainability. The following question is asked:
    - ○ What are the knowledge and skills of each team member, and how can they be leveraged to the project's advantage?
- Characterize and understand the development environment: The observation and understanding of the environment, as well as its trends in the past that may have had some kind of impact on the generation of our final plan, allow us to carry out projects with greater guarantees. The determination of these external factors, in the form of indicators facilitates the establishment of mechanisms and strategies for implementing the "business". Their determination is addressed by answering the following question:
    - ○ What external characteristics, in the form of indicators, can alter future decisions and strategies on sustainability innovation?
    - ○ How can digital technologies, such as IoT and AI, contribute to optimizing port logistics and reducing environmental impact?
- Macro-environmental analysis: The study of the macro-environment, also known as the generic environment, involves obtaining technological, socio-cultural, economic, political and environmental factors that may alter or intervene in the success of the project. That is why, this section focuses on the indicators that condition the objective of the study:
    - ○ What characterizes the observed environment and can influence us to favor the project?
    - ○ What policies and investments are needed to support the transition to sustainable port infrastructure and low-carbon transport networks?

3.2.1. Definition of Indicators

The selection of indicators plays a crucial role in evaluating the sustainability of port infrastructure and its impact on transport systems. These indicators provide a structured framework for assessing key dimensions such as economic viability, environmental impact, technological adoption, and social engagement. By analyzing these factors, decision-makers can develop strategic plans that align with global sustainability objectives while ensuring the operational efficiency of maritime transport.

After gathering all this information, we extract the possible indicators that generate some kind of impact on the "business" we want to analyze.

From the study of the work scenario, 75 indicators for the BOT were obtained, separated into the different groups proposed by the tool. Below, we will learn about them in greater detail, the reason for their choice, their identification, and their subsequent definition. All this will be synthesized through the elaboration of tables with indicators, which will respond to the different factors, depending on the scenario in which we find ourselves, that intervene in innovation in sustainability within the Spanish Port System.

The selection of indicators was based on a comparative analysis of best practices in port sustainability, utilizing reports from major European and global ports. This allowed for the identification of 75 indicators that were categorized under the BOT framework. These indicators were further validated through the Delphi method with experts from the maritime sector, ensuring that the selected criteria were relevant and applicable to the Spanish port context.

The expert panel consisted of 12 professionals specializing in port sustainability, logistics, and infrastructure management, each with at least 10 years of experience in the sector. Their evaluations were collected through structured interviews and the Delphi panel discussions. The weight assignment process followed an iterative approach where experts independently provided their assessments, and subsequent rounds were conducted to refine the weights. The process continued until a consensus threshold of ±5% deviation between expert responses was reached, ensuring reliability and minimizing bias in the final weights.

For the construction of the BOT after the study of the environment, the following were determined for each group:

- Motivations and resources to move forward: six indicators.
- Formation of working groups: nine indicators were developed, three in reference to Knowledge, three in reference to Skills and three in relation to Group work.
- Characterize and understand the development environment: thirteen indicators were developed, four in relation to History, six in reference to the port as a Place, and three in reference to People as the agents involved.
- Macro-environment analysis: This last scenario will be the one that includes the largest number of indicators because the object of study of this work will be innovation in sustainability, developing a total of forty-seven indicators, including six in the Technological field, nine in the Socio-Cultural field, eleven in the Economic, eight in the Political, and thirteen in the Environmental field.

To ensure a robust framework for evaluating innovation and sustainability in the Spanish Port System, an extensive benchmarking analysis was conducted. This process involved the review of previous studies on port sustainability, including reports from European ports recognized for their advanced sustainability strategies (e.g., Rotterdam, Hamburg, and Barcelona). Key dimensions such as energy efficiency, digitalization, and operational resilience were analyzed to identify best practices. As a result, 75 indicators were selected, grouped according to the BOT methodology. These indicators were classified based on their impact on port performance and sustainability, taking into account existing evaluation models such as PESTEL and circular economy frameworks. The selection process was further validated through expert consultations, ensuring relevance and applicability to the Spanish port context.

This represents a total of 75 indicators with which the database was developed. Below is an example of allocation and indicators (Table 1).

Table 1. Assigning indicators to the BOT (selection). Source: own elaboration.

Identifier	Indicator	Definition
M_3	Economic Profitability	Increasing profitability in ports will help recover large investments.
R_3	Previous Experiences	Recent innovations in sustainability.
C_2	Information	Information from involved agents on sustainability strategy, machinery, and equipment.
G_2	Communication	Communication between different members and across all levels of the organization.
HI_3	Change in Trade	Globalization of port markets. High service costs and port fees.
L_1	Logistics Chain Link	Seeking alternative financing due to investment shortages.
T_1	Digitalization and Automation	Seeking alternative financing due to investment shortages.
SC_3	Union Support	Union support in ports for a model shift towards sustainability.
A_2	Water Quality	Port water may become contaminated due to port operations.

For the construction of the BOT, a total of 75 indicators were identified and categorized into four main groups according to the affinity matrix methodology: (1) motivations and resources to move forward (6 indicators), (2) formation of working groups (9 indicators), (3) characterization and understanding of the development environment (13 indicators), and (4) macro-environment analysis (47 indicators, covering Technological, Socio-Cultural, Economic, Political, and Environmental aspects).

However, Table 1 presents a selected subset of indicators from different groups, chosen to illustrate the most representative aspects within each category. The full distribution of all indicators is detailed in the study framework but is not exhaustively included in the table for conciseness and clarity

Each indicator in the BOT framework follows a systematic coding structure, where prefixes denote thematic categories, followed by a numerical value indicating its placement within the hierarchy of assessed factors. This nomenclature ensures a structured approach to indicator classification, as exemplified in Table 1.

To ensure a structured classification of the indicators within the BOT framework, a specific nomenclature has been adopted. Each indicator is assigned an uppercase letter corresponding to its category in Spanish, followed by its specific descriptor in English. The categories are as follows: M (Motivaciones-Motivations), R (Recursos-Resources), C (Conocimiento-Knowledge), G (Grupo de Trabajo-Work Group), HI (Historia-History), L (Lugar-Place), P (Personas-People), T (Tecnológico-Technological), SC (Socio-Cultural), and A (Ambiental-Environmental) Table 1. This system allows for a clear and logical representation of the indicators while maintaining their connection to the original classification methodology.

Table 1 presents a selection of representative indicators from the total set of 75, illustrating how different categories contribute to the overall analysis. The full distribution of indicators is available within the study framework but is not exhaustively included in the table to maintain conciseness and clarity. This approach ensures that the methodology remains comprehensive while highlighting the most relevant aspects within each category.

These indicators serve as the foundation for the Business Observation Tool (BOT), enabling a structured analysis of sustainability challenges and opportunities within the Spanish Port System. By evaluating each indicator through measurement variables and scoring methodologies, the BOT provides actionable insights that guide decision-making.

This structured approach ensures that innovation efforts are aligned with sustainability goals, enhancing the resilience and efficiency of port operations within the broader framework of sustainable transport infrastructure.

3.2.2. Measurement of Indicators

Once the indicators have been obtained that, at a current level, can influence the understanding of the reality of Spanish ports in terms of innovation in sustainability, they are measured.

The measurement will consist of two parts, a first where a weight is assigned to the indicator according to its importance within the Spanish Port System, and a second where we will try to measure the indicator using measurement variables with current data.

All this allows us to evolve beyond the observed reality, proposing possible objectives and strategies for the future, which will help to achieve sustainability within Spanish ports.

MEASUREMENT 1. WEIGHT ALLOCATION

In the first measurement, a weight was assigned to each indicator, depending on its importance for innovation in sustainability within the Spanish Port System.

The weighting of the selected indicators was carried out through a qualitative–quantitative mixed approach. Initially, a panel of port industry experts was consulted to determine the perceived importance of each indicator within the sustainability framework. The Delphi method was employed to achieve a consensus among the experts, ensuring that subjective biases were minimized. In parallel, quantitative data from sustainability reports of Spanish ports were analyzed to correlate the impact of each indicator with real operational performance metrics. The combination of expert-based qualitative assessments and empirical quantitative analysis enabled a structured and objective assignment of weights, thus enhancing the reliability and validity of the evaluation framework.

The qualitative–quantitative analysis in this study consisted of two phases: First, a panel of experts from the port industry assessed each indicator's importance using the Likert scale methodology. This was followed by a statistical validation of these indicators using empirical data extracted from port sustainability reports. The weighting process was further refined using the Delphi technique, ensuring a balanced assessment that integrated expert judgment with quantitative performance metrics.

Following the selection of a group of experts, a qualitative–quantitative analysis was carried out on the basis of the selected indicators. Depending on the degree of importance of the indicator considered within the innovation in sustainability of the Spanish Port System, each indicator was scored qualitatively and quantitatively. As seen when defining these indicators, the scores were determined by differentiating them into the different groups that comprise the tool used, BOT, which has been classified by colors. These scores were classified into three groups depending on the degree of importance of the indicator, which are 1, 2.5 and 5; the value 0 was discarded because these indicators have been proposed for the purpose of influencing innovation in sustainability, even if to a minimal extent. The measurements are as follows:

- Low Importance "LOW" >1
- Medium degree of importance "MEDIUM" >2 to 5
- High degree of importance "HIGH" >5

The weight allocation methodology employed in this study follows a well-established multi-criteria decision-making approach, which provides reliability and reproducibility in sustainability assessments. Alternative methods, such as fuzzy logic or AHP (Analytic Hierarchy Process), were considered but were deemed less suitable given the dataset characteristics and expert availability. Future research may explore the integration of AI-based weighting models to enhance automation and precision in sustainability evaluations.

An example of the measurement results can be seen in Figure 3.

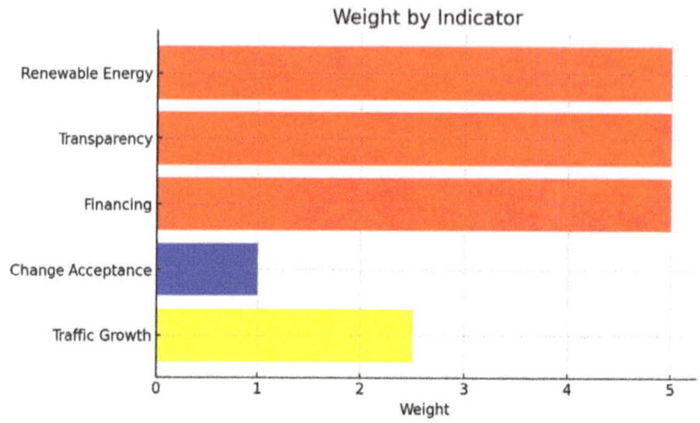

Figure 3. Examples of measurement 1. Source: own elaboration.

After the development of measurement 1 and the assignment of the values, the results are summarized in Figure 4.

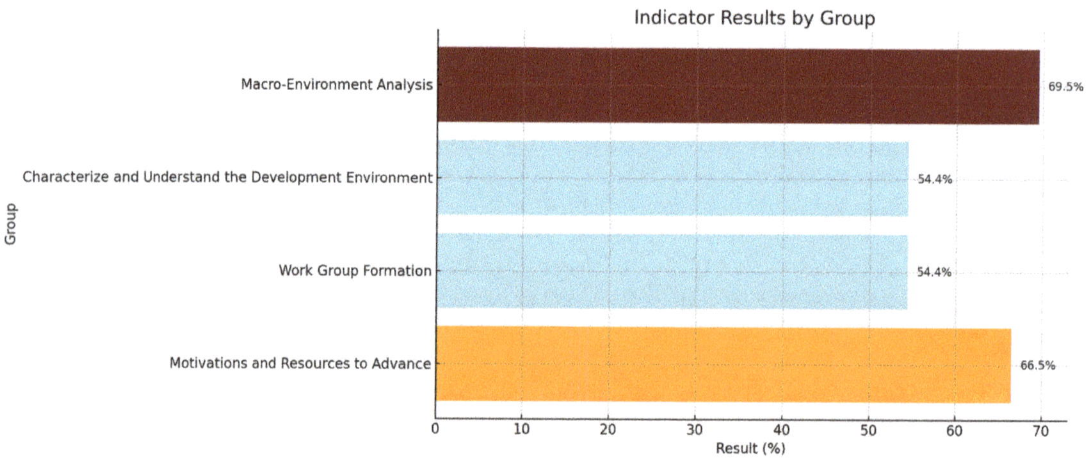

Figure 4. Measurement results 1. Source: own elaboration.

MEASUREMENT 2. ASSIGNING VALUES TO VARIABLES

In the second measurement, the indicators were measured according to their value within the Spanish Port System. This was achieved through the definition of a measurement variable, which was assigned to each of them to help understand the current situation within Spanish ports compared to ports worldwide.

The objective of this measurement was to assign values to the measurement variables of each indicator, allowing us to measure each one based on its value within the Spanish Port System and in comparison with the rest of the world's ports. The measurement was conducted in the same way as the first, differentiating the indicators by groups and colors according to the study tool, BOT. The scores were classified into three groups, depending on the situation of each indicator within the Spanish Port System compared to ports worldwide. These scores are as follows:

- Bottom "LOW" > 1
- Equivalent "MEDIUM" > 2.5
- Superior "HIGH" > 5

The classification of certain qualitative indicators into LOW, MEDIUM, and HIGH categories was conducted based on expert consensus, ensuring a structured assessment of factors such as 'Previous Experiences' and 'Information'. This approach, while simplified, facilitates comparability and standardization across multiple case studies. Further refinements, including ordinal scaling or fuzzy logic applications, may enhance the robustness of future assessments.

An example of the results of measurement 1 can be seen in Figure 5.

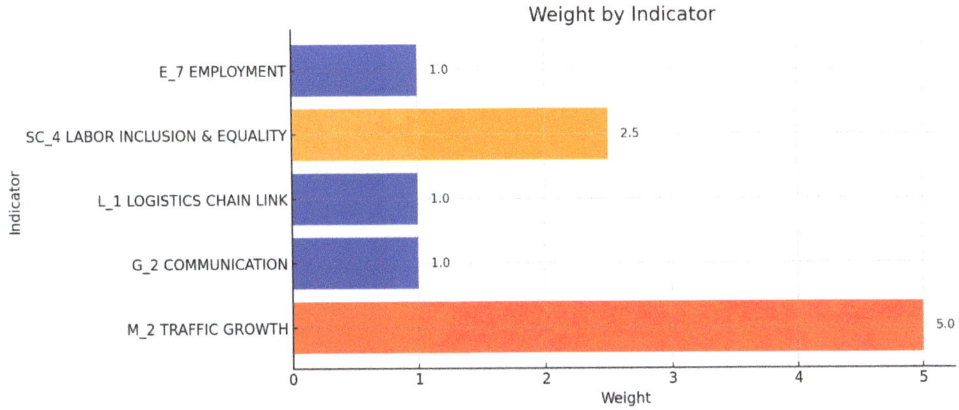

Figure 5. Measurement example 2. Source: own elaboration.

After the development of measurement 2 and the assignment of the values, the results are summarized in Figure 6.

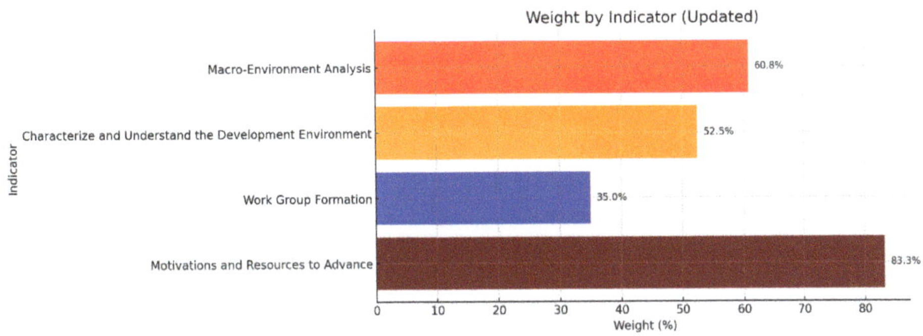

Figure 6. Measurement results 2. Source: own elaboration.

4. Analysis of Results

Analysis of results is the final and conclusive part of an investigation. In this section, all information from our study will be processed, presented in an orderly and understandable way, and analyzed to draw conclusions based on the measurements and their data.

The application of the BOT framework allowed for the identification of critical areas where innovation is most needed to enhance sustainability in Spanish ports. The results indicate that technological advancements, particularly in energy management and automation, have the highest potential impact. Ports with higher levels of digitalization (e.g., Barcelona and Valencia) exhibited better sustainability performance metrics compared to less automated ports. Specifically, the use of renewable energy sources and electrification of port operations were identified as key drivers for reducing carbon emissions. The results were further compared with international benchmarks, demonstrating that Spanish ports still lag behind in full-scale automation but show promising advancements in green energy implementation. These findings underscore the necessity for targeted investments in smart port technologies and intermodal transport solutions to bridge the sustainability gap.

To determine a future scenario, we will have to address past results and know where we want to go. After carrying out the first measurement, which identified the factors or scenarios within the BOT that were most important to achieve innovation in sustainability, it will become evident that the macro-environment group is the most relevant. This is because it includes sub-groups such as economic, political, socio-cultural, or environmental, which are fundamental to achieving sustainability. Therefore, to achieve our objective, we must formulate strategies that allow us to strengthen this scenario, always keeping in mind the motivations and resources necessary for their implementation.

Another aspect, which is clear here, is the irrelevance of work teams when it comes to achieving innovation in sustainability, leading to not having to take measures in this regard in the near future, but making it clear that in the long run, and for it to be achieved optimally, special emphasis will have to be placed on the training of workers, since most of the innovative measures will have a high technological degree that port agents must handle appropriately. A reconversion of the labor fabric will have to be carried out, since the staff does not have the necessary qualifications and experience for the new developments.

After the second measurement, clearer conclusions can be drawn about which aspects should be enhanced and which will act as driving forces for others. At first, the limitations of the technological pillar will become evident, in relation to digitalization processes, integration of information technologies and innovative information systems, making this the critical point at present. This situation will have to be reversed in the short term to avoid reduction in efficiency and guarantee future growth compared to other international ports.

On the economic side, we will be on the right track to achieve the objectives, but there will be factors with ample room for improvement in the short term, such as increase in employment or the search for new financing techniques. It is also worth mentioning here that significant improvement has been achieved by Spanish ports in environmental matters in recent years, obtaining numerous recognitions from renowned institutions in the field of sustainability at European and global levels. In this regard, the strategy for the future will be to follow the same line as before.

Finally, but the most important according to the results collected, is the political scenario. This indicates that the Spanish Port System, through its different institutions, is fully involved with sustainability and innovation, developing new proposals and ensuring that sustainability objectives are met. The strength of this scenario means that it can act as a driving force in achieving the sustainability innovation objectives for the other parties.

The results obtained through the application of the Business Observation Tool (BOT) to the Spanish Port System provide valuable insights into the current state of innovation in sustainability. The findings highlight key strengths, weaknesses, and potential opportunities to enhance the sustainability of port infrastructure and operations. This section will discuss the implications of the two measurement phases, analyze trends, and propose strategic actions to optimize the implementation of sustainability-driven innovation.

Key Findings from Measurement 1: Weight Allocation

The first measurement assigned weights to various indicators based on their perceived importance within the Spanish Port System. The macro-environmental factors emerged as the most critical, with a weight of 69.5%, followed by motivations and resources (66.5%), observation (54.4%), and work group formation (45.5%).

The high weight assigned to macro-environmental factors highlights their fundamental role in achieving sustainability in ports. Economic, political, socio-cultural, and environmental elements shape the feasibility and effectiveness of sustainable innovation. Notably, political support emerged as a key enabler, with Spanish institutions demonstrating a strong commitment to sustainability through policies and funding mechanisms. This finding underscores the importance of aligning port innovation with national and European sustainability policies to ensure regulatory and financial support.

The relatively low weight assigned to work group formation suggests that workforce development is not perceived as an immediate priority for implementing sustainability innovations. However, this could present a long-term challenge, as many innovative solutions require advanced technical expertise. Thus, while immediate focus should be placed on technological and policy innovations, a parallel effort is needed to ensure that port professionals are adequately trained to operate and manage these new systems.

Key Findings from Measurement 2: Indicator Value Assignment

The second measurement assessed the actual value of indicators within the Spanish Port System, comparing them to international benchmarks. The findings revealed significant disparities in different sustainability dimensions.

One of the most critical findings is the relatively low technological maturity of Spanish ports compared to global benchmarks. The automation of container terminals remains limited, with only two semi-automated terminals (Barcelona and Algeciras). Digitalization processes and the adoption of smart port technologies, such as IoT and AI-driven logistics, require accelerated investment to prevent efficiency losses and maintain competitiveness.

The socio-cultural and political indicators scored high, reflecting Spain's strong institutional support for sustainability initiatives. Spanish ports have demonstrated leadership in labor inclusion, regulatory compliance, and stakeholder engagement. However, the frequent occurrence of labor strikes highlights the need for improved labor relations and collaborative approaches to balance worker rights with operational efficiency.

The economic performance of Spanish ports presents mixed results. While connectivity and trade facilitation are strong (e.g., maritime highways), employment growth in the sector has stagnated. Financing mechanisms, such as the Ports 4.0 fund, have been successful in promoting innovation, but further diversification of funding sources is needed. On the environmental front, Spain has made significant progress, particularly in water quality management, emission control, and waste management. However, air pollution remains a concern, especially in high-traffic ports like Barcelona and Palma de Mallorca.

The Business Observation Tool (BOT) has been previously utilized in the strategic analysis of automation and sustainability in port terminals, as demonstrated in prior research [16]. The systematic evaluation approach of BOT enables its adaptation to various port environments, facilitating data-driven decision-making for sustainable infrastructure and operational efficiency.

While Barcelona and Valencia exhibit strong digitalization and green energy initiatives, Bilbao presents a distinct profile, characterized by its strategic investment in industrial sustainability. The Port of Bilbao has focused on offshore wind energy projects and carbon capture technologies, positioning itself as a leader in renewable energy integration within the Spanish Port System. However, its slower adoption of full-scale automation compared to Barcelona and Valencia highlights an opportunity for targeted innovation in logistics and operational efficiency.

Strategic Implications and Future Directions

Based on the results, several strategic actions can be proposed to enhance sustainability innovation in Spanish ports.

- Accelerating Technological Adoption
  - Increase investments in smart port technologies, including AI, digital twins, and automation.
  - Expand digitalization initiatives, ensuring interoperability between port systems and logistics networks.
  - Strengthen cybersecurity measures to protect digital infrastructure.
- Enhancing Policy and Institutional Support
  - Continue leveraging European sustainability initiatives to secure funding and regulatory support.
  - Foster collaboration between public and private stakeholders to enhance policy implementation.
  - Promote standardization in sustainability reporting and performance monitoring.
- Addressing Economic and Workforce Challenges
  - Develop financial incentives for green investments in port infrastructure and operations.
  - Strengthen employment policies to balance labor rights with efficiency improvements.
  - Implement training programs for port workers to adapt to new technologies.
- Strengthening Environmental Initiatives
  - Expand the electrification of docks and promote shore power connections.
  - Enhance pollution monitoring systems to address air and water quality issues.
  - Foster circular economy initiatives, including waste-to-energy projects.

The application of the BOT methodology has provided a comprehensive understanding of the factors influencing sustainability innovation in the Spanish Port System. While significant progress has been made in regulatory and environmental efforts, technological and economic challenges remain. By prioritizing digital transformation, enhancing policy frameworks, and fostering economic and workforce resilience, Spanish ports can strengthen their sustainability performance and position themselves as global leaders in green maritime transport.

5. Conclusions

This study provides a structured approach to assessing sustainability in the Spanish Port System through the application of the Business Observation Tool (BOT). The results highlight progress in adopting green energy solutions and digitalization strategies, while also revealing significant challenges in automation, intermodal connectivity, and regulatory alignment needed to foster innovation. The integration of smart technologies, such as IoT-based monitoring systems and AI-driven logistics optimization, should be prioritized to bring Spanish ports closer to international best practices.

The findings emphasize that technological advancements play a crucial role in improving efficiency and reducing emissions. However, digital transformation remains a challenge, and many ports still face barriers to full automation. Achieving sustainability requires coordinated efforts across multiple sectors, as governance, regulatory frameworks, and economic incentives shape the adoption of innovation. The macro-environment, including economic, social, and environmental factors, is a determining element in the sustainability of port operations, requiring a balanced approach that does not compromise long-term development.

From a methodological perspective, this study contributes to the field by integrating the BOT framework with multi-criteria decision analysis, providing a structured and replicable model for evaluating sustainability in port operations. While the research is centered on the Spanish Port System, the methodology can be adapted to other European ports facing similar challenges. The cases of Rotterdam, Hamburg, and Antwerp serve as relevant examples where this approach could be applied to systematically assess sustainability strategies and measure innovation-driven improvements.

Looking ahead, Spanish ports are expected to continue their transition toward smart port models, where technology plays a fundamental role in optimizing logistics, reducing environmental impact, and ensuring more efficient and resilient operations. Key areas for development include further investment in shore-side electrification, the use of AI-driven predictive maintenance for port equipment, and the implementation of blockchain solutions to improve transparency and efficiency in maritime logistics.

Despite significant progress made in recent years, achieving a fully sustainable port system will require greater collaboration among port authorities, policymakers, and industry stakeholders. Future research should explore comparative analyses between Spanish and other European ports, further refining best practices and fostering innovation in sustainable port management.

Author Contributions: Methodology, J.V.-C., J.D.R. and N.G.-C.; Software, J.V.-C.; Validation, N.G.-C.; Formal analysis, N.G.-C.; Investigation, J.V.-C., J.D.R., N.G.-C. and A.C.-O.; Writing – original draft, N.G.-C.; Writing – review & editing, J.V.-C., J.D.R., N.G.-C. and A.C.-O.; Supervision, J.V.-C. All authors have read and agreed to the published version of the manuscript.

Funding: This research received no external funding.

Institutional Review Board Statement: Ethical review and approval were waived for this study because this research consists of a questionnaire on port exploitation indicators, with no collection of sensitive personal data or interventions affecting participants. The participants are port industry professionals, not a vulnerable population. According to Spanish research regulations, including the General Data Protection Regulation (GDPR) and institutional policies, this type of study is exempt from ethical review requirements.

Informed Consent Statement: Patient consent was waived because according to the research regulations of the Universidad Politécnica de Madrid (UPM), our study does not require written informed consent from participants. The UPM's ethical guidelines specify that informed consent is only mandatory when sensitive personal data are collected or when the study involves interventions that could affect participants. Our research does not fall into any of these categories, as it only involves a focus group with professionals from the maritime sector, without the collection of any personal or sensitive data. Additionally, the UPM Ethics Committee only requires ethical review and consent documentation for studies involving human subjects under specific conditions, such as medical, psychological, or personally intrusive research. Since our study does not involve such elements, it is exempt from formal ethical review and informed consent requirements.

Data Availability Statement: The data used in this study is not publicly available due to confidentiality agreements with the participating port terminals. Aggregated or anonymized data may be provided upon reasonable request.

Acknowledgments: The authors would like to thank the Polytechnic University of Madrid in general and particularly the Department of Transport, Territory and Urban Planning Engineering of the Polytechnic University of Madrid.

Conflicts of Interest: The authors declare no conflicts of interest.

References

1. Hristov, I.; Chirico, A. The Role of Sustainability Key Performance Indicators (KPIs) in Implementing Sustainable Strategies. *Sustainability* **2019**, *11*, 5742. [CrossRef]
2. Fratila (Adam), A.; Gavril (Moldovan), I.A.; Nita, S.C.; Hrebenciuc, A. The Importance of Maritime Transport for Economic Growth in the European Union: A Panel Data Analysis. *Sustainability* **2021**, *13*, 7961. [CrossRef]
3. Notteboom, T.E.; Parola, F.; Satta, G. The relationship between transhipment incidence and throughput volatility in North European and Mediterranean container ports. *J. Transp. Geogr.* **2019**, *74*, 371–381. [CrossRef]
4. González-Cancelas, N.; Molina Serrano, B.; Soler-Flores, F.; Camarero-Orive, A. Using the SWOT Methodology to Know the Scope of the Digitalization of the Spanish Ports. *Logistics* **2020**, *4*, 20. [CrossRef]
5. Alamoush, A.S.; Ballini, F.; Ölçer, A.I. Revisiting port sustainability as a foundation for the implementation of the United Nations Sustainable Development Goals (UN SDGs). *J. Shipp. Trade* **2021**, *6*, 19. [CrossRef]
6. Caliskan, A. Seaports participation in enhancing the sustainable development goals. *J. Clean. Prod.* **2022**, *379*, 134715. [CrossRef]
7. Othman, A.; El-gazzar, S.; Knez, M. A Framework for Adopting a Sustainable Smart Sea Port Index. *Sustainability* **2022**, *14*, 4551. [CrossRef]
8. Di Vaio, A.; Zaffar, A.; Balsalobre-Lorente, D.; Garofalo, A. Decarbonization technology responsibility to gender equality in the shipping industry: A systematic literature review and new avenues ahead. *J. Shipp. Trade* **2023**, *8*, 9. [CrossRef]
9. Brunila, O.-P.; Kunnaala-Hyrkki, V.; Inkinen, T. Sustainable small ports: Performance assessment tool for management, responsibility, impact, and self-monitoring. *J. Shipp. Trade* **2023**, *8*, 14. [CrossRef]
10. Del Giudice, M.; Di Vaio, A.; Hassan, R.; Palladino, R. Digitalization and new technologies for sustainable business models at the ship–port interface: A bibliometric analysis. *Marit. Policy Manag.* **2022**, *49*, 410–446. [CrossRef]
11. Ranking Measures to Improve the Sustainability of Mediterranean Ports Based on Multicriteria Decision Analysis: A Case Study of Souda Port, Chania, Crete | Environment, Development and Sustainability. Available online: https://link.springer.com/article/10.1007/s10668-021-01711-7 (accessed on 13 December 2023).
12. Zhao, D.; Wang, T.; Han, H. Approach towards Sustainable and Smart Coal Port Development: The Case of Huanghua Port in China. *Sustainability* **2020**, *12*, 3924. [CrossRef]
13. Klar, R.; Fredriksson, A.; Angelakis, V. Digital Twins for Ports: Derived from Smart City and Supply Chain Twinning Experience. *IEEE Access* **2023**, *11*, 71777–71799. [CrossRef]
14. Chlomoudis, C.; Pallis, P.; Platias, C. Environmental Mainstreaming in Greek TEN-T Ports. *Sustainability* **2022**, *14*, 1634. [CrossRef]
15. Gonzalez, J.N.; Camarero-Orive, A.; González-Cancelas, N.; Guzman, A.F. Impact of the COVID-19 pandemic on road freight transportation—A Colombian case study. *Res. Transp. Bus. Manag.* **2022**, *43*, 100802. [CrossRef] [PubMed]
16. Santos Martín, A.E.; González-Cancelas, N.; Serrano, B.M.; Soler-Flores, F. Towards the sustainability of the Spanish port system through the business observation tool. *Proc. Inst. Civ. Eng. Marit. Eng.* **2021**, *174*, 53–61. [CrossRef]
17. Iris, Ç.; Lam, J.S.L. A review of energy efficiency in ports: Operational strategies, technologies and energy management systems. *Renew. Sustain. Energy Rev.* **2019**, *112*, 170–182. [CrossRef]
18. Puig, M.; Azarkamand, S.; Wooldridge, C.; Selén, V.; Darbra, R.M. Insights on the environmental management system of the European port sector. *Sci. Total Environ.* **2022**, *806*, 150550. [CrossRef]
19. Spengler, T.; Tovar, B. Potential of cold-ironing for the reduction of externalities from in-port shipping emissions: The state-owned Spanish port system case. *J. Environ. Manag.* **2021**, *279*, 111807. [CrossRef]
20. Haezendonck, E.; Van den Berghe, K. Patterns of Circular Transition: What Is the Circular Economy Maturity of Belgian Ports? *Sustainability* **2020**, *12*, 9269. [CrossRef]
21. Acciaro, M.; Renken, K.; El Khadiri, N. Technological Change and Logistics Development in European Ports. In *European Port Cities in Transition: Moving Towards More Sustainable Sea Transport Hubs*; Carpenter, A., Lozano, R., Eds.; Strategies for Sustainability; Springer International Publishing: Cham, Switzerland, 2020; pp. 73–88. ISBN 978-3-030-36464-9.
22. Notteboom, T.; van der Lugt, L.; van Saase, N.; Sel, S.; Neyens, K. The Role of Seaports in Green Supply Chain Management: Initiatives, Attitudes, and Perspectives in Rotterdam, Antwerp, North Sea Port, and Zeebrugge. *Sustainability* **2020**, *12*, 1688. [CrossRef]
23. Tola, F.; Mosconi, E.M.; Marconi, M.; Gianvincenzi, M. Perspectives for the development of a circular economy model to promote ship recycling practices in the European context: A systemic literature review. *Sustainability* **2023**, *15*, 5919. [CrossRef]

sustainability

MDPI

Article

BIM Uses for Mitigating Deficiencies in Road Scheduling Planning

Karen Castañeda [1,2], Omar Sánchez [3], Carlos A. Peña [4,5], Rodrigo F. Herrera [1,*] and Guillermo Mejía [2]

1 School of Civil Engineering, Pontificia Universidad Católica de Valparaíso, Valparaíso 2374631, Chile; karen.castaneda@pucv.cl or karen2178721@correo.uis.edu.co
2 Department of Civil Engineering, Universidad Industrial de Santander, Bucaramanga 680002, Colombia; gmejia@uis.edu.co
3 Department of Civil Engineering, Pontificia Universidad Javeriana, Bogotá D.C. 111711, Colombia; omar_sanchezr@javeriana.edu.co
4 Department of Civil Engineering, Universitat Politècnica de València, 46022 Valencia, Spain; carpeso1@upv.edu.es or carlosalbertops@ufps.edu.co
5 Department of Civil Engineering, Universidad Francisco de Paula Santander, Cúcuta 540003, Colombia
* Correspondence: rodrigo.herrera@pucv.cl

Abstract: Efficient scheduling is essential for successful and sustainable road construction projects. However, conventional planning approaches often lack adaptability, visualization, and integration with modern technologies, leading to schedule deficiencies and cost overruns. Therefore, this paper studies the influence of BIM uses in mitigating planning deficiencies in road construction scheduling. A four-stage research method was employed: (1) identification of BIM uses relevant to road projects, (2) determination of principal causes contributing to scheduling deficiencies, (3) cross-impact analysis to quantify BIM's influence on these deficiencies, and (4) expert interviews to characterize key BIM uses. Results highlight that 4D construction planning, quantity take-off and cost estimation, and traffic management planning positively impact road scheduling, reducing errors in work breakdown structures, resource allocation, and coordination between project stakeholders. BIM streamlines communication, supports proactive risk management, and enables real-time schedule adjustments by providing an integrated digital environment. This research shows the advantages of applying BIM in road construction to improve project planning, curtail scheduling setbacks, and encourage adopting innovative practices. The results provide information that planners can utilize to enhance the use of BIM in scheduling road construction projects.

Keywords: BIM; scheduling; construction planning; road infrastructure; deficiencies; cross impact matrix

1. Introduction

Road infrastructure is essential for the development and progress of any country [1]. However, this industry faces several obstacles, such as managing limited resources, meeting tight deadlines, and dealing with environmental and social impacts. Planning the road construction schedules is crucial to ensuring these projects' efficient and timely completion [2]. However, the methods for preparing such schedules face significant challenges, compromising accuracy and efficiency [3]. Furthermore, the growing need for sustainable road networks has prompted the industry to seek innovative procedures to transform this field [4]. The process's complexity is compounded by the coordination of various teams and tasks, uncertainty in the availability of materials, and variable weather conditions, among others [5]. Considering these challenges, it is essential to implement methodologies and technologies that support the scheduling planning process and ensure the timely delivery of road projects while meeting budget, quality, and sustainability goals [6].

Scheduling road projects involves various traditional methods that depend on the project's complexity, available resources, and specific characteristics of each construction site. The most common methods include Gantt charts for visual planning, the Critical Path Method (CPM), and the Project Evaluation and Review Method (PERT) for time management [7–9]. These methods are essential in project management as they provide visibility of activities, optimize resources, and help identify the critical path. However, traditional approaches have limitations like the need for constant updates due to changes in requirements, their rigidity, which makes them less adaptable for dynamic projects, and lack of integration with 3D models, which can limit the visualization of the physical development of the project and make it challenging to identify potential problems during construction [9]. Therefore, traditional methods require technological and methodological adaptations to overcome limitations and maximize their usefulness in project management.

Technological advancements, such as augmented reality, geographic information systems, the Internet of Things, artificial intelligence, digital twins, 3D printing, and Building Information Modeling (BIM), are transforming the planning and management of construction projects [10]. Among these, BIM has shown remarkable benefits in complementing traditional scheduling methods in construction projects. BIM enables the creation of three-dimensional digital models of a project's physical and functional characteristics, offering a detailed visualization of the construction process before its commencement [11]. This significantly improves planning and schedule management by integrating time and cost data with the 3D model, providing real-time construction simulations and work sequence analysis [12]. Visualizing the project's development over time helps identify potential conflicts, cost overruns, or delays before they occur, improving decision-making and coordination between teams. Furthermore, BIM can support resource optimization and risk management by providing a deeper understanding of the project, facilitating more accurate and efficient planning [13].

Research in the construction sector's adoption of BIM for schedule planning has made significant progress [14], particularly in the field of buildings. Predictive models based on BIM have been developed to enable more integrated and collaborative project management, leading to more informed and efficient decision-making [15]. These studies have utilized advanced technologies such as genetic algorithms and multi-objective optimization systems to address scheduling challenges and optimize the balance between time and cost [16,17]. Implementing these models has improved the accuracy and efficiency of schedule generation, with several case studies demonstrating reduced planning time and improved cost management. Additionally, hybrid systems that integrate BIM with multi-objective-based techniques (MOFBI) and multi-criteria decision-making (MCDM) to balance resources in project scheduling have been explored, showing results with uniform distributions of solutions [18]. Other studies highlight using BIM to facilitate communication and collaboration, implement decision support systems, and improve the accuracy of cost and schedule estimates. Further research demonstrates the effectiveness of BIM in coding and breaking down construction activities, managing progress through three-dimensional and PDCA models, and automating the sequence of construction activities [19]. This has led to reduced planning times and improved accuracy. These studies highlight how BIM can significantly contribute to solving construction project management challenges by optimizing schedule, cost, and collaboration between participants.

Despite significant progress in schedule planning in the construction industry, notable challenges remain, such as project delays and cost overruns [20–22]. These challenges are mainly related to the deficiencies observed during the planning phase of road projects, where managing a vast amount of information to create work breakdown structures is critical to planning and controlling the construction process effectively. Several shortcomings have been identified that directly impact schedules, including those related to the quantification of materials, integration with traffic management plans, overestimation of equipment and personnel productivity, and problems in assigning precedence relationships, among others [23]. BIM has been proven to offer considerable benefits in schedule management, highlighting its ability to mitigate many deficiencies in road project planning [24,25]. Despite this, the adoption of BIM in this sector is still limited, underscoring the need for focused research into how this technology could mitigate various factors that negatively affect schedules.

The integration of emerging technologies in the planning of road construction schedules has been little explored. This presents an opportunity for further research to adapt and expand advanced technologies to address the unique challenges involved in road infrastructure construction. BIM can be used to plan and manage schedules and has the potential to help mitigate some of the issues that influence the planning of construction schedules for road projects [10,11,23]. Therefore, this study aims to (1) estimate the influence of using BIM to mitigate road project scheduling deficiencies and (2) characterize the BIM uses required to mitigate road project scheduling deficiencies. Thus, the development of this research provides an understanding of effective BIM integration and how it can lead to significant improvements in schedule planning, directly contributing to schedule deficiencies that affect road construction. The results provide a valuable perspective for improving road infrastructure while highlighting the need to adopt innovative practices in project planning and execution.

The innovation of this study lies in the systematic methodological approach that directly connects specific deficiencies in road schedule planning with the most suitable use of BIM to address those deficiencies. Unlike previous research, which has primarily focused on buildings or partial scheduling methods, this study explores the use of BIM in road infrastructure. Using a cross-impact matrix, it quantifies how each BIM use contributes to alleviating the identified planning deficiencies. Additionally, semi-structured interviews are conducted to complement the quantitative analysis and better understand the practical implementation of these BIM uses in real-world contexts. This combination of methods (expert questionnaires, assessments of deficiencies in road construction projects, and qualitative interviews) creates a framework that researchers could adopt in other contexts. It identifies the most impactful BIM uses and provides concrete guidelines for improving scheduling accuracy and efficiency in road projects. Therefore, this study contributes to the field of civil engineering by providing empirical evidence and practical strategies for enhancing road construction planning through BIM adoption.

2. Literature Review

BIM implementation in construction projects has remarkably improved schedule planning [11,12,24]. Through BIM, the creation of detailed digital replicas of projects is facilitated, allowing the execution of simulations of the construction process before the physical start of construction. This anticipation capability substantially improves decision-making processes by enabling the analysis and comparison of alternative construction procedures [14]. In this way, project managers can select the option that maximizes efficiency and profit based on a thorough and detailed analysis facilitated by BIM [26]. The benefits of BIM in planning construction project schedules have led to several studies in the field. Nguyen et al. [18] developed a hybrid system that integrates BIM, multi-objective forensics (MOFBI), and multi-criteria decision-making (MCDM) to facilitate resource balancing in project scheduling. To demonstrate the efficiency and effectiveness of the proposed framework, they applied it to three building project scheduling problems. The study found that the framework can generate result curves with highly uniform resource distribution of solutions and provide visual analysis to aid decision-making. Guo et al. [27] have created a system for coding and breaking down construction activities, which can be used for automated scheduling with BIM implementation. The system helps calculate each construction activity's duration, allowing for the automatic generation of component-level schedules for the most commonly used hybrid concrete structures in building projects. This provides a time sequence for construction simulation and automation for intelligent construction equipment.

Rehman et al. [28] conducted a literature review to investigate the main factors that generate delays in construction projects and how BIM can help address these issues. They supplemented their research with primary data collection and a case study to evaluate the effectiveness of the aspects studied. The authors concluded that BIM substantially impacts construction projects by providing an effective solution for schedule management and mitigating delay risks. In schedule control, Zhang et al. [19] presented a study to improve the use of technology in engineering projects by focusing on construction progress management. They used a parametric modeling technique to create a three-dimensional BIM model, which was integrated into the Glodon BIM5D platform to develop a four-dimensional model. The PDCA (plan-implement-check-act) cycle theory was then applied to the project. Through a case study, the proposed methodology was tested, and it was found that integrating BIM and PDCA clarified schedule control objectives and optimized the management process. This resulted in a 60% reduction in scheduled simulations and a 3.5% advancement in the construction period. Roa'a et al. [15] examined the valuable contributions of project management in managing the cost and schedule of construction projects by providing higher levels of precision and reliability. Online surveys and interviews were implemented to determine the importance of BIM technology in managing the cost and schedule of construction projects. A study case of a building is used to compare in terms of accuracy, effort, and quantity budgeting between manual results and REVIT software (https://www.autodesk.com/products/revit/overview, accessed on 30 January 2025) estimates. The findings indicate that applying project management theories significantly reduces budget, time, and effort in construction projects and is essential to prevent financial losses and problems with cost overruns or delays. They also show that using modern methods such as BIM improves the accuracy of cost and schedule estimates, helping to avoid errors.

Singh et al. [17] proposed a method to automate the installation of pipes and optimize the schedule using 4D BIM technology. The technique extracts information from 3D models and facilitates time planning. The method establishes specific rules for automatically matching piping components with installation activities and uses constraint analysis with sequential rules to coordinate piping systems efficiently. A case study was conducted, which demonstrated that this approach could generate optimal installation sequences and schedules, reducing planning time by 96% to 97% and significantly improving accuracy compared to traditional methods. ElMenshawy and Marzouk [16] have developed a model that automatically creates schedules for construction projects using BIM. The model primarily focuses on solving the time-cost dilemma by providing user-variable scenarios. The model exports quantities from BIM and generates construction activities and their durations, applying logical sequences to connect them. The study uses the NSGA-II genetic algorithm for multi-objective optimization, aiming to balance project duration and costs by selecting the most appropriate scenario and exporting it to Primavera. This approach allows the schedule to be adjusted based on different performance parameters, such as the number of crews or construction methods, offering various solutions for different combinations of duration and cost.

Furthermore, the implementation of BIM in road projects has developments that enable 3D modeling combined with time and cost analysis [29–32]. By integrating new technologies such as UAV equipment, LIDAR, and remote sensing, it has been possible to obtain models of road projects for monitoring and project management [33–36]. However, there are challenges to implementing BIM in highway projects, where researchers report studies about improving interoperability between different cost and schedule software or avoiding data loss when using data interchange formats [37,38]. The development of BIM tools for highways is lower with respect to buildings. Therefore, BIM adoption on highways is limited [39–41]. Significant progress has been made in generating new entities and property sets for roadway elements in the Industry Foundation Classes (IFC) standard [42–46]. Studies have developed methodologies and tools for BIM integration and scheduling, but they need to be adopted at the industrial level to improve project management [47–51]. Recent studies have focused on machine learning that allows the integration of computer data analysis for generating or evaluating automated models [52–56]. This enables the integration of BIM models and artificial intelligence (AI), finding recent applications for the different stages of a construction project life cycle [57–61]. Similarly, research related to BIM has been developed for road infrastructure projects [62,63]. Although there has been progress in the use of BIM in construction schedule planning, most studies have only focused on building projects. This lack of attention to BIM in road project planning highlights a research gap. It is essential to assess the potential of BIM to improve schedule planning in road infrastructure projects. Additionally, there is a weakness in studies investigating how BIM can help mitigate factors that adversely affect construction schedules. This observation emphasizes the need for more targeted and in-depth research in these critical areas to expand understandings of BIM and its functional application in the construction industry.

3. Research Method

This study examined how BIM uses can help reduce deficiencies in road schedule planning. The research method used four steps (See Figure 1). First, a questionnaire was created to identify the main BIM uses that can mitigate schedule planning problems. It was sent to construction planning experts knowledgeable in BIM technology and road engineering. Second, the critical deficiencies affecting road construction scheduling were selected from previous research [64]. Third, a second questionnaire was designed to estimate the effect of BIM uses on the improvement of the schedule planning deficiencies groups, according to the experience and knowledge of the professionals. Fourth, a cross-impact matrix was created to visualize and quantify the interactions between the different variables. This helped identify the BIM uses with the most relevant potential effect on mitigating specific schedule problems. Fifth, semi-structured meetings were conducted with professionals with expertise in road construction process planning and BIM. These interviews focused on obtaining information on how BIM helps mitigate the deficiencies that affect schedule planning. Finally, the results are reported in chapter four.

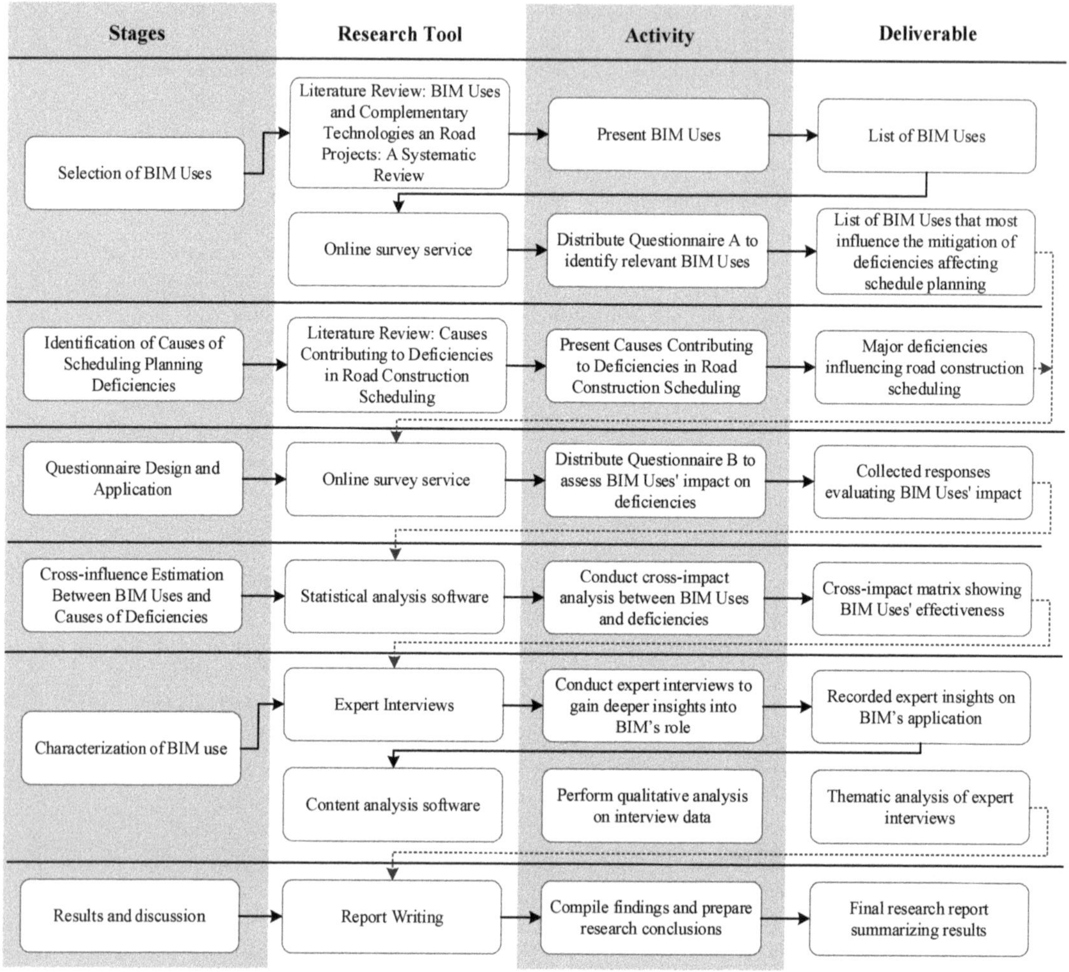

Figure 1. Methodological framework.

3.1. Selection of BIM Uses

This study is grounded in the work of Castañeda et al. [64], which presents a recent (2024) systematic review of BIM uses in road infrastructure projects, offering a comprehensive and up-to-date foundation for the research. From the 39 BIM uses identified by Castañeda et al. [64], (see Table 1), it was proposed to conduct a questionnaire to detect the main BIM uses aligned to the mitigation of shortcomings in planning construction processes of road projects (see Supplementary Material S1). The questionnaire gathered opinions from an expert group with experience in road project construction process planning and BIM adoption. The questionnaire consisted of two parts. The first part was oriented to collect information on five categorical variables: (1) profession, (2) academic degree, (3) country of experience, (4) role in the construction sector, and (5) years of experience. The second part identifies the BIM uses aligned with mitigating deficiencies in planning construction processes for roadway projects. Each BIM use was evaluated using a scale with three options for ranking: (1) Yes = 2, in case the BIM use contributes to schedule deficiency mitigation; (2) Indirectly = 1, in case the BIM use indirectly contributes to schedule deficiency mitigation; and (3) No = 0, in case the BIM use does not contribute to schedule deficiency mitigation. The questionnaire was hosted on the Survey Monkey platform, facilitating its distribution to practitioners through a URL link.

Table 1. BIM uses for road projects.

Category	BIM Uses
Road design	3D Existing Conditions Modeling, Alignment Optimization, Clash Analysis, Design and Evaluation of Roadside Facilities, Design Documentation, Design Review, Geometric Design, Modularization for Prefabrication.
Traffic analysis	Traffic Management Plan, Traffic Monitoring, and Traffic Analysis in Design.
Soil aspects	Analyzing Earthmoving Operations, Geological and Geotechnical Analysis.
Road safety	Construction Safety Analysis, Driving Simulation, Road Lighting Analysis, Road Safety Analysis.
Environmental issues	Environmental Impact, Solar Radiation Analysis, Sustainability Analysis.
Other engineering analysis	Drainage Analysis, Pavement Analysis, Road Acoustic Analysis, Structural Analysis, Transmission Lines Analysis, Underground Utility Analysis, Vulnerability Analysis.
Construction planning and analysis	Equipment and Material Planning, Space Use Planning on the Site, Workforce/Labor Planning and Training, 4D Construction Process Impact Analysis, 4D Construction Planning, Constructability Analysis, Maintenance Plan, and Schedule Estimation.
Cost analysis	5D Cost Analysis, Quantity Take-Off, and Cost Estimation.
Construction monitoring and control	4D and 5D as-built—as Planned Comparison and Monitoring, Tracking Onsite Construction Progress.

A preliminary survey was sent to five professionals with over five years of experience in road projects and BIM adoption. The objective was to evaluate the clarity of the content. After analyzing the answers and observations obtained, the authors adjusted the questionnaire. A revised version was submitted to the same group of experts for a final validation process. This concluded with the approval of the questionnaire's final version for use in the study. Questionnaire A was sent to 20 professionals with experience in road construction process planning and BIM adoption in the architecture, engineering, and construction (AEC) industry (see Supplementary Material S1). The purpose of Questionnaire A was to identify the most relevant BIM uses for mitigating scheduling deficiencies in road construction projects by gathering expert evaluations and calculating a Relative Influence Index (RII) to rank the 39 BIM uses based on their perceived impact. Out of the 20 responses, 12 were analyzed for the final sample (see Appendix A).

The data collected was used to conduct a Relative Influence Index (RII) analysis to identify BIM uses that have the most impact on mitigating road infrastructure project scheduling deficiencies. This analysis was centered on a questionnaire by sector professionals, who assigned weights to each of the three classification options allowed for each BIM use. Thus, the RII was calculated using Equation (1).

$$RII = \left(\frac{\sum_{i=1}^{N} w_i}{N * w_{max}} \right) \quad (1)$$

where w_i is the classification weights, assigned by the evaluator i; N is the total number of valid questionnaires ($N = 12$); and w_{max} is the maximum possible weights for classification ($w_{max} = 2$), based on the evaluation scale in the questionnaire (0, 1, 2).

The BIM uses were evaluated based on their RII, sorted from highest to lowest. This helped identify the uses that have the most impact in mitigating the causes of deficiencies in schedule planning. After analyzing the results, seven BIM uses were selected with the most significant influence in mitigating such deficiencies.

3.2. *Identification of Causes of Scheduling Planning Deficiencies*

Identifying deficiencies in road construction scheduling was based on a systematic literature review to consolidate the causes most frequently identified in previous studies. The study by Castañeda et al. [65] served as a key reference, as it synthesized findings from multiple sources and structured them into a comprehensive set of 34 deficiencies. Thus, the selected deficiencies came from a single study and represented a broader consensus within the existing literature. In addition, Castañeda et al. [65] applied exploratory factor analysis to classify these deficiencies into seven components (see Table 2), which were subsequently used to assess the impact of the selected BIM uses in the first phase of the analysis. These seven components were used to determine the impact of each of the seven BIM uses selected in the first stage of the analysis.

Table 2. Causes and components contributing to schedule planning deficiencies.

Principal Components	Causes Contributing to Deficiencies in the Scheduling
Project Preparation and Contextual Conditions	Inadequate planning of the articulation with neighboring areas Inadequate exploration of existing site conditions Inadequate estimation of permit acquisition durations Deficient project location Poor understanding of the project by the planner Lack of political/social considerations in planning Deficient estimation of material supply (stocks)
	Inadequate selection of construction methods
Scope and Schedule Planning	Failures in the definition of successor and predecessor relationships between activities Failures in the configuration of the work schedule (working days, holidays, working hours, seasons of the year) Omission or deficiency in the definition of work packages (WBS) Inadequacy of planning software adopted
Traffic Management Planning	Omission of road markings in planning Poor zoning of the construction process based on traffic conditions Failures in on-site traffic estimation Poor planning of machinery routes
Project Complexity	Poor estimation of workforce performance Deficient coordination with public utilities (water, sanitary, electrical, and other surface or subway networks) High complex access to the project area
Resource Estimation	Lack of planning of storage and parking sites
	Deficient estimation of material quantities
	Poor estimation of workforce quantity
Considering External Factors	Inadequate estimation of construction document approval times Poor identification of project stakeholders Request for project start time extensions (applies in cases of rescheduling requests)
Planning Accuracy and Operational Complexity	Lack of verification during the schedule preparation Inadequate estimation and consideration of waiting times (setting time or other waiting times between activities) High complexity of construction work

3.3. Questionnaire Design and Application

The quantitative estimation of the influence of BIM uses in mitigating deficiencies in road schedule planning was carried out for the seven main components of deficiencies in road schedule planning, which were identified in the study by Castañeda et al. [65], and the seven BIM uses selected in the first stage from the RII. A questionnaire was created to collect the data, which was carried out by gathering opinions from a group of experts with experience in road construction planning and BIM implementation (see Supplementary Material S2). The questionnaire consisted of two parts. The first part was oriented to collect information on five categorical variables: (1) profession, (2) academic degree, (3) country of experience, (4) role in the construction sector, and (5) years of experience. The second part is centered on collecting the impact of BIM uses on each of the seven significant causes affecting schedule planning on road projects. Each effect of BIM uses on the components was determined using a five-point Likert-type scale, where 1 = very low, 3 = medium, and 5 = very high. Questionnaire B was conducted on the Survey Monkey platform, facilitating its distribution to practitioners via a URL link.

The same group of professionals used in the first validation of Section 3.1 was used again to validate the draft version of the questionnaire. The purpose of this validation was to assess the clarity of the content. After analyzing the obtained answers and comments, the authors adjusted the questionnaire. This process was iterative and concluded with the approval of the questionnaire's final version for use in the study. The final version of Questionnaire B was then sent to 42 architecture, engineering, and construction (AEC) professionals with experience in road construction planning and BIM implementation (see Supplementary Material S2). The purpose of Questionnaire B was to quantitatively assess the influence of the BIM uses (selected in Section 3.1) on mitigating seven key components of scheduling deficiencies in road construction projects (presented in Section 3.2) by collecting expert evaluations through a five-point Likert scale and applying a cross-impact analysis to identify the main BIM uses. After a thorough review, 12 incomplete responses were discarded, and 30 were included in the final analysis (see Appendix B).

To ensure the validity of the research instrument, a Cronbach's alpha analysis was conducted to assess the internal consistency of the responses obtained in the impact section of the questionnaire (see Equation (2)). In quantitative research utilizing questionnaires as a data collection tool, Cronbach's alpha serves as a key metric for evaluating the reliability of measurement scales. This coefficient ranges from 0 to 1, where $\alpha \geq 0.9$ indicates excellent internal consistency, $0.8 \leq \alpha < 0.9$ is considered good, $0.7 \leq \alpha < 0.8$ is acceptable, $0.6 \leq \alpha < 0.7$ is questionable, $0.5 \leq \alpha < 0.6$ is poor, and $\alpha < 0.5$ is deemed unacceptable. The coefficient is computed using the following equation:

$$\alpha = \frac{K}{K-1}\left(1 - \frac{\sum_{i=1}^{k} S_i}{S_t}\right) \quad (2)$$

where K is the number of combinations between causes of deficiencies of scheduling planning and BIM uses ($K = 49$), Si is the variance of the scores assigned to the influence of each combination, and St is the total variance of the influence.

Cronbach's alpha analysis for responses concerning the impact of BIM uses on the identified causes yielded a coefficient of 0.92, indicating excellent internal reliability. This result suggests a high level of consistency in responses, reinforcing the validity of the research instrument employed in this study.

3.4. Cross-Influence Estimation Between BIM Uses and Causes of Deficiencies

After expert consultation, a cross-impact analysis was conducted to evaluate the influence of BIM uses in mitigating deficiencies in road construction scheduling. This method systematically quantifies the effect of BIM uses on scheduling deficiencies, allowing for a structured evaluation of interactions between variables, highlighting key cause–effect relationships, and guiding decision-making based on their potential impact. Three main steps were identified for the analysis. Firstly, the cross-impact matrix was structured, with seven main components of causes of deficiencies in road construction scheduling arranged in the rows, and selected BIM uses placed in the columns. Secondly, the collected survey data were used to calculate the arithmetic mean (see Equation (3)) and a measure of central tendency such as the median (see Equation (4)). This measure of central tendency provides a better representation of the perceived "typical impact" without being distorted by extremely low or high responses. It also provides a clear picture of the trend or tilt of respondents' overall opinions. Thus, a matrix of seven columns and seven rows was prepared to analyze the impact of BIM uses and deficiencies in schedule planning for road projects.

$$\overline{X} = \frac{1}{n} \times \sum_{i=1}^{n} x_i = \left(\frac{x_1 + x_2 + \cdots + x_n}{n} \right) \tag{3}$$

where n is the total number of valid questionnaires ($n = 30$); and x_i is the score assigned by the experts consulted (very low = 1 to very high = 5).

$$M_e = \frac{x_{\frac{n}{2}} + x_{\frac{n}{2}+1}}{2} \tag{4}$$

where n is the total number of valid questionnaires ($n = 30$), $x_{\frac{n}{2}}$ is the score found at position $\frac{n}{2}$, and $x_{\frac{n}{2}+1}$ is the score found at position $\frac{n}{2} + 1$.

From the cross-impact analysis scores, a color scale was used to visually identify the influence of interactions between BIM uses and causes of deficiencies. The matrix obtained helped to identify its application limitations and propose solutions oriented towards best practices based on the experiences, perceptions, and opinions of those with deep knowledge in the field.

Therefore, the selection of the main BIM uses followed a two-stage process combining expert evaluation and cross-impact analysis. First, an initial list of 39 BIM uses, identified by Castañeda et al. [30], was evaluated through a questionnaire distributed to 20 professionals with experience in road project planning and BIM adoption, of which 12 responses were validated and analyzed. Each professional assessed the contribution of each BIM use to mitigate scheduling deficiencies using a three-point scale (2 = contributes directly, 1 = contributes indirectly, 0 = does not contribute). Using these responses, a Relative Influence Index (RII) was calculated to rank the BIM uses from highest to lowest relevance. The BIM uses were identified based on their RII scores (top five rank). In the second stage, these BIM uses were further evaluated through a cross-impact analysis, conducted using a second questionnaire distributed to 42 professionals, with 30 valid responses included in the analysis. These professionals rated the influence of each BIM use on the mitigation of seven key components of scheduling deficiencies identified in previous research by Castañeda et al. [31], using a five-point Likert scale (1 = very low to 5 = very high). From these ratings, both the arithmetic mean and the median were calculated, forming a cross-impact matrix that visually illustrates the relative influence of each BIM use on each deficiency component. Based on this matrix, the three main BIM uses were selected.

3.5. Characterization of BIM Use

An interview was conducted to synthesize the main BIM uses (selected in the previous stage) in road projects to manage the lack of uses in the scheduling planning of construction processes. The researchers conducted semi-structured interviews with eight experts with more than five years of experience in road project planning, management, and

BIM adoption. The eight experts were interviewed and selected based on their extensive experience in the construction of road infrastructure projects and specialized knowledge of BIM adoption in road infrastructure projects (see Appendix C). The purpose of the interviews was to gain deeper insights into the practical application of BIM in addressing the root causes of scheduling deficiencies in construction projects. The semi-structured interview had three parts, which are detailed in Supplementary Material S3. Firstly, informed consent was obtained from the experts to conduct and record the interview, and their training and experience were assessed. Secondly, the experts gave a presentation to contextualize the research stages that led to the interview. Thirdly, the experts were asked how each of the main BIM uses can mitigate the causes of shortcomings in the scheduling of construction procedures. Each session lasted between 15 and 30 min and aimed to collect information about existing processes, implementation deficiencies, opportunities for improvement, and required resources, among others. These interviews provided valuable information centered on the professionals' experience and in-depth comprehension of the field of interest, forming the basis for describing BIM uses required to mitigate road project scheduling deficiencies.

Finally, a qualitative analysis was conducted based on the transcripts of the interview recordings. This type of analysis began with a detailed reading of the responses to identify common themes, patterns, and trends. In addition, it was complemented with the use of NVIVO software (https://lumivero.com/products/nvivo/, accessed on 30 January 2025), which allowed for the categorizing of responses, identification of patterns, drawing of key conclusions, and understanding of how BIM can contribute to mitigating deficiencies in the planning of construction schedules. Therefore, the analysis was structured in six main steps: (1) transcribing the recordings, (2) cleaning up the text, (3) breaking the text into smaller units, (4) eliminating conjunctions, (5) categorizing the text, and (6) content analysis. This approach sought to capture the complexity and multidimensionality of BIM implementation in road infrastructure construction directly from the professionals' perspective. In addition, the combination of qualitative analysis with NVIVO allowed a more detailed interpretation of the information collected, thus, facilitating the development of practical solutions to the challenges identified.

4. Results and Discussion

4.1. BIM Uses to Mitigate Schedule Deficiencies

Table 3 presents the main BIM uses that address issues with schedule planning based on Questionnaire A responses of 12 experts in road infrastructure construction. An analysis was conducted using a Relative Influence Index (RII) to establish a ranking of the BIM uses that professionals identified as having the most significant influence in mitigating the causes of deficiencies. The seven main BIM uses that contribute to the mitigation of planning deficiencies are the following: (1) Quantity Take-Off and Cost Estimation: This has the highest contribution in relationships of cost and time estimation and its variation during the project lifecycle, allowing the determination of possible cost and schedule overruns due to changes in the project [66]. (2) Equipment and Material Planning: This refers to project planning, availability, and supply of equipment and materials, which has been identified as one of the most critical factors in previous studies related to BIM [67]. (3) 4D Construction Planning: Integrating time data into a 3D BIM model is essential, enabling risk reduction and resource optimization [68]. (4) Schedule Estimation: Integrating the time dimension in BIM facilitates change management and project version control [69]. (5) Traffic Management Plan: This is relevant in road projects as is associated with coordination, risk management, construction processes, and user interaction [70]. (6) Space Use Planning on the Site: BIM allows visualization, analysis, and simulation of conditions and scenarios that may occur before construction, coordinating the design with equipment and personnel [71]. (7) 5D Cost Analysis: This improves the project's financial planning and contributes to decision-making throughout the life cycle [72].

Table 3. Relative influence index of BIM uses that contribute to mitigating road schedule deficiencies.

Id	BIM Use	RII	Rank
U_{37}	Quantity Take-Off and Cost Estimation	1.00	1
U_{28}	Equipment and Material Planning	0.96	2
U_{32}	4D Construction Planning	0.92	3
U_{35}	Schedule Estimation	0.92	3
U_9	Traffic Management Plan	0.79	5
U_{29}	Space Use Planning on the Site	0.79	5
U_{36}	5D Cost Analysis	0.79	5
U_{30}	Workforce/Labor Planning and Training	0.75	8
U_{33}	Constructability Analysis	0.75	8
U_{34}	Maintenance Plan	0.75	8
U_1	3D Existing Conditions Modeling	0.71	11
U_3	Clash Analysis	0.71	11
U_8	Modularisation for Prefabrication	0.71	11
U_{31}	4D Construction Process Impact Analysis	0.71	11
U_{39}	Tracking Onsite Construction Progress	0.71	11
U_{12}	Analyzing Earthmoving Operations	0.67	16
U_{38}	4D and 5D as-built—as Planned Comparison and Monitoring	0.63	17
U_4	Design and Evaluation of Roadside Facilities	0.58	18
U_5	Design Documentation	0.58	18
U_{14}	Construction Safety Analysis	0.58	18
U_{15}	Driving Simulation	0.58	18
U_{17}	Road Safety Analysis	0.58	18
U_{26}	Underground Utility Analysis	0.58	18

4.2. Impact Between BIM Uses and Schedule Planning Deficiencies

Table 4 displays the cross-impact matrix, where the columns present the top seven BIM uses that help mitigate schedule planning issues identified in Section 4.1. The rows contain the main components and related deficiencies in planning the road construction schedule. Based on expert opinion, a color-coding scheme has been implemented to determine the most influential uses. Dark green represents very high values; light green corresponds to high, yellow to medium, orange to low, and red to very low. The values obtained, as mentioned in Section 3.4, are the mean and median for each cross. According to Questionnaire B responses conducted with experts in road infrastructure construction, a new ranking was established based on the total score obtained. It was determined that the top three most influential BIM uses in mitigating the causes of deficiencies analyzed are (1) 4D Construction Planning, (2) Traffic Management Plan, and (3) Schedule Estimation, out of the seven analyzed. Similarly, the components most relevant by BIM for their mitigation were organized, and it was observed that the components of the causes of deficiencies most influenced by BIM for their mitigation, according to the total score, are (1) Scope and Schedule Planning, (2) Resource Estimation, and (3) Traffic Management Planning.

Table 4. The cross-impact matrix between BIM uses and causes of deficiency in road schedule planning.

		BIM Uses X (Me)								
Id	**Main Components**	**4D Construction Planning**	**Traffic Management Plan**	**Schedule Estimation**	**Space Use Planning on the Site**	**Equipment and Material Planning**	**Quantity Take-Off and Cost Estimation**	**5D Cost Analysis**	**Total Score**	**Rank**
C_1	Scope and Schedule Planning	4.5 (5.0)	4.0 (4.0)	4.4 (5.0)	3.8 (4.0)	4.1 (4.0)	4.3 (5.0)	3.8 (4.0)	28.97	1
C_2	Resource Estimation	4.2 (4.0)	3.9 (4.0)	3.9 (4.0)	3.9 (4.0)	4.0 (4.5)	4.4 (5)	3.4 (3.5)	27.73	2
C_3	Traffic Management Planning	4.3 (5.0)	4.4 (5.0)	3.9 (4.0)	4.1 (4.0)	4.0 (4.0)	3.5 (3.5)	3.1 (3.0)	27.47	3
C_4	Planning Accuracy and Operational Complexity	4.2 (4.5)	4.0 (4.0)	4.1 (4.0)	3.7 (4.0)	3.8 (4.0)	3.9 (4.0)	3.5 (3.0)	27.27	4
C_5	Project Complexity	4.0 (4.0)	3.4 (3.5)	3.2 (3.0)	3.7 (4.0)	3.4 (3.0)	3.4 (3.0)	3.3 (3.00)	24.53	5
C_6	Considering External Factors	3.5 (3.5)	3.45 (3.5)	3.7 (3.0)	3.1 (3.0)	3.0 (3.0)	2.8 (3.0)	3.1 (3.0)	22.7	6
C_7	Project Preparation and Contextual Conditions	3.0 (3.0)	3.3 (3.0)	3.0 (3.0)	3.4 (3.0)	2.9 (3.0)	2.5 (2.0)	2.7 (3.0)	20.83	7
	Total score X	27.77	26.73	26.17	25.8	25.23	24.9	22.9	179.5	
	Rank	1	2	3	4	5	6	7		

Table 4 indicates that some BIM uses are more closely related to the main components of schedule deficiencies than others. Some BIM uses share a relationship with other uses, while others do not. BIM uses such as 4D Construction Planning, Traffic Management Planning, and Schedule Estimation have a high relationship with the first four main components: Scope and Schedule Planning, Resource Estimation, Traffic Management Planning, and Planning Accuracy and Operational Complexity. These BIM uses in the road domain are aligned with the corresponding deficiencies in planning. Previous studies have shown that BIM uses can mitigate scheduling deficiencies [73–75]. BIM could have a high impact on mitigating deficiencies such as a lack in the definition of work packages (WBS), failures in setting the work schedule, shortcomings defining successor and predecessor relationships between activities, lack of planning of storage and parking locations, and deficient zoning of the construction process according to traffic conditions.

The BIM Space Use Planning on the Site significantly influences mitigating deficiencies related to Traffic Management Planning, Resource Estimation, and Scope and Scheduling Planning. Therefore, some of the causes of deficiencies where the BIM use Space Use Planning on the Site shows high influence are poor planning of machinery routes, shortcomings in storage planning and parking sites, and inadequacy of planning software adopted. Previous research has shown that BIM adoption can provide valuable contributions in these aspects within the construction sector [76,77]. Similarly, this research confirms its applicability and relevance in road project planning. In addition, the adoption of advanced technologies for spatial analysis [78] has enhanced detailed data-driven road planning using tools such as a Geographic Information System (GIS) [79] and temporal elements by integrating spatial planning with BIM 4D [80,81]. This synergy between 4D BIM and other technologies effectively addresses several of the shortcomings in planning previously mentioned, demonstrating how BIM can complement and enrich the process. Although the use of BIM for site spatial planning does not stand out as having the highest impact compared to other BIM uses discussed in the study, its significant association with key components positions it as a relevant element in improving the planning and execution of roadway projects.

Equipment and Material Planning has a moderate relationship with some significant components. The three components with which it is most closely related are Scope and Schedule Planning, Resource Estimating, and Traffic Management Planning. Some of the shortcomings associated with this use of BIM include failures to establish successor and predecessor relationships between activities and deficient estimation of material quantities. Although it has been proven in other studies [82,83] that the use of Equipment and Material Planning shows good performance and can cover the deficiencies generated by these components [84], Equipment and Material Planning using BIM still faces challenges. This may be one of the reasons why experts have shown a slight preference for other uses of BIM on road projects. Specifically, the BIM use of 4D Construction Planning has effectively mitigated scheduling and resource management deficiencies. It provides smoother integration and advanced visualization that allows potential problems to be anticipated and resolved before they become significant obstacles in the construction process. Therefore, while equipment and materials planning with BIM are beneficial, its impact is less pronounced than other uses of BIM in road construction planning.

The Quantity Take-Off and Cost Estimation are closely related to Scope and Scheduling Planning, Resource Estimation, Planning Accuracy, and Operational Complexity. Some issues associated with using BIM in this context include a lack of definition of work packages (WBS), deficient estimation of material quantities, and lack of verification during the schedule preparation. Previous research in the construction industry has shown a direct link between the use of BIM and these factors [85–87], and this study expands on those findings in the context of roadway projects. While specific components are highly correlated with material quantity planning and cost estimating using BIM, others have a moderate or low relationship with this use of BIM. This indicates that while material quantity and cost estimation are crucial to certain aspects of road planning and management, their impact may be more limited in other phases of the construction process. The Quantity Take-Off and Cost Estimation enhances the accuracy and efficiency of project planning, providing a solid foundation for informed decision-making. BIM's capacity to integrate data and visualize potential scenarios enables professionals to identify and rectify deficiencies before they significantly impact project progress. Therefore, despite some components showing a weaker relationship with the use of BIM, its application remains vital for the overall optimization of road construction planning.

The 5D Cost Analysis has a medium to low relationship with most of the components analyzed, except Scope and Schedule Planning, with which it has a high relationship. Some of the deficiencies associated with this use of BIM include shortcomings in configuring the work Schedule and Omission or failures in the definition of work packages (WBS). 5D Cost Analysis through BIM integrates costs into the dimensional model, providing a powerful tool for project financial management [88]. This integration has improved the alignment between schedule and budget planning [89], ensuring that planning decisions are informed by full cost considerations and applied to infrastructure projects [90,91]. The ability of 5D BIM to provide a detailed view of costs throughout the project life cycle facilitates more effective management of financial resources, contributing to risk mitigation and greater predictability in project execution. However, applying this use of BIM has proven to be more significant in road projects' construction and operation phases [92]. During these phases, 5D cost analysis allows for the continuous and accurate assessment of actual versus planned costs, providing critical data that can influence strategic decision-making and operational process optimization. This ability to monitor and adjust costs in real time is crucial to keeping projects within budget and maximizing financial efficiency.

4.3. Practical Experiences of Using BIM in the Mitigation of Road Schedule Deficiencies

The analysis of interviews with eight AECO industry professionals experienced in using BIM for road infrastructure projects provided valuable information on how BIM can help address construction process schedule planning deficiencies. The focus was on understanding the practical experiences of these professionals and gaining insight into the strategies and methods used to resolve schedule planning shortcomings in the context of road infrastructure construction (see Supplementary Material S4).

4.3.1. Four-Dimensional Construction Planning

Based on insights from professionals, it is clear that using the 4D Construction Plan has significant potential to reduce errors in defining the relationships between project activities. BIM allows for visualizing activities in a timeline relative to their physical space, combining the project's 3D model with the construction schedule. This visualization makes it easier to identify and rectify errors in activity sequences before they impact actual construction. Additionally, the preview feature provided by 4D simulation helps ensure that all activities are planned based on logical and physical dependencies, thus, preventing construction delays and errors. One of the professionals explains that during a 4D simulation, it is possible to observe whether the animation of an asphalt binder appears before the animation of the granular layers. This type of visualization allows the detection of problems in the definition of precedence relationships, making it possible to correct them and adjust activities to ensure they are executed in the correct order. Additionally, improving the accuracy of the sequence of activities contributes to increased safety on the construction site. Another professional mentioned that with 4D simulation, one could foresee and visualize earth movements and other activities that may compromise the safety of personnel and equipment. This way, 4D planning optimizes project efficiency and ensures a safer working environment. This BIM use also provides a clear and detailed visualization of the schedule and the 3D model, improving communication and coordination for more efficient and effective project execution.

The detailed, time-dependent visualization of the work schedule, including specific dates and times, makes it possible to identify problems that may be overlooked in a traditional schedule, such as incorrectly scheduled activities on holidays or at inappropriate times. One interviewee mentioned that integrating the work schedule into the simulation allows planners to ensure that all activities are scheduled at appropriate times, considering labor and seasonal constraints. This could improve schedule accuracy, optimize resource allocation, and prevent additional costs associated with out-of-hours work or adverse weather conditions. Four-dimensional simulation enables planners to identify activities not included in the WBS early on. An interviewed professional explains, "If an important activity, such as the construction of abutments, is forgotten in the planning of a bridge construction process, it will become evident during the simulation because the simulation of the abutments will not be visible. This allows us to correct the error before it affects the project". This early detection capability is crucial to ensure that all necessary activities are executed as planned.

BIM offers an integrated platform compatible with various tools and data formats. One practitioner explains that "by linking data to the BIM model, planners can identify format incompatibilities or missing information, which may indicate inappropriate software selection. For instance, if schedule data does not integrate correctly with the 3D model, it could be due to format incompatibilities, suggesting that the wrong software has been chosen". Early detection of these problems enables planners to switch software or adjust data exchange formats to ensure seamless integration. Four-dimensional planning also allows for assessing whether the chosen software meets the specific project requirements. One interviewed professional mentioned that "4D planning can determine whether the software is capable of effectively handling time and cost data and whether it is compatible with other software used in the project". Compatibility with standards such as IFC is crucial for realizing the exchange of information between different platforms and ensuring the accuracy and availability of all project data. One of the professionals stresses the importance of choosing software compatible with the project's needs and capable of handling 4D planning. They emphasize that selecting the right software is vital to the project's success. In a complex road infrastructure project, such as constructing a new road with bridges and tunnels, choosing planning software that handles geometry and construction schedules and integrates seamlessly with other cost and human resource management systems is crucial. Working with standard data formats could help identify compatibility issues and avoid inadequate software selection early on.

4.3.2. Quantity Take-Off and Cost Estimation

This BIM use accurately calculates the materials required for each project phase. Knowing these quantities makes it possible to plan the volumes of material that will need to be stored at the construction site. One practitioner explains that "if a project requires a significant amount of granular subbase, BIM can help determine the space needed to store this material before its use, ensuring that there is sufficient space designated for it and avoiding congestion on the site". Additionally, BIM's ability to integrate logistics planning with site scheduling could support planning when and where certain materials and equipment would be needed. This allows for efficient storage and parking space management, as materials can be delivered and stored just before use, minimizing unnecessary space occupation and reducing the risk of damage or loss of materials due to prolonged storage. Furthermore, this BIM use facilitates communication and coordination between project stakeholders. Another professional emphasizes that the detailed and centralized information in the BIM model can be shared with suppliers and contractors, allowing for more accurate planning of material deliveries and machinery usage. This includes identifying suitable sites for unloading materials and organizing parking vehicles and equipment spaces, especially in urban areas or projects with limited space. The detailed digital replications of a project with a high level of development (LOD) allow for accurate estimates, reflecting changes in the BIM model due to design modifications. This can address poor material estimates, reducing the risk of under- or over-estimates that may impact on the project budget and schedule. Integrating job scheduling with resource management through BIM enables optimal labor planning and allocation. Professionals believe that using BIM methodology for calculating quantities and costs could ensure resource availability for each project phase, improve team coordination, and streamline scenario simulations, among other benefits.

4.3.3. Traffic Management Plan

This BIM use helps mitigate the omission of roadway markings by supporting the creation of digital twins, which make it possible to include the affected intersections and routes. These virtual models can represent in detail the operation and traffic flow in a road project, thus, providing a virtual replica of the environment that could help foresee and plan the construction phases, including road markings. One of the interviewees states that "digital traffic simulation, supported by BIM, makes it possible to identify the need and location of road signs with high accuracy. By visualizing traffic in a virtual environment, engineers can determine what type of signs are needed, how many are required, and where they should be placed to maintain traffic safety and flow". To address issues related to poor zoning, planners can integrate traffic flow simulation with the work of other specialties. This will help identify optimal work areas that minimize traffic obstruction. For example, suppose a construction phase requires closing a section of road. In that case, BIM simulation can demonstrate how this closure will affect traffic flow and enable planners to adjust zoning to minimize delays and congestion. By combining these simulations, a zoning plan can be created that reduces traffic disruptions and optimizes the use of space. Additionally, this BIM use allows for vehicle flow simulations calibrated with actual site data, which helps to improve the accuracy of these estimates. Planners can analyze critical parameters such as operating speed, queue lengths at traffic lights, and travel times. These analyses allow the virtual models to be adjusted to accurately reflect actual traffic conditions in the project area. By calibrating these models with data observed at the site, it is possible to predict with greater certainty how traffic will behave during the different phases of construction. This improves the planning of machine routing activities, which is critical to avoid interference with traffic and ensure efficient operation. Planners can simulate and visualize machinery routes within the construction environment, allowing them to identify potential traffic conflicts and optimize routes to minimize disruptions during project execution. According to professionals, a BIM simulation can show the best route for transporting heavy materials, minimize traffic impact, and avoid congested areas. Adjusting machinery routes based on changing traffic conditions and project needs could improve more efficient coordination and reduce delays and additional costs.

4.3.4. Schedule Estimation

This BIM use supports the mitigation of verification gaps during schedule development by providing an integrated and consistent platform for the planning and scheduling of road infrastructure projects. Because it allows for creating a digital replica of the project, which facilitates simulation analysis and continuous data integration, this ensures that time estimates are more realistic and reliable, reducing the risk of unexpected deviations. One practitioner posits, "In bridge construction, BIM can break down the project into specific tasks such as foundation, pier, and deck construction, assigning precise times to each activity". This level of detail and accuracy in planning allows the schedule to be continuously verified and adjusted based on updated site data, which improves precision and confidence in the final schedule. In addition, the ability to perform automated checks of resources and activity durations through creating code schedules, according to the practitioners, could minimize human error and ensure that the schedule is more accurate to the construction plan.

BIM addresses the inadequate estimation and consideration of waiting times by precisely simulating construction sequences, which optimizes coordination between different teams and reduces times, which do not generate value for the project; one practitioner states, "On a road project, BIM can simulate the time a dump truck needs to wait while a backhoe loads material, or the waiting time due to traffic crossings on congested road corridors". Another professional noted, "BIM allows you to look at different scenarios and adjust activities as needed, which is crucial to mitigate risks and maintain planning throughout the project development and schedule". These simulations allow forecasting and planning of the impact of traffic and other disruptions on machine waiting time, ensuring that activities are coordinated efficiently. In addition, the ability to dynamically adjust waiting times according to actual project conditions and identify bottlenecks through the thorough decomposition of activities allows for more accurate and effective management of waiting times. This improves project performance, optimizes the use of resources, and minimizes costs associated with downtime. Finally, managing the high complexity of work in road infrastructure projects is significantly enhanced with the use of BIM by allowing the decomposition of the project into manageable sub-activities and the creation of accurate models representing the elements involved "in the construction of a tunnel, BIM can model each phase of the process with a high LOD to understand the complexity of the elements and thus the project, which could be seen from excavation to lining and installation of ventilation and lighting systems". This detailed modeling capability facilitates the identification and resolution of conflicts before they occur on-site, which improves the coordination and planning of construction activities. In addition, BIM enables the integration of multiple disciplines into a single model, ensuring that all teams manage the same up-to-date information, which is crucial on complex projects. The simulations and scenario analysis that BIM provides help to foresee the challenges associated with the complexity of the works and develop strategies to manage them. Continuous monitoring of project progress and the ability to update the model with real-time data ensures that planners can respond quickly to unforeseen changes and challenges, optimizing project execution and minimizing risks.

4.4. Discussion, Limitations, and Future Research

The findings of this study highlight the importance of implementing BIM uses to mitigate deficiencies in the scheduling of road construction projects. Four-dimensional Construction Planning, Quantity Take-Off and Cost Estimation, and Constructability Analysis were identified as the most influential ways to improve schedule planning in road projects. These results align with previous studies demonstrating that integrating BIM into construction planning enhances activity sequences' visualization, analysis, and optimization, thereby reducing risks and improving schedule accuracy [12,19,24].

This study's key contribution was identifying specific interactions between BIM uses and the principal components contributing to scheduling deficiencies. The cross-impact matrix enabled the mapping of these relationships, revealing that 4D Construction Planning has a significant impact on reducing deficiencies related to scope definition and activity sequencing, which is consistent with previous research highlighting the role of 4D planning in integrating activities and enabling early detection of conflicts [17,68]. Furthermore, the results confirmed the importance of the Traffic Management Plan in mitigating deficiencies associated with traffic control during construction, which is in line with prior studies demonstrating its effectiveness in optimizing site logistics and traffic flow management in road projects [70,75].

The strong relationship between Quantity Take-Off and Cost Estimation and the resource estimation component reinforces findings from previous research, which emphasizes BIM's ability to improve accuracy in material and cost estimation, thereby reducing planning uncertainty and improving decision-making processes [85,90]. However, the study also revealed that the impact of BIM is not uniform across all components. For example, 5D Cost Analysis showed a lower influence on mitigating operational planning deficiencies, suggesting that its primary contribution relates to financial management rather than detailed activity sequencing, consistent with prior studies' findings [88,89].

Despite these benefits, the literature also highlights important challenges associated with BIM adoption in road projects, several of which were confirmed in this study. First, high initial implementation costs remain a significant barrier, particularly for small and medium-sized firms with limited technology budgets [73]. Second, the learning curve and the need for specialized training to ensure effective BIM use are critical factors frequently reported as obstacles to widespread adoption [66,87]. Lastly, regulatory and legal barriers can constrain the implementation of BIM in road infrastructure projects, as some countries still lack comprehensive policies and regulations mandating or encouraging BIM use in public works contracts [19,22].

Another important aspect requiring further exploration is the empirical validation of BIM's impact on schedule performance in road projects. Although this study relied on expert opinions, no historical project data were used to quantify how BIM reduces scheduling errors or improves planning efficiency. Prior research underscores the importance of combining expert input with real-world data analysis to strengthen the empirical basis for BIM's benefits, providing quantifiable metrics on time and cost savings [28,83]. Moreover, recent studies have begun exploring the integration of emerging technologies that could amplify BIM's impact on schedule planning. For instance, combining BIM with artificial intelligence (AI) and machine learning (ML) has shown promise in improving schedule prediction accuracy and identifying potential delays in advance [14,18]. Similarly, real-time data from digital twins and IoT sensors is increasingly used to dynamically update project schedules, enhancing responsiveness and real-time decision-making during construction [78,80]. Future research should investigate how these advanced technologies can complement BIM, enabling more proactive and adaptive schedule management in road projects.

This study presents several limitations that should be considered when interpreting the findings. First, the analysis was based exclusively on the participation of professionals with expertise in road construction projects and prior experience using BIM. This sector-specific focus limits the generalizability of the results to other types of construction projects, such as building projects, urban infrastructure, or energy-related developments, where the nature of planning deficiencies and the effectiveness of BIM uses may differ. Second, the study concentrated exclusively on applying BIM technologies to mitigate scheduling deficiencies without incorporating complementary methodologies, such as Lean Construction, which could provide valuable process-focused strategies for improving schedule planning. By emphasizing BIM as the primary tool, the study potentially overlooks the synergistic benefits of integrating BIM with alternative process improvement approaches. Another limitation stems from the estimation of cross-impact relationships between BIM uses and causes of scheduling deficiencies based on expert perceptions rather than empirical project data. While expert input provides valuable insights, the absence of validation using historical data from real projects restricts the capacity to quantitatively assess the actual magnitude of BIM's contribution to mitigating planning deficiencies. Consequently, future research should complement expert-based assessments with longitudinal project data to enhance the robustness and external validity of the findings. Finally, the causes of scheduling deficiencies analyzed in this study were derived from prior research on road construction projects. As such, the identified deficiencies and their mitigation through BIM uses may not fully capture the challenges in other infrastructure projects. Expanding the analysis to include diverse project typologies would help refine the understanding of BIM's role across a broader spectrum of construction scenarios.

Future research should pursue several avenues to build on the findings of this study and address the identified limitations. One important direction involves empirically validating the cross-impact relationships identified between BIM uses and causes of scheduling deficiencies, considering the different realities around the world, i.e., including country, environment, and people, among other factors. Conducting case studies on real-world road projects, where BIM's influence on scheduling performance can be systematically measured and compared against projects using conventional planning approaches, would provide valuable quantitative evidence to support or refine the expert-based insights obtained in this research. Another promising avenue is the extension of the analysis to other construction sectors, such as building projects, transportation hubs, and large-scale urban developments. Examining how BIM uses interact with planning deficiencies in different project types would enhance the generalizability of the findings and reveal whether the critical BIM uses identified in road projects hold similar importance in other contexts. This cross-sectoral comparison could also highlight sector-specific challenges and opportunities for BIM adoption. Additionally, future research could explore the integration of BIM with complementary methodologies, particularly Lean Construction, to develop a more holistic framework for mitigating scheduling deficiencies. Combining BIM's technological capabilities with Lean Construction's emphasis on process efficiency and waste reduction could yield innovative solutions for improving schedule reliability in road projects. Emerging technologies also represent a valuable research opportunity. Integrating BIM with artificial intelligence (AI) and machine learning (ML) could enable predictive analytics for schedule forecasting, risk identification, and adaptive rescheduling. Furthermore, using real-time data from sensors and digital twins could enhance the dynamic updating of schedules, improving responsiveness to on-site conditions and unforeseen events. Finally, future research should investigate the policy and regulatory environment necessary to foster wider BIM adoption in road infrastructure projects. Comparative studies across countries with differing regulatory frameworks could identify the best practices and effective incentives that promote the use of BIM in public sector projects. This policy-oriented research would provide valuable insights for governments and industry bodies aiming to accelerate BIM implementation in transportation infrastructure projects.

5. Conclusions

This study advances the theoretical understanding of how BIM can mitigate schedule deficiencies in road project planning by identifying and ranking the BIM uses with the greatest influence, based on input from 12 AEC professionals experienced in road projects and BIM implementation. Through the application of the Relative Influence Index (RII), the study highlights the seven most relevant BIM uses: (1) Quantity Take-Off and Cost Estimation, (2) Equipment and Material Planning, (3) 4D Construction Planning, (4) Schedule Estimation, (5) Traffic Management Plan, (6) Space Use Planning on the Site, and (7) 5D Cost Analysis.

The study also contributes by applying a cross-impact matrix to analyze and map the interactions between these BIM uses and the principal components of schedule deficiencies, providing a systematic framework for understanding the role of BIM in addressing specific planning vulnerabilities. This matrix-based approach offers theoretical insights into how the influence of BIM is distributed across different dimensions of project planning, from scope definition to operational complexity. Furthermore, the study enhances the theoretical framework by incorporating qualitative data from semi-structured interviews supported by NVIVO analysis. This mixed-method approach enriches theoretical understanding by capturing both quantitative relationships and qualitative insights, enabling a more comprehensive view of how BIM uses contribute to mitigating planning deficiencies in practice. Thus, this study offers practical guidance for road infrastructure project teams by identifying which BIM uses contribute most effectively to mitigating deficiencies in schedule planning. This allows decision-makers to prioritize specific BIM tools and processes, maximizing their benefits. BIM uses such as 4D Construction Planning and 5D Cost Analysis enhance project visualization, allowing real-time plan adjustments and better stakeholder coordination. These capabilities reduce the risk of delays and cost overruns by improving activity sequencing, resource management, and progress tracking.

Additionally, integrating BIM in space use planning and traffic management provides critical support for logistics and operational efficiency, addressing site constraints and public traffic interference challenges. By using BIM for Quantity Take-Off and Schedule Estimation, project teams gain more reliable and traceable data, reducing errors and improving planning accuracy. Identifying interactions between BIM uses and deficiency causes also supports the development of more targeted risk management strategies, enabling practitioners to address vulnerabilities in the planning process proactively. The focus on road projects and the exclusive participation of professionals experienced in this sector limits generalizability to other project types. Additionally, the study emphasizes BIM-based solutions, excluding complementary methodologies. Future research could expand to different project types and explore the combined application of BIM and Lean Construction to develop comprehensive strategies for mitigating schedule planning deficiencies, leveraging the strengths of both approaches.

Supplementary Materials: The following supporting information can be downloaded at: https://www.mdpi.com/article/10.3390/su17062729/s1, Supplementary Material S1. Questionnaire A; Supplementary Material S2. Questionnaire B; Supplementary Material S3. Informed consent to participate in an interview for the development of a research project; Supplementary Material S4. Description of the potential of how BIM can help to mitigate the causes of deficiencies in the planning of construction process schedules in road projects.

Author Contributions: Conceptualization, K.C., O.S., C.A.P., R.F.H. and G.M.; methodology, K.C., O.S., C.A.P., R.F.H. and G.M.; results analysis, K.C., O.S., C.A.P., R.F.H. and G.M.; writing—original draft preparation, K.C., O.S., C.A.P. and R.F.H.; writing—review and editing, K.C., O.S., C.A.P., R.F.H. and G.M.; visualization, K.C., C.A.P. and O.S.; supervision, K.C., O.S., C.A.P., R.F.H. and G.M. All authors have read and agreed to the published version of the manuscript.

Funding: Omar Sánchez gratefully acknowledges the financial support from the Pontificia Universidad Javeriana, Colombia, through "Apoyo a proyectos de investigación liderados por profesores que se encuentran en su primera etapa 2023" with the project entitled "Planificación de la construc-ción de proyectos viales, mediante la integración de simulaciones digitales BIM nD, líneas de balance y Last Planner System" (ID 20673). Rodrigo F. Herrera acknowledge the financial support from ANID FONDECYT Iniciación 2023 No. 11230455.

Institutional Review Board Statement: Ethical review and approval were waived for this study because it involved only interviews to gather the opinions of a group of experts who provided informed consent prior to participation. All measures were implemented to ensure participant anonymity and safeguard their data in compliance with relevant policies.

Informed Consent Statement: Informed consent was obtained from all subjects involved in the study.

Data Availability Statement: The original contributions presented in this study are included in the article. Further inquiries can be directed to the corresponding author.

Acknowledgments: Karen Castañeda thanks the Technology, Innovation, Management, and Sustainability in Civil Engineering (TIMS) research group of the Pontificia Universidad Católica de Valparaíso for the support received during her research internship.

Conflicts of Interest: The authors declare no conflicts of interest.

Appendix A. Professional Profiles of Experts Who Participated in the Questionnaire

Id	Profession (Grade)	Country	Rol	Years of Experience
1	Civil Engineer, M.Sc.	Chile	Contractor	>15
2	Civil Engineer, M.Sc., Ph.D.	Colombia	Consultant	>15
3	Civil Engineer, Esp.	Chile	Owner	>15
4	Civil Engineer, M.Sc.	Colombia	Designer	>15
5	Civil Engineer, M.Sc., Ph.D.	Colombia	Academic/Researcher	>15
6	Civil Engineer, Esp.	Chile	Designer	>10

Id	Profession (Grade)	Country	Rol	Years of Experience
7	Civil Engineer, M.Sc., Ph.D.	Chile	Designer	>10
8	Civil Engineer, Esp.	Chile	Contractor	>10
9	Civil Engineer, M.Sc., Ph.D.	Colombia	Academic/Researcher	>10
10	Civil Engineer, M.Sc.	Colombia	Academic/Researcher	>5
11	Civil Engineer, M.Sc.	Colombia	Contractor	>5
12	Civil Engineer, M.Sc., Ph.D.	Colombia	Academic/Researcher	>5

Appendix B. Characterization of Professional Roles and Years of Experience

Professional Roles \ Years of Experience	Less Than 5 Years	From 5 to 10 Years	From 10 to 15 Years	More Than 15 Years	Total
Designer	--	1	2	--	3
Consultant/Auditor	1	--	1	1	3
Contractor	1	5	1	6	13
Owner	1	--	--	--	1
Academic/Researcher	--	4	2	4	10
Other	--	--	--	--	--
Total	3	10	6	11	30

Appendix C. Professional Profiles of Experts Who Participated in the Questionnaire

Id	Profession (Grade)	Country	Rol	Years of Experience
1	Civil Engineer, M.Sc., Ph.D.	Spain	Academic/Researcher	>15
2	Civil Engineer, M.Sc., Ph.D.	Chile	Consultant/Designer	>15
3	Architect, Ph.D.	Spain	Academic/Researcher	>15
4	Civil Engineer, M.Sc., Ph.D.	Spain	Academic/Researcher	>15
5	Civil Engineer, M.Sc., Ph.D.	Colombia	Academic/Researcher	>10
6	Civil Engineer, M.Sc., Ph.D.	Colombia	Academic/Researcher	>10
7	Architect, M.Sc., Ph.D.	Spain	Academic/Designer	>10
8	Civil Engineer, M.Sc., Ph.D.	Colombia	Academic/Researcher	>7
9	Civil Engineer, M.Sc.,	Colombia	Consultant/Designer	>7
10	Civil Engineer, Ph.D.	Chile	Researcher/Designer	>5

References

1. Koike, A.; Tavasszy, L.; Sato, K.; Monma, T. Spatial Incidence of Economic Benefit of Road-Network Investments: Case Studies under the Usual and Disaster Scenarios. *J. Infrastruct. Syst.* **2012**, *18*, 252–260. [CrossRef]
2. Miralinaghi, M.; Davatgari, A.; Seilabi, S.E.; Labi, S. Contract Bundling Considerations in Urban Road Project Scheduling. *Comput. -Aided Civil. Infrastruct. Eng.* **2022**, *37*, 427–450. [CrossRef]
3. Castañeda, K.; Sánchez, O.; Herrera, R.F.; Mejía, G. Highway Planning Trends: A Bibliometric Analysis. *Sustainability* **2022**, *14*, 5544. [CrossRef]
4. Giunta, M. Assessment of the Environmental Impact of Road Construction: Modelling and Prediction of Fine Particulate Matter Emissions. *Build. Environ.* **2020**, *176*, 106865. [CrossRef]
5. Mejía, G.; Sánchez, O.; Castañeda, K.; Pellicer, E. Stakeholders' Issues as a Source of Project Delays: A Meta-Analy-Sis between Building and Road Projects. *Rev. Constr.* **2023**, *22*, 51–73. [CrossRef]
6. Hosseininasab, S.M.; Shetab-Boushehri, S.N.; Hejazi, S.R.; Karimi, H. A Multi-Objective Integrated Model for Selecting, Scheduling, and Budgeting Road Construction Projects. *Eur. J. Oper. Res.* **2018**, *271*, 262–277. [CrossRef]

7. Kim, K.; De La Garza, J.M. Evaluation of the Resource-Constrained Critical Path Method Algorithms. *J. Constr. Eng. Manag.* **2005**, *131*, 522–532. [CrossRef]
8. Lu, M.; Li, H. Resource-Activity Critical-Path Method for Construction Planning. *J. Constr. Eng. Manag.* **2003**, *129*, 412–420. [CrossRef]
9. Olivieri, H.; Seppänen, O.; Alves, T.D.C.L.; Scala, N.M.; Schiavone, V.; Liu, M.; Granja, A.D. Survey Comparing Critical Path Method, Last Planner System, and Location-Based Techniques. *J. Constr. Eng. Manag.* **2019**, *145*, 04019077. [CrossRef]
10. Lozano-Ramírez, N.E.; Sánchez, O.; Carrasco-Beltrán, D.; Vidal-Méndez, S.; Castañeda, K. Digitalization and Sustainability in Linear Projects Trends: A Bibliometric Analysis. *Sustainability* **2023**, *15*, 15962. [CrossRef]
11. Sigalov, K.; König, M. Recognition of Process Patterns for BIM-Based Construction Schedules. *Adv. Eng. Inform.* **2017**, *33*, 456–472. [CrossRef]
12. Chen, C.; Tang, L. BIM-Based Integrated Management Workflow Design for Schedule and Cost Planning of Building Fabric Maintenance. *Autom. Constr.* **2019**, *107*, 102944. [CrossRef]
13. Nusen, P.; Boonyung, W.; Nusen, S.; Panuwatwanich, K.; Champrasert, P.; Kaewmoracharoen, M. Construction Planning and Scheduling of a Renovation Project Using Bim-Based Multi-Objective Genetic Algorithm. *Appl. Sci.* **2021**, *11*, 4716. [CrossRef]
14. Kim, S.; Peavy, M.; Huang, P.; Kim, K. Development of BIM-Integrated Construction Robot Task Planning and Simulation System. *Autom. Constr.* **2021**, *127*, 103720. [CrossRef]
15. Abdulwahhab, R.; Naimi, S.; Abdullah, R. Managing Cost and Schedule Evaluation of a Construction Project via BIM Technology and Experts' Points of View. *Math. Model. Eng. Probl.* **2022**, *9*, 1515–1522. [CrossRef]
16. ElMenshawy, M.; Marzouk, M. Automated BIM Schedule Generation Approach for Solving Time–Cost Trade-off Problems. *Eng. Constr. Archit. Manag.* **2021**, *28*, 3346–3367. [CrossRef]
17. Singh, J.; Cheng, J.C.P.; Anumba, C.J. BIM-Based Approach for Automatic Pipe Systems Installation Coordination and Schedule Optimization. *J. Constr. Eng. Manag.* **2021**, *147*, 04021143. [CrossRef]
18. Nguyen, D.T.; Chou, J.S.; Tran, D.H. Integrating a Novel Multiple-Objective FBI with BIM to Determine Tradeoff among Resources in Project Scheduling. *Knowl. Based Syst.* **2022**, *235*, 107640. [CrossRef]
19. Yuzhuo, Z.; Hyuk, J.; Qinghe, W.; Xinlong, Z. Research on Schedule Control of Large Scale Complex Project Based on BIM and PDCA Theory. *J. Archit. Civil. Eng.* **2022**, *39*, 183–191. [CrossRef]
20. Mahamid, I.; Bruland, A.; Dmaidi, N. Causes of Delay in Road Construction Projects. *J. Manag. Eng.* **2012**, *28*, 300–310. [CrossRef]
21. Herrera, R.F.; Sánchez, O.; Castañeda, K.; Porras, H. Cost Overrun Causative Factors in Road Infrastructure Projects: A Frequency and Importance Analysis. *Appl. Sci.* **2020**, *10*, 5506. [CrossRef]
22. Amoatey, C.; Okanta, A. Exploring Critical Road Project Delay Factors in Ghana. *J. Facil. Manag.* **2017**, *15*, 110–127. [CrossRef]
23. Castañeda, K.; Sánchez, O.; Herrera, R.F.; Pellicer, E.; Porras, H. BIM-Based Traffic Analysis and Simulation at Road Intersection Design. *Autom. Constr.* **2021**, *131*, 103911. [CrossRef]
24. Oreto, C.; Massotti, L.; Biancardo, S.A.; Veropalumbo, R.; Viscione, N.; Russo, F. BIM-Based Pavement Management Tool for Scheduling Urban Road Maintenance. *Infrastructures* **2021**, *6*, 148. [CrossRef]
25. Zhang, S.; Wang, X. Quantifying Schedule Delay Risk in Construction Projects: A Data-Driven Approach with BIM and Probabilistic Reliability Analysis. *Adv. Civil. Eng.* **2023**, *2023*, 5525655. [CrossRef]
26. Kehily, D.; Underwood, J. Embedding Life Cycle Costing in 5D BIM. *J. Inf. Technol. Constr.* **2017**, *22*, 145–167.
27. Guo, H.; Ye, X.; Ren, Q.; Luo, Z. Automatic Generation of Construction Schedules Based on BIM and Rule Reasoning. *Qinghua Daxue Xuebao/J. Tsinghua Univ.* **2022**, *62*, 189–198. [CrossRef]
28. Sami Ur Rehman, M.; Thaheem, M.J.; Nasir, A.R.; Khan, K.I.A. Project Schedule Risk Management through Building Information Modelling. *Int. J. Constr. Manag.* **2022**, *22*, 1489–1499. [CrossRef]
29. Lamas, D.; Justo, A.; Soilán, M.; Riveiro, B. 3D Point Cloud to Bim: Automated Application to Define IFC Alignment and Roadway Width Entities from Mls-Acquired Lidar Data of Mountain Roads. *ISPRS Ann. Photogramm. Remote Sens. Spat. Inf. Sci.* **2022**, *10*, 169–176. [CrossRef]
30. Dolson, G.A. Scheduling and Management Improvements Accelerate Roadway Design. *Transp. Res. Rec.* **1999**, *2*, 92–97. [CrossRef]
31. Castañeda, K. Marco Metodológico para la Integración de Planes de Manejo de Tráfico en la Planificación de Procesos Constructivos de Intersecciones Viales, a Través de BIM y LEAN. Ph.D. Thesis, Industrial University of Santander, Bucaramanga, Colombia, 2024.
32. Mirboland, M.; Smarsly, K. BIM-Based Description of Intelligent Transportation Systems for Roads. *Infrastructures* **2021**, *6*, 51. [CrossRef]
33. Soilán, M.; Justo, A.; Sánchez-Rodríguez, A.; Riveiro, B. 3D Point Cloud to BIM: Semi-Automated Framework to Define IFC Alignment Entities from MLS-Acquired LiDAR Data of Highway Roads. *Remote Sens* **2020**, *12*, 2301. [CrossRef]
34. Cafiso, S.; Di Graziano, A.; Pappalardo, G.; Severino, A. Bim of an Existing Road Using Advanced Monitoring Systems. *Int. J. Eng. Appl.* **2021**, *9*, 243–250. [CrossRef]

35. Tang, F.; Ma, T.; Zhang, J.; Guan, Y.; Chen, L. Integrating Three-Dimensional Road Design and Pavement Structure Analysis Based on BIM. *Autom. Constr.* **2020**, *113*, 103152. [CrossRef]
36. Justo, A.; Soilán, M.; Sánchez-Rodríguez, A.; Riveiro, B. Scan-to-BIM for the Infrastructure Domain: Generation of IFC-Complaint Models of Road Infrastructure Assets and Semantics Using 3D Point Cloud Data. *Autom. Constr.* **2021**, *127*, 103703. [CrossRef]
37. Han, C.; Han, T.; Ma, T.; Tong, Z.; Wang, S.; Hei, T. End-to-End BIM-Based Optimization for Dual-Objective Road Alignment Design with Driving Safety and Construction Cost Efficiency. *Autom. Constr.* **2023**, *151*, 104884. [CrossRef]
38. Li, C.; Ding, L.; Zhong, B. Highway Planning and Design in the Qinghai–Tibet Plateau of China: A Cost–Safety Balance Perspective. *Engineering* **2019**, *5*, 337–349. [CrossRef]
39. Toca, C.; Bastos Costa, D. Increasing Production Efficiency through the Reduction of Transportation Activities and Time Using 4D BIM Simulations. *Eng. Constr. Andarchitectural Manag.* **2022**, *28*, 2222–2247. [CrossRef]
40. Costin, A.; Adibfar, A.; Hu, H.; Chen, S.S. Building Information Modeling (BIM) for Transportation Infrastructure—Literature Review, Applications, Challenges, and Recommendations. *Autom. Constr.* **2018**, *94*, 257–281. [CrossRef]
41. Biancardo, S.A.; Capano, A.; de Oliveira, S.G.; Tibaut, A. Integration of BIM and Procedural Modeling Tools for Road Design. *Infrastructures* **2020**, *5*, 37. [CrossRef]
42. Ait-Lamallam, S.; Sebari, I.; Yaagoubi, R.; Doukari, O. IFCInfra4OM: An Ontology to Integrate Operation and Maintenance Information in Highway Information Modelling. *ISPRS Int. J. Geoinf.* **2021**, *10*, 305. [CrossRef]
43. Eslahi, M.; El Meouche, R.; Doukari, O.; Ruas, A. BIM and Contribution to IFC-Bridge Development: Application on Raymond Barre Bridge. In *Advances in Building Information Modeling, Proceedings of the First Eurasian BIM Forum, EBF 2019, Istanbul, Turkey, 31 May 2019*; Communications in Computer and Information Science Series CCIS; Springer: Berlin/Heidelberg, Germany, 2020; Volume 1188, pp. 182–194.
44. Esser, S.; Aicher, K. IfcBridge Model. Generation Using. Visual Programming. In Proceedings of the Forum Bauinformatik, Berlin, Germany, 11–13 September 2019.
45. Ding, L.; Li, K.; Zhou, Y.; Love, P.E.D. An IFC-Inspection Process Model for Infrastructure Projects: Enabling Real-Time Quality Monitoring and Control. *Autom. Constr.* **2017**, *84*, 96–110. [CrossRef]
46. Report, U.M.L.M.; Elements, R. *IFC Road Conceptual Model Report*; buildingSMART: Hertfordshire, UK, 2020.
47. Esmaeili, B.; Hallowell, M. Integration of Safety Risk Data with Highway Construction Schedules. *Constr. Manag. Econ.* **2013**, *31*, 528–541. [CrossRef]
48. Sharma, H.; McIntyre, C.; Gao, Z.; Nguyen, T.-H. Developing a Traffic Closure Integrated Linear Schedule for Highway Rehabilitation Projects. *J. Constr. Eng. Manag.* **2009**, *135*, 146–155. [CrossRef]
49. Sukumaran, P.; Bayraktar, M.E.; Hong, T.; Hastak, M. Model for Analysis of Factors Affecting Construction Schedule in Highway Work Zones. *J. Transp. Eng.* **2006**, *132*, 508–517. [CrossRef]
50. Yin, Y.; Choi, K.; Jeong, H.D.; Touran, A. Integration of Schedule, Cost, Traffic, and Risk into the Determination of Construction Contract Times. *J. Manag. Eng.* **2022**, *38*, 04022061. [CrossRef]
51. Li, Y.; Al-Haddad, S.; Taylor, T.R.B.; Goodrum, P.M.; Sturgill, R.E. Impact of Utilizing Construction Engineering and Inspection Consultants on Highway Construction Project Cost and Schedule Performance. *Transp. Res. Rec.* **2019**, *2673*, 716–725. [CrossRef]
52. Xia, Y.; Zhang, C.; Wang, C.; Liu, H.; Sang, X.; Liu, R.; Zhao, P.; An, G.; Fang, H.; Shi, M.; et al. Prediction of Bending Strength of Glass Fiber Reinforced Methacrylate-Based Pipeline UV-CIPP Rehabilitation Materials Based on Machine Learning. *Tunn. Undergr. Space Technol.* **2023**, *140*, 105319. [CrossRef]
53. Liu, B.; Lu, W.; Olofsson, T.; Zhuang, X.; Rabczuk, T. Stochastic Interpretable Machine Learning Based Multiscale Modeling in Thermal Conductivity of Polymeric Graphene-Enhanced Composites. *Compos. Struct.* **2024**, *327*, 117601. [CrossRef]
54. Liu, B.; Vu-Bac, N.; Zhuang, X.; Fu, X.; Rabczuk, T. Stochastic Integrated Machine Learning Based Multiscale Approach for the Prediction of the Thermal Conductivity in Carbon Nanotube Reinforced Polymeric Composites. *Compos. Sci. Technol.* **2022**, *224*, 109425. [CrossRef]
55. Liu, B.; Vu-Bac, N.; Rabczuk, T. A Stochastic Multiscale Method for the Prediction of the Thermal Conductivity of Polymer Nanocomposites through Hybrid Machine Learning Algorithms. *Compos. Struct.* **2021**, *273*, 114269. [CrossRef]
56. Liu, B.; Vu-Bac, N.; Zhuang, X.; Lu, W.; Fu, X.; Rabczuk, T. Al-DeMat: A Web-Based Expert System Platform for Computationally Expensive Models in Materials Design. *Adv. Eng. Softw.* **2023**, *176*, 103398. [CrossRef]
57. Lim, C.P.; Yussof, M.M.; Sopian, S.F.; Jaganathan, J. Review on the Non-Destructive Test (NDT) Approaches on Concrete Crack Mapping Prediction. *J. Adv. Res. Appl. Mech.* **2025**, *131*, 172–184. [CrossRef]
58. Naji, K.; Gowid, S.; Ghani, S. AI and IoT-Based Concrete Column Base Cover Localization and Degradation Detection Algorithm Using Deep Learning Techniques. *Ain Shams Eng. J.* **2023**, *14*, 102520. [CrossRef]
59. Van Giang, H.; Anh, N.T.; Nam, L.H.; Nam, T.P.; Tu, T.T. Challenges on application of digital technology in building operation and maintenance of high-rise buildings in vietnam. *Malays. Constr. Res. J.* **2024**, *23*, 131–149.
60. Waqar, A.; Bheel, N.; Tayeh, B.A. Modeling the Effect of Implementation of Artificial Intelligence Powered Image Analysis and Pattern Recognition Algorithms in Concrete Industry. *Dev. Built Environ.* **2024**, *17*, 100349. [CrossRef]

61. Zou, X.; Yan, G.; Mohammed, K.J.; Suhatril, M.; Khadimallah, M.A.; Marzouki, R.; Almirante, H.; Escorcia-Gutierrez, J. Advancements in Nano-Enhanced Steel Structures for Earthquake Resilience: Integrating Metallic Elements, AI, and Sensor Technology for Engineering Disasters Mitigation in Steel Buildings. *Steel Compos. Struct.* **2024**, *53*, 443–460. [CrossRef]
62. Uaissova, M.; Zharlykassov, B. Application of AI Techniques for Asphalt Concrete Mix Production Optimization. *J. Eur. Des Syst. Autom.* **2024**, *57*, 353–361. [CrossRef]
63. Ramadan, S.; Kassem, H.; ElKordi, A.; Joumblat, R. Incorporating Artificial Intelligence Applications in Flexible Pavements: A Comprehensive Overview. *Int. J. Pavement Res. Technol.* **2024**. [CrossRef]
64. Castañeda, K.; Sánchez, O.; Herrera, R.F.; Gómez-Cabrera, A.; Mejía, G. Building Information Modeling Uses and Complementary Technologies in Road Projects: A Systematic Review. *Buildings* **2024**, *14*, 563. [CrossRef]
65. Castañeda, K.; Sánchez, O.; Herrera, R.F.; Mejía, G. Deficiencies Causes in Road Construction Scheduling: Perspectives from Construction Professionals. *Heliyon* **2024**, *11*, e41514. [CrossRef] [PubMed]
66. Aladağ, H.; Demirdöğen, G.; Demirbağ, A.T.; Işık, Z. Understanding the Perception Differences on BIM Adoption Factors across the Professions of AEC Industry. *Ain Shams Eng. J.* **2023**, *14*, 102545. [CrossRef]
67. Yu, Q.; Li, K.; Luo, H. A BIM-Based Dynamic Model for Site Material Supply. *Procedia Eng.* **2016**, *164*, 526–533. [CrossRef]
68. Rashidi, A.; Yong, W.Y.; Maxwell, D.; Fang, Y. Construction Planning through 4D BIM-Based Virtual Reality for Light Steel Framing Building Projects. *Smart Sustain. Built Environ.* **2023**, *12*, 1153–1173. [CrossRef]
69. Essam, N.; Khodeir, L.; Fathy, F. Approaches for BIM-Based Multi-Objective Optimization in Construction Scheduling. *Ain Shams Eng. J.* **2023**, *14*, 102114. [CrossRef]
70. Malcher, A.; Saliba, A.; Wolgast, E. Herausforderungen Und Perspektiven Bei Der BIM-Planung von Brücken—Das Beispiel Der Geh- Und Radwegüberführung Riederwald. *Bautechnik* **2023**, *100*, 396–405. [CrossRef]
71. Xinting, G.; Weimin, Z. Apply Digital-Twin Model to Optimize the Planning of Equipment Pipeline System in the Laboratory Campus. In Proceedings of the International Conference on Computational Design and Robotic Fabrication, Shanghai, China, 24–25 June 2023; Springer: Singapore, 2024. Part F2072. pp. 260–272. [CrossRef]
72. Aranda, J.Á.; Santonja, M.M.; Saurí, M.G.; Peris-Fajarnés, G. Minimizing Shadow Area in Mountain Roads for Improving the Sustainability of Infrastructures. *Sustainability* **2021**, *13*, 5392. [CrossRef]
73. Al-Roumi, H.; Al-Sabah, R. Exploring the Rate of Adoption and Implementation Depth of Building Information Modeling (BIM): A Case of Kuwait. *J. Eng. Res.* **2023**, *12*, 86–99. [CrossRef]
74. Kovvuri, A.R. Web-Based Visual Project Progress (4D BIM) to Plan and Monitor in a More Lean and Efficient Way. In Proceedings of the Indian Lean Construction Conference, Hyderabad, India, 13–16 December 2022; Springer: Singapore, 2024; pp. 407–412. [CrossRef]
75. Schlosser, T.; Schlosser, P.; Zuzulová, A.; Cápayová, S.; Bálint, G. Traffic Planning as a Preliminary Tool in Project Documentation for BIM. *Buildings* **2022**, *12*, 381. [CrossRef]
76. RazaviAlavi, S.R.; AbouRizk, S. Construction Site Layout Planning Using a Simulation-Based Decision Support Tool. *Logistics* **2021**, *5*, 65. [CrossRef]
77. Le, P.L.; Dao, T.M.; Chaabane, A. BIM-Based Framework for Temporary Facility Layout Planning in Construction Site: A Hybrid Approach. *Constr. Innov.* **2019**, *19*, 424–464. [CrossRef]
78. Wang, F.; Zou, Y.; del Rey Castillo, E.; Ding, Y.; Xu, Z.; Zhao, H.; Lim, J.B.P. Automated UAV Path-Planning for High-Quality Photogrammetric 3D Bridge Reconstruction. *Struct. Infrastruct. Eng.* **2022**, *20*, 1595–1614. [CrossRef]
79. Zavari, M.; Shahhosseini, V.; Ardeshir, A.; Sebt, M.H. Multi-Objective Optimization of Dynamic Construction Site Layout Using BIM and GIS. *J. Build. Eng.* **2022**, *52*, 104518. [CrossRef]
80. Hosny, A.; Nik-Bakht, M.; Moselhi, O. Physical Distancing Analytics for Construction Planning Using 4D BIM. *J. Comput. Civil. Eng.* **2022**, *36*, 04022012. [CrossRef]
81. Caldart, C.W.; Scheer, S. Construction Site Design Planning Using 4D BIM Modeling. *Gest. Prod.* **2022**, *29*, e5312. [CrossRef]
82. Kontrimovičius, R.; Ustinovichius, L. BIM-Based Prototype of a Mathematical Model of Construction Planning. *J. Civil. Eng. Manag.* **2023**, *29*, 1–14. [CrossRef]
83. Krasovskaya, O.A.; Vyaznikov, V.E.; Mamaeva, A.I. Application of BIM Technologies as IT Projects for Digital Transformation in Industry. In Proceedings of the International Scientific and Practical Conference Digital and Information Technologies in Economics and Management, Online, 2 November 2021; Lecture Notes in Networks and Systems LNNS; Springer: Cham, Switzerland, 2022; Volume 432, pp. 104–116. [CrossRef]
84. Magill, L.J.; Jafarifar, N.; Watson, A.; Omotayo, T. 4D BIM Integrated Construction Supply Chain Logistics to Optimise On-Site Production. *Int. J. Constr. Manag.* **2022**, *22*, 2325–2334. [CrossRef]
85. Chamikara, K.L.; Perera, B.A.K.S.; Kurukulasooriya, I.N. Use of building information modelling to mitigate cost overruns in design and build projects. *World Constr. Symp.* **2023**, *1*, 1104–1115. [CrossRef]
86. Elyano, M.R. Yuliastuti Analysis of Clash Detection and Quantity Take-off Using BIM for Warehouse Construction. *IOP Conf. Ser. Earth Environ. Sci.* **2021**, *794*, 012012. [CrossRef]

87. Sampaio, A.Z.; Azevedo, G.; Gomes, A. BIM Manager Role in the Integration and Coordination of Construction Projects. *Buildings* **2023**, *13*, 2101. [CrossRef]
88. Rohani, M.; Shafabakhsh, G.; Haddad, A.; Asnaashari, E. Strategy Management of Construction Workspaces by Conflict Resolution Algorithm and Visualization Model Strategy Management of Construction Workspaces. *Eng. Constr. Archit. Manag.* **2018**, *25*, 1053–1074. [CrossRef]
89. Hui, S.; Jing, K.T.; Yee, H.C.; Siyao, C.; Zihao, D.; Yangle, S. The Impacts of 5D Building Information Modelling towards Cost Management in the Construction Industry. *J. Adv. Res. Appl. Sci. Eng. Technol.* **2023**, *32*, 471–487. [CrossRef]
90. Baldrich Aragó, A.; Roig Hernando, J.; Llovera Saez, F.J.; Coll Bertran, J. Quantity Surveying and BIM 5D. Its Implementation and Analysis Based on a Case Study Approach in Spain. *J. Build. Eng.* **2021**, *44*, 103234. [CrossRef]
91. Hussain, O.A.I.; Moehler, R.C.; Walsh, S.D.C.; Ahiaga-Dagbui, D.D. Minimizing Cost Overrun in Rail Projects through 5D-BIM: A Conceptual Governance Framework. *Buildings* **2024**, *14*, 478. [CrossRef]
92. Trochymiak, W.; Krygier, A.; Stachura, M.; Jaworski, J. The BIM 5D Model of the Bridge Built Using the Incremental Launching Method. *Arch. Civil. Eng.* **2023**, *69*, 157–172. [CrossRef]

sustainability

MDPI

Article

Towards More Sustainable Planning Decisions Around Airports: Investigating Global Airport Classifications and Proposing a Four-Tiered System for Australia

Isara Khanjanasthiti [1,*], Bhishna Bajracharya [2] and Daniel O'Hare [2]

1 Department of Geography and Planning, University of New England, Parramatta, NSW 2150, Australia
2 Faculty of Society & Design, Bond University, Gold Coast, QLD 4226, Australia; bbajrach@bond.edu.au (B.B.); dohare@bond.edu.au (D.O.)
* Correspondence: ikhanjan@une.edu.au

Abstract: Airports differ vastly in purpose, scale, land use and activities. Although several criteria and definitions have been used to classify airports worldwide, no universally accepted classification currently exists. The lack of standardisation means an airport could fall under multiple overlapping categories, creating confusion and obscuring its actual role. Australia also lacks an official system that reflects the diverse nature of its airports. This paper addresses that gap by proposing a holistic and integrated four-tiered classification framework for Australian airports. Drawing on an international literature review and comparative policy analysis, the classification is based on four criteria: airport location, governance, network function and passenger profile. To illustrate its practical application, the system is applied to case studies of selected airports representing each tier, as well as airports that overlap between tiers. The proposed classification has the potential to enable more strategic, coordinated and sustainable infrastructure investment and land use decision-making around airports.

Keywords: airport classification; international airports; domestic airports; regional airports; remote airports; Australia

1. Introduction

Airports vary in purpose, scale, land use, activities and relationships with their surrounding regions. Given such complexity, national governments and international agencies employ airport classification for several purposes, including slot allocation (designated time intervals for an aircraft to land or depart at an airport), network management, public allocation funding, competition assessment, security regulations or setting use charges [1]. However, no universally accepted classification of airports currently exists [2]. There are many ways to categorise an airport, using a variety of factors and circumstances, including the airport's historical context, functions, political power and other geographical factors. Todd et al. [2] (p. 143) observe that relying on these broad factors has led to "a simplification of the true role of any airport, trading sophistication for clarity". They argue, however, that airport classification is useful as it allows some degree of categorising airport infrastructure, thus allowing generalisation. Additionally, due to the lack of standardised classification, an airport could fall under several categories from various authorities worldwide, all of which employ different terms and criteria.

There is no official Australian airport classification that comprehensively captures the characteristics and context of airports despite the critical role of airports in connecting cities and towns across a geographically vast country. The only official Australian classification currently applies to its international airports. While the Australian Airports Association (AAA) offers a tiered classification system, it lacks detailed explanations for how each application of each tier. This can lead to inconsistent interpretations and limits its usefulness for decision-making. For example, Gold Coast Airport and Cairns Airport are categorised as 'regional' despite having higher passenger volumes than some capital city airports. Similarly, the existing paradigm of grouping all privatised airports as 'major airports' can misrepresent their actual activity levels. These gaps highlight the need for a clearer, more consistent classification that holistically considers airport characteristics and contexts.

To address these issues, the aim of this paper is to develop a conceptual classification for civil airports in Australia that holistically considers their key characteristics. Three objectives underpin this aim: (1) to explore how airports are classified in the literature and globally, (2) to review classifications currently in use for airports in Australia and (3) to develop a classification for Australian airports using the findings from the two prior objectives. To achieve the aim and objectives, the paper first details the methodology of the study. It then reviews the literature on how airports are classified. Subsequently, the paper analyses relevant policy documents, including official and unofficial classifications and categories for airports worldwide and in Australia. The literature review and policy analysis findings are then synthesised into a four-tiered classification system for Australian airports. The classification is also applied to selected airports to demonstrate its practical use. The paper then explores various ways in which the system can support government decision-making, ultimately promoting economic, environmental and social sustainability. The paper concludes with a discussion of the potential applicability of the conceptual classification to other countries, as well as an outline of the limitations and future research directions for the proposed system.

2. Methodology

To develop the conceptual classification for Australian airports, this study employed a qualitative research design comprising four scaffolded steps: (1) a literature review, (2) a comparative policy analysis, (3) development of the proposed framework and (4) application of the proposed framework. This approach enabled a comprehensive exploration of airport classification frameworks across jurisdictions and facilitated the development of a context-specific classification for Australia.

Firstly, the literature review examined peer-reviewed publications on airport classifications, with an emphasis on identifying commonly used classification criteria. Secondly, the comparative policy analysis examined airport classification systems in the United States, the United Kingdom, Europe and China. These jurisdictions were selected due to the maturity, diversity and policy innovation of their airport networks. Existing airport classifications in Australia were also evaluated to determine their application and limitations, as well as to explore the diverse nature of Australian airports. Thirdly, key criteria for an Australian airport classification system were established, based on the international review of airport classification systems and the characteristics of Australian airports. This led to the development of a four-tiered framework for classifying Australian airports based on geographical and operational characteristics. Lastly, to illustrate the practical applicability of the proposed framework, it was applied to a set of airport case studies representing each tier. These case studies were chosen to portray the diversity of airport types captured across the four proposed tiers, including examples from capital cities, major regional centres, regional towns and remote communities. The case studies showcase the classification system's ability to distinguish airport characteristics in a manner that is both consistent and policy-relevant. This application highlights the classification framework's utility in supporting planning, regulatory and investment decisions. Figure 1 provides a visual summary of the four steps employed in the study's methodology.

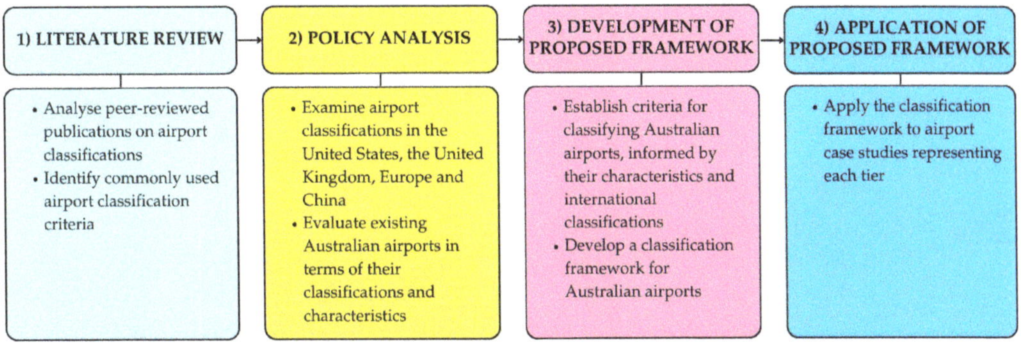

Figure 1. Diagram illustrating the qualitative methodology employed in this study. Source: Authors.

3. Literature Review: Diverse Perspectives in Airport Classification

3.1. Airport Classification Studies

Past studies have employed different definitions of airport types [3]. According to Chen et al. [4], airport classification studies are highly varied—an unsurprising trend given their diverse geographical and geopolitical factors. They propose that airport classification studies have fallen into three categories. The first category involves using single or single-type indicators—such as the airport's function as a hub [5]. The second type of study utilises multiple indicators, capturing airport properties and geographical factors. The third approach involves developing a 'synthetic cross-network indicator system', combining aspects of the first two techniques and including other modes of non-air transport [4]. Chen et al. [4] highlight that basic airport classification studies employing single or single-type indicators tend to rely on passenger volumes or types of airport services and produce results that are broad but limited in nuance. Nevertheless, more recent studies have increasingly employed the second approach, combining simple indicators with various geographical factors and regional economic influence. Table 1 summarises the principal criteria employed in selected airport classifications.

Table 1. Airport classification criteria employed in the literature.

Source	Scope	Classification Criterion/Criteria
[6]	Worldwide	Portion of passenger traffic in the same 'multiairport system'
[7]	The United Kingdom and the United States	(1) Geographical scale of the airport (e.g., national, regional or international airport) (2) The airport's role (e.g., origin and destination or hub airport)
[8]	United States	Three categories of airports in the same metropolitan area (i.e., a 'primary' airport, other 'core' airports and 'fringe' airports), based on the following attributes: (1) Portion of passenger traffic in the metropolitan area (2) Distance to the primary airport, which determines the 'core' or 'fringe' status
[1]	United States	Three-tier hierarchy based on the airport's hubbing activity indicators, including traffic generation and connectivity

Table 1. *Cont.*

Source	Scope	Classification Criterion/Criteria
[9]	Worldwide	Freight activity indicators, including the following: (1) Annual freight volume (2) The proportion of freight volume compared to all airport activities (3) The proportion of freight aircraft compared to all commercial aircraft (4) The proportion of international freight compared to all freight
[10]	Worldwide	Identification of multiple airport regions (MARs) based on the following factors: (1) Airport passenger traffic (2) Temporal distance between airports
[3]	Worldwide	Five-tier hierarchy, with each tier containing up to two types of airports, based on the following characteristics: (1) Continental aircraft traffic shares (2) Proximity to other airports in the same network (3) Hubbing potential based on connectivity level
[4]	China	(1) Interaction between air and high-speed rail (2) Airport's properties, including the number of seats, number of carriers, airport competitors, routes and destinations

As illustrated in Table 1, multiple studies have employed various criteria in their airport classifications, with passenger traffic as a commonly used criterion. Several studies have also utilised the relationship between nearby airports—grouped into one 'region', 'system' or 'area'—as an essential consideration in their classification approach.

3.2. International Classification Systems

Internationally, two agencies play a significant role in governing commercial aviation operations: the International Civil Aviation Organisation (ICAO) and the International Air Transport Association (IATA). The ICAO's Aerodrome Reference Code, included in the organisation's Annex 14—a document establishing Standards and Recommended Practices (SARPs) for airport design and operations—can be used to categorise airports [11] (The Aerodrome Reference Code specifies the maximum size of aircraft for which an airport is designed based on three criteria: runway length, wingspan and outer main gear wingspan). In the same document, the International Civil Aviation Organization [12] also employs peak-hour aircraft traffic volume to classify airport traffic density into light, medium and heavy categories. Meanwhile, the IATA employs the Worldwide Airport Slot Guidelines to allocate airport slots by designating airports as Level 1, Level 2 or Level 3 depending on their potential congestion level [13].

Having explored how airports have been classified in past studies, the paper next discusses airport classifications currently employed in different regions, including the United States, the United Kingdom (UK), Europe and China.

3.3. Airport Classifications: Policy Examples from the United States, the United Kingdom, Europe and China

3.3.1. United States

In the United States, airports are classified by the Federal Aviation Administration (FAA), which publishes and annually updates the National Plan of Integrated Airport Systems (NPIAS). The NPIAS assists the FAA in coordinating and funding the system of 3287 public-use airports nationwide under the federally run Airport Capital Improvement Program over a five-year timeframe. The document groups these airports into two major categories: primary and nonprimary [14]. Most (2904 or 88.3%) of the public-use airports covered by the NPIAS are nonprimary airports. The document also employs statute categories, defined in Title 49 of the United States Code and outlined in Table 2, to categorise the airports.

Table 2. Statute categories of public-use airports in the United States [15].

Category	Description
	Primary Airports
Commercial Service Airports	• Publicly owned airports that facilitate more than 10,000 passenger boardings annually and receive scheduled passenger service • Further classified into four 'hub' categories (large hub, medium hub, small hub and nonhub primary) based on their percentage of national passenger boardings
	Nonprimary Airports
Commercial Service Airports	• Publicly owned airports that facilitate between 2500 and 10,000 passenger boardings annually and receive scheduled passenger service • Also referred to as 'nonhub nonprimary'
Reliever Airports	• Help relieve congestion at a commercial service airport and facilitate general aviation access for the community
General Aviation Airports	• Airports with no scheduled passenger service or with scheduled service that facilitates less than 2500 passenger boardings yearly

In addition to those listed in Table 2, the Federal Aviation Administration [16] also specifies another category, 'Cargo Service Airports', comprising airports served by cargo aircraft carrying no passengers, with a total landed weight exceeding 100 million pounds. An airport can be concurrently categorised as both a Commercial Service Airport and a Cargo Service Airport. Following cooperation with aviation stakeholders and a national review of General Aviation Airports, the FAA published General Aviation Airports: A National Asset in 2012. The document provides a further policy-based categorisation of nonprimary airports, classifying them as 'national', 'regional', 'local', 'basic' and 'unclassified' based on their existing activity, geographic factors and public interest functions [14]. Notably, ownership is a common criterion determining the applicable category for an airport across both the statute and policy-based airport categories employed by the FAA.

3.3.2. United Kingdom

The UK Government's White Paper on Airports Policy specifies four airport categories: 'gateway international airports', 'regional airports', 'local airports' and 'general aviation airports' [17]. All the Gateway International Airports, except Manchester Airport and Glasgow Prestwick Airport, were in London [18]. Unlike the other three airport categories, Gateway International Airports served international flights. Following the White Paper publication, the UK Government encouraged additional investment in major regional airports to meet local demand and "shift the burden away from the London area airports" [19]. The government also promoted the further development of Manchester Airport as the principal Gateway International Airport outside the London area [19]. However, in the 1980s, the UK Government abolished this official classification, shifting towards market-driven approaches and believing strict categorisation would hinder airports from developing efficiently in response to passenger demand [20].

More recently, Butcher [21] employs the term 'regional airport' to refer to a UK airport outside South East England, a commonly employed approach in airport-related publications in the UK. Likewise, the Environment, Transport and Regional Affairs Committee [22] refers to all UK airports other than the five 'London airports'—including Heathrow, Gatwick, Stansted, Luton and London City airports—as 'regional airports'. Similarly, the Civil Aviation Authority [23] uses the term 'regional airport' in its Aviation Trends report for the third quarter of 2018 to refer to all airports that do not fall under its 'London airports' category.

The six 'London airports', as of January 2023 [24], are not the six busiest airports in the UK in terms of terminal passenger volume—although four of them, namely Heathrow, Gatwick, Stansted and Luton, are in the top five. All six airports, however, are located in the London metropolitan area and are the only international airports in the region. The 'London airports' comprise the 'London Airport System', the busiest network of nearby airports worldwide by passenger volume [25]. London City and Southend airports service a significantly smaller number of passengers than the other 'London airports'—3.2 million and 89,000, respectively, in comparison to Heathrow and Gatwick airports, which served 64.5 and 34.2 million passengers, respectively, in January 2023 [24]. As such, passenger volume at individual airports is a less important criterion in the UK's classification of 'London airports' and 'regional airports'. Instead, the cumulative number of passengers of the larger airport 'system' or 'network' that an airport is part of could be seen as a more relevant criterion for the country's airport classification.

The cluster of 'London airports' resides within the expansive metropolitan vicinity of London and collectively handled 60.7% of the nation's air passenger traffic between February 2022 and January 2023 [24]. Consequently, these pivotal airports play a momentous role within the UK aviation sector, reflecting the metropolitan primacy structure wherein London substantially surpasses all other cities, including Birmingham, in terms of population. Thus, the six 'London airports' are classified based on three primary criteria, namely: (1) the spatial location of the airport, i.e., whether it falls within or outside the London metropolitan area, (2) the airport's function as an international airport and (3) the magnitude of passenger traffic cumulatively managed by the airport system that the airport is part of.

Nevertheless, one major downside of the UK's unofficial airport classification system is the large degree of simplification applied to 'regional airports', referring to all other airports in the country that are not 'London Airports'. Several regional airports manage a large passenger volume, e.g., Manchester, Edinburgh and Birmingham airports, which serviced 24.2 million, 11.8 million and 10 million passengers, respectively, between February 2022 and January 2023 [24]. In contrast, other regional airports handle a significantly smaller number of passengers, e.g., Southampton and Tiree airports, which serviced 658,000 and 11,000 passengers, respectively, over the same period. Furthermore, many regional airports also handle a larger passenger volume than two of the six London airports: London City and Southend airports.

3.3.3. Europe

No integrated and complete airport classification uniformly applies to European airports [26,27]. The European Parliament, European Council, European Commission and Committee of the Regions each employ their own classification systems. However, these classifications consider only passenger volume as the sole criterion for airport categorisation [26]. Interestingly, European airport classification has received considerable attention from scholars in recent years (e.g., [26–29]), possibly attributable to the lack of an integrated classification for the whole continent.

3.3.4. China

The Civil Aviation Administration of China (CAAC) classifies four types of airports in the Civil Aviation [2018] No. 18 document as a basis for setting airport charges [30]:

- Primary Airport, class 1—including three airports (Beijing, Shanghai Pudong and Guangzhou);
- Primary Airport, class 2—including three airports (Shenzhen, Chengdu and Shanghai Hongqiao);
- Secondary Airport—including 20 airports (e.g., Kunming, Hangzhou, Wuhan and Qingdao);
- Tertiary Airport—including all other airports.

Similarly to Europe's airport classifications, the primary criterion for the CAAC's airport classification appears to be passenger volume. This observation is grounded in the fact that all six Primary Airports are amongst the country's top ten busiest airports by departing seats as of May 2025 [31].

4. Australian Airport Classifications and Their Limitations

Australia's official airport classification is limited to international airports only and considers border facilities and services at these airports. Although a new airport categorisation was recently introduced, its scope is only intended for determining each airport's requirements for security screening measures [32]. The international airport classification system is specified under Section 9 of the Air Navigation Act 1920. Depending on the range of border facilities and onsite services, an airport is designated as an 'international airport' with one of the categories outlined in Table 3.

Table 3. Categories of international airports in Australia under the Air Navigation Act 1920 [33].

Category	Description
Major International Airport	An airport of entry and departure, open to both scheduled and non-scheduled flights, where all procedures related to customs, immigration and quarantine are available
Restricted Use International Airport	An airport of entry and departure where all procedures related to customs, immigration and quarantine are available on a restricted basis only to flights with prior approval
Alternate Airport	An airport specified in the flight plan at which a flight may land if it is inadvisable to land at the originally designated airport
International Non-Scheduled Flight Airport	An airport where international non-scheduled flights may be granted with approval and no other form of international operation is permitted
External Territory International Airport	An airport located in an Australian External Territory where international flights can enter and depart, and all procedures related to customs, immigration and quarantine are available

To enhance the country's aviation security, the Federal Government recently introduced a risk categorisation framework for Australian airports through the Aviation Transport Security Amendment (Security Controlled Airports) Regulations 2019, which was in effect from December 2019 to December 2020. The seven-tiered classification considers the overall risk profile of airports, evaluating their regional attributes, operating context and existing threat and security environments [32].

Australia's airport sector is represented by the AAA, a peak national body with more than 340 airports and aerodromes as member organisations. According to Merkert [34], the AAA categorises its member airports and aerodromes into the following seven tiers:

- Tier 1 Capital City Airports;

- Tier 2 Non-Capital International Gateway Airports;
- Tier 3 Major Regional Airports with direct interstate services;
- Tier 4 Major Regional RPT airports without direct interstate services, servicing more than 20,000 passengers (RPT denotes Regular Public Transport, referring to scheduled commercial flights "on which seats and/or cargo space is available to the general public" [35]);
- Tier 5 Regional Airports without direct interstate services, servicing less than 20,000 passengers;
- Tier 6 Regional Airports without Regular Passenger Transport (RPT) services;
- Tier 7 Remote Community Aerodromes (used for medical emergencies).

The AAA [36] also suggests that Australia's airport industry comprises four broad categories, outlined in Table 4.

Table 4. The Australian Airports Association's categories of airports in Australia [36].

Category	Description
Capital city and major privatised airports	Airports located in the capital city of a state/territory, e.g., Sydney and Melbourne, all of which have been privatised under the Airports Act 1996
Regional airports	Airports located in a regional area, the majority of which were owned and operated by the Federal Government, primarily for the military before being transferred to the management and ownership of their respective local governments
Department of Defence airports	Airports owned and operated by the Department of Defence
Completely privately owned airports and air strips	Airports and air strips that are privately owned

Most airports in Australia are 'regional airports', which play a significant role in the Australian aviation industry due to the beneficial outcomes they facilitate for regional communities such as connecting regional and rural areas with major cities, emergency medical services, aerial firefighting, agricultural activities, pilot training and aerial photography [36] (p. 7). These diverse activities position regional airports as important economic and social assets for regional Australia. Regional airports are estimated to account for 45% of Australia's tourism industry and facilitate 12% of annual domestic trips [37].

Donehue and Baker [38] employ the acronym RRR to refer to remote, rural and regional airports in their investigation of challenges faced by airports in non-metropolitan Australia. They utilised the Australian Standard Geographical Classification (ASGC) Remoteness Structure, illustrated in Figure 2, to categorise RRR airports, most of which are owned and managed by local governments, into the following three sub-categories:

- Regional airports, located in 'Inner Regional Australia' and providing RPT to capital cities, 'Outer Regional Area' and 'Remote Australia';
- Rural airports, located in 'Outer Regional Australia' and facilitating both RPT services and general aviation activities; and
- Remote airports, located in 'Remote Australia' and 'Very Remote Australia' and facilitating RPT, general aviation and charters.

Australia's aviation industry lacks an official classification system that encapsulates both domestic and international airports. The AAA's airport categories, which are not officially implemented in the country's aviation sector, do not clearly define the roles, functions and characteristics of airports. This gap signals the need for a classification system that comprehensively considers a range of airport attributes and the unique context of Australia's aviation sector. The following section develops such a classification system based on an analytical discussion of the airport classifications outlined above.

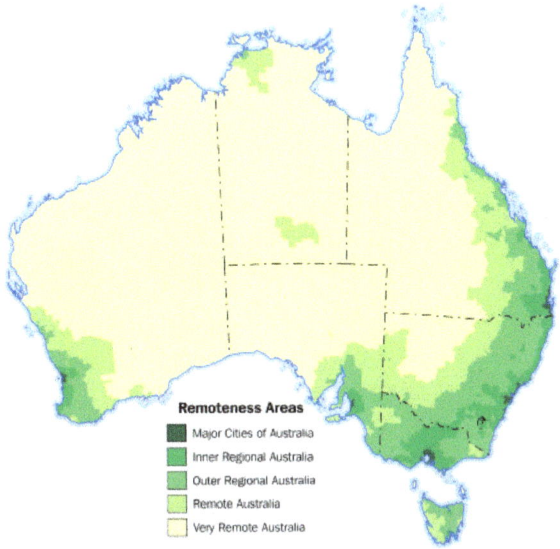

Figure 2. The Australian Standard Geographical Classification Remoteness Structure map, with the dash lines illustrating state and territory boundaries. This figure illustrates a map of ASGS Edition 3 Remoteness Areas for Australia, published by the Australian Bureau of Statistics [39] and reproduced in this paper under the CC BY 4.0 license [40].

5. Developing a Classification System for Australian Airports

Australia's international airport classification only considers facilities and services related to overseas passengers at these airports. Meanwhile, the recently introduced (and now repealed) security classification employs risk-based criteria to evaluate the risk level in the operational environment of airports. The latter system is more comprehensive in classification criteria and applies to both domestic and international airports. The AAA utilises three criteria in categorising their member airports, namely airport location (in a capital city or a regional area), airport governance (managed and operated by a private organisation, a local government or the Federal Government) and airport ownership (private or public).

Notably, unlike the rest of the world, Australian airport classifications—including official ones and the unofficial system employed by the AAA—do not incorporate passenger-related measurements. However, passenger volume is arguably an essential criterion for classifying airports in Australia, given that several airports normally categorised as 'regional airports' by the AAA still service a relatively large volume of passengers, thus playing a significant role in the Australian aviation industry. In 2018–19, Gold Coast Airport and Cairns Airport were Australia's sixth and seventh busiest airports, servicing more passengers than three out of the eight airports categorised by the AAA as 'Tier 1 Capital City Airports' [41]. Yet, under the AAA's categorisation system, these airports would be considered 'regional airports' due to their location outside their state's capital city, Brisbane. However, this category would inaccurately reflect the role and function of these airports as the sixth and seventh busiest airports in Australia.

Privatised airports in Australia are often collectively referred to 'major airports' (e.g., [42,43]) (21 of the federally owned airports in Australia are 'privatised' under the Airports Act 1996 [44]). Through this arrangement, these airports are managed and operated by private organisations). However, not all of these 'major airports' serve a large number of passengers [41], with privatised airports comprising only 11 of the country's 20 busiest airports in 2023–24 [45]; as such, simply referring to privatised airports as 'major airports' is not an accurate reflection of the actual passenger volume of these airports. This approach could create a public misconception that all privatised airports play a significant role in the country's aviation sector regarding passenger volume.

The AAA's seven-tiered system categorises its member airports by the following five factors related to airport characteristics:

1. Geographical location (e.g., capital city, non-capital city, regional area or remote community);
2. Operational scope (i.e., whether the airport primarily serves domestic or international flights);
3. Passenger volume;
4. Availability of direct interstate flights;
5. Facilitation of RPT services.

Although this classification framework appears comprehensive concerning the criteria considered, the AAA has not provided an exhaustive exposition for each airport category. Consequently, the lack of detailed elaboration may result in diverse interpretations of the seven-tiered categorisation among industry stakeholders.

The proposed definition of RRR airports by Donehue and Baker [38] offers a valuable framework for categorising the diverse range of airports in non-metropolitan Australia. Such a classification is necessary given the significant variations in their roles, functions, and attributes. They utilised the ASGC Remoteness Structure to categorise airports outside major urban centres into regional, rural, and remote airports. This approach adeptly incorporates the contextual nuances of these airports, acknowledging that their characteristics can markedly differ based on the level of remoteness. Moreover, the ASGC Remoteness Structure also delineates the 'Major Cities of Australia' category, encompassing capital cities and significant regional centres. This classification is particularly pertinent for major airports outside capital cities, as it enables their placement based on their location within a major urban centre.

6. Airport Tiers: A Proposed Airport Classification System for Australia

The authors have developed a conceptual classification encompassing four distinct categories for Australian airports, drawing on the above discussion on airport classification practices worldwide and a comprehensive understanding of Australian airport attributes. The classification, which employs the approach of combining multiple indicators related to airport properties and geographical factors [4], defines the role, function and characteristics of all Australian airports using four criteria: (1) location, (2) governance, (3) network function and (4) passenger profile.

While existing airport classification systems provide useful foundations, they vary in their focus on operational, spatial and institutional characteristics. The criterion related to passenger numbers, such as those in the United States, Europe and China, offers quantifiable thresholds but may prioritise volume over context. This approach can obscure the unique roles of regional or remote airports, which may be critical to network connectivity and regional sustainability despite relatively low traffic. Therefore, other relevant criteria are required. The governance criterion reflects the regulatory and ownership structures shaping airport management. Australia's privatisation of airports, which resembles models in the United States and the United Kingdom, has shaped several airports and, as such, should be reflected in the classification. Meanwhile, the network function of airports, including hub-versus-spoke models or role-based typologies, allows the nuanced nature of airports' activities and operations to be captured. In Australia, many regional airports serve diverse roles—from emergency services to pilot training to FIFO operations—that form the broader 'general aviation' activities.

The proposed airport classification combines structural attributes (governance and passenger profile) with contextual factors (location and network function), aiming to reflect not only what an airport is but also what it does, and for whom. This integrated approach recognises the differentiated nature of airports across Australia's vast and varied landscape.

The classification is based on a 'tier' system with four tiers. The notion of 'tier' within the proposed airport classification model finds its foundation in the definition of "one of several layers or levels" [46]. Considering the wide-ranging variations in airports' nature, location, function and operation, it is desirable to categorise them into distinct 'tiers' based on shared criteria, enabling their differentiation across various levels. The term 'tier' has also been widely used in the aviation industry and the literature. For instance, the AAA's classification system for its member airports is based on seven tiers.

Additionally, de Neufville [47] employs the term to describe airports such as Memphis and Louisville as 'second-tier airports'. Similarly, Grubesic et al. [48] offer a tier-based classification for airports worldwide, considering factors such as traffic flow and dominance over other airports, with Tier 1 airports exerting the highest influence over their counterparts. These different usages of 'tier' exemplify the term's utility in facilitating a coherent approach to comprehending the diverse airport landscape. The four airport categories within the proposed classification are First-Tier Airports, Second-Tier Airports, Third-Tier Airports and Fourth-Tier Airports.

6.1. First-Tier Airports

First-Tier Airports, having been privatised through the Airports Act 1996, currently operate under the management of private entities. Positioned within the confines of a state/territory's capital city, these airports fall within the 'Major Cities of Australia' zone of the ASGC Remoteness Structure (as depicted in Figure 2). Possessing pivotal significance, these airports play a paramount role in facilitating the seamless movement of both passengers and freight, serving as central hubs for multiple airlines. These airports also fulfil the vital function of being the principal international gateways for their respective state/territory. As primary facilitators of air travel domestically and internationally, First-Tier Airports command the highest proportion of passengers served amongst all airports within their respective state/territory.

6.2. Second-Tier Airports

Second-Tier Airports are located in major regional cities within the boundaries of 'Inner Regional Australia' in the ASGC Remoteness Structure. These airports, which may fall under the management of private entities or local governments, play a pivotal role in facilitating the movement of passengers and freight. Serving as prominent hubs for some airlines, Second-Tier Airports not only cater to a substantial portion of passengers relative to all airports within their respective state/territory but also may serve international flights. Owing to the array of on-site services, these airports can serve as alternate landing venues for First-Tier Airports in aircraft congestion or adverse weather conditions. The flexibility provided by Second-Tier Airports enhances the overall efficiency and reliability of air travel throughout the country.

6.3. Third-Tier Airports

Third-Tier Airports, located in regional areas within either the 'Inner Regional Australia' or 'Outer Regional Australia' zone of the ASGC Remoteness Structure, are owned and managed by local councils—with a few exceptions. Functioning as 'spokes' for both First-Tier and Second-Tier airports, these airports primarily facilitate domestic flights and cater to a relatively limited number of passengers. Depending on the scope of available facilities, Third-Tier Airports can also act as reliever airports for their higher-tier counterparts and facilitate a wide range of general aviation activities. In the Australian context, these airports play a vital role in connecting regional communities with capital cities and major regional centres, mitigating the challenges posed by 'the enormous distances that often exist between ... towns and cities' in Australia [38]. Third-Tier Airports contribute significantly to fostering connectivity and accessibility within the country's regional landscapes.

6.4. Fourth-Tier Airports

Fourth-Tier Airports predominantly comprise airports in communities located within the 'Remote Australia' or 'Very Remote Australia' classification in the ASGC Remoteness Structure. Similarly to Third-Tier Airports, most of these airports are owned and managed by local governments. Operating primarily to facilitate general aviation activities, Fourth-Tier Airports serve as key facilitators of pilot training, aerial work (e.g., aerial photography and agriculture), emergency medical services, and recreational flying pursuits. Given their remote locations and the corresponding constraints, Fourth-Tier Airports typically offer a relatively limited range of facilities, restricting access primarily to smaller aircraft compared to higher-tier airports. As a result, RPT and chartered flights servicing Fourth-Tier Airports may be available but remain limited in frequency. Nevertheless, Fourth-Tier Airports play an indispensable role in supporting aviation activities within their respective remote communities, fostering crucial services and contributing to the overall connectivity of such isolated regions.

6.5. Exclusion of Freight Dimension

The proposed classification excludes the freight aspect of airport operations, an important activity for many airports. The discussion below outlines the rationale behind this omission, which should be recognised as a potential limitation of the framework's applicability.

Despite their smaller size, remote location, and limited passenger traffic, certain Australian airports handle substantial volumes of valuable freight. Toowoomba Wellcamp Airport is an exemplary illustration. Situated in the agricultural region of Toowoomba, this Third-Tier Airport serves as more than just a passenger facility; it focuses on fostering robust freight activities for the region. The airport was strategically built as 'the only regional airport in Australia truly focused on growing a strong freight airline base rather than simply a passenger service facility' [49]. As the state of Queensland's sole airport with a dedicated 747-8F international freighter service, it has facilitated weekly international cargo flights since 2016 [50]. These flights transport agricultural products to Hong Kong, a gateway to the broader markets of China.

Meanwhile, certain major Australian airports within capital cities, which service substantial passenger volumes, handle minimal quantities of high-value freight and undertake limited freight operations. According to the Bureau of Infrastructure and Transport Research Economics BITRE [51], in 2018/19, three First-Tier Airports, namely Darwin, Canberra and Hobart airports, processed a relatively limited international freight volume of 475, 83 and 0 tonnes, respectively. However, they each served between two and three million total passengers. In contrast, although Toowoomba Wellcamp Airport, a Third-Tier Airport, serviced 110,000 passengers, its international freight totalled 1213 tonnes in the same period—a notably larger volume than that processed by the three First-Tier Airports mentioned here. These statistics portray the diversity in freight operations amongst Australian airports. Despite serving many passengers, higher-tier airports may not exhibit substantial freight activities. Conversely, lower-tier airports could surpass higher-tier airports in terms of freight handling. These characteristics underscore the impracticality of incorporating freight volume as a criterion for the proposed Australian airport classification. Therefore, future studies could develop a classification based on freight characteristics of Australian airports, expanding on the system proposed by Mayer [9] and reflecting the nuances of freight activities across Australian airports.

In Australia, freight activities are critical to many regional airports, particularly in supporting local industries and enabling remote community access to goods and services [52]. However, this classification framework prioritises indicators that apply consistently across the full range of Australian airports. Freight volumes are highly variable, often airport-specific, and not systematically reported across all airports, making them less suitable as a criterion in the proposed categorisation. Nevertheless, freight activities could form the basis for a future extension of the proposed classification or the development of an alternative categorisation based primarily or entirely on freight activities.

A summary of the proposed classification system is outlined in Table 5.

Table 5. Proposed classification for Australian airports. Source: Authors.

Location	Governance	Network Function	Passenger Profile
		First-Tier Airport	
Capital city of a state/territory	Privatised	• Main international gateway for the state/territory • Key hub for multiple airlines • Serve the highest passenger proportion amongst all airports in the state/territory	• Domestic and international passengers
		Second-Tier Airport	
Major regional city or centre	Privatised or owned and managed by a local government agency	• Hub for some airlines • Spoke and reliever airport for First-Tier Airports • Serve a large proportion of passengers in the state/territory	• Domestic and international passengers • Lower proportion of international passengers than First-Tier Airports
		Third-Tier Airport	
Regional city or town	Most are owned and managed by a local government agency	• Spoke and reliever airport for higher-tier airports • Connects regional communities with capital cities • Facilitates general aviation activities	• Only domestic passengers
		Fourth-Tier Airport	
Remote communities	Most are owned and managed by a local government agency	• Primarily facilitate general aviation activities • May occasionally service commercial flights	• Only domestic passengers and in limited numbers

6.6. Application to Australian Airports

Figure 3 provides a broad yet non-exhaustive list of Australian airports (represented by their International Air Transport Association (IATA) codes) categorised across the four tiers under the proposed classification system. All First-Tier Airports are fully listed, while airports in other tiers are selectively included. The figure also identifies selected airports exhibiting overlaps between adjacent tiers.

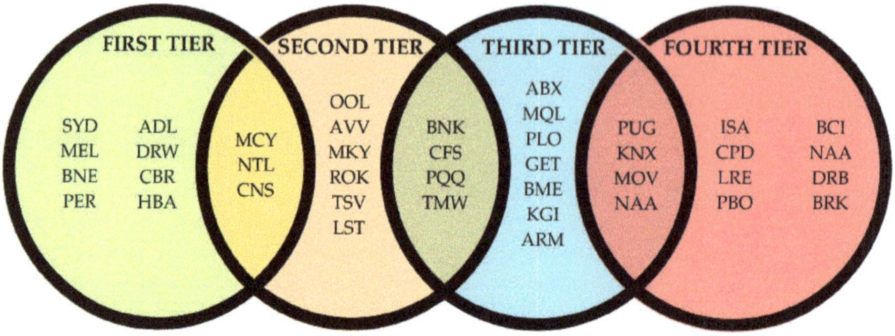

Figure 3. List of selected airports by IATA codes across the proposed four-tiered airport classification. Source: Authors.

Table 6 provides the full names of the airports shown in Figure 3.

Table 6. Full names of the airports listed in Figure 3.

IATA Code		Airport
	First Tier	
SYD		Sydney Kingsford Smith Airport
MEL		Melbourne Tullamarine Airport
BNE		Brisbane Airport
PER		Perth Airport
ADL		Adelaide Airport
DRW		Darwin International Airport
CBR		Canberra Airport
HBA		Hobart International Airport
	First/Second Tier	
MCY		Sunshine Coast Airport
NTL		Newcastle Airport
CNS		Cairns Airport
	Second Tier	
OOL		Gold Coast Airport
AVV		Avalon Airport
MKY		Mackay Airport
ROK		Rockhampton Airport
TSV		Townsville Airport
LST		Launceston Airport
	Second/Third Tier	
BNK		Ballina Byron Gateway Airport
CFS		Coffs Harbour Airport
PQQ		Port Macquarie Airport
TMW		Tamworth Airport
	Third Tier	
ABX		Albury Airport
MQL		Mildura Airport
PLO		Port Lincoln Airport
GET		Geraldton Airport
BME		Broome International Airport
KGI		Kalgoorlie-Boulder Airport
ARM		Armidale Airport
	Third/Fourth Tier	
PUG		Port Augusta Airport
KNX		Kununurra Airport
MOV		Moranbah Airport
NAA		Narrabri Airport

Table 6. *Cont.*

IATA Code	Airport
Fourth Tier	
ISA	Mount Isa Airport
CPD	Coober Pedy Airport
LRE	Longreach Airport
PBO	Paraburdoo Airport
BCI	Barcaldine Airport
NAA	Narrabri Airport
DRB	Derby Airport
BRK	Bourke Airport

Table 7 applies the classification to selected Australian airports, elaborating on one airport per tier.

Table 7. Application of the proposed classification to selected Australian airports.

Airport (IATA Code)	Location, State	Tier Attributes
First-Tier Airport		
Sydney Kingsford Smith Airport (SYD)	Sydney, New South Wales	• **Capital City Location:** The airport is located in Sydney, the capital city of the state of New South Wales, and the 'Major Cities of Australia' zone [53]. • **Significant Passenger Volume:** The airport is Australia's busiest airport, servicing a total of 41.5 million passengers in 2024 [45]. The airport handles a substantial share of passengers at both the state and national levels. In comparison, Newcastle Airport—the second busiest in the state—served 1.2 million passengers, while Melbourne Airport—the second busiest in the country—handled 35.5 million, both significantly lower than Sydney Kingsford Smith Airport's traffic in 2024 [45]. • **International Gateway:** In 2024, 40% of the airport's passengers are international passengers [45]. It plays a pivotal role as the primary international gateway for both the state and Australia. • **Privatised:** The airport has been operated by Sydney Airport Corporation Limited since its privatisation in 2002. • **Airline Hub:** The airport serves as an operational hub for Qantas Airways and a transit hub between regional New South Wales, other Australian cities and other countries. It is also a strategic hub for major domestic airlines, including Virgin Australia and Jetstar Airways [54].
Second-Tier Airport		
Gold Coast Airport (OOL)	Gold Coast, Queensland	• **Location in a Major City and Mega Region:** The airport is located across the City of Gold Coast and Tweed Shire within the 'Major Cities of Australia' zone [53]. The City of Gold Coast, which receives the majority of the airport's passenger traffic [55], is a major regional city located in proximity to Brisbane, the capital city of the state of Queensland. The city is also part of the '200 km city', a mega-region centred around Brisbane [56]. • **Significant Passenger Volume:** In 2024, the airport was Australia's sixth busiest airport, serving 6.2 million passengers [45]. As the second busiest airport in the state behind Brisbane Airport, which served 23.3 million passengers in 2024 [45], Gold Coast Airport is an important hub connecting regional communities in South East Queensland and Northern New South Wales with major Australian Cities and a few countries, including New Zealand, the United States and Canada [57,58]. The recently completed expansion of the airport's southern terminal has also increased the airport's capacity to handle passengers, expected to grow to 13 million by 2044 [58]. • **Hub for Airlines:** Domestically, the airport serves as a key spoke airport for Qantas Airways, Jetstar Airways and Virgin Australia, with daily flights to several First-Tier Airports. • **Low International Passenger Proportion:** While the airport services international passengers, its share of such passengers (10% in 2024) is significantly lower than that of First-Tier Airports. For example, Sydney Kingsford Smith Airport, Melbourne Airport, and Brisbane Airport recorded international passenger proportions of 40%, 32%, and 26%, respectively, in the same year [45]. Nevertheless, international travellers are a notable presence in the airport's passenger profile. Before the COVID-19 pandemic, the airport operated several direct international flights to other destinations, including Japan, Hong Kong, Singapore and Fiji [59]. In 2018–19, international flights accounted for 15% of the airport's total passenger volume (6.4 million) [45]. • **Privatised:** In 1998, the airport was privatised and its operating rights were transferred to Queensland Airports Limited. • **Diversion Destination for First-Tier Airport:** While the airport does not officially function as a reliever airport, it has served as a diversion destination for flights bound for Brisbane Airport in events of severe weather or runway closures or aircraft [60,61].

Table 7. *Cont.*

Airport (IATA Code)	Location, State	Tier Attributes
		Third-Tier Airports
Albury Airport (ABX)	Albury, New South Wales	• **Regional City Location:** The airport is located in Albury, a city adjacent to the state border of New South Wales and Victoria. Owned and operated by Albury City Council, the airport serves the Albury-Wodonga region, which spans the state border and falls within 'Inner Regional Australia' [53]. • **Spoke Function:** The airport connects residents of the region to Sydney, Melbourne and Brisbane. Thus, it serves only domestic flights and operates as a spoke airport for First-Tier Airports, servicing approximately 276,000 passengers in 2023–24 [45]. • **Facilitates General Aviation:** The airport plays an important role in facilitating general aviation activities, servicing 32,700 general aviation aircraft movements in 2016–17. This figure is expected to steadily increase to 41,500 by 2031–32 [62].
		Fourth-Tier Airport
Mount Isa Airport (ISA)	Mount Isa, Queensland	• **Remote City Location:** The airport is situated in Mount Isa, a remote city in North West Queensland within 'Remote Australia' [53]. • **Limited Commercial Flights:** According to Queensland Airports Limited [63], which has operated the airport since 2005, more than 82% of the airport's passengers are attributable to chartered flights bringing in fly-in fly-out (FIFO) workers. Nonetheless, the airport facilitates RPT flights to Brisbane and other major cities in the state, recording 232,000 domestic RPT passengers in 2023–24 [45]. • **Facilitates General Aviation:** With dedicated general aviation facilities, the airport enables general aviation activities encompassing helicopter operations, flight training and aerial surveying conducted by several businesses [64].

To highlight the shortcomings of how Australian airports are currently categorised, Sydney Kingsford Smith Airport, Gold Coast Airport and Mount Isa Airport, due to their privatised nature, would all be referred to as 'major airports'. Such a perspective does not accurately account for the airports' vastly different attributes, as illustrated in Table 6. Nevertheless, certain airports, due to their distinct characteristics, may overlap between two adjacent tiers, as illustrated in Table 8.

Table 8. Selected airports exhibiting overlaps between adjacent tiers of the proposed classification system. Source: Authors.

Airport (IATA Code)	Location (City, State)	Higher Tier Attributes	Lower Tier Attributes
		First-Tier/Second-Tier Airport	
Newcastle Airport (NTL)	Newcastle, New South Wales	• **Newcastle's Close Relationship to Sydney:** The airport serves Newcastle, a city situated within 'Major Cities of Australia' [53] and in the proposed 'Six Cities Region' centred around Greater Sydney [65] and its surroundings. • **Significant Passenger Volume:** As the second busiest airport in the state and the fifteenth busiest airport nationally, the airport served 1.2 million domestic passengers in 2024 [45]. • **Robust Links to Capital Cities, Regional Cities and Remote Towns**: The airport provides RPT flights to ten destinations, including capital cities, regional cities and remote towns nationwide. With frequent daily flights to First-Tier Airports and Second-Tier Airports and weekly flights to Third- and Fourth-Tier Airports, Newcastle Airport functions as a key spoke airport in the national flight network. • **Increasing Focus on International Flights:** With frequent daily flights to Brisbane and Melbourne Airports—where passengers can connect to international services—the airport has been actively positioning itself as an alternative to Sydney Kingsford Smith Airport, which offers limited accessibility to regional communities. Its long-term vision includes offering "direct international flights to key trading and tourism destinations" by 2036, supported by ongoing investment in a new international terminal [66] (p. 15).	• **Non-Capital City Location:** The airport is located in 'Inner Regional Australia', and Newcastle is not the capital city of New South Wales. • **Public Ownership:** The airport is owned and operated by Newcastle Airport Pty Limited, a joint venture between the City of Newcastle and Port Stephens Council. • **No International Flights:** Despite its emphasis on international connections for passengers, the airport currently facilitates only domestic flights, aligning it with the Third Tier.

Table 8. *Cont.*

Airport (IATA Code)	Location (City, State)	Higher Tier Attributes	Lower Tier Attributes
		Second-Tier/Third-Tier Airport	
Ballina Byron Gateway Airport (BNK)	Ballina, New South Wales	• **Location in Regional City:** The airport is situated in Northern New South Wales, within Inner Regional Australia. It serves Ballina, Byron Bay—a major tourism destination—and other surrounding cities and towns. • **Significant Passenger Volume:** With approximately 628,000 domestic passengers served in 2024, the airport is the seventeenth busiest airport in Australia and the third busiest airport in the state (or fourth if counting Gold Coast Airport) [45]. • **Spoke for First-Tier Airports:** The airport offers daily RPT services to First-Tier Airports, including Sydney Kingsford Smith Airport and Melbourne Airport.	• **Public Ownership:** Ballina Byron Gateway Airport is owned and operated by Ballina Shire Council. • **Serves Only Domestic Flights:** The airport facilitates only domestic flights, aligning it with the Third Tier. • **Facilitates General Aviation:** The airport offers facilities for multiple general aviation businesses, including hangars, aircraft maintenance areas, a clubhouse and office spaces, fostering a vibrant aviation community [67].
		Third-Tier/Fourth-Tier Airport	
Port Augusta Airport (PUG)	Port Augusta, South Australia	• **Location in Regional Town:** The airport is situated in Port Augusta, which lies within 'Outer Regional Australia' [53]. • **Emergency Services Hub:** Since 2001, the airport has been a major base for the Royal Flying Doctor Service (RFDS) (The RFDS is **Australia's** leading nonprofit aeromedical organisation, delivering emergency and primary healthcare services to rural and remote communities across the country) [68]. Its significance as an emergency services hub effectively positions it at a higher Tier. • **Facilitates General Aviation:** Port Augusta Airport hosts the Port Augusta Aero Club and a range of general aviation facilities. Between 2015 and 2035, the airport's annual general aviation aircraft movements are expected to grow from 5427 to up to 9802. The robust general aviation activities at the airport suggest placement at a higher Tier.	• **Proximity to Remote Communities:** Port Augusta lies on the edge of the Outer Regional Australia zone and is bordered by 'Remote Australia' and 'Very Remote Australia' at distances of approximately 12 km and 70 km, respectively, to the north, northeast, east, and southeast [53]. The proximity of the airport to remote communities aligns with the Fourth Tier's location criterion. • **No Commercial Flights:** The airport currently services no RPT flights. The airport handles around 16,000 passengers via charters. These passengers are FIFO workers for the mining industry across the north of the state [69].

Airports in the overlapping areas between the proposed tiers, particularly between the first and second tiers and between the second and third tiers, offer the opportunity to prevent or mitigate the encroachment of incompatible developments around airports—such as tall structures and residential development. Government agencies can employ the proposed classification to identify airports with robust potential for future growth, development and expansion. This identification can assist in proactive land-use planning around these airports, allowing authorities to prohibit incompatible uses or reduce their intensity through mechanisms such as zoning regulations or land protection strategies.

7. Contribution to Sustainability

The proposed airport classification can help government agencies make strategic planning decisions that promote economic, social and environmental sustainability. Table 9 demonstrates how the classification could enhance government decision-making concerning airports in relation to the three pillars of sustainability: economy, environment and society [70,71].

Table 9. The potential sustainability contribution of the proposed Australian airport classification to government decision-making.

Contribution	Explanation	Examples
	Economic Sustainability	
Efficient Resource Allocation	The distinct categorisation of airports enables national and state/territory government agencies to prioritise investments in infrastructure and services based on the specific needs of each tier. For instance, First-Tier and Second-Tier Airports may require more funding towards transport infrastructure to support their large number of passengers, whereas Third- and Fourth-Tier Airports may necessitate subsidies for emergency medical services.	• Melbourne Airport, as a First-Tier Airport, is one of the busiest airports in the country. To support the airport's projected growth, a Melbourne Airport Rail project, expected to be completed in 2030, will connect the airport to the regional and metropolitan train network [72]. • Under the 2025–27 Regional Airports Development Scheme, the Western Australian Government is seeking grant applications for projects enhancing the state's regional air services and safety. A major project supported in a previous round of the program involves upgrading of regional airstrips in Indigenous communities to enable more efficient emergency evacuations by the RFDS [73].
Regional Economic Development	The classification can assist government agencies in identifying potential airports with untapped potential for regional economic development through trade or tourism. For example, Second-Tier and Third-Tier Airports, with their strong linkages with capital cities, could be strategically marketed as gateways to surrounding natural attractions, boosting tourism and retail activities.	• In late 2023, Mackay Airport established a partnership with Harrup Park, a local sporting club, to attract major sporting events and tourists to Mackay. This initiative seeks to capitalise on the airport's recently expanded flight routes and the host region's rich availability of sporting facilities and nature attractions [74].
	Environmental Sustainability	
Land Use Efficiency	The classification can be used to identify and preserve key areas for biodiversity protection or productive farmland around Third- and Fourth-Tier Airports or to create business parks within or around First- and Second-Tier Airports.	• Leigh Creek Airport is a Fourth-Tier Airport located in the remote town of Leigh Creek, approximately 500 kilometres north of Adelaide, the capital of South Australia. The airport lies near Aroona Sanctuary, a legally protected nature reserve under state legislation [75]. With a resort and a town in its catchment, the airport can also be promoted as a base for chartered flights supporting scientific research and tourism, thus contributing to the conservation efforts linked to the sanctuary. • In July 2024, MAB Corporation launched Haystone Business Park, a 38-hectare precinct to house warehouses and manufacturing facilities that benefit from the proximity to Melbourne Airport [76].
Tailored Environmental Initiatives	Lower-tier airports could serve as testing grounds for environmental initiatives with high potential impact, such as trialling new solar technologies at a Fourth-Tier Airport that benefits from year-round solar access.	• To capitalise on the local availability of feedstock, Toowoomba Wellcamp Airport is developing a Sustainable Aviation Fuel facility in partnership with Boeing [77]. • Gold Coast Airport is trialing autonomous shuttles to reduce the airport's environmental footprint. It is also transitioning fleets to low- to zero-emission technologies as part of its Net Zero targets [58]. These environmental initiatives represent potential areas for collaborative research and development between the airport operators and external agencies.

Table 9. *Cont.*

Contribution	Explanation	Examples
	Social Sustainability	
Development Equity	The classification system can assist in advocating for more equitable resource distribution, where lower-tier airports are not overlooked in favour of higher-tier airports.	• The Australian Government's Regional Airports Program and Remote Airstrip Upgrade Program are national competitive grant initiatives with the intention to support regional and remote airports [78]. As such, the proposed classification's criteria for Third-Tier and Fourth-Tier Airports can assist the programs in identifying eligible airports.
Essential Services	Lower-tier airports, particularly those in the Fourth Tier, play a critical role in providing healthcare and supplies to remote communities. The classification can be used to highlight these airports' social functions to ensure ongoing support from government agencies and the private sector.	• Mildura Regional Airport, a Third-Tier Airport, receives 10 RFDS flights daily, emphasising the airport's important role in facilitating healthcare services for residents of Mildura [79].

8. Conclusions

Airports vary significantly in purpose, scale, land use, activities and relationships with their surroundings and host region and country. Agencies worldwide classify airports for decision-making purposes. However, there is no standardised classification of airports globally. Similarly, Australia lacks a comprehensive airport classification that considers domestic and international airports and captures the manifold characteristics and diverse contexts of airports. This paper has developed a conceptual classification for Australian airports to address this gap. Integrating indicators related to airport properties and geographical factors, the tier-based classification defines Australian airport roles, functions and characteristics based on four criteria: location, governance, network function and passenger profile.

The proposed classification has several potential real-world applications and benefits. Allowing an airport's role and characteristics to be clearly defined can assist national, state and local governments in strategic planning and decision-making for and around airports, including land use planning and infrastructure planning and funding. On the regulatory side, by distinguishing airports by their internal and external characteristics, the Federal Government can employ the classification to determine the applicable standards and regulations for all airports concerning such matters as safety, operation and development controls. Notably, the classification can assist government agencies in their decision-making towards more sustainable outcomes. For instance, recognising the important regional service role of airports in the second and third tiers could support more targeted investment in these airports and their associated infrastructure, while designation in an overlap between two adjacent tiers may imply the need for stricter land use controls to manage potential urban encroachment. Aligning planning and investment decisions with an airport's tier can help prioritise sustainability goals in a more context-sensitive and efficient manner.

The conceptual classification also has potential applicability to airports in other countries. However, as countries vary in geographical contexts, the specific nature, roles and characteristics of airports outside Australia differ significantly from those of Australian airports. As such, future comparative research could investigate the characteristics of airports in another country, the findings of which can be used to revise the proposed classification by adjusting the specifics of each criterion and adding or reducing the number of tiers. Future studies could also investigate the relationship between the proposed airport tiers and the existing tier-based city classification systems. Since the proposed classification does not account for freight activities, future research on Australian airport classification could focus on developing a system that explicitly categorises airports based on freight volume and characteristics.

This paper has outlined key findings from the initial qualitative phase of the authors' Australian airport classification study, highlighting limitations associated with the existing classification and proposing a new framework based on a literature review and policy analysis. The future plan for this study involves further development of the proposed four-tiered Australian airport classification, including quantitative validation and application to categorise all Australian airports.

Author Contributions: Conceptualization, I.K., B.B. and D.O.; methodology, I.K., B.B. and D.O.; software, I.K.; validation, I.K.; formal analysis, I.K.; investigation, I.K.; resources, I.K., B.B. and D.O.; data curation, I.K.; writing—original draft preparation, I.K.; writing—review and editing, I.K., B.B. and D.O.; visualization, I.K.; supervision, B.B. and D.O.; project administration, I.K., B.B. and D.O.; funding acquisition, I.K., B.B. and D.O. All authors have read and agreed to the published version of the manuscript.

Funding: This research was supported by the Australian Government's Australian Postgraduate Scholarship (now entitled Research Training Program).

Institutional Review Board Statement: Not applicable.

Informed Consent Statement: Not applicable.

Data Availability Statement: No new data were created or analyzed in this study. Data sharing is not applicable to this article.

Conflicts of Interest: The authors declare no conflicts of interest.

References

1. Suau-Sanchez, P.; Voltes-Dorta, A.; Rodríguez-Déniz, H. Regulatory Airport Classification in the US: The Role of International Markets. *Transp. Policy* **2015**, *37*, 157–166. [CrossRef]
2. Todd, D.; Maquire, J.; Steenhuis, H.-J. Airports. In *The Global Commercial Aviation Industry*; Eriksson, S., Steenhuis, H.-J., Eds.; Routledge: Oxford, UK, 2016; pp. 138–169.
3. Wong, W.-H.; Zhang, A.; Cheung, T.K.-Y.; Chu, J. Examination of Low-Cost Carriers' Development at Secondary Airports Using a Comprehensive World Airport Classification. *J. Air Transp. Manag.* **2019**, *78*, 96–105. [CrossRef]
4. Chen, Y.; Fuellhart, K.; Zhang, S.; Witlox, F. Airport Classification in Chinese Multi-Airport Regions. *Eur. J. Transp. Infrastruct. Res.* **2022**, *22*, 1–21. [CrossRef]
5. Derudder, B.; Devriendt, L.; Witlox, F. Flying Where You Don't Want to Go: An Empirical Analysis of Hubs in the Global Airline Network. *Tijdschr. Voor Econ. Soc. Geogr.* **2007**, *98*, 307–324. [CrossRef]
6. Bonnefoy, P.A.; de Neufville, R.; Hansman, R.J. Evolution and Development of Multiairport Systems: Worldwide Perspective. *J. Transp. Eng.* **2010**, *136*, 1021–1029. [CrossRef]
7. Derudder, B.; Devriendt, L.; Witlox, F. A Spatial Analysis of Multiple Airport Cities. *J. Transp. Geogr.* **2010**, *18*, 345–353. [CrossRef]
8. Brueckner, J.K.; Lee, D.; Singer, E. City-Pairs Versus Airport-Pairs: A Market-Definition Methodology for the Airline Industry. *Rev. Ind. Organ.* **2014**, *44*, 1–25. [CrossRef]
9. Mayer, R. Airport Classification Based on Cargo Characteristics. *J. Transp. Geogr.* **2016**, *54*, 53–65. [CrossRef]
10. Sun, X.; Wandelt, S.; Hansen, M.; Li, A. Multiple Airport Regions Based on Inter-Airport Temporal Distances. *Transp. Res. Part E Logist. Transp. Rev.* **2017**, *101*, 84–98. [CrossRef]
11. Kato, H.; Murakami, J. Economic Impacts of Airport Upgrading on the Regional Economy: Case Study of the People's Republic of China. In *Developing Airport Systems in Asian Cities: Spatial Characteristics, Economic Effects, and Policy Implications*; Kato, H., Murakami, J., Eds.; Asian Development Bank: Manila, Philippines, 2022.
12. International Civil Aviation Organization. *Annex 14—Aerodromes—Volume I—Aerodromes Design and Operations*; International Civil Aviation Organization: Montreal, QC, Canada, 2018.
13. Airports Council International; International Air Transport Association; Worldwide Airport Coordinators Group. *Worldwide Airport Slot Guidelines (WASG)*; International Air Transport Association: Montreal, QC, Canada, 2020.

14. Federal Aviation Administration. *National Plan of Integrated Airport Systems (NPIAS) 2023—2027*; Secretary of Transportation: Washington, DC, USA, 2022.
15. Federal Aviation Administration. *National Plan of Integrated Airport Systems (NPIAS) 2023–2027: Appendix C—Statutory and Policy Definitions; Data Sources and NPIAS Process*; Secretary of Transportation: Washington, DC, USA, 2022.
16. Federal Aviation Administration. Airport Categories. Available online: https://bit.ly/3DgCbwO (accessed on 15 May 2025).
17. Secretary of State for Trade. *Airports Policy (CMND: 7084)*; Secretary of State for Trade: London, UK, 1978.
18. Maltby, D.A.; White, H.P. The Transport System of the United Kingdom. In *Transport in the United Kingdom*; Maltby, D., White, H.P., Eds.; Palgrave: London, UK, 1982; pp. 39–78.
19. Secretary of State for Transport. *Airports Policy (Volume 4: Debated on Wednesday 13 May 1981)*; Secretary of State for Transport: London, UK, 1981.
20. Twomey, J.; Tomkins, J. Development Effects at Airports: A Case Study of Manchester Airport. In *Transport and Urban Development*; Banister, D., Ed.; Routledge: London, UK, 2003; pp. 187–211.
21. Butcher, L. *Briefing Paper: Regional Airports*; House of Commons Library: London, UK, 2016.
22. Environment, Transport and Regional Affairs Committee. *Environment, Transport and Regional Affairs—Eighth Report*; House of Commons: London, UK, 1998.
23. Civil Aviation Safety Authority. *Aviation Trends: Quarter 3 2018*; Civil Aviation Safety Authority: London, UK, 2018.
24. Civil Aviation Safety Authority. Table 1: Size of Reporting Airports February 2022–January 2023. Available online: https://bit.ly/3K3UQQj (accessed on 15 May 2025).
25. Egeland, J. *Political Economy of Airport Capacity Planning in the London Airport System*; Washington, DC, USA, 2018. Available online: https://www.itf-oecd.org/sites/default/files/docs/airport-capacity-planning-london.pdf (accessed on 14 April 2025).
26. Kazda, A.; Turiak, M.; Gőtz, K. Airport Typology for LCC Policy Changes: A European Perspective. *Aviation* **2020**, *24*, 90–98. [CrossRef]
27. Turiak, M. Selected Aspects of Airport Typology Formation. In Proceedings of the XXI International Scientific-Technical Conference, Varna, Bulgaria, 1–2 July 2013; pp. 82–85.
28. Madas, M.A.; Zografos, K.G. Airport Capacity vs. Demand: Mismatch or Mismanagement? *Transp. Res. Part A Policy Pract.* **2008**, *42*, 203–226. [CrossRef]
29. Malighetti, P.; Paleari, S.; Redondi, R. Airport Classification and Functionality within the European Network. *Probl. Perspect. Manag.* **2009**, *6*, 183–196.
30. Caisen, G.; Qi, Y. The Slot Auction Sharing Mechanisms of the Airport and the Air Traffic Control—The Chinese case. *Aviat. Space J.* **2019**, *Year XVIII n. 1*, 2–14.
31. OAG Aviation Worldwide Limited. China Aviation Market: Flight Data for China. Available online: https://bit.ly/3OtlepA (accessed on 15 May 2025).
32. Office of Parliamentary Counsel. Aviation Transport Security Amendment (Security Controlled Airports) Regulations 2019. Available online: https://bit.ly/43UaGGp (accessed on 15 May 2025).
33. Airservices Australia. *Aeronautical Information Publication Australia*; Airservices Australia: Canberra, Australia, 2023.
34. Merkert, R. Air Transport in Regional, Rural and Remote Areas. In *Air Transport Management: An International Perspective*, 2nd ed.; Budd, L., Ison, S., Eds.; Routledge: Oxon, UK, 2020.
35. Civil Aviation Safety Authority. Glossary. Available online: https://bit.ly/4iOX2uC (accessed on 15 May 2025).
36. Australian Airports Association. *Australian Airports Association (AAA)—Submission to Commonwealth Budget 2019–2020*; Australian Airports Association: Canberra, Australia, 2019.
37. ACIL Allen Consulting. *Regional Airport Infrastructure Study: Economic Contribution and Challenges of Regional Airports in Australia*; Department of the Treasury: Canberra, Australia, 2016.
38. Donehue, P.; Baker, D. Remote, Rural, and Regional Airports in Australia. *Transp. Policy* **2012**, *24*, 232–239. [CrossRef]
39. Australian Bureau of Statistics. Remoteness Areas: Australian Statistical Geography Standard (ASGS) Edition 3. Available online: https://bit.ly/3rS5DXx (accessed on 15 May 2025).
40. Creative Commons. CC BY 4.0: Attribution 4.0 International (Deed). Available online: https://bit.ly/30lGkfD (accessed on 15 May 2025).
41. Khanjanasthiti, I. Planning for Economic Development around a Second-Tier Airport: A Case Study of Gold Coast Airport. Ph.D. Thesis, Bond University, Gold Coast, Australia, 2021.
42. Donnet, T.; Ryler, T.; Lohmann, G.; Spasojevic, B. Developing a Queensland (Australia) Aviation Network Strategy: Lessons from Three International Contexts. *J. Air Transp. Manag.* **2018**, *73*, 1–14. [CrossRef]
43. United Voice. *United Voice Submission: Rural and Regional Affairs and Transport References Committee—Inquiry into Airport and Aviation Security*; Parliament of Australia: Canberra, Australia, 2015.
44. Australian Government. Airports. Available online: https://bit.ly/3y89MWM (accessed on 15 May 2025).
45. Australian Government. Airport Traffic Data. Available online: https://bit.ly/3iWBnkV (accessed on 15 May 2025).

46. Cambridge University Press. TIER | English Meaning—Cambridge Dictionary. Available online: https://bit.ly/2HsfagR (accessed on 15 May 2025).
47. de Neufville, R. Low-Cost Airports for Low-Cost Airlines: Flexible Design to Manage the Risks. *Transp. Plan. Technol.* **2008**, *31*, 35–68. [CrossRef]
48. Grubesic, T.H.; Matisziw, T.C.; Zook, M.A. Spatio-Temporal Fluctuations in the Global Airport Hierarchies. *J. Transp. Geogr.* **2009**, *17*, 264–275. [CrossRef] [PubMed]
49. .id (Informed Decisions) Consulting. *Wellcamp Airport: Helping Toowoomba Take Off*; .id (Informed Decisions) Consulting: Toowoomba, Australia, 2016.
50. Toowoomba Regional Council. *Toowoomba Region Transport and Logistics Profile 2022*; Toowoomba Regional Council: Toowoomba, Australia, 2022.
51. Australian Government. *Australian Infrastructure and Transport Statistics: Yearbook 2021*; Australian Government: Canberra, Australia, 2021.
52. Deloitte Access Economics. *Taking Flight: The Economic and Social Contribution of Australia's Airports*; Australian Airports Association: Canberra, Australia, 2023.
53. Australian Bureau of Statistics. Change in the ASGS Remoteness Classification: ASGS Edition 2 (2016) to ASGS Edition 3 (2021). Available online: https://bit.ly/3GEf84f (accessed on 15 May 2025).
54. Sydney Airport Corporation Limited. *Sydney Airport Master Plan 2039*; Sydney Airport Corporation Limited: Sydney, Australia, 2019.
55. Khanjanasthiti, I.; O'Hare, D.; Bajracharya, B. Cross-Border Planning for Promoting Economic Development: A Case Study of Gold Coast Airport, Australia. *Sustainability* **2025**, *17*, 1672. [CrossRef]
56. Spearritt, P. The 200 KM City: Brisbane, The Gold Coast, and Sunshine Coast. *Aust. Econ. Hist. Rev.* **2009**, *49*, 87–106. [CrossRef]
57. Queensland Airports Limited. Gold Coast Airport: Our Destinations. Available online: https://bit.ly/3bdUMKh (accessed on 15 May 2025).
58. Gold Coast Airport Pty Ltd. *Preliminary Draft Master Plan 2024*; Gold Coast Airport Pty Ltd.: Gold Coast, Australia, 2024.
59. Gold Coast Airport Pty Ltd. *2017 Master Plan: Our Blueprint for the Future of Gold Coast Airport*; Gold Coast Airport Pty Ltd.: Gold Coast, Australia, 2017.
60. Barnsley, W. Brisbane Airport Flight Delays, Diversions Due to Thick Fog over City. Available online: https://bit.ly/42G7COT (accessed on 15 May 2025).
61. Bowman, J. Large Parts of Queensland Blanketed by Fog, Causing Visibility Issues and Flight Diversions. Available online: https://ab.co/4kkOY5d (accessed on 15 May 2025).
62. The Airport Group. *Albury Airport Master Plan 2018*; Albury City Council: Albury, Australia, 2018.
63. Queensland Airports Limited. Mount Isa Airport. Available online: https://bit.ly/3ShJ1ty (accessed on 15 May 2025).
64. Queensland Airports Limited. General Aviation. Available online: https://bit.ly/43duFAI (accessed on 15 May 2025).
65. Greater Cities Commission. *Discussion Paper: The Six Cities Region*; New South Wales Government: Sydney, Australia, 2022.
66. Newcastle Airport Pty Limited. *2036 Newcastle Airport Vision: Delivering the Airport the Region Deserves*; Newcastle Airport Pty Limited: Newcastle, Australia, 2018.
67. Airport Master. About Ballina Byron Gateway Airport. Available online: https://bit.ly/4iMJGho (accessed on 15 May 2025).
68. Port Augusta City Council. *Port Augusta Laurie Wallis Aerodrome Master Plan 2016–2036*; 68. Port Augusta City Council: Port Augusta, Australia, 2016.
69. Port Augusta City Council. Port Augusta Airport. Available online: https://bit.ly/4m0vp3E (accessed on 15 May 2025).
70. Kajikawa, Y. Research Core and Framework of Sustainability Science. *Sustain. Sci.* **2008**, *3*, 215–239. [CrossRef]
71. Purvis, B.; Mao, Y.; Robinson, D. Three Pillars of Sustainability: In Search of Conceptual Origins. *Sustain. Sci.* **2019**, *14*, 681–695. [CrossRef]
72. Victorian Government. Melbourne Airport Rail Overview. Available online: https://bit.ly/41yStwZ (accessed on 15 May 2025).
73. Western Australian Government. Regional Airports Development Scheme Grants Now Available. Available online: https://bit.ly/3E9aTga (accessed on 15 May 2025).
74. Mackay Airport Pty Ltd. Key Partnership to Grow Mackay Sport, Tourism. Available online: https://bit.ly/3XBEtRZ (accessed on 15 May 2025).
75. Government of South Australia. *The South Australian Government Gazette*; Government of South Australia: Adelaide, Australia, 1995.
76. MAB Corporation. MAB Provides Businesses a Gateway to Melbourne with the Launch of Its Latest Business Park. Available online: https://bit.ly/4hXrdyu (accessed on 15 May 2025).
77. Alsop, E. Wagners, Boeing to produce SAF at Wellcamp. Available online: https://bit.ly/4j0l3yb (accessed on 15 May 2025).

78. Australian Government. Regional and Remote Aviation. Available online: https://bit.ly/3RuoUb7 (accessed on 15 May 2025).
79. Arcadis Australia Pacific Pty Ltd. *Mildura Regional Airport Master Plan (2017–2037)*; Mildura Airport Pty Ltd.: Mildura, Australia, 2017.

sustainability

MDPI

Article

Sustainable Railway Infrastructure: Modernization Strategies for Integrating 1520 mm and 1435 mm Gauge Systems

Iryna Bondarenko

Institute of Mechanical Engineering and Transport, Lviv Polytechnic National University, 79000 Lviv, Ukraine; iryna.o.bondarenko@lpnu.ua

Abstract

This article examines the modernization of railway systems with a focus on sustainable infrastructure development, aligning with the European Commission's strategy for integrating 1520 mm gauge railways into the European 1435 mm gauge network. A key challenge lies in addressing the technical aspects of the railway infrastructure that are not explicitly detailed in the European strategy but have evolved through the parallel historical development of two distinct railway engineering systems. An analysis of calculation methodologies highlights that the primary difference in determining technical parameters for 1435 mm and 1520 mm tracks stems from the selection of the primary classifier based on functional purpose and strength requirements. Furthermore, the existing concept of mechanical system motion presents limitations in harmonizing the technical aspects of railway systems with different track gauges. To bridge this gap, two potential solutions are proposed. The first suggests expanding the conventional mechanical system motion framework by incorporating principles from the theory of relativity, while the second explores the application of elastic wave propagation theory as a novel conceptual model for railway system dynamics. The choice of modernization strategy will play a crucial role in ensuring long-term sustainability of the railway infrastructure, requiring a balanced approach that accounts for the operational intensity, infrastructure wear, and specific technical requirements of track elements in different railway gauge systems.

Keywords: sustainable infrastructure; railway systems; technical aspects; mechanical system; elastic wave; classification; dynamic load

1. Introduction

Is it possible to integrate railway systems with 1520 mm and 1435 mm gauges without a comprehensive strategy, or does such an approach inevitably lead to a technological and logistical dead end? This question becomes particularly relevant in the context of railway system modernization aligned with the European Commission's strategy [1], which plays a strategic role in developing trade, transit potential, and competitiveness; improving transport and logistics infrastructure; expanding railway corridors and the network of intermodal terminals; and enhancing control systems, security, and international cooperation.

According to the European Commission strategy, a plan has been developed for implementing a 1435 mm gauge railway network, which will be operated alongside the existing 1520 mm gauge network, based on the following principles:

- "The 1435 mm gauge system would focus on higher-speed transportation (international passenger, IC, and container/platform wagon freight), with the 1520 mm system catering for lower-speed transport (local and regional passenger traffic and heavy bulk).
- The development of a 1435 mm backbone network in the two countries would be implemented in a phased manner, from West to East, with the largest urban agglomerations eventually connected to the new 1435 mm corridors to support future economic development.
- For each line, the required configuration is defined (1435 mm only, 1520 mm only, or both). This is based on a number of factors including:
    a. Cost and combined network-wide operational aspects (1435 mm and 1520 mm systems).
    b. Time for project development and implementation.
    c. Ensuring the capacity of the current 1520 mm gauge system during the implementation and operation of the new lines."

The European Commission strategy applies an advanced planning methodology that takes into account the existing data limitations and uncertainties inherent to the current situation. Although the European Commission strategy considers the economic and infrastructure aspects of integration, the successful implementation of the project largely depends on the technical compatibility of railway infrastructure systems. Unfortunately, the tendency to write scientific articles is focused on the discussion of tactical decisions in engineering tasks, rather than strategic decisions. However, the article focuses on the technical strategy of integrating 1520 mm and 1435 mm gauges. Therefore, this article examines the feasibility of implementing this strategy from an engineering perspective.

The challenge lies in considering technical aspects that are not reflected in statistical data, but rather are the result of the parallel historical development of two distinct railway engineering systems: 1435 mm and 1520 mm.

The fundamental differences between these engineering systems stem from the selected railway classification frameworks, the conceptual approaches used to evaluate track performance, and the modeling methodologies that were applied during the development of each system.

Thus, the goal of this study is to justify the need to consider not only the efficiency of modernizing railway engineering systems but also the historical development of their concepts, while integrating modern technical requirements.

The main question addressed in this study is as follows:

Should railway infrastructure be developed through the introduction of new engineering concepts, or should existing technologies be adapted to new conditions?

To comprehensively present the current state of railway engineering development, this article is structured as follows:

Section 2 examines the technical aspects defined by the European Commission strategy and presents a comparative analysis of the key approaches to the development of railway track engineering systems with 1435 mm and 1520 mm gauges. Section 3 discusses the challenges and solutions that define the technical aspects of railway tracks. This section sets out an approach to modernizing the description of the physical interaction processes between the track and rolling stock. Two potential solutions are proposed. The first suggests expanding the conventional mechanical system motion framework by incorporating principles from the theory of relativity, while the second explores the application of elastic wave propagation theory as a novel conceptual model for railway system dynamics. Section 4 presents a discussion on the feasibility of implementing the European Commission strategy. Section 5 provides the conclusions and recommendations based on the conducted research.

2. Materials and Methods

2.1. Technical Aspects Formed in the «Strategy for the EU Integration of the Ukrainian and Moldovan Rail Systems»

1. The European Commission strategy characterizes the analysis of the existing 1520 mm railway track system through the lens of the 1435 mm railway network as follows:

Section 4.1.1.3 «The Ukrainian network mainly comprises Rail R65 and R50 types. Standard rail length ranges between 12.5 m and 25 m, with joints and two joint bars or fishplates bolted through the web of the rail. Rail type R50 is not compliant with TEN-T requirements regarding axle load. With 51.50 kg/m weight, it is used on some mainline sections with 200 m length, but mainly on siding areas and marshalling yards. Rail type R65 complies with the TEN-T axle load standards.».

Section 4.1.1.4 «Main issues in relation to poor track condition can be classified as follows:

a. Inadequate and/or absence of drainage leading to flooding.
b. Track components in poor condition, with geometry in good condition (alignment and level defects).
c. The profile of tracks not maintained (rail's corrugations, rail head's surface defects with cracks, intrusion, shelling, etc.).
d. Sleepers (old and second hand) and/or not well adjusted (squareness of the ties). Heterogeneity in track support (e.g., use of mixed wooden and concrete sleepers on a given section) can also affect track condition and lead to speed restrictions.
e. Ballast polluted and/or missed.
f. Elastic fastening missing, and crampon not adequate for modern tracks.
g. Absence of uniform track gauge.
h. Joints low and in poor condition.
i. Rail with excessive wear and/or important side wear, becoming worn down.
j. Weak rails and sleepers.
k. Turnouts in poor condition.
l. Rail life span is conditioned by the overall tonnage catered for by the different lines and determined by established rules.»

Section 4.4 Maintenance

a. «Significant deterioration of rail infrastructure affects provision of high-quality and safe freight and passenger transportation services. The situation leads to numerous speed restrictions and lower average speed of transportation of cargo and passengers»;
b. «Speed restrictions mainly apply to regional lines; however, it is estimated that about 50% of the network is operated with a limitation in the order of 60 km/h».

Section 4.9 SWOT analysis:

c. Strengths: (i) the existing rail infrastructure provides sufficient capacity and coverage. (ii) With investments into modern interlocking systems, track protection, traffic management and communication, level crossings, and limited alignment corrections, an operational speed of 200 km/h on certain parts of the network are possible. (iii) With the high axle loads of 25 t and a longer train development length of approximately 1000 m, the capacities for heavy bulk transport are comparatively high;
d. Weaknesses: (i) low technological level of the rail system. With the introduction of modern interlocking and traffic management systems (ERTMSs), the performance and therefore the competitiveness of the rail system can be significantly improved. (ii) Different track gauge than in the neighboring EU countries. The rail system is currently not compatible with the EU rail system;

e. Opportunities: as part of the future revised TEN-T regulation there will be opportunities for enhanced coordination in the medium and long-term development of the lines.

Section 5.1.1 Technical and operational integration of the Ukrainian and Moldovan rail systems into the EU, including the ERTMS: one of the key elements to achieve integration is ensuring rail interoperability. In that regard, compliance with the technical specification for interoperability (TSI) is essential.

Section 5.2.10 Opening up the rail markets to up-to-date technologies is essential for future development and the EU integration of both systems. In addition to safety and security improvements, the use of new technologies will increase the capacity of the system and its operational flexibility.

2. The main changes in the reconstruction of 1520 mm gauge railway systems are characterized in study [1] by the following concepts:

Section 6.1.1 Investments that allow for the continuous and effective operation of the 1520 mm system will be crucial. This could include, for example, the modernization of damaged and outdated infrastructure, but also investments at border-crossing areas to improve their effective throughput capacity. Nevertheless, bearing in mind the expected integration with the EU and the planned transition to 1435 mm track gauge, it is recommended that when investing in the 1520 mm network, such investments already consider a future potential transformation into the 1435 mm gauge system by, for example, using polyvalent sleepers able to accommodate both gauges. This approach has been applied successfully in similar situations (e.g., Spain), with the difference in the cost of such sleepers being insignificant.

Section 6.1.4 presented options for the reconstruction of railway infrastructure systems, from which four options were selected, shown in Figures 1–4. In Section 6.1.5, a cost assessment of these options was conducted, demonstrating the reconstruction of the following elements of the railway track system:

a. «Track substructure.
b. Drainage-main line.
c. Track superstructure.
d. Turnouts.
e. Level crossings.
f. Sidings.
g. Supply and installation of fencing (for high speed only).
h. Small bridges.
i. Signalling-ETCS-interlocking-telecommunications.
j. Overhead Catenary System (OCS).
k. Power supply.»

Section 6.2 defined the defined the network functionalities:

1. «1435 mm network functionalities:
    i. Passenger International;
    ii. Passenger National;
    iii. Freight Fast.
2. 1520 mm network functionalities:
    i. Passenger Local/Regional;
    ii. Passenger Night trains;
    iii. Freight Slow/heavy.»

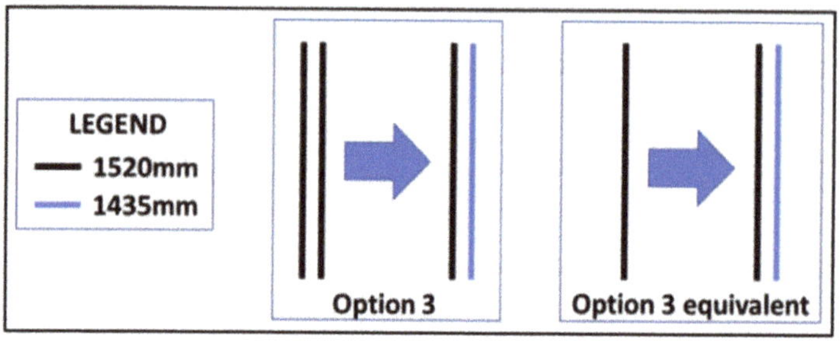

Figure 1. Option 3: 1520 mm single-track upgrading; 1520 mm to 1435 mm single-track transformation [1].

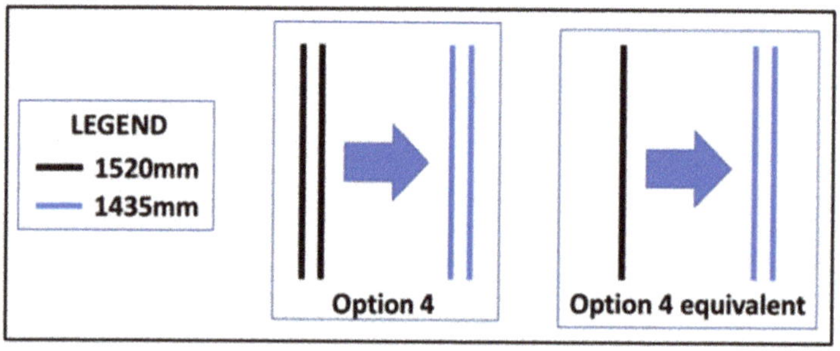

Figure 2. Option 4: 1520 mm to 1435 mm double-track transformation [1].

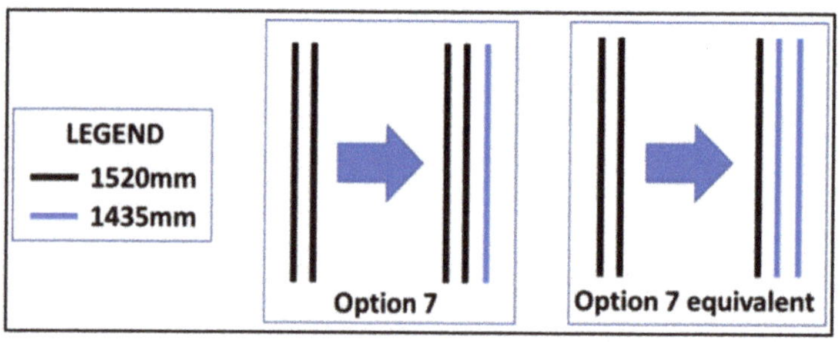

Figure 3. Option 7: 1520 mm double-track upgrading; 1435 mm single-track parallel new construction [1].

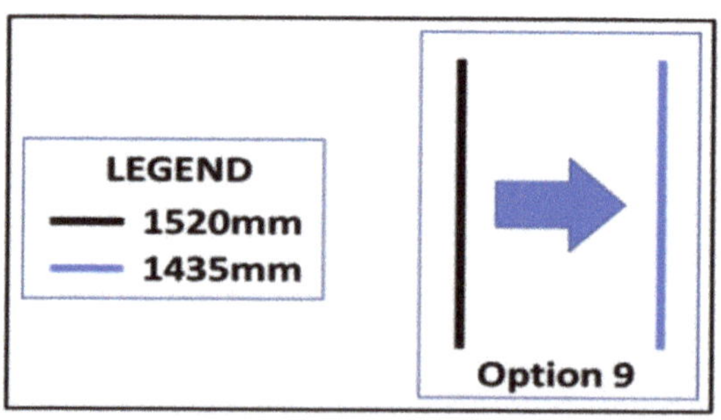

Figure 4. Option: 1520 mm single-track to 1435 mm transformation [1].

Section 6.7 «Parameters that are not subject to a specific options analysis in this study For certain parameters, there is no need to develop an options analysis as they are either already defined by policy requirements (e.g., TSI), or they should be developed to be compatible to the existing network (interoperability). For other parameters, the options analysis should be developed in the Feasibility Studies for each corridor.»

Section 6.7.1 «Axle load and structure gauge. The axle load is considered a "Hard" parameter in the TSI and thus, its minimum value is defined there, depending on the traffic category of each rail line.

For rail lines with a track gauge of 1520 mm, the minimum axle load is 22.5 t for dedicated passenger lines and 25 t for freight or mix traffic lines. In both cases, as a minimum the structure gauge S defined in the TSI should be applied.

Regarding the future 1435 mm lines, most of them will be envisaged for mixed traffic. Consequently, recommendation is to develop them with an axle load of 22.5 t and structure gauge GB or GC, depending on the envisaged traffic category. If in the future some 1435 mm rail lines are planned to be developed for passenger use only, the axle load and structure gauge requirements should be assessed in a Feasibility Study, depending on the traffic category.»

Thus, the aforementioned technical aspects will be further used for a comparative analysis of the key approaches to the formation of technical aspects of railway track engineering systems with 1435 mm and 1520 mm gauges, their modern implementation, and interchangeability. Additionally, the nuances of the possible expansion of the accumulated operational experience of both systems are discussed, taking into account the modernization of concepts regarding the physical nature of the interaction processes between the track and rolling stock by supplementing existing concepts with the principles of elastic wave theory.

2.2. Technical Aspects of Railway Track Classification

To solve most techno-economic challenges in transport process management, a classification system for railway lines and sections is required.

In many countries, the concept of "railway track classification" is not established as an independent regulatory standard; however, the use of track classes, categories, or groups serves as a foundational element in shaping the regulatory framework for railway infrastructure management. This classification is crucial for standardizing design and operational practices, as well as for ensuring consistent safety and maintenance requirements.

Railway classification varies across countries based on several key parameters. Table 1 presents the main characteristics used in international practice for infrastructure evaluations.

Table 1. Railway track classifiers.

Country	Axle Load	Max Axle Load	Freight Traffic Density	Theoretical Transport Volume	Max Speed	Freight/ Passenger Train Speeds	Structure Gauge	Usable Length of Platform/Train Length
Germany		x	x		x			
France				x				
USA			x		x			
Canada			x			x		
European Union (TSI INF)	x				x		x	x

For 1435 mm gauge railways, different countries employ single-, dual-, and triple-parameter classification systems [2–5] (see Table 1). In the context of 1520 mm gauge railways, two classification systems have historically evolved [6]. The first of these systems is centered on design standards, the purpose of which is to regulate geometric alignment, track profiles, and substructure specifications. The second system is centered on operational criteria, the purpose of which is to define the structure of the track and its maintenance requirements.

As shown in Table 1, the TSI INF (Technical Specifications for Interoperability relating to the "Infrastructure" subsystem of the EU rail system) proposes a classification system based on four key classifiers.

The classification of railways in different countries has evolved due to historical development and functional differences in the railway infrastructure. The fundamental criteria, validated through decades of railway operation, that influence classification choices include the track strength [7–12], stability [13–18], and safety [19–24].

Engineering and Structural Considerations

Currently, the selection of track structure characteristics is carried out according to regulatory documents. However, it is important to note that initial classification criteria were based on track strength calculations.

Track strength calculations aimed to answer whether certain structural elements could be used within the track system by establishing stress criteria in specific locations. These stress values were determined based on extensive experiments and operational experience.

Key Findings from Comparative Methodology Analysis

1. Researchers studying both 1435 mm and 1520 mm gauge track systems based their track strength calculations on the prevailing knowledge of physics and mathematics of their time. These calculations were grounded in several well-established principles: the differential equation of rail bending, treating the rail as a beam on a continuous elastic foundation under a vertical static load; dynamic process modeling using Gaussian probability distributions to account for the influence of multiple independent factors; and the inclusion of horizontal forces [25,26], drawing on the foundational work of Timoshenko [27].
2. Mixed passenger and freight traffic—including both empty and heavily loaded trains—is a defining feature of the 1520 mm railway system, with operational safety for this diverse flow as its primary objective. As a result, a dynamic factor analysis emphasized the evaluation of force variations (dynamic loads). The calculated values for key forces—such as (i) the maximum equivalent load for stress calculations in rails, (ii) the maximum equivalent load for deflection calculations of rails, stresses, and forces in the sub-rail foundation elements, and (iii) the dynamic force exerted by the rail on its support—account for multiple influencing factors, including the structural design of rolling stock bogies, train speed, the presence of isolated and continuous wheel irregularities, and the characteristics of track geometry deviations.
3. For 1435 mm railways, the narrower range of axle loads and emphasis on track maintenance cost optimization led to a focus on the Dynamic Amplification Factor (DAF), which measures the track condition based on track–vehicle interaction dynamics without specifying which aspect of the impact is considered.

Accounting for Thermal Forces in Stability Calculations

The track stability assessment methodology, particularly concerning thermal forces (track stress caused by temperature fluctuations), revealed a common dependence between rail elongation and temperature variations [28–30]. However, differences in track strength calculation methodologies for different gauge widths resulted in distinct applications of this dependency.

Uniformity of Track Elevation Calculation Methodology

The method for calculating rail elevation in curved track sections is consistent across different track gauges, as it is based on universal physical principles. This approach considers factors such as unbalanced acceleration, train speed, and curve radius and applies a unified formula to determine the required rail elevation.

Similarly, derailment risk assessment principles remain consistent for both 1435 mm and 1520 mm gauge tracks [31,32].

Engineering Approaches in Track Design and Unification Challenges

Historical railway engineering developments show that both 1435 mm and 1520 mm gauge railway infrastructure systems were built using similar principles available at the time.

The primary difference between track parameter calculation methodologies for 1435 mm and 1520 mm tracks arose from the choice of the main classification factor, which was based on the functional purpose of the tracks and structural strength requirements.

Efforts to standardize railway requirements have been made through international standards such as TSI INF. However, differences in the operational conditions, infrastructure, and economic models make it impractical to create a universal classification or to transfer design solutions directly from one gauge system to another.

3. Results

3.1. Challenges in Defining the Technical Aspects of Railway Tracks

When calculating railway tracks, three key challenges arise: (1) the uncertainty of the interaction mass, (2) the heterogeneity of the track structure, and (3) the need to account for dynamic loads in different planes. Below, we examine how these challenges relate to traditional mechanical concepts.

To begin with, let us consider the challenges in the conceptual description of mechanical systems.

Firstly, historically, there have been two stages in the development of mechanical system descriptions:

1. Analysis of uniform rectilinear motion using Newton's laws, where
    a. The nature of motion was determined by Newton's First Law, stating that inertial reference frames record rectilinearity and uniformity of motion in the absence of interactions or when the resultant force equals zero.
    b. The parameters of mechanical motion were formulated by Newton's Second Law, defining the relationship among (i) the resultant force applied to the body, (ii) the rate of velocity change in the direction of the applied force, and (iii) the mass involved in the interaction.
    c. Newton's Third Law established equilibrium in mechanical systems by considering forces/reactions from each interacting body.
2. Analysis of accelerated/decelerated rectilinear and rotational motion, where
    a. The nature of motion was defined by introducing non-inertial reference frames.
    b. The parameters of mechanical motion were formulated using D'Alembert-Lagrange's principle and its further developments [33–36].
    c. To create a balanced closed system, "fictitious" forces were introduced, such as inertial forces, centrifugal force, and Coriolis force [37–43].

Railway track calculations, as part of mechanical systems, fully utilize these principles [44–50]. However, railway tracks have three inconvenient properties:

1. Their dimensions exceed the local interaction zones with rolling stock, meaning that the interaction mass is unknown.
2. Their structure is heterogeneous both in length and depth, necessitating the introduction of stiffness characteristics in all computational planes, at various levels, and

along the track length. This significantly affects the velocity variation of track and rolling stock components during the interaction.

3. For railway tracks to be considered a balanced system, the interaction of active and passive forces in different planes must be accounted for. This complicates calculations as it becomes difficult to determine the exact resultant force impact.

In all dynamic track calculations, three unknown parameters remain, making it necessary to conduct experiments to identify interaction patterns between rolling stock and rails.

Secondly, all railway track calculations rely on a geometric assessment of the localized force distribution, independent of time. The concept of a "time step" is used, which determines the position of structural track and rolling stock components for force distribution under given interaction conditions. It is assumed that all forces and reactions are applied simultaneously [50–57].

This approach requires probabilistic consideration of the possible simultaneous impacts of various forces, with a specific correlation of force magnitudes, obtained through experimental studies that determine "equivalent" or "reduced" interaction masses

Thirdly, rolling stock movement on railway tracks is characterized by complex oscillations, caused by the following: (i) changes in the movement regime along track sections, (ii) heterogeneities in the track structure, alignment, and profile, (iii) design features of rolling stock, and (iv) atmospheric and climatic conditions.

These physical aspects of the track–rolling stock interaction must be considered in geometric descriptions by analyzing non-uniform curvilinear motion.

A key feature of this motion's geometric description is the distinction between "normal" and "tangential" interaction components. This distinction is entirely artificial, introduced to describe observed motion trends in curves, and does not reflect the underlying physical processes occurring during curvilinear motion.

In rectilinear motion, the force direction fully determines the velocity change direction of a body/system after impact.

However, oscillations caused by curvilinear motion exhibit different origins and force application locations. The determination of force magnitudes requires solving separate problems, each with distinct reference frames.

For instance, the "fictitious" centrifugal force can be incorporated into calculations in various forms, including as the Newtonian centrifugal force, D'Alembert's centrifugal force, or as the normal component of the transport inertia force within the equations of relative motion.

The lack of understanding of fictitious forces requires extensive experimental research to determine the conditions under which they occur and their effects. For this purpose, research centers have been established to conduct a series of experiments on railways to evaluate the effect of fictitious forces on the stability of trains on curves.

However, developing experimental methodologies is essential, as they must account for the causes of the studied effects. Any methodology is based on a specific concept, which is missing for these forces. This is why fictitious forces have been introduced based on observed effects rather than on their physical origins.

Thus, the existing regulatory framework for both 1435 mm and 1520 mm track systems was developed through extensive experimental and operational experience.

Consequently, under the current calculation concept, transferring requirements from one system to another is impossible without large-scale experimental studies. Attempts to transfer TSI INF standards to 1520 mm railway tracks have shown that without large-scale experiments, adapting these standards requires substantial modifications [2].

If such conditions had been met, a unified calculation system for railway maintenance would likely already be in place—capable of simulating the real behavior of track sections under actual rolling stock operations, while accounting for long-term climatic influences specific to each location. To achieve this, calculation models must incorporate time-dependent geometric changes resulting from track–rolling stock interactions, variations in stiffness over time, and track geometry monitoring data that enable modeling under different train configurations, speeds, and load conditions.

Thus, overcoming the challenges in defining railway track technical aspects requires a new calculation concept for modeling the railway track and rolling stock interaction. The most promising concept today is the theory of elastic waves.

Since this study is part of a broader research direction, some aspects of elastic wave theory application for railway track modeling and monitoring have already been addressed in studies [58–61]. This article will focus on demonstrating the potential of this theory for determining the technical aspects of railway tracks, applicable to both 1435 mm and 1520 mm gauges, based on the known properties of waves.

As mentioned earlier, the main limitations of the existing concept are related to the following:

1. The lack of time-dependent modeling of mechanical system motion.
2. The lack of understanding of force origins in curvilinear motion modeling.
3. The inability to model stiffness processes.

The following sections will address these issues in detail.

3.2. Overcoming Challenges in Defining the Technical Aspects of Railway Tracks

3.2.1. Considering Time in the Modeling of Mechanical System Motion

Time is one of the seven fundamental physical quantities, forming the basis for understanding the process of recording what is referred to as "change."

The concepts of "statics" and "dynamics" are inherently relative and are interpreted in each context based on two key factors: the time scale over which changes are observed and the capabilities of the measuring instruments used by the observer.

Any change is considered "static" if, during the measurement period, the measuring instruments do not detect a transition from one fixed state to another due to external influences. Thus, the concept of statics and dynamics depends not only on the physical process itself but also on the method of its observation and measurement. If different states are recorded, the change is considered "dynamic."

In the natural sciences, changes are typically described through scalar quantities, which represent specific states, and vectors, which characterize the transitions between these states.

Mathematical descriptions of "change" can be constructed without directly incorporating time in vector representations. This approach is widely used but is limited by the difficulty of deriving analytical dependencies, leading to the broad application of numerical methods. Furthermore, various vector formulations exist depending on their functional use in modeling.

Without incorporating time, the description of natural phenomena using vectors becomes incomplete. Therefore, modern physics increasingly considers processes not in a three-dimensional spatial system but in a four-dimensional space–time framework. This approach accounts for both the geometric properties of a system and its evolution over time, which is particularly important for analyzing dynamic processes.

The consideration of time in mechanical systems is especially crucial when analyzing the propagation of elastic waves, as their dynamics are directly linked to changes in energy over time.

In elastic wave theory, there are two primary types of waves: longitudinal and transverse waves. All other types of waves are specific cases that arise under particular propagation conditions.

Although waves have a well-defined propagation speed, this parameter is often neglected in practical problem-solving. This characteristic is widely used in structural monitoring (e.g., non-destructive testing), yet it is rarely included in mathematical models. However, the fact that waves inherently propagate at a specific speed makes them highly suitable for use as vectors. As is well known, waves serve as carriers of energy, making them physical vectors of energy transmission through space and time. Due to the fundamental properties of waves, it is always possible to determine the following:

1. The direction in which the wave will propagate.
2. The speed at which it moves.
3. The way in which its amplitude will change as it propagates.

Elastic wave propagation is a natural phenomenon, which, according to the theory of relativity, remains invariant in any reference frame. This property allows elastic waves to be used to describe the motion of mechanical systems, complementing classical concepts of interactions. Therefore, incorporating elastic waves into the description of mechanical system motion offers an expanded interpretation of Newton's laws.

For example, Newton's first law can be reinterpreted through the lens of elastic wave theory as follows: any change in momentum P(t) results in a change in the state of a mechanical system, and conversely, altering the system's state requires an impulse. While classical mechanics focuses on the motion of objects, it does not explain how an impulse propagates through a medium—this gap is addressed by elastic wave theory, which describes motion as a time-dependent process of energy transmission.

Newton's second law can be extended by incorporating the time-dependent nature of its fundamental components, including the momentum P(t), force F(t), interacting mass m(t), and velocity u(t), thereby providing a more dynamic and realistic representation of mechanical interactions over time.

At any given moment t over a time period T, these components are related as follows: $P(t) = F(t)T = m(t)u(t)$.

This allows changes in the system to be considered not only in terms of applied forces but also in terms of the distribution of energy across all points within the system over time. Thus, the second law of Newton gains an additional interpretation through energy distribution processes in space and time.

Impulse propagation in a system is based on the wave property of forming sequences of incidence–transmission–refraction when encountering barriers, such as at the transition between different structural elements or the presence of defects. This property of waves demonstrates the law of energy conservation, allowing for the superposition of impulses at any location while accounting for changes in the impulse magnitude over time.

Additionally, the superposition of waves within a mechanical system, such as incident and reflected waves, provides a complete understanding of dissipation processes. This further supports the law of energy conservation and formalizes thermodynamic laws by specifying energy magnitudes, types, and transformation directions.

Moreover, the superposition of refracted waves (the interaction of waves as they transition between different media) provides a complete description of system inertia, including all components of the second law of Newton (momentum, velocity, displacement) during motion in response to an applied impulse.

The third law of Newton, in its original form, explains interactions between bodies that were not previously in contact. However, it struggles to explain reaction forces in systems that were already in contact within the existing mechanical system motion framework.

The phenomenon of incidence–transmission–refraction in wave propagation is fully described by the laws of energy, momentum, and angular momentum conservation [58,59]. This principle explains energy transmission within a system, eliminating the need for fictitious forces that were historically introduced to balance mechanical equations.

This effect is evident in the propagation of elastic waves through a multi-element rotating structure, where waves reflect at the boundaries between elements, transmit into adjacent media, and undergo amplitude changes depending on the material properties at the contact interfaces.

This insight allows previously unexplained effects to be understood without introducing fictitious forces (such as centrifugal force or Coriolis force). By employing elastic wave theory with time-dependent effects, the need for fictitious forces disappears, and interactions within a system can be naturally described.

As mentioned earlier, the traditional mechanical motion model considered time only in terms of discrete steps ("time steps"). In this framework, time was treated logically, based on observed amplitude–phase relationships obtained through experiments.

For example, after numerous experiments, sensor characteristics were established for measuring deflections and stresses in track structures under the impact of rolling stock. These studies relied on an amplitude–frequency analysis. Although recorded oscillograms of passing trains appeared visually different in length at different speeds, this aspect was not considered relevant in traditional data processing.

A similar approach is used in geometric wave representation, where logical time is applied, and amplitude–frequency characteristics are evaluated based on wavelength.

However, the fact that humans can recognize speech or music regardless of location indicates that sound waves propagate identically across different points. This suggests that impulse propagation should be analyzed not only along wave propagation directions (spatial dimensions) but also at each local point, incorporating time-dependent factors.

By integrating time into elastic wave theory, the description of wave processes from impulse impacts becomes more precise, incorporating superposition effects and the ability to analyze vibration amplitudes at any spatial location.

The specified capability enables an expanded description of many dynamic processes. For example, the following sections illustrate how modeling can be extended for fracture mechanics processes and phenomena, such as shock waves. Both processes are associated with considering amplitude variations at specific locations.

For a comprehensive study of fracture mechanics in explosions, it is necessary to analyze the propagation of elastic and electromagnetic waves while accounting for the structural composition of the medium. This is because interactions between surfaces in mechanical systems involve complex energy transfer processes similar to the propagation of elastic waves. For instance, considering an explosive wave, the explosion impulse can be represented as a function consisting of two parts. The first part is a steeply rising segment, where the impulse magnitude increases rapidly over a short period, representing the rate of impulse change over time. The second part is a decaying curve, which lasts significantly longer than the first part, while the rate of impulse variation is considerably lower. The intensity characteristics of both parts have a crucial impact on the destructive effect and the alteration of material properties, defining zones of irreversible changes.

For elastic waves, this is determined by the relationship between the rate of explosion impulse change and the propagation speed of this impulse in the materials affected by the explosion. The boundaries of irreversible changes due to elastic wave impact will be present if the following condition is met: the rate of explosion impulse change is greater than the propagation speed of the elastic wave in the medium.

Thus, irreversible processes under an elastic wave influence are primarily determined by the first part of the explosion impulse. However, when considering all aspects of fracture mechanics, the general criterion for defining the boundary of irreversible changes under an impulse impact is as follows: the energy directed toward the realization of the explosion amplitude is greater than the energy required to maintain bonds within the medium.

Shock wave phenomena can be more intuitively understood through aircraft motion.

The impulse intensity function resulting from an aircraft's passage at each point along its trajectory is determined by both the aircraft's geometry and its velocity. Since an aircraft has a specific configuration at a constant speed, every cross-section along its path will experience the same impulse intensity over time, meaning that the process of the aircraft passing through a given section remains consistent. However, variations in impulse characteristics depend on the aircraft's speed. As the speed increases, the time required to traverse a given section decreases. Therefore, on one hand, increasing the speed shortens the duration of passage or effectively reduces the aircraft's perceived length, while on the other hand, it alters the characteristics of the impulse effect. Consequently, modeling an aircraft's flight involves analyzing the superposition of waves over time, generated by the aircraft's impulse at each trajectory point, considering both the aircraft's velocity and wave propagation in the medium.

Regarding the application of elastic wave theory with time considerations in the "rolling stock–railway track" system, this approach provides two key advantages.

First, it enables modeling based on real impulse sources, such as wheel–rail contacts, while accounting for their location and duration of action. Second, it allows for the use of experimental recording tools that consider the parameters of these impulses. This approach facilitates the adaptation of theoretical studies to experimental data while accounting for different data update rates. This is because theoretical modeling updates data based on wave propagation speeds, whereas experimental data updates depend on sensor characteristics and their inertia. However, analyzing experimental data obtained in this manner enables a comprehensive assessment of the impact of all components within the "rolling stock–railway track" system, ensuring a well-informed interpretation of recorded data.

This approach not only improves the diagnostic accuracy but also enables the refinement of existing mathematical models based on real-world data. Incorporating time into elastic wave theory allows for the recording of the amplitude–frequency characteristics of the system while also enabling an inverse analysis—determining the impulse source based on its response. Table 2 presents a comparative analysis based on elastic wave theory, drawing on data from sources [58–61].

Table 2. Comparison of classical mechanics and elastic wave theory.

Criterion	Classical Mechanics	Elastic Wave Theory
Main Object of Analysis	Rigid and elastic bodies in space	Wave propagation in a medium
Role of Time	Describes object motion but does not account for energy propagation over time	Time is a key parameter that determines energy propagation
Momentum Transfer	Momentum is transferred only upon contact	Momentum is transferred through the medium as wave processes
Interaction Description	Uses forces (Newtonian and fictitious) to explain motion	Interaction is described through wave reflection, refraction, and superposition

Table 2. *Cont.*

Criterion	Classical Mechanics	Elastic Wave Theory
Fictitious Forces	Introduced to explain motion in non-inertial systems (centrifugal force, Coriolis force)	Not required, as interaction dynamics are explained by wave propagation
Energy Conservation	Considered as the sum of potential and kinetic energy	Energy is transmitted through elastic waves, allowing for dissipation and inertia effects
Dynamic System Analysis	Describes object motion under applied forces	Describes momentum and energy changes over time and space
Application in Calculations	Used for analyzing strength, stability of structures, and motion of bodies	Used in defectoscopy, vibration modeling, and dynamic process simulations
Application in Railway Mechanics	Calculates track strength, stability, and dynamic loads on tracks	Enables real-time analysis of track and rolling stock behavior

3.2.2. The Nature of Rotational Forces in the Modeling of Curvilinear Motion

It is well known that a perpendicular force or distributed load applied to the surface of an object induces normal stresses in the section under the applied impulse, and calculating these stresses has never posed a difficulty. Issues arise, however, when the force is applied outside the considered section. In such cases, concepts such as the moment of force, tangential stresses, and tangential forces come into play. This force application can cause rotation, and in engineering terminology, there are various names for this effect, including torque, rotational moment, twisting moment, and bending moment.

The existing concept of mechanical motion accepts this effect as a given and does not focus on the nature of its emergence. However, the propagation of elastic waves in space and time fully explains this process.

It is commonly known that

1. The point of force application, perpendicular to its axis, divides any space into two zones: compression and tension.
2. Pure longitudinal waves always propagate in the direction of the applied force.
3. Pure transverse waves always propagate perpendicular to the direction of the applied force.
4. In all other directions, both longitudinal and transverse waves propagate.
5. Thus, the point of force application divides the surrounding space into two regions:
    a. Radially converging towards the force application point.
    b. Radially diverging from it.
6. Within the force field:
    a. Any point along the radial direction aligned with the force experiences either compression or tension, depending on its position relative to the force application point.
    b. Any point along the radial direction perpendicular to the force experiences pure shear stress in the force direction, regardless of its position relative to the force application point.
    c. Any point along other radial directions, except for those perfectly aligned or perpendicular to the force, experiences rotation:
        i. Clockwise rotation if the point is to the left of the force direction.
        ii. Counterclockwise rotation if the point is to the right of the force direction.

The force axis and the perpendicular direction can be considered time axes, since the propagation speeds of longitudinal and transverse waves are known. Thus, the ellipsoidal surfaces with semi-axes in equal time intervals demonstrate identical phases of medium/particle oscillation under force action in time and space.

Since elastic waves are spherical by nature, the amplitude of oscillations decays inversely proportional to the distance from the source. Therefore, every point in the radial space experiences the same impulse effect in terms of duration and nature but with a time shift and a decreasing impact depending on its location relative to the force application point.

Previously, we examined the incidence–reflection–refraction wave chain phenomenon as the basis for force formation within a system. Next, we will analyze the nature of rotational force emergence using a specific example.

Example: The Effect of a Tangential Force on a Disk

Let us consider the effect of a tangential force on a disk, illustrated in Figure 5.

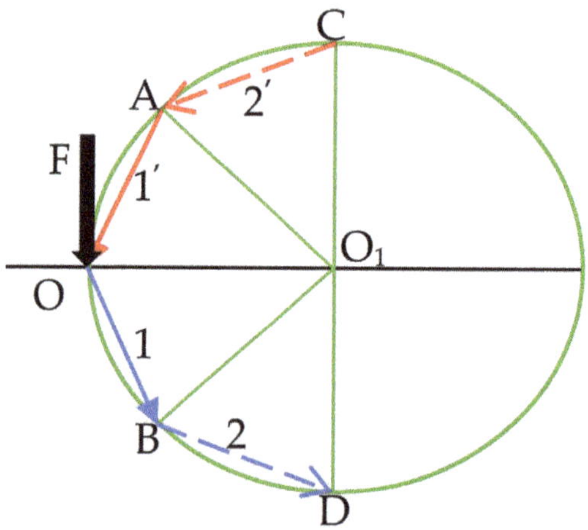

Figure 5. Diagram explaining the nature of circular body rotation under force *F*.

To do so, we will use the explanation of a single force F applied at point *O*, as described earlier. For simplicity, we assume a disk centered at O_1, where both points *O* and O_1 lie on the same straight line. We analyze the case where the disk is fixed such that it can only rotate about its center O_1.

When the force *F* is applied at point *O*, the straight line OO_1 divides space into the following:

1. A region radially converging toward the force application point.
2. A region radially diverging away from it.

At all times, there exists a pair of vectors (converging and diverging) lying in the same plane and positioned at the same angle to force F.

As an example, Figure 5 illustrates two such diagonal vectors, 1 and $1'$.

Since the process occurs within the same body, the wave propagation speed is identical.

Therefore, these vectors will simultaneously reach the body boundaries at points A and B, where the incidence–reflection–refraction wave chain phenomenon occurs.

In this case, we do not analyze wave motion outside the disk, so we limit our scope to the emergence of reflection waves 2 and $2'$. It is well known that reflection waves are oriented at the same angle to the normal as the incident waves. Hence, the wave characteristics of these vectors are identical, but their directions differ. When they simultaneously reach points C and D, the process continues.

This process will occur for various pairs of vectors within the disk plane. Since the vectors tend to rotate in the same direction, their effect can be interpreted as a force moment. Additionally, there will be diametrically opposite vectors, which means that the action of force F can be represented using the conventional force-pair approach.

3.2.3. Modeling of Stiffness Processes

In classical mechanics, the stiffness of an object is defined by its elasticity and resistance to external loads. However, in dynamic systems, stiffness is determined not only by static characteristics but also by energy transmission processes within the material, which is especially important for the propagation of elastic waves. In this context, stiffness can be considered through the dynamic characteristics of the medium and the conditions of its contact with other objects.

It is well known that an object's stiffness depends not only on the characteristics of the applied impulse but also on three fundamental factors: the material's physical and mechanical properties, the object's geometric features, and the conditions of its interaction with surrounding elements—typically expressed as boundary conditions in computational models.

Previously in this article, the phenomenon of the incidence–reflection–refraction chain was discussed. Below, we examine how the conditions for stiffness formation are considered through this wave propagation phenomenon over time.

The consideration of the physical and mechanical properties of materials in stiffness formation has been extensively studied from the perspective of elastic wave propagation in time.

In classical mechanics, stiffness is determined based on the static parameters of the material. However, in dynamic systems, wave propagation plays a key role in stiffness formation since the wave velocity is directly related to the material properties.

The speed of wave propagation in any medium is determined by three fundamental properties: the material's density, its elastic modulus, and Poisson's ratio.

Therefore, the question of wave propagation in space and time does not pose significant difficulties in terms of theoretical modeling.

In addition, materials exhibit specific critical angles related to the behavior of longitudinal and transverse waves, as defined by Snell's law. The first critical angle marks the minimum incidence angle at which a longitudinal wave ceases to penetrate the second medium, generating head waves. The second critical angle corresponds to the threshold at which the refracted transverse wave no longer enters the second medium, leading to the formation of Rayleigh surface waves. The third critical angle defines the minimum incidence angle for a transverse wave at which no reflected longitudinal wave is produced.

These angles are crucial in calculating structural stiffness, as they determine how waves transmit energy within a material. For example, if the incident wave angle exceeds the second critical angle, Rayleigh surface waves are generated in the material, significantly affecting local structural stiffness.

Thus, the incidence–reflection–refraction chain phenomenon accounts for the individuality of each material and provides the broadest explanation of how force interactions propagate within objects.

The consideration of the geometric characteristics of an object is well-researched and thoroughly described in the scientific literature.

When waves reach the boundaries of objects, impact, reflection, and refraction processes occur. These wave interactions, over time, determine the distribution of stresses within the object, influencing its response to external loads.

However, stiffness is determined not only by the internal properties of an object but also by how it interacts with other structural elements. In real-world conditions, all mechanical systems interact with one another, making the consideration of contact properties critically important.

Contact Stiffness Considerations

The description of contact conditions has been the subject of extensive research, with contact stiffness defined by a combination of factors: the material properties of the interacting objects, their geometric configurations, and the geometry of the contact surfaces.

For example, it is well known that the direction of wave reflection changes when bodies with different densities come into contact. This phenomenon is closely related to the concept of acoustic impedance, which is defined as the product of a medium's density and the speed of sound propagation within it. At the same time, the speed of sound itself depends not only on the material's density but also on its elastic properties, creating either a hard or soft acoustic medium.

Although contact stiffness is fundamentally determined by the material properties and geometry of the objects involved, in practice, it is also influenced by additional factors such as the contact pressure, friction, and surface characteristics of the materials. In elasticity theory, these influences are accounted for by modeling impulse variations across both time and space.

Stiffness as a Function of Elastic Waves

Examining stiffness through the lens of elastic waves allows for the connection of the static and dynamic characteristics of a system.

Traditional calculations typically separate stiffness into static and dynamic components; however, wave-based methods enable its representation as a continuous time-dependent function. By accounting for various impulse characteristics through wave superposition, stiffness can be accurately modeled under a wide range of conditions. This approach is especially valuable in engineering applications such as analyzing vibrational loads on rails and evaluating the dynamic response of bridge structures, ultimately contributing to improved travel comfort and greater environmental sustainability [62,63].

Unlike traditional methods, which consider only static stiffness, wave-based modeling allows for a time-dependent analysis of structural behavior.

This significantly improves the accuracy of calculations and enhances the prediction of potential structural failures.

4. Discussion

This section analyzes the feasibility of harmonizing the technical requirements of railway systems with track gauges of 1520 mm and 1435 mm within the framework of the European Commission's concept. It also examines the technical viability of the European Commission's proposals [1] for integrating post-Soviet railways with a 1520 mm gauge into the European railway system with a 1435 mm gauge. The analysis aims to identify potential challenges and possible solutions.

4.1. Background Analysis

In previous sections, it was demonstrated that the experience gained through numerous experiments and railway track operations has been instrumental in establishing technical assessment criteria for tracks with gauges of 1435 mm and 1520 mm.

As shown in Table 1, various countries have extensive experience in classifying railway tracks based on different parameters for the 1435 mm gauge. This facilitates the harmonization of technical aspects under alternative parameters. However, such experience is entirely lacking for the 1520 mm gauge railway track, as, during the formation of classification criteria, this railway system belonged to a single country. Following the dissolution into separate nations, the classification approach remained unchanged. Consequently, there is a complete absence of unified classification experience that incorporates both design and maintenance aspects, as observed for the 1435 mm gauge.

Therefore, the direct transfer of methodologies from the 1435 mm system to the 1520 mm system is impossible. To ensure reliability, additional experiments are required to establish appropriate classification criteria.

From this perspective, the existing concept requires adaptation and further research to harmonize the technical aspects of the two railway infrastructure systems with different track gauges. This process is both time- and resource-intensive.

To reduce time and resource expenditure, two conceptual modifications are possible. Based on the analysis of challenges and their solutions in defining the technical aspects of railway tracks, the following conceptual approaches for mechanical system motion are proposed.

4.2. Considerations for Technical Aspects of Railway Tracks

For evaluating the technical aspects of railway tracks, it is essential to consider the principles of mechanics that form the basis for stability and strength calculations. Historically, the concept of mechanical system motion was developed based on identifying the correlation between critical technical states of a mechanical system and its operating conditions. In other words, the feasibility of the mechanical system functioning under specific operational conditions was examined by defining the critical state as a limiting factor for its operation. To achieve this, the motion process of the mechanical system itself needed to be analyzed. Therefore, the process observed in experiments was described.

In traditional approaches to mechanical system calculations, the primary focus was on geometric changes in structures and strength assessments. This justified the use of idealized models that ignored dynamic factors such as wave processes and external influences. In most cases, the decisive factor in choosing a mechanical system option was the economic feasibility of its manufacturing, construction, and operation.

A key challenge in modern operational conditions lies in understanding and managing the interrelationship among three critical technical aspects: the functional purpose of the railway track, the maintenance and servicing system, and the structural characteristics of the track elements and construction.

Under this approach, the primary selection criterion is the functional safety of the railway track, both from the user's perspective and in terms of environmental impact.

Thus, contemporary advancements in human understanding necessitate consideration not only of the movement process of the mechanical system under load but also of the natural phenomena accompanying this process. Consequently, it is essential to account for the rate of changes in the stress–strain state of the mechanical system itself and the speed of natural phenomena within it.

As is known, there is a fundamental contradiction from this perspective that prevents the application of the currently accepted concept of mechanical system motion:

- The classical law of velocity addition has a limited scope of application.

This limitation arises because, on the one hand, discussing velocity without specifying a reference system is meaningless within the existing concept of mechanical system motion. On the other hand, the invariance of elastic and electromagnetic wave velocity in media with certain characteristics is an established fact and is supported by the general principles of relativity theory.

4.3. Possible Solutions for Harmonizing Technical Aspects of 1435 mm and 1520 mm Railways

Based on the above considerations, there are two possible solutions to the problem of harmonizing the technical aspects of railway tracks with 1435 mm and 1520 mm gauges.

The first option involves introducing dependency coefficients for various parameters into the existing concept of mechanical system motion by accounting for the following:

- The property of waves to decrease in amplitude proportionally to their length.
- The interaction time between wheels and rails at different train speeds.
- The relationships of mass, time, and length in accordance with the theory of relativity.

The second option involves applying the theory of elastic wave propagation in space and time as the concept of mechanical system motion [58–61].

Thus, each option provides a solution to the problem but with opposite potential capabilities in describing the analytical dependence between influencing factors and the results of changes. Table 3 presents a comparative analysis based on elastic wave theory, drawing on data from sources [26,58–61].

Table 3. Comparison of proposed options.

Criterion	Option 1: Coefficients	Option 2: Elastic Waves
Method Basis	Introduction of coefficients to adjust calculations	Complete rethinking of the model based on wave processes
Ease of Implementation	Easier, as it adjusts existing calculations	Requires a new methodology and experiments
Required Data	Depends on empirical coefficients	Operates with the physical characteristics of the medium
Accuracy	Limited by the correctness of the coefficients	Potentially more accurate, as it accounts for the dynamics of the path

The use of coefficients (Option 1) allows for adapting the existing concept; however, it remains dependent on empirical data, which reduces the accuracy of forecasts. At the same time, transitioning to a model based on elastic waves (Option 2) requires significant computational resources but potentially provides a more accurate description of the dynamics of the railway track.

4.4. Technical Feasibility Analysis of Integrating the 1520 mm Railway Network into the 1435 mm European System: Challenges and Critical Factors

Below is an analysis of the technical feasibility of the proposals introduced by the European Commission [1] for integrating the post-Soviet railway network with a track gauge of 1520 mm into the European railway system with a track gauge of 1435 mm.

The characteristics of the railway system with a 1520 mm gauge, presented in Sections 4.1.1.3, 4.1.1.4, 4.4, 4.9, 5.1.1, and 5.2.10, demonstrate a lack of understanding of the discrepancies between the technical aspects and the classification of requirements for railway infrastructure systems with gauges of 1435 mm and 1520 mm.

It has previously been shown that the primary difference between the calculation methods for determining the technical parameters of railway tracks with gauges of 1435 mm and 1520 mm is due to the functional purpose of the tracks.

For instance, in the railway system with a 1520 mm gauge, there are no established correlations between the force, the load-bearing area, stresses, or deformation modulus of granular materials and their thickness. There is also no classification of fastenings operating under specific force and stiffness characteristics. Moreover, there are no limitations on the magnitude of force impacts from rolling stock on the track, neither in classification nor in experimental studies.

At the same time, for example, on the railways of Ukraine, a system of the multiple reuse of rail and sleeper grid elements is employed. On sections with lower traffic intensity, so-called "old stock materials" are used—materials that have reached their usage limits according to established criteria for more intensive operating conditions. Furthermore, in an effort to accommodate the maximum amount of freight, railway systems with a 1520 mm gauge consider the impact of repeated maximum dynamic forces. However, while stress limitations are considered, displacement restrictions are not. Therefore, the elasticity modulus of the track in the vertical plane was previously unrestricted.

For clarity, Table 4 presents the stiffness characteristics of the track. The research was conducted between 1998 and 2001 by staff members of the "Track and Track Facilities" department and the track testing laboratory of DIIT [64]. The track stiffness was measured using a modular railcar. Table 4 demonstrates the moduli of elasticity of the sub-rail foundation in the vertical direction (MPa) under axial loads of 22–25 kN and train speeds of up to 140 km/h for passenger trains and 90 km/h for freight trains under summer conditions.

Table 4. Track structure characteristics and module values (MPa).

Track Structure Type	Module Values (MPa)		
	Average	Maximum	Minimum
P65[1](6)[2]1440[3](HBZ)[4]SCH[5]	24	46	9
P65(6)1600(HBZ)SCH	25	48	9.5
P65(6)1840(HBZ)SCH	26	50	10
P65(6)2000(HBZ)SCH	26.8	52	12
P50(6)1440(IA,IB)SCH	22	40	8
P50(6)1600(IA,IB)SCH	22.5	42	8.5
P50(6)1840(IA,IB)SCH	23	44	9
P50(6)2000(IA,IB)SCH	24	48	9.5

[1]—Type of rails; [2]—the wear value (in mm) is shown in parentheses; [3]—sleeper spacing, pcs/km; [4]—designation of reinforced concrete sleepers with KB fastening; [5]—designation of crushed stone ballast.

As seen in Table 2, the average values of actual elasticity modules are far from the modern requirements. The variation between the minimum and maximum values ranges from 4.3 to 5 times. Therefore, considering subpoints (ii) and (iii) in Section 4.9 as a strong aspect is illogical, as this assessment is made from the perspective of meeting the criteria used in the railway system with a track gauge of 1435 mm. A railway section with such subgrade characteristics cannot be certified without a complete reconstruction, while in [1] the proposed reconstruction options, the highest costs for upgrading the substructure amount to a maximum of 5%; see Figure 6.

Easy profile cost distribution per component	Option 1	Option 2	Option 3	Option 4	Option 5	Option 6	Option 7	Option 8	Option 9
TRACK SUBSTRUCTURE	2%	5%	4%	4%	3%	2%	5%	5%	4%
DRAINAGE- MAIN LINE	1%	1%	1%	1%	1%	1%	1%	1%	1%
TRACK SUPERSTRUCTURE	34%	31%	26%	24%	33%	35%	29%	26%	22%
TURNOUTS	1%	1%	1%	1%	1%	2%	1%	1%	1%
LEVEL CROSSINGS	0%	0%	1%	1%	1%	0%	1%	1%	2%
SIDINGS	2%	1%	4%	1%	2%	1%	2%	2%	4%
SUPPLY AND INSTALLATION of FENCING (for high speed only)	0%	0%	0%	0%	0%	0%	0%	0%	0%
SMALL BRIDGES	3%	1%	1%	1%	1%	1%	2%	2%	1%
SIGNALLING - ETCS - INTERLOCKING - TELECOMMUNICATIONS	16%	21%	26%	32%	25%	28%	25%	25%	33%
OCS	10%	11%	9%	9%	9%	8%	10%	9%	8%
POWER SUPPLY	16%	12%	10%	10%	8%	6%	8%	11%	9%
OVERHEAD COSTS	17%	17%	17%	17%	17%	17%	17%	17%	17%
TOTAL	100%	100%	100%	100%	100%	100%	100%	100%	100%

Figure 6. Flat terrain cost distribution per component. Source: prepared by Egis for JASPERS [1].

Thus, replacing the upper part of the subgrade and the superstructure of the track, as indicated in Section 6.1.1, will not yield operational benefits even in the medium term when using such a foundation under the influence of rolling stock moving on one track at 200 km/h and on the other with axle loads of 25 tons.

Furthermore, the main problem lies in the fact that the intensity of use of the railway system with a 1520 mm track gauge has historically been much higher than that of the 1435 mm gauge. As of today, the restoration measures proposed based on the experience of rehabilitating worn-out but non-aging track structures are unsuitable. Therefore, it is necessary to consider the current state of railway infrastructure systems with a 1520 mm gauge, which differs from that of railway tracks with a 1435 mm gauge.

Moreover, the gradual reconstruction options presented in Section 6.1.4 imply the operation of rolling stock with different functional purposes in the same direction on an unprepared foundation. In other words, using two tracks of different gauges, where rolling stock moves simultaneously as on single-track lines, ensures the possibility of trains running in the same direction.

Neither the methodology for calculating the impact of temperature changes considers the influence of longitudinal forces from rolling stock on track stability, nor does any railway system impose restrictions based on this criterion. Therefore, assuming that the effect of longitudinal forces can be neglected in cases where longitudinal displacements originate from two sources with different amplitude–frequency characteristics is inadequate, even on a prepared foundation. This is even more critical for a trackbed that has been in operation for over 100 years and was built according to the requirements in place at the time of construction.

Considering all the above, the following conclusions can be drawn:

1. The implementation of the European Commission's concept requires an adaptation of the methodology, as the 1435 mm and 1520 mm systems have historically been developed independently, and their operational characteristics differ significantly. Without accounting for these differences, it is impossible to ensure the safety and durability of the railway infrastructure.
2. A direct transfer of technical requirements is impossible due to the absence of unified assessment criteria for 1520 mm tracks.
3. For successful harmonization, either the gradual introduction of corrective coefficients or the development of a fundamentally new model based on wave processes is required.

5. Conclusions

The study has shown that the successful integration of 1520 mm and 1435 mm railway systems requires not only the adaptation of regulatory frameworks but also a rethinking of the methodology for calculating strength, stability, and safety. The article analyzes existing challenges and proposes alternative approaches to addressing them.

When modernizing the railway infrastructure, it is crucial to consider not only the efficiency of new solutions but also the evolution of existing engineering systems, as well as contemporary operational requirements.

The work emphasizes the importance of considering technical aspects in the creation of a unified classification of railway infrastructure systems. This requires the harmonization of existing operational and design regulatory frameworks, which are based on different methodologies for calculating strength, stability, and safety.

The analysis presented in this study on the harmonization possibilities of technical aspects of railway infrastructure systems with different track gauges has shown that existing methods are insufficient. To overcome the identified difficulties, it is necessary either to adapt current concepts to new factors or to apply a new paradigm based on the wave modeling methods of railway systems. Both approaches offer different solutions to the problem, requiring further research and experimentation.

The proposed paradigm for railway system modernization is based on modern scientific concepts and takes into account dynamic processes in the "track–rolling stock" system. Unlike traditional methods, the new approach considers the propagation of loads over time, taking into account the interaction of elastic waves.

This paradigm relies on the concept of elastic wave propagation over time, which allows for the consideration of dynamic characteristics of the track and rolling stock, such as the vibration amplitude, load variation speed, and stress distribution effects. Unlike traditional methods, this approach describes not only static parameters but also the influence of time-dependent factors on track stability.

The conceptual implementation of railway system modernization depends on the chosen strategy. Emphasizing the introduction of new concepts requires significant investment in research and testing but ensures maximum accuracy in calculations and improved operational reliability. At the same time, expanding existing technologies reduces costs and the adaptation time but limits the implementation of fundamentally new approaches to load and track stability management. The final choice of strategy should consider the intensity of railway operations, the state of infrastructure, and the specifics of track design in each railway system.

Therefore, the successful modernization of railway infrastructure systems requires a comprehensive approach that integrates theoretical analysis, experimental data, and a strategically selected method of adaptation. In this context, the author intends to further develop models of dynamic interactions between the track and rolling stock based on the principles of elastic wave theory.

Funding: This research received no external funding.

Institutional Review Board Statement: Not applicable.

Informed Consent Statement: Not applicable.

Data Availability Statement: The original contributions presented in this study are included in the article. Further inquiries can be directed to the corresponding author.

Conflicts of Interest: The author declares no conflicts of interest.

References

1. Strategy for the EU Integration of the Ukrainian and Moldovan Rail Systems. European Commission 11.07.23. 159p. Available online: https://transport.ec.europa.eu/system/files/2023-07/Integration_of_the_UAMD_railway_system_into_the_EU_transport_system.pdf (accessed on 14 May 2025).
2. Infrastructure TSI. Available online: https://eur-lex.europa.eu/eli/reg/2014/1299/2023-09-28 (accessed on 14 May 2025).
3. A. Classes of Track: Operating Speed Limits. Available online: https://tc.canada.ca/en/rail-transportation/rules/2022-2023/rules-respecting-track-safety/part-ii-track-safety-rules (accessed on 14 May 2025).
4. Track Classifications. Available online: https://www.trains.com/trn/train-basics/abcs-of-railroading/track-classifications/ (accessed on 14 May 2025).
5. Analysis of the Basic Parameters for Maintaining the Technical and Operational Compatibility of the 1 520-mm Gauge Rail System at the CIS-EU Border. Subsystem: Infrastructure. Permanent Way and Track Facilities. Available online: https://www.era.europa.eu/system/files/2022-11/Analysis%201520%20INF%20(EN).pdf (accessed on 14 May 2025).
6. *DBN V.2.3-19-2018–Kyiv*; State Building Standards of Ukraine. Transport Structures. Railways with a Gauge of 1520 mm. Ministry of Regional Development of Ukraine: Kyiv, Ukraine, 2018; 129p.
7. Kukulski, J.; Jacyna, M.; Gołębiowski, P. Finite Element Method in Assessing Strength Properties of a Railway Surface and Its Elements. *Symmetry* **2019**, *11*, 1014. [CrossRef]

8. Chawla, S.; Shahu, J.T.; Kumar, S. Analysis of cyclic deformation and post-cyclic strength of reinforced railway tracks on soft subgrade. *Transp. Geotech.* **2021**, *28*, 100535. [CrossRef]
9. Avsievich, A.; Avsievich, V.; Avsievich, N.; Ovchinnikov, D.; Ivaschenko, A. Railway Track Stress–Strain Analysis Using High-Precision Accelerometers. *Appl. Sci.* **2021**, *11*, 11908. [CrossRef]
10. Claudet, B.; Duhamel, D.; Foret, G.; Hoang, T.; Sabatier, F. Wave finite element method for computing the dynamic response of periodic structures with transition zones and subjected to moving loads: Application to railways tracks with damaged or reinforced zones. *Transp. Geotech.* **2025**, *52*, 101572. [CrossRef]
11. Heydari, H.; Zakeri, J.A.; Noori, A.S. Material-oriented maintenance of railway track using under sleeper pad (USP). *Int. J. Transp. Sci. Technol.* **2025**. [CrossRef]
12. Lee, S.Y.; Truong, K.D.; Do, D.D.; Minh Le, T.H. Improving Railway Track Durability: Evaluating Sand and Emulsion Effects on Cement Asphalt Mortar. *Transp. Res. Procedia* **2025**, *85*, 120–127. [CrossRef]
13. Agustin, D.; Wu, Q.; Ngamkhanong, C. A review on railway track buckling prediction methods. *Constr. Build. Mater.* **2025**, *466*, 140295. [CrossRef]
14. Zhai, Y.; Jia, H.; Xu, L.; Zeng, Q.Z.; Zhao, C.; Zheng, S. Damage analysis and failure mechanism of high-speed railway bridge-track system considering site conditions under mainshock-aftershock sequences. *Eng. Struct.* **2025**, *334*, 120270. [CrossRef]
15. Donaire-Ávila, J.; Aceituno, J.F.; Suárez, F. Cumulative effects of cracking in monoblock multipurpose precast concrete sleepers on railway tracks. *Eng. Struct.* **2025**, *326*, 119491. [CrossRef]
16. Rempelos, G.; Ognibene, G.; Le Pen, L.; Blainey, S.; Preston, J.; Powrie, W. Railway Track Deterioration Models: A Review of the State of the Art. *Transp. Geotech.* **2024**, *49*, 101377. [CrossRef]
17. Ramos, A.; Gomes Correia, A.; Nasrollahi, K.; Nielsen, J.C.O.; Calçada, R. Machine Learning Models for Predicting Permanent Deformation in Railway Tracks. *Transp. Geotech.* **2024**, *47*, 101289. [CrossRef]
18. Powrie, W. Soil mechanics principles for modelling railway track performance. *Transp. Geotech.* **2024**, *47*, 101265. [CrossRef]
19. Neves Costa, J.; Ambrósio, J.; Andrade, A.R.; Frey, D. Safety assessment using computer experiments and surrogate modeling: Railway vehicle safety and track quality indices. *Reliab. Eng. Syst. Saf.* **2023**, *229*, 108856. [CrossRef]
20. Niu, L. Automatic identification method of foreign body intrusion in railway transportation track based on improved LeaderRank identification of key points. *Egypt. Inform. J.* **2025**, *30*, 100682. [CrossRef]
21. Kedia, N.K.; Wangtawesap, R.; Ngamkhanong, C. Dynamics responses of railway bridges influenced by flooded ballasted tracks subjected to high-speed trains. *Transp. Eng.* **2025**, *20*, 100334. [CrossRef]
22. Poyyamozhi, M.; Devadharshini, A.S.K.; Murugesan, B.; Novak, T.; Mlcak, T.; Narayanamoorthi, R. Enhancing railway track maintenance with real-time ultrasonic and moisture sensing: Proactive and zone-specific management strategies. *Results Eng.* **2024**, *24*, 103472. [CrossRef]
23. Koohmishi, M.; Kaewunruen, S.; He, X.; Guo, Y. Advancing railway sustainability: Strategic integration of circular economy principles in ballasted track systems. *J. Clean. Prod.* **2025**, *490*, 144713. [CrossRef]
24. Burdzik, R.; Nowak, B.; Rozmus, J.; Słowiński, P.; Pankiewicz, J. Safety in the railway industry. *Arch. Trans.* **2017**, *44*, 15–24. [CrossRef]
25. Ershkov, O.P. *Calculations of Transverse Horizontal Forces in Curves*; All-Union Research Institute of Railway Transport: Moscow, Russia, 1966; pp. 3–236. Available online: https://biblioserver.usurt.ru/cgi-bin/irbis64r_13/cgiirbis_64.exe?LNG=uk&Z21ID=&I21DBN=KN&P21DBN=KN&S21STN=1&S21REF=&S21FMT=fullwebr&C21COM=S&S21CNR=10&S21P01=0&S21P02=1&S21P03=A=&S21STR=%D0%95%D1%80%D1%88%D0%BA%D0%BE%D0%B2,%20%D0%9E.%20%D0%9F (accessed on 14 May 2025).
26. Shakhunyants, G.M. Railway Route—Moscow: Transport, 1987. pp. 32–45. Available online: https://f.eruditor.link/file/1486741/ (accessed on 14 May 2025).
27. Timoshenko, S.P. *Stability of Elastic Systems*; Vlasov, V.Z., Ed.; Snitko, I.K., Translator; Gostekhizdat: Leningrad, Russia; Moscow, Russia, 1955; 532p.
28. Pershin, S.P. Method of Calculating the Stability of Continuous Welded Track. Proceedings of the MIIT, Issue 147: Track and Track Facilities. Moscow, Russia, 1962; pp. 28–97. Available online: https://studwood.net/1925899/tehnika/kompleksnyy_raschet_besstykovogo_puti_prochnost_ustoychivost#google_vignette (accessed on 14 May 2025).
29. Thermal Stress. Available online: https://en.wikipedia.org/wiki/Thermal_stress (accessed on 14 May 2025).
30. Banerjee, T.; Moharana, S.; Parida, L. Early detection of thermal instability in railway tracks using piezo-coupled structural signatures. *J. Infrastruct. Intell. Resil.* **2023**, *2*, 100063. [CrossRef]
31. Gholamizadeh, K.; Pamucar, D.; Moslem, S.; Basiri, P.; Esztergár-Kiss, D.; Mohammadfam, I. Decoding rail derailments: Unraveling the weighted factors influencing safety and sustainability using the best-worst method. *Results Eng.* **2024**, *23*, 102539. [CrossRef]
32. Zhang, D.; Ishihara, T. A comparative study on the crosswind stability of the railway vehicle considering distinct national standards. *J. Wind Eng. Ind. Aerodyn.* **2024**, *254*, 105901. [CrossRef]

33. Cruz, P.; Batista, P.; Sanyal, A. Design and analysis of attitude observers based on the Lagrange-d'Alembert principle applied to constrained three-vehicle formations. *Adv. Space Res.* **2022**, *69*, 4001–4012. [CrossRef]
34. Hentosh, O.E.; Prykarpatsky, Y.A.; Blackmore, D.; Prykarpatski, A.K. Lie-algebraic structure of Lax–Sato integrable heavenly equations and the Lagrange–d'Alembert principle. *J. Geom. Phys.* **2017**, *120*, 208–227. [CrossRef]
35. Sumin, M.I. Regularized Lagrange Principle and Pontryagin Maximum Principle in Optimal Control and Inverse Problems. *IFAC-PapersOnLine* **2018**, *51*, 871–876. [CrossRef]
36. Abdel Aal, O.F.; Viola, J.; Chen, Y. Fractional Order Euler-Lagrange Model for Accelerated Gradient Methods. *IFAC-PapersOnLine* **2024**, *58*, 466–471. [CrossRef]
37. Giusteri, G.G.; Marcuzzi, F.; Rinaldi, L. Replacing voids and localized parameter changes with fictitious forcing terms in boundary-value problems. *Results Appl. Math.* **2023**, *20*, 100402. [CrossRef]
38. Gutierrez-Prieto, E.; Gomez, M.; Reis, P.M. Harnessing centrifugal and Euler forces for tunable buckling of a rotating elastica. *Extrem. Mech. Lett.* **2024**, *72*, 102246. [CrossRef]
39. Tajima, M.; Yamada, T. Topology optimization with accessibility constraint from multiple bi-directions using fictitious anisotropic diffusion equation based on coupled fictitious physical model. *Comput. Methods Appl. Mech. Eng.* **2024**, *429*, 117169. [CrossRef]
40. Bura, E.; Torabi, A.R.; Seweryn, A. Fracture prediction in flat PMMA notched specimens under tension - effectiveness of the equivalent material concept and fictitious material concept. *Theor. Appl. Fract. Mech.* **2024**, *130*, 104273. [CrossRef]
41. Cao, M.; Zhou, A. Fictitious pile method for fixed-head pile groups with dissimilar piles subjected to horizontal loading. *Soils Found.* **2022**, *62*, 101212. [CrossRef]
42. Schäfers, K.; Finkenrath, J.; Günther, M.; Knechtli, F. Hessian-free force-gradient integrators. *Comput. Phys. Commun.* **2025**, *309*, 109478. [CrossRef]
43. Carapellese, F.; Faedo, N. Mechanical interactions modeling of inertial wave energy converters. *Int. J. Mech. Sci.* **2024**, *284*, 109731. [CrossRef]
44. Fu, C.; Zhu, Y.; Tong, D. Stiffness Assessment of Cracked Reinforced Concrete Beams Based on a Fictitious Crack Model. *KSCE J. Civ. Eng.* **2021**, *25*, 516–528. [CrossRef]
45. de Oliveira Barbosa, J.M.; Fărăgău, A.B.; van Dalen, K.N. A lattice model for transition zones in ballasted railway tracks. *J. Sound Vib.* **2021**, *494*, 115840. [CrossRef]
46. Catena, M.; Gesualdo, A.; Lisi, S.; Penta, F.; Pucillo, G.P. A continuous model for the railway track analysis in the lateral plane. *Procedia Struct. Integr.* **2018**, *12*, 538–552. [CrossRef]
47. Malmborg, J.; Persson, P.; Persson, K. Numerical investigation of railway subgrade stiffening: Critical speed and free-field vibrations. *Transp. Geotech.* **2022**, *34*, 100748. [CrossRef]
48. Silva, R.; Costa, C.; Arêde, A. Numerical methodologies for the analysis of stone arch bridges with damage under railway loading. *Structures* **2022**, *39*, 573–592. [CrossRef]
49. Bettinelli, L.; von der Thannen, B.; Stollwitzer, A.; Fink, J. Comparison of different approaches for considering vehicle-bridge-interaction in dynamic calculations of high-speed railway bridges. *Eng. Struct.* **2022**, *270*, 114897. [CrossRef]
50. Charoenwong, C.; Connolly, D.P.; Odolinski, K.; Alves Costa, P.; Galvín, P.; Smith, A. The effect of rolling stock characteristics on differential railway track settlement: An engineering-economic model. *Transp. Geotech.* **2022**, *37*, 100845. [CrossRef]
51. Lingamanaik, S.N.; Thompson, C.; Nadarajah, N.; Ravitharan, R.; Widyastuti, H.; Chiu, W.K. Using Instrumented Revenue Vehicles to Inspect Track Integrity and Rolling Stock Performance in a Passenger Network During Peak Times. *Procedia Eng.* **2017**, *188*, 424–431. [CrossRef]
52. Moliner, E.; Museros, P.; Allahvirdizadeh, R. Track–bridge interaction effects in the acceleration and displacement response of high-speed railway bridges: Simplified vs. refined modelling. *Eng. Struct.* **2024**, *314*, 118304. [CrossRef]
53. Latvala, J.; Kolisoja, P.; Luomala, H. The cyclic loading resistance of old railway track sub-ballast materials at different water contents. *Transp. Geotech.* **2022**, *35*, 100772. [CrossRef]
54. Yang, F.; Wei, Z.; Sun, X.; Shen, C.; Núñez, A. Wheel-rail rolling contact behavior induced by both rail surface irregularity and sleeper hanging defects on a high-speed railway line. *Eng. Fail. Anal.* **2021**, *128*, 105604. [CrossRef]
55. Fazilova, Z.T.; Egereva, E.N.; Chuprynin, A.Y. The Simulation of the Stress-Strain Track Structure Condition on the Section with Variable Stiffness. *Transp. Res. Procedia* **2023**, *68*, 712–716. [CrossRef]
56. Knap, S.; Suchánek, A.; Harušinec, J. Calculation of the geometric transit through the track curves of a two-section platform wagon. *Transp. Res. Procedia* **2021**, *55*, 853–860. [CrossRef]
57. Fomin, O.; Prokopenko, P.; Medvediev, Y.; Degtyareva, L. Determination of safety indicators of the freight wagons by mobile systems. *Procedia Struct. Integr.* **2024**, *59*, 516–522. [CrossRef]
58. Bondarenko, I.; Lukoševičius, V.; Neduzha, L. Novel 'Closed'-System Approach for Monitoring the Technical Condition of Railway Tracks. *Sustainability* **2024**, *16*, 3180. [CrossRef]
59. Bondarenko, I.; Severino, A.; Olayode, I.O.; Campisi, T.; Neduzha, L. Dynamic Sustainable Processes Simulation to Study Transport Object Efficiency. *Infrastructures* **2022**, *7*, 124. [CrossRef]

60. Bondarenko, I.; Keršys, R.; Neduzha, L. Modeling the Dynamism of Complex Dynamic Systems Using the Universal Laws of the Wave Theory. In *Transport Means 2023, Proceedings of the 27th International Conference, Palanga, Lithuania, 4–6 October 2023*; Kaunas University of Technology: Kaunas, Lithuania; Part II, pp. 639–644. Available online: https://crust.ust.edu.ua/server/api/core/bitstreams/94f7a23c-6c92-4ad9-bf2f-8d609602ec1f/content (accessed on 14 May 2025).
61. Bondarenko, I.; Campisi, T.; Tesoriere, G.; Neduzha, L. Using Detailing Concept to Assess Railway Functional Safety. *Sustainability* **2023**, *15*, 18. [CrossRef]
62. Sivilevičius, H.; Maskeliūnaitė, L.; Petkevičienė, B.; Petkevičius, K. The model of evaluating the criteria, describing the quality of organization and technology of travel by international train. *Transport* **2012**, *27*, 307–319. [CrossRef]
63. Maskeliūnaitė, L.; Sivilevičius, H. The model for evaluating the criteria describing the quality of the trip by international train. *Technol. Econ. Dev. Econ.* **2014**, *20*, 484–506. [CrossRef]
64. Taturevich, A.P. Results of Determining Actual Values of Track Rigidity for Research of Track and Rolling Stock Interaction. Science and Progress of Transport. *Bull. Dnepropetr. Natl. Univ. Railw. Transp.* **2003**, *2*, 95–100. Available online: https://cyberleninka.ru/article/n/rezultaty-opredeleniya-fakticheskih-znacheniy-zhestkosti-puti-dlya-issledovaniy-vzaimodeystviya-puti-i-podvizhnogo-sostava (accessed on 14 May 2025).

sustainability

MDPI

Article

Choosing Sustainable and Traditional Public Transportation Alternatives Using a Novel Decision-Making Framework Considering Passengers' Travel Behaviors: A Case Study of Istanbul

Pelin Büşra Şimşek [1], Akın Özdemir [1,*], Selahattin Kosunalp [2] and Teodor Iliev [3]

1 Department of Industrial Engineering, Ondokuz Mayıs University, Samsun 55139, Türkiye; 19060546@stu.omu.edu.tr
2 Department of Computer Technologies, Gönen Vocational School, Bandırma Onyedi Eylül University, Bandırma 10200, Türkiye; skosunalp@bandirma.edu.tr
3 Department of Telecommunications, University of Ruse, 7017 Ruse, Bulgaria; tiliev@uni-ruse.bg
* Correspondence: akin.ozdemir@omu.edu.tr

Abstract

A public transportation system consists of complex processes and requires comprehensive planning activities for a city when dealing with the travel behavior decisions of passengers. Travel behavior decisions are important in selecting suitable transportation alternatives for passengers. In the literature, little attention has been paid to prioritizing the criteria and ranking the alternatives for assessing sustainable and traditional public transportation modes when considering the travel behavior decisions of passengers. In this paper, a five-phased novel decision analysis framework, including Analytic Hierarchy Process (AHP), Technique for Order of Preference by Similarity to Ideal Solution (TOPSIS), and VIekriterijumsko KOmpromisno Rangiranje (VIKOR) techniques, is proposed to evaluate the alternatives. In addition, to the best of our knowledge, the novel decision-making framework in this paper has not been employed before to assess sustainable transportation alternatives dealing with the travel behavior decisions of passengers. Next, the thirteen criteria are specified, including economics, safety, travel quality, and environmental and health aspects, to analyze the travel behavior decisions of passengers with regard to the experts' notions, published reports, and papers. Then, the seven public transportation alternatives are determined, including sustainable and traditional transportation modes. A case study was carried out in Istanbul, Türkiye. Based on the results, service frequency, the vehicle type and its mechanism, and ease of accessibility were found to be the top three significant criteria that affect travel behavior decisions. Furthermore, metro, Marmaray, and metrobus are the top three public transportation alternatives. In addition, the results were verified. Moreover, managerial and theoretical recommendations are provided to policymakers. Lastly, sustainable development goals 11.2 and 11.b can be achieved by designing an accessible, affordable, environmentally friendly, safe, and sustainable public transportation system when analyzing the travel behavior decisions of passengers.

Keywords: public transportation; travel behavior analysis; transportation policy; sustainability; multi-criteria decision analysis; sustainable development goals

1. Introduction

Nowadays, people spend an increasing amount of time in traffic, especially in metropolitan cities, which affects travel behavior decisions. The most effective method that municipalities can implement to prevent traffic problems and stampedes that grow with the time people spend on the road is offering public transportation options. For example, Istanbul, one of the metropolitan cities in the world, has a wide coverage of the public transportation network, serving a population of 15,701,602 people [1]. Also, 18.3% of the population of Türkiye lives in Istanbul [1]. Moreover, the population of Istanbul increased by 45,678 people compared to the previous year [1]. Based on this awareness, increasing public transportation options and making them available will benefit cities by preventing traffic problems. Also, the use of public transportation may deter people from using individual vehicles because of heavy traffic. Moreover, different public transportation modes can provide more options and lead more people to use public transportation, decreasing traffic accidents and time spent in traffic. In addition, environmental damage, such as carbon footprint and exhaust emissions, which increases with the increased use of individual vehicles, can be prevented using sustainable transportation modes.

A number of public transportation modes can provide access to more people and meet the transportation needs of more people from a social perspective. Another benefit of several public transportation modes is that public transportation will contribute to city planning. Moreover, better and more organized public transportation networks will contribute to the sustainable and effective planning of cities. When examining all the reasons, the increase in public transportation usage and the diversity of public transportation open up many opportunities to make life easier. So, evaluating sustainable and traditional public transportation modes is a crucial issue when analyzing the travel behavior decisions of passengers. In the literature, there is a research gap when it comes to addressing this issue dealing with passengers' travel behaviors. Based on the framework of this study, municipalities may carry out various projects to expand, improve, and tighten their public transportation networks for sustainable cities.

1.1. Literature Review

The relevant studies of sustainable and public transportation mode selections are reviewed in this subsection as follows. Particularly, Yedla et al. [2] investigated the effect of involving qualitative criteria for the choice of transportation alternatives in Delhi. In addition, 4-stroke 2-wheelers, CNG cars, and CNG buses were ranked as the three transportation options associated with the six criteria: adaptability, barrier, cost, energy, environment, and technology. Next, Aydın and Kahraman [3] specified the criteria under three main economic, technological, and social categories to choose between nine buses with different features. They used the analytic hierarchy process (AHP) and VIekriterijumsko KOmpromisno Rangiranje (VIKOR) methods for this particular purpose. Along the same lines, Camargo Pérez et al. [4] provided a comprehensive review of published papers between 1982 and May 2014 dealing with the decision-making framework for the urban passenger transportation system, including design and operation issues. Then, Bai et al. [5] introduced a sustainable transportation fleet assessment framework dealing with economic, environmental, and different vehicle performance criteria. Also, they presented a hybrid approach to evaluate and select the sustainable transportation vehicle when integrating an interval grey number with a three-parameter using a rough set approach and the VIKOR technique. Afterward, Barbosa et al. [6] introduced a multi-criteria approach to evaluate urban public transportation based on the perceptions of users. Moreover, the objective and subjective parameters were specified associated with the notions of users for the integrated public transportation system in Florianópolis and other cities in Brazil. Next, Ercan et al. [7]

examined the transportation mode selection behavior for public transportation ridership. They also highlighted that mode choice behavior was affected by gasoline fuel sale taxes. It should be noted that they did not use the decision-making theory in their study.

Sustainable transportation alternatives are significant for environmental and social aspects. Notably, Büyüközkan et al. [8] presented a research study focusing on the choice of public bus technologies for sustainable transportation options. They also applied an intuitionistic fuzzy Choquet integral with a decision-making technique. Next, Saplıoğlu and Aydın [9] implemented the AHP technique to examine survey data from four hundred sixty participants and determined the efficient parameters of safe and serviceable bicycle routes when integrating cycling and public transport. On the other hand, researchers have paid attention to advanced transportation modes. In particular, Lee [10] presented a decision-making framework to rank the alternative advanced transportation modes, such as bus rapid transit, light rail transit, and bimodal tram, for new towns in Korea. Furthermore, the insights were provided for urban public transportation in Korea. Then, Jasti and Ram [11] used the AHP with fuzzy logic framework and direct weighting techniques to integrate a priority-related weighting system into benchmarking the Mumbai metro rail system. Moreover, the performance of the Mumbai metro rail system was investigated for multimodal integration. After that, Nalmpantis et al. [12] used the AHP technique to determine the weights of the three criteria: feasibility, innovativeness, and utility for public transportation. Furthermore, they drew a conclusion about which of the innovations were important and acquired recommendations for more attractive public transport.

The integrated public transit system is a crucial parameter to enhance public transit efficiency and usage. Thus, Errampalli et al. [13] presented a methodology to evaluate three policies when combining metro and bus services in Delhi. They also examined sustainability indicators to calculate the integration level. Next, Seker and Aydin [14] used interval-valued intuitionistic fuzzy AHP and combinative distance-based assessment (CODAS) techniques to evaluate public transportation options, such as automated guideway transit, battery electric buses, personal rapid transit, and trams. Also, they presented a case study on campus for 36,000 users. Along the same lines, Alkharabsheh et al. [15] introduced a grey theory-based AHP to rank the criteria of supply quality of the public transportation system for a case study of Amman City, Jordan, when dealing with non-expert respondents. Also, fare, service quality, tractability, and transportation quality were specified as the four main criteria. Afterward, Dahlgren and Ammenberg [16] applied the multi-criteria method to various bus technologies in the Swedish case and evaluated buses based on biomethane, diesel, electricity, ethanol, hydrotreated vegetable oil, fatty acid methyl ester, and natural gas. Furthermore, Görçün [17] integrated the criteria importance through intercriteria correlation (CRITIC) and the evaluation based on distance from average solution (EDAS) techniques to assess choosing urban rail vehicles for the part of public transportation. Then, Romero-Ania et al. [18] combined the ELECTRE TRI and the DELPHI algorithms in order to evaluate and classify public buses. The alternatives were determined as diesel, diesel hybrid, GNC, induction electric, and plug-in electric, where the costs and emissions criteria were specified. Along the same lines, Canbulut et al. [19] studied the tramway selection problem for a public transportation firm in Türkiye. For this purpose, they used the AHP to obtain the weights of the specified nine criteria and the grey relationship analysis to determine the best alternative among eight options.

Recently, Çelikbilek et al. [20] adopted a grey model of the best–worst method, AHP, and multi-objective optimization ratio analysis for evaluating the Budapest public transportation system. Also, they concluded that acquiring new buses was the first alternative to enhance urban transportation quality. Then, Derbel and Boujelbene [21] developed an assessment model to enhance the competitiveness of public transportation operators. In addition, economic efficiency, effectiveness, financial efficiency, pertinence, and service quality were denoted as the performance criteria by the authors. Further, Zhang et al. [22] employed a unified prior evaluation technique for guiding public transportation planning activities without using observational data. Along the same lines, Borghetti et al. [23] proposed an integrated method with AHP, ELimination Et Choix Traduisant la REalitè I (ELECTRE I), and a simple weighted sum model (SWSM) for choosing alternative fuels for a bus fleet. They also consider cost, environment, and lifecycle criteria when presenting an Italian case study. Next, Dilian et al. [24] investigated the effects of public transportation on the health of older adults. Furthermore, Kundu et al. [25] suggested fuzzy-based best–worst and fuzzy-based multi-attribute ideal–real comparative analysis methods to assess six transportation alternatives, such as bus rapid transport, commuter trains, light rail trams, metro, public buses, and trams associated with eleven specified criteria when consulting experts. Moreover, Zehra and Wong [26] systematically reviewed the four areas, such as infrastructure, transportation modes, networks, and planning, for considering wildfire evacuations. Then, Aghaabbasi and Sabri [27] reviewed the potential of digital twin approaches to obtain a holistic travel behavior representation for supporting planning activities and policymaking. Further, Alhassan and Anciaes [28] examined the literature to evaluate the effects of public transportation investments or disinvestments based on economic and social levels. Next, Lyu et al. [29] analyzed the data from two hundred fifteen Chinese cities between 2013 and 2018 to investigate whether urban sprawl may impact carbon emissions and public transportation efficiency. Lastly, Nilsson et al. [30] examined the decision-making mobility of senior citizens dealing with public transportation in three metropolitan areas in Sweden, and they observed that senior citizens want to see less car traffic.

Table 1 summarizes the relevant literature studies. As observed in Table 1, this study covers the thirteen criteria, involving economics, safety, travel quality, and environmental and health aspects, when evaluating the seven sustainable and traditional public transportation modes. Moreover, the presented five-phased technique in this study is comprehensive and novel, including AHP, TOPSIS, and VIKOR.

Table 1. Brief literature review on public transportation selection.

Studied by	Criteria	Alternatives	Techniques
Yedla et al. [2]	Both qualitative and quantitative criteria	4-stroke 2-wheelers, CNG cars, and CNG buses	Multi-criteria approach
Aydın and Kahraman [3]	Economic, technological, and social categories	Nine buses with different features	AHP and VIKOR
Bai et al. [5]	Economic, environmental, and different vehicle performance criteria	Transport vehicle fleet (ten vehicles with different attributes)	A rough set approach and VIKOR
Büyüközkan et al. [6]	Economical, technical, environmental, and social	Public bus technologies	Intuitionistic fuzzy Choquet integral with a decision-making technique
Saplıoğlu and Aydın [9]	Parameters for bicycle route selection	Safe and serviceable bicycle routes when integrating cycling and public transport	AHP
Lee [10]	Provider and user as main criteria for metropolis and small- and medium-scale cities	A railway-type public transport, a dual mode of tram and bus, and a bus transit system	A multi-criteria approach
Jasti and Ram [11]	Performance indicators	Metro rail system	AHP and fuzzy logic

Table 1. *Cont.*

Studied by	Criteria	Alternatives	Techniques
Nalmpantis et al. [12]	Feasibility, utility, and innovativeness	Four lists of innovations were derived and ranked.	AHP
Errampalli et al. [13]	Economic, social, and environmental	Metro rail and bus	Multi-criteria analysis
Seker and Aydin [14]	Economic, environmental, usage, safety, and technical conditions	Automated guideway transit, battery electric buses, personal rapid transit, and trams	Interval-valued intuitionistic fuzzy AHP and CODAS
Alkharabsheh et al. [15]	The criteria of supply quality of the public transportation system	Rank the criteria of supply quality of the public transportation system.	A grey theory-based AHP
Dahlgren and Ammenberg [16]	The costs and emissions criteria	Bus technologies	A multi-criteria method
Görçün [17]	Performance indicators for twenty-two criteria	Urban rail vehicles	CRITIC and EDAS
Romero-Ania et al. [18]	The costs and emissions criteria	Classify public buses.	ELECTRE TRI and DELPHI
Canbulut et al. [19]	Performance indicators for nine criteria	Tramway selection	AHP and grey relationship analysis
Çelikbilek et al. [20]	Performance indicators	Buses	A grey model of the best–worst method, AHP, and multi-objective optimization ratio
Borghetti et al. [23]	Cost, environment, and lifecycle criteria	Alternative fuels for a bus fleet	AHP, ELECTRE I, and SWSM
Kundu et al. [25]	Eleven specified criteria	Bus rapid transport, commuter trains, light rail trams, metro, public buses, and trams	Fuzzy-based best–worst and fuzzy-based multi-attribute ideal–real comparative analysis methods
This study	The thirteen specified criteria, including economics, safety, travel quality, and environmental and health aspects	The seven public transportation alternatives, including sustainable and traditional transportation modes	A five-phased novel decision analysis framework, including AHP, TOPSIS, and VIKOR

1.2. Research Contributions and Objectives

Public transportation is a comprehensive and extensive process that involves many factors and requires much planning effort. This study aims to select public transportation modes based on the travel behavior decisions of passengers when providing efficiency and enhancing planning activities. Moreover, the contributions of this study are summarized as follows. First, most relevant research studies about public transportation use the AHP technique to prioritize the criteria involving economic and non-polluting aspects. This paper proposes an integrated multi-criteria decision-making process that includes AHP, Technique for Order of Preference by Similarity to Ideal Solution (TOPSIS), and VIKOR techniques. The AHP part is used to determine the weights of the criteria based on the travel behavior decisions of passengers for the public transportation mode selection problem. Also, the TOPSIS and VIKOR methods are applied to rank the alternative transportation modes. The comparison analyses are conducted. Secondly, the thirteen criteria are specified based on published reports, papers, and expert opinions to analyze passengers' travel behavior decisions. The seven alternative public transportation options, including sustainable and traditional vehicles, are considered. Third, this study is performed in a megacity, including different types of transportation modes when discussing sustainable development goals (SDGs). In addition, theoretical and managerial implications are provided for the policymakers.

The remainder of this work is presented in the following way. First, the integrated decision-making holistic framework is proposed in Section 2 with the problem definition, criteria, and alternatives. Next, a case study of Istanbul, Türkiye, is presented in Section 3. Then, the managerial and theoretical implications of the public transportation mode selection problem are provided in Section 4. Lastly, concluding remarks and potential future studies are summarized in Section 5.

2. Problem Definition, Criteria, and Proposing an Integrated Multi-Criteria Decision-Making Method

2.1. Problem Definition

Cities are complicated spatial structures with infrastructures such as transportation systems. Significant transportation problems take place when transportation systems may not appropriately fulfill the necessities of urban mobility. Notably, public transportation systems are either over- or underused because of the passengers' demand associated with the public transportation alternatives. For example, crowdedness may cause discomfort for passengers and affect travel behavior and the selection of transportation modes during peak hours. Hence, adequate transit infrastructure should be provided that is associated with passengers' expectations. Moreover, the public transportation system may not generate adequate income to keep the operations and costs financially sustainable without considering passengers' expectations and travel behavior decisions. Also, the financial burden should be considered by municipalities when investing in a public transportation system. For all these reasons, the selection of public transportation alternatives is a crucial problem in providing efficiency, decreasing costs, and reducing the environmental effect and traffic congestion when meeting the passengers' needs. So, the decision-making framework is useful for selecting suitable transportation alternatives when investigating the travel behavior decisions of passengers.

2.2. Specified Criteria and Alternatives

When examining the travel behavior decisions of passengers to specify the criteria in this paper, the published reports and papers in the literature [2–30] were reviewed, and experts' notions (five decision-makers) were asked. Thirteen criteria are specified in this paper when considering the published reports, papers, and experts' notions. Also, Table 1 shows the differences between the previously published works and this study. In this paper, the decision-makers consist of five engineers working on transportation planning. Each decision-maker has at least two years of work experience. Also, each decision-maker has a bachelor of science degree.

The thirteen criteria are explained as follows:

Pricing (C1): This criterion is related to the economic aspect. Each public transportation alternative has a different fee, and the pricing varies according to the distance. In addition, the monthly subscription opportunity offered for vehicles used with Istanbul Card reduces the fee for each ride in the case study of this paper.

Speed (C2): This criterion is associated with the aspect of time management. Passengers want to reach a location in a minimum time, so the speed of the vehicle is an important criterion. Notice that some transportation options may be slow due to the technical requirements or transportation infrastructure.

Ease of accessibility (C3): This criterion is based on the infrastructure aspect of public transportation. Passengers desire to experience ease of accessibility and save time while using public transportation.

Number of transfers (C4): This criterion is related to the aspect of passengers' travel behavior. Passengers desire to complete their ride with the fewest number of transfers while traveling. Although transfers sometimes provide benefits in terms of time, they may not be preferred in terms of cost.

Crowdedness (C5): This criterion is related to the aspect of passengers' travel behavior. During peak hours, the efficiency and quality of public transportation decrease due to crowdedness. Operating vehicles at optimum capacity ensures better service.

Security (C6): This criterion is associated with the safety concern. Since public transportation brings many people together, passengers want to complete their trip safely against possible problems. Also, sufficient security measures can reduce crime rates in public transportation.

Air conditioning (C7): This criterion is relevant to the technical conditions of public transportation vehicles that affect the travel behavior of passengers. Air conditioning in public transportation includes heating, cooling, and ventilation operations. Air conditioning provides passengers with a comfortable environment, cleans the air inside, and provides a healthier travel opportunity.

Vehicle type and its mechanism (C8): This criterion denotes a technical concern of public transportation alternatives that impact the travel behavior of passengers. Different types of vehicles can be used in public transportation. Some passengers may not prefer to travel on buses or minibuses because of sudden braking or shaking, which causes nausea and inability to stand. Due to environmental awareness, some passengers may prefer sustainable transportation alternatives.

Service Quality (C9): This criterion is directly related to the quality of the service for passengers. Making the necessary announcements in the vehicle and informing about the next stop or the stops where transfers can be made increases the service quality.

Service frequency (C10): This criterion is related to the services that directly affect the passengers' decisions. High service frequencies make public transportation more accessible for passengers by reducing waiting time.

Noise (C11): This criterion is related to health and environmental concerns. The noise of vehicles is not desirable for passengers. Also, high noise levels negatively affect human health and cause environmental noise pollution.

Service comfort (C12): This criterion is based on the service of the transportation alternative. Passengers desire to increase their comfort levels when using public transportation. For instance, seated passengers can travel safely in the case of sudden stops or movements of the vehicle. In addition, public transportation alternatives with seating areas for pregnant, elderly, and disabled individuals increase the accessibility of these individuals to public transport.

Phobia (C13): This criterion specifies a psychological effect of passengers' travel behavior decisions. Some people may fear one of the public transportation modes, such as sea services, metro, etc. For these reasons, people may turn to different public transportation options.

Sustainable and traditional public transportation modes can be determined as various alternatives for evaluating the public transportation systems of cities when investigating the travel behavior decisions of passengers. In this case study, the seven alternatives—bus, ferry, Marmaray, metro, metrobus, minibus, and tram—are specified based on the serving population of the metropolitan city. Notice that a few alternatives, such as ferry and minibus, in the case study may not be valid in other urban cities. Moreover, it is essential to determine the appropriate alternatives for evaluating the public transportation system used by people in a city.

2.3. An Integrated Multi-Criteria Decision-Making Method

An integrated decision-making framework consists of five main phases: (1) planning, (2) AHP, (3) TOPSIS, (4) VIKOR, and (5) concluding phases. First of all, the planning phase includes determining the experts, criteria, and alternatives for the problem. Second, the AHP technique is applied to prioritize the specified criteria. The geometric mean is utilized as an aggregation operator in the AHP technique. Indeed, this aggregation approach is suitable for the meaning of judgments and priorities [31]. Also, surveys are conducted for passengers, asking about their preferences. Third, the TOPSIS method is employed to rank the specified alternatives. Fourth, the VIKOR method is also applied to rank the determined alternatives. Finally, the TOPSIS and VIKOR methods are compared, and managerial recommendations are drawn for policymakers. The details of the five main phases are provided below.

Phase 1: The planning phase. This phase consists of five sub-phases, and it includes defining the problem, determining the group of experts, determining the criteria, and determining the alternatives.

Phase 1.1. Start the planning phase.

Phase 1.2. Define the problem. The choice of public transportation modes is a significant problem in increasing efficiency, decreasing costs, and eliminating adverse environmental effects and traffic congestion when satisfying the passengers' requirements.

Phase 1.3. Determine the group of experts. The decision-makers should be knowledgeable and experienced in the studied problem.

Phase 1.4. Determine the criteria based on the published papers, reports, and experts' notions. The specified criteria reflect the economic, safety, travel quality, environmental, and health aspects of the problem to investigate the travel behavior decisions of passengers.

Phase 1.5. Specify the alternatives. The alternatives are determined based on public transportation modes serving the population of a city. Then, go to Phase 2.

Phase 2: The AHP and data collection phase. The second phase involves seven sub-phases, and it is related to how the AHP procedure is executed to obtain weights of the specified criteria. Moreover, data is collected by surveying passengers.

Phase 2.1. Begin the AHP method.

Phase 2.2. Generate pairwise comparison matrices (*PCMs*) for the *k*th decision-maker (DM) using Equation (1). Each decision-maker is asked about the preferences individually.

$$PCM_k = \begin{pmatrix} 1 & a_{12} & \ldots & a_{1n} \\ a_{21} = 1/a_{12} & 1 & \ldots & a_{2n} \\ \ldots & \ldots & \ldots & \ldots \\ a_{n1} = 1/a_{1n} & a_{n2} = 1/a_{2n} & \ldots & 1 \end{pmatrix}_{n \times n} \quad (1)$$

where a_{ij} denotes the scale of importance between the *i* and *j* criteria.

Phase 2.3. Obtain an aggregated pairwise comparison matrix (*APCM*) using the geometric mean as follows.

$$APCM = \begin{pmatrix} ag_{11} = (1_1 \times \ldots \times 1_k)^{1/k} & ag_{12} = (a_{12_1} \times \ldots \times a_{12_k})^{1/k} & \ldots & ag_{1n} = (a_{1n_1} \times \ldots \times a_{1n_k})^{1/k} \\ ag_{21} = (1/a_{12_1} \times \ldots \times 1/a_{12_k})^{1/k} & ag_{22} = (1_1 \times \ldots \times 1_k)^{1/k} & \ldots & ag_{2n} = (a_{2n_1} \times \ldots \times a_{2n_k})^{1/k} \\ \ldots & \ldots & \ldots & \ldots \\ ag_{n1} = (1/a_{1n_1} \times \ldots \times 1/a_{1n_k})^{1/k} & ag_{n2} = (1/a_{2n_1} \times \ldots \times 1/a_{2n_k})^{1/k} & \ldots & ag_{nn} = (1_1 \times \ldots \times 1_k)^{1/k} \end{pmatrix}_{n \times n} \quad (2)$$

where ag_{ij} represents the geometric mean of each decision-maker's scale of importance between the *i* and *j* criteria.

Phase 2.4. Normalize the *APCM* in the following way.

$$ag'_{ij} = ag_{ij} / \sum_{i=1}^{n} ag_{ij} (i, j = 1, 2, \ldots, n) \quad (3)$$

Phase 2.5. Find the *i*th criterion weight, wgt_i, by taking the sum of each row of the normalized APCM and dividing it by the matrix size.

$$wgt_i = (1/n) \sum_{i=1}^{n} ag'_{ij} (i, j = 1, 2, \ldots, n) \quad (4)$$

Phase 2.6. Check the consistency ratio (*CR*) as follows:

$$CR = CI/RI \quad (5)$$

$$CI = (\text{Principal Eigen Value} - n)/(n-1) \quad (6)$$

where *CI* and *RI* denote the consistency index and random index, respectively. If the value of *CR* is less than or equal to 0.10, the APCM is consistent. Notice that Tavana et al. [32] expressed the *RI* values for the AHP technique.

Phase 2.7. Survey the passengers who frequently use public transportation, asking for public transportation preferences. Next, go to Phase 3.

Phase 3: The TOPSIS phase. This phase includes seven sub-phases. The mathematical equations and details of the TOPSIS procedure are provided to rank the alternatives.

Phase 3.1. Begin the TOPSIS method.

Phase 3.2. A multi-criteria decision analysis matrix (*MCDAM*) is obtained by using the collected data from the surveys as follows:

$$MCDAM = [f_{ij}]_{n \times m} \quad (7)$$

where f_{ij} denotes a collected value for the *i*th alternative on the *j*th criterion.

Phase 3.3. The normalized *MCDAM* is calculated as follows:

$$fnormalized_{ij} = \frac{f_{ij}}{\sqrt{\sum_{i=1}^{n} f_{ij}^2}} \quad (8)$$

where $fnormalized_{ij}$ denotes the normalized value.

Phase 3.4. The positive, abbreviated as *Pideal*, and negative, abbreviated as *Nideal*, ideal solutions are found in the following way.

$$\begin{aligned} &Pideal = [fnormalized_1^+, \ldots, fnormalized_m^+] \\ &Nideal = [fnormalized_1^-, \ldots, fnormalized_m^-] \\ &\text{where } fnormalized_j^+ = \begin{cases} \max fnormalized_{ij} \text{ for benefit} \\ \min fnormalized_{ij} \text{ for cost} \end{cases} \\ &fnormalized_j^- = \begin{cases} \min fnormalized_{ij} \text{ for benefit} \\ \max fnormalized_{ij} \text{ for cost} \end{cases} \end{aligned} \quad (9)$$

Phase 3.5. The weighted Euclidean distances are obtained associated with the positive and negative ideal solutions, abbreviated as $Epos_i^+$ and $Epos_i^-$, respectively, below.

$$\begin{aligned} Epos_i^+ &= \sqrt{\sum_{j=1}^{n} wgt_j \left(f_{ij} - fnormalized_j^+\right)^2} \\ Epos_i^- &= \sqrt{\sum_{j=1}^{n} wgt_j \left(f_{ij} - fnormalized_j^-\right)^2} \end{aligned} \quad (10)$$

Phase 3.6. The *i*th total performance score (TPS_i) is calculated for the *i*th alternative as follows:

$$TPS_i = {\left(Epos_i^-\right)}/{\left(Epos_i^- + Epos_i^+\right)} \quad (11)$$

Phase 3.7. The alternatives are ranked based on the TPSs. Then, go to Phase 4.

Phase 4: The VIKOR phase. This phase consists of six sub-phases. The mathematical equations and details of the VIKOR procedure are presented to rank the alternatives.

Phase 4.1. Begin the VIKOR method.

Phase 4.2. Use the MCDAM in Equation (7).

Phase 4.3. *Pideal* and *Nideal* are found for the VIKOR technique as follows.

$$Pideal_i = \max f_{ij} \text{ and } Nideal_i = \min f_{ij} \ (i = 1, 2, \ldots, n \text{ and } j = 1, 2, \ldots, m) \quad (12)$$

where *Pideal* and *Nideal* represent positive and negative ideal solutions, respectively.

Phase 4.4. *Sj* and *Rj* values are calculated in the following way.

$$\begin{aligned} S_j &= \sum_{i=1}^{n} wgt_i (Pideal_i - f_{ij}) / (Pideal_i - Nideal_i) \\ R_j &= \max_i wgt_i (Pideal_i - f_{ij}) / (Pideal_i - Nideal_i) \end{aligned} \quad (13)$$

Phase 4.5. *Qj* values are computed as follows:

$$\begin{aligned} &Q_j = \frac{q(S_j - S^+)}{(S^- + S^+)} + \frac{(1-q)(R_j - S^+)}{(R^- + R^+)} \\ &\text{where } S^+ = \min_i S_j, S^- = \max_i S_j, R^+ = \min_i R_j, R^- = \max_i R_j, \text{ and } 0 \le q \le 1 \end{aligned} \quad (14)$$

Phase 4.6. The specified alternatives are ranked based on the ascending order of the values. Next, go to Phase 5.

Phase 5: The concluding phase. This phase involves four sub-phases. A conclusion is drawn in this phase when comparing the TOPSIS and VIKOR methods. Moreover, the results are validated by previous works and experts' notions. Next, a sensitivity analysis is carried out to assess the stability and robustness of the presented case study. Furthermore, managerial insights are provided for policymakers.

Phase 5.1. Draw a conclusion when comparing the TOPSIS and VIKOR method results. This phase is important for providing theoretical recommendations about the methods because the different multi-criteria methods may give different results for a case study.

Phase 5.2. Verify the results from the published reports, papers, and experts' notions.

Phase 5.3. Perform the sensitivity analysis. A sensitivity analysis is conducted to check the stability and robustness of the case study.

Phase 5.4. Provide managerial recommendations associated with the results for policymakers.

3. A Case Study of Istanbul

3.1. The Planning Phase of the Case Study

The public transportation network in Istanbul, Türkiye, is spread over an area of 5712 km^2. It has a wide coverage area, including trams, metros, buses, minibuses, metrobuses, ferries, and Marmaray, serving a population of 15,701,602 people [1]. In this paper, a case study is performed for the public transportation in Istanbul, Türkiye. The five experts' opinions are asked individually in online meetings to generate PCMs in the first week of January 2024, as shown in Tables 2–6. Also, surveys were conducted voluntarily and individually between January and August 2024 through online meetings with one thousand five hundred people living in Istanbul, Türkiye, who use public transportation. Figure 1 shows the flowchart of the proposed integrated decision-making framework. Moreover, Figure 2 shows the map of Istanbul, Türkiye [33].

Table 2. The PCM for the first DM.

	Criteria												
DM1	**1**	**2**	**3**	**4**	**5**	**6**	**7**	**8**	**9**	**10**	**11**	**12**	**13**
1	1.000	2.000	0.500	0.500	0.500	2.000	1.000	1.000	2.000	0.500	2.000	1.000	2.000
2	0.500	1.000	0.333	0.333	0.333	1.000	0.500	0.500	1.000	0.333	1.000	0.500	1.000
3	2.000	3.000	1.000	1.000	1.000	3.000	2.000	2.000	3.000	1.000	3.000	2.000	3.000
4	2.000	3.000	1.000	1.000	1.000	3.000	2.000	2.000	3.000	1.000	3.000	2.000	3.000
5	2.000	3.000	1.000	1.000	1.000	3.000	2.000	2.000	3.000	1.000	3.000	2.000	3.000
6	0.500	1.000	0.333	0.333	0.333	1.000	0.500	0.500	1.000	0.333	1.000	0.500	1.000
7	1.000	2.000	0.500	0.500	0.500	2.000	1.000	1.000	2.000	0.500	2.000	1.000	2.000
8	1.000	2.000	0.500	0.500	0.500	2.000	1.000	1.000	2.000	0.500	2.000	1.000	2.000
9	0.500	1.000	0.333	0.333	0.333	1.000	0.500	0.500	1.000	0.333	1.000	0.500	1.000
10	2.000	1.000	1.000	1.000	1.000	3.000	2.000	2.000	3.000	1.000	3.000	1.000	3.000
11	0.500	1.000	0.333	0.333	0.333	1.000	0.500	0.500	11.000	0.333	1.000	0.500	1.000
12	1.000	2.000	0.500	0.500	0.500	2.000	1.000	1.000	2.000	1.000	2.000	1.000	2.000
13	0.500	1.000	0.333	0.333	0.333	1.000	0.500	0.500	1.000	0.333	1.000	0.500	1.000

Table 3. The PCM for the second DM.

	Criteria												
DM2	**1**	**2**	**3**	**4**	**5**	**6**	**7**	**8**	**9**	**10**	**11**	**12**	**13**
1	1.000	1.000	1.000	1.000	5.000	5.000	9.000	5.000	7.000	1.000	7.000	7.000	7.000
2	1.000	1.000	1.000	1.000	5.000	5.000	9.000	5.000	7.000	1.000	7.000	7.000	7.000
3	1.000	1.000	1.000	1.000	5.000	5.000	9.000	5.000	7.000	1.000	7.000	7.000	7.000
4	1.000	1.000	1.000	1.000	1.000	5.000	9.000	5.000	7.000	1.000	7.000	7.000	7.000
5	0.200	0.200	0.200	1.000	1.000	1.000	5.000	1.000	3.000	0.200	3.000	3.000	3.000
6	0.200	0.200	0.200	0.200	1.000	1.000	5.000	1.000	3.000	0.200	3.000	3.000	3.000
7	0.111	0.111	0.111	0.111	0.200	0.200	1.000	0.200	0.333	0.111	0.333	0.333	0.333
8	0.200	0.200	0.200	0.200	1.000	1.000	5.000	1.000	3.000	0.200	3.000	3.000	3.000
9	0.143	0.143	0.143	0.143	0.333	0.333	3.000	0.333	1.000	0.143	1.000	1.000	1.000
10	1.000	1.000	1.000	1.000	5.000	5.000	9.000	5.000	7.000	1.000	7.000	7.000	7.000
11	0.143	0.143	0.143	0.143	0.333	0.333	3.000	0.333	11.000	0.143	1.000	1.000	1.000
12	0.143	0.143	0.143	0.143	0.333	0.333	3.000	0.333	1.000	0.143	1.000	1.000	1.000
13	0.143	0.143	0.143	0.143	0.333	0.333	3.000	0.333	1.000	0.143	1.000	1.000	1.000

Table 4. The PCM for the third DM.

	Criteria												
DM3	**1**	**2**	**3**	**4**	**5**	**6**	**7**	**8**	**9**	**10**	**11**	**12**	**13**
1	1.000	1.000	0.250	0.200	4.000	0.250	0.333	0.200	0.200	0.200	0.200	0.333	0.250
2	1.000	1.000	0.250	0.200	4.000	0.250	0.333	0.200	0.200	0.200	0.200	0.333	0.250
3	4.000	4.000	1.000	0.500	7.000	1.000	2.000	0.500	0.500	0.500	0.500	2.000	1.000
4	5.000	5.000	2.000	1.000	8.000	2.000	3.000	1.000	1.000	1.000	1.000	3.000	2.000
5	0.250	0.250	0.143	0.125	1.000	0.143	0.167	0.125	0.125	0.125	0.125	0.167	0.143
6	4.000	4.000	1.000	0.500	7.000	1.000	2.000	0.500	0.500	0.500	0.500	2.000	1.000
7	3.000	3.000	0.500	0.333	6.000	0.500	1.000	0.333	0.333	0.333	0.333	1.000	0.500
8	5.000	5.000	2.000	1.000	8.000	2.000	3.000	1.000	1.000	1.000	1.000	3.000	2.000
9	5.000	5.000	2.000	1.000	8.000	2.000	3.000	1.000	1.000	1.000	1.000	3.000	2.000
10	5.000	2.000	2.000	1.000	8.000	2.000	3.000	1.000	1.000	1.000	1.000	3.000	2.000
11	5.000	5.000	2.000	1.000	8.000	2.000	3.000	1.000	11.000	1.000	1.000	3.000	2.000
12	3.000	3.000	0.500	0.333	6.000	0.500	1.000	0.333	0.333	0.333	0.333	1.000	0.500
13	4.000	4.000	1.000	0.500	7.000	1.000	2.000	0.500	0.500	0.500	0.500	2.000	1.000

Table 5. The PCM for the fourth DM.

	Criteria												
DM4	**1**	**2**	**3**	**4**	**5**	**6**	**7**	**8**	**9**	**10**	**11**	**12**	**13**
1	1.000	2.000	1.000	1.000	1.000	1.000	1.000	1.000	1.000	1.000	1.000	4.000	6.000
2	0.500	1.000	0.500	0.500	0.500	0.500	0.500	0.500	0.500	0.500	0.500	3.000	5.000
3	1.000	2.000	1.000	1.000	1.000	1.000	1.000	1.000	1.000	1.000	1.000	4.000	6.000
4	1.000	2.000	1.000	1.000	1.000	1.000	1.000	1.000	1.000	1.000	1.000	4.000	6.000
5	1.000	2.000	1.000	1.000	1.000	1.000	1.000	1.000	1.000	1.000	1.000	4.000	6.000
6	1.000	2.000	1.000	1.000	1.000	1.000	1.000	1.000	1.000	1.000	1.000	4.000	6.000
7	1.000	2.000	1.000	1.000	1.000	1.000	1.000	1.000	1.000	1.000	1.000	4.000	6.000
8	1.000	2.000	1.000	1.000	1.000	1.000	1.000	1.000	1.000	1.000	1.000	4.000	6.000
9	1.000	2.000	1.000	1.000	1.000	1.000	1.000	1.000	1.000	1.000	1.000	4.000	6.000
10	1.000	1.000	1.000	1.000	1.000	1.000	1.000	1.000	1.000	1.000	1.000	4.000	6.000
11	1.000	2.000	1.000	1.000	1.000	1.000	1.000	1.000	11.000	1.000	1.000	4.000	6.000
12	0.250	0.333	0.250	0.250	0.250	0.250	0.250	0.250	0.250	0.250	0.250	1.000	3.000
13	0.167	0.200	0.167	0.167	0.167	0.167	0.167	0.167	0.167	0.167	0.167	0.333	1.000

Table 6. The PCM for the fifth DM.

	Criteria												
DM5	**1**	**2**	**3**	**4**	**5**	**6**	**7**	**8**	**9**	**10**	**11**	**12**	**13**
1	1.000	5.000	5.000	6.000	5.000	4.000	2.000	1.000	5.000	3.000	5.000	4.000	5.000
2	0.200	1.000	1.000	2.000	1.000	0.500	0.250	0.200	1.000	0.333	1.000	0.500	1.000
3	0.200	1.000	1.000	2.000	1.000	0.500	0.250	0.200	1.000	0.333	1.000	0.500	1.000
4	0.167	0.500	0.500	1.000	0.500	0.333	0.200	0.167	0.500	0.250	0.500	0.333	0.500
5	0.200	1.000	1.000	2.000	1.000	0.500	0.250	0.200	1.000	0.333	1.000	0.500	1.000
6	0.250	2.000	2.000	3.000	2.000	1.000	0.333	0.250	2.000	0.500	2.000	1.000	2.000
7	0.500	4.000	4.000	5.000	4.000	3.000	1.000	0.500	4.000	2.000	4.000	3.000	4.000
8	1.000	5.000	5.000	6.000	5.000	4.000	2.000	1.000	5.000	3.000	5.000	4.000	5.000
9	0.200	1.000	1.000	2.000	1.000	0.500	0.250	0.200	1.000	0.333	1.000	0.500	1.000
10	0.333	3.000	3.000	4.000	3.000	2.000	0.500	0.333	3.000	1.000	3.000	2.000	3.000
11	0.200	1.000	1.000	2.000	1.000	0.500	0.250	0.200	11.000	0.333	1.000	0.500	1.000
12	0.250	2.000	2.000	3.000	2.000	1.000	0.333	0.250	2.000	0.500	2.000	1.000	2.000
13	0.200	1.000	1.000	2.000	1.000	0.500	0.250	0.200	1.000	0.333	1.000	0.500	1.000

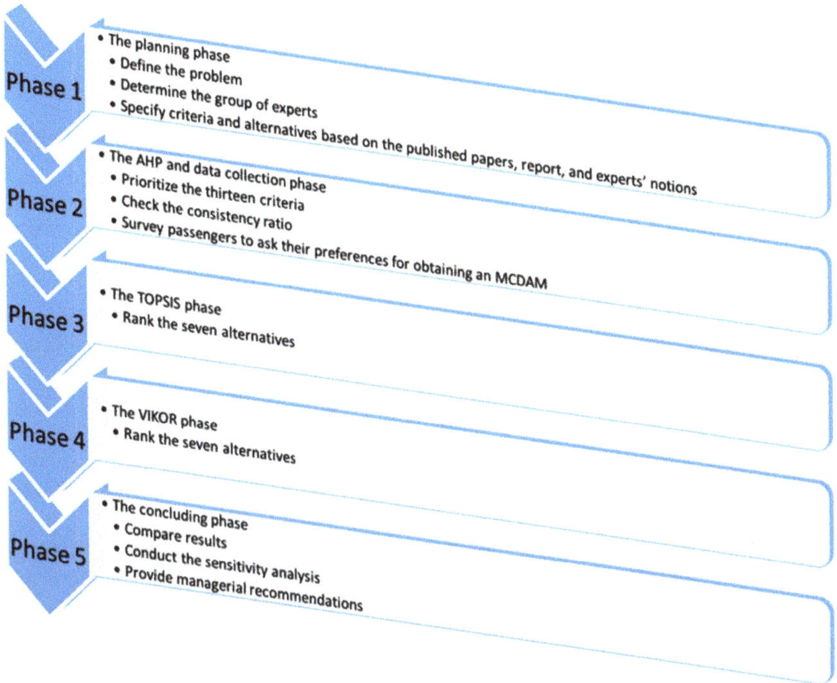

Figure 1. The flowchart of the proposed integrated decision-making framework.

Figure 2. Map of Istanbul, Türkiye [33].

The seven alternatives as part of the planning phase of the proposed methodology are determined for the case study in this paper as follows:

Bus (A1): It denotes one of the public transportation alternatives when using buses.

Ferry (A2): It represents passengers carried by a boat on a body of water.

Marmaray (A3): It denotes a 76.6-kilometre-long commuter rail line in Istanbul, Türkiye.

Metro (A4): It denotes rapid transit, a type of high-capacity public transport.

Metrobus (A5): It denotes a 52 km (32.3 mi) bus rapid transit route in Istanbul, Türkiye.

Minibus (A6): It represents one of the public transportation modes constructed to carry between 9 and 16 passengers.

Tram (A7): It denotes a type of urban rail transit.

Marmaray, metro, and tram provide sustainability and offer energy-efficient, low-emission alternatives while decreasing traffic congestion and air pollution. So, Marmaray, metro, and tram are defined as sustainable transportation modes. On the other hand, in this paper, buses, ferries, metrobuses, and minibuses are specified as traditional transportation modes.

3.2. Results and Discussion for Prioritizing the Thirteen Criteria Employing the AHP and Data Collection Phase

The AHP results are obtained using Phases 2.1–2.6 to prioritize the thirteen criteria. The PCMs are presented in Tables 2–6 for each decision-maker (DM) using Equation (1).

Next, the APCM in Equation (2) and the normalized APCM in Equation (3) are shown in Tables 7 and 8. Further, Table 9 shows the weights of the thirteen criteria using Equation (4). The CR value was found to be 0.036 when employing Equations (5) and (6). The APCM is consistent since the CR value is less than 0.10. Hence, the obtained weights in Table 9 are valid for the thirteen criteria.

Table 7. The APCM.

	Criteria												
APCM	**1**	**2**	**3**	**4**	**5**	**6**	**7**	**8**	**9**	**10**	**11**	**12**	**13**
1	1.000	1.821	0.910	0.903	2.187	1.585	1.431	1.000	1.695	0.786	1.695	2.063	2.537
2	0.549	1.000	0.530	0.582	1.272	0.792	0.715	0.549	0.931	0.407	0.931	1.118	1.543
3	1.099	1.888	1.000	1.000	2.036	1.496	1.552	1.000	1.600	0.699	1.600	2.237	2.631
4	1.108	1.719	1.000	1.000	1.320	1.585	1.609	1.108	1.600	0.758	1.600	2.237	2.631
5	0.457	0.786	0.491	0.758	1.000	0.735	0.839	0.549	1.024	0.384	1.024	1.149	1.505
6	0.631	1.262	0.668	0.631	1.361	1.000	1.108	0.574	1.246	0.441	1.246	1.644	2.048
7	0.699	1.398	0.644	0.621	1.191	0.903	1.000	0.506	0.977	0.517	0.977	1.320	1.516
8	1.000	1.821	1.000	0.903	1.821	1.741	1.974	1.000	1.974	0.786	1.974	2.702	3.245
9	0.590	1.074	0.625	0.625	0.977	0.803	1.024	0.506	1.000	0.437	1.000	1.246	1.644
10	1.272	1.431	1.431	1.320	2.605	2.268	1.933	1.272	2.290	1.000	2.290	2.787	3.764
11	0.590	1.074	0.625	0.625	0.977	0.803	1.024	0.506	11.000	0.437	1.000	1.246	1.644
12	0.485	0.894	0.447	0.447	0.871	0.608	0.758	0.370	0.803	0.359	0.803	1.000	1.431
13	0.394	0.648	0.380	0.380	0.665	0.488	0.660	0.308	0.608	0.266	0.608	0.699	1.000

Table 8. The normalized APCM.

	Criteria												
Normalized APCM	**1**	**2**	**3**	**4**	**5**	**6**	**7**	**8**	**9**	**10**	**11**	**12**	**13**
1	0.101	0.108	0.093	0.092	0.120	0.107	0.092	0.108	0.063	0.108	0.101	0.096	0.093
2	0.056	0.059	0.054	0.059	0.070	0.054	0.046	0.059	0.035	0.056	0.056	0.052	0.057
3	0.111	0.112	0.103	0.102	0.111	0.101	0.099	0.108	0.060	0.096	0.096	0.104	0.097
4	0.112	0.102	0.103	0.102	0.072	0.107	0.103	0.120	0.060	0.104	0.096	0.104	0.097
5	0.046	0.047	0.050	0.077	0.055	0.050	0.054	0.059	0.038	0.053	0.061	0.054	0.055
6	0.064	0.075	0.069	0.064	0.074	0.068	0.071	0.062	0.047	0.061	0.074	0.077	0.075
7	0.071	0.083	0.066	0.063	0.065	0.061	0.064	0.055	0.037	0.071	0.058	0.062	0.056
8	0.101	0.108	0.103	0.092	0.100	0.118	0.126	0.108	0.074	0.108	0.118	0.126	0.120
9	0.060	0.064	0.064	0.064	0.053	0.054	0.066	0.055	0.037	0.060	0.060	0.058	0.061
10	0.129	0.085	0.147	0.135	0.143	0.153	0.124	0.138	0.086	0.137	0.137	0.130	0.139
11	0.060	0.064	0.064	0.064	0.053	0.054	0.066	0.055	0.411	0.060	0.060	0.058	0.061
12	0.049	0.053	0.046	0.046	0.048	0.041	0.048	0.040	0.030	0.049	0.048	0.047	0.053
13	0.040	0.039	0.039	0.039	0.036	0.033	0.042	0.033	0.023	0.037	0.036	0.033	0.037

Table 9. Weights of the thirteen criteria.

Criterion	wgt_i
Pricing (C1)	0.0987
Speed (C2)	0.0548
Ease of accessibility (C3)	0.1001
Number of transfers (C4)	0.0986
Crowdedness (C5)	0.0538
Security (C6)	0.0677
Air conditioning (C7)	0.0624
Vehicle type and its mechanism (C8)	0.1078
Service quality (C9)	0.0581
Service frequency (C10)	0.1293
Noise (C11)	0.0868
Service comfort (C12)	0.0460
Phobia (C13)	0.0359

In Table 9, service frequency (C10) is the most important criterion when analyzing the travel behavior decisions of passengers. Based on the results, the passengers desired more frequent services for public transportation. Indeed, the finding is consistent with the previous studies by Mouwen [34] and Brechan [35]. However, Kundu et al. [25] stated that frequency is the fifth most crucial criterion among the other eleven criteria in their work. The results show that the passengers' travel behaviors are significant in evaluating the public transportation system to satisfy passengers' needs. Next, the vehicle type and its mechanism (C8) is the second important criterion from the passengers' perspective. Note that passengers have paid attention to using sustainable transportation alternatives. Following this, ease of accessibility (C3) is the third significant criterion. Everyone can easily access the public transportation system. Also, the finding of the importance of the ease of accessibility criterion is in agreement with the study of Lee [10], Friman et al. [36], and Zhang et al. [22]. The pricing and the number of transfers are the fourth and fifth significant criteria, respectively. As indicated in the previous research by Brechan [35], the price reduction did not substantially affect the usage of the public transportation system. As shown in Table 9, the pricing criterion is not as important as the service frequency criterion. Also, the transfers should not be ignored in public transportation modes, as indicated by Errampalli et al. [13], and the weight of C4 verifies this statement. After this, noise is the sixth significant criterion in Table 9. Noise annoyance is a critical concern, and it should be reduced to increase the convenience of passengers. In addition, the finding of the vehicle noise is consistent with the previous study by Seker and Aydin [14] in terms of environmental awareness. Furthermore, security and air conditioning are the seventh and eighth crucial criteria, respectively. Then, service quality, speed, and crowdedness are obtained as the ninth, tenth, and eleventh substantial criteria. Also, these three criteria account for approximately 17% of the total weight and should not be ignored to satisfy passengers' needs. Lastly, service comfort and phobia are found to be the least significant criteria for the public transportation system when dealing with passengers' priorities.

3.3. Results and Discussion for Ranking the Seven Alternatives Using the TOPSIS and VIKOR Phases of the Proposed Methodology

In this paper, MCDAM using Equation (7) is obtained using the data collected from the participants. The participants were asked about their preferences for all criteria separately. For example, based on the first criterion, 310 participants prefer A1, 87 participants prefer A2, 121 participants prefer A3, 588 participants prefer A4, 302 participants prefer A5, 40 participants prefer A6, and 52 participants prefer A7. These values indicate the performance values for alternatives based on the criteria. Then, a 7 × 13 MCDAM was obtained with values.

The TOPSIS technique in Phases 3.1–3.7 was applied to rank the alternatives. For comparison purposes, the VIKOR technique was also utilized to sort the specified alternatives when employing Phases 4.1–4.6. Tables 10 and 11 show the weighted normalized MCDAMs for the TOPSIS employing Equations (8)–(10) and VIKOR techniques using Equation (12), respectively. Furthermore, Tables 12 and 13 present the results of the TOPSIS phase using Equation (11) and the VIKOR phase employing Equations (13) and (14).

Table 10. The weighted normalized MCDAM for the TOPSIS technique.

	Criteria												
	1	**2**	**3**	**4**	**5**	**6**	**7**	**8**	**9**	**10**	**11**	**12**	**13**
A1	0.041	0.003	0.064	0.029	0.013	0.015	0.015	0.031	0.026	0.025	0.008	0.008	0.009
A2	0.012	0.001	0.001	0.014	0.021	0.004	0.039	0.006	0.027	0.026	0.005	0.001	0.018
A3	0.016	0.030	0.016	0.029	0.027	0.039	0.032	0.064	0.018	0.070	0.055	0.007	0.017
A4	0.078	0.040	0.058	0.075	0.037	0.051	0.031	0.077	0.024	0.098	0.063	0.036	0.023
A5	0.040	0.022	0.038	0.045	0.014	0.013	0.012	0.023	0.030	0.026	0.019	0.026	0.008
A6	0.005	0.003	0.026	0.004	0.004	0.005	0.004	0.004	0.012	0.012	0.003	0.002	0.001
A7	0.007	0.003	0.010	0.013	0.005	0.003	0.005	0.009	0.008	0.005	0.009	0.002	0.001

Table 11. The weighted normalized MCDAM for the VIKOR technique.

	Criteria													S_i	R_i
	1	**2**	**3**	**4**	**5**	**6**	**7**	**8**	**9**	**10**	**11**	**12**	**13**		
A1	0.050	0.053	0.000	0.064	0.039	0.051	0.043	0.068	0.010	0.102	0.081	0.036	0.023	0.620	0.102
A2	0.090	0.055	0.100	0.085	0.026	0.065	0.000	0.106	0.008	0.100	0.084	0.046	0.007	0.772	0.106
A3	0.084	0.015	0.077	0.064	0.016	0.017	0.013	0.019	0.030	0.039	0.011	0.038	0.010	0.434	0.084
A4	0.000	0.000	0.009	0.000	0.000	0.000	0.014	0.000	0.015	0.000	0.000	0.000	0.000	0.039	0.015
A5	0.052	0.026	0.041	0.043	0.038	0.053	0.048	0.080	0.000	0.100	0.063	0.013	0.024	0.582	0.100
A6	0.099	0.053	0.061	0.099	0.054	0.064	0.062	0.108	0.048	0.120	0.087	0.045	0.036	0.935	0.120
A7	0.097	0.053	0.086	0.087	0.052	0.068	0.061	0.101	0.058	0.129	0.079	0.044	0.035	0.951	0.129

Table 12. The results of the TOPSIS technique for the seven alternatives.

Alternative	$Epos_i^+$	$Epos_i^-$	TPS_i	Rank
A1	0.138	0.088	0.390	4
A2	0.176	0.053	0.232	5
A3	0.104	0.124	0.544	2
A4	0.012	0.197	0.945	1
A5	0.128	0.084	0.397	3
A6	0.189	0.026	0.119	6
A7	0.190	0.014	0.069	7

According to Table 12, the metro is the first transportation system alternative based on the choices of passengers. Then, the second alternative is Marmaray. The first and second alternatives in Table 12 validate the importance of the rail transportation network when decreasing adverse environmental effects and traffic congestion. These findings are also verified by the study by Güzel and Alp [37] when considering environmental impacts, such as greenhouse gas (GHG) emissions. Also, some metro lines are operated at a service frequency of approximately every two and a half minutes [38]. Due to the high frequency of service, it is observed that the metro is selected as the first choice by the passengers as well. Next, the service frequency is approximately every two minutes for the metrobus lines [39]. Also, there are some advantages of using metrobuses over buses. Hence, the third alternative is the metrobus, as per the selection of passengers. Furthermore, the fourth alternative is the bus. Afterward, the ferry is the fifth public transportation alternative. Furthermore, the minibus is the sixth alternative. Finally, the tram is the last alternative to the public transportation system in this study.

Table 13. The results of the VIKOR technique for the seven alternatives.

	S_j	Rank	R_j	Rank	Q_j	Rank	Alternative
	0.620	4	0.102	4	0.701	4	A1
	0.772	5	0.106	5	0.798	5	A2
	0.434	2	0.084	2	0.520	2	A3
	0.039	1	0.015	1	0.000	1	A4
	0.582	3	0.100	3	0.669	3	A5
	0.935	6	0.120	6	0.948	6	A6
	0.951	7	0.129	7	1.000	7	A7
S^+ and R^+	0.039		0.015				
S^- and R^-	0.951		0.129				

Table 13 shows the VIKOR results for ranking the seven alternatives. Based on the VIKOR results, the alternatives are ranked as follows: metro (A4), Marmaray (A3), metrobus (A5), bus (A1), ferry (A2), minibus (A6), and tram (A7). Notice that VIKOR and TOPSIS rankings are the same for the alternatives. Therefore, the ranking of sustainable and traditional public transportation alternatives is verified by the methods, published reports, and papers.

3.4. Sensitivity Analysis Using the Concluding Phase of the Proposed Methodology

The sensitivity analysis in this subsection is conducted with respect to the variation of the criteria weight [40,41]. The weight change limits, Δx, are calculated between −0.129 and 0.871. Table 14 shows the new criteria weight with six different scenarios.

Table 14. The six different scenarios obtained from the sensitivity analysis.

Scenario	S1	S2	S3	S4	S5	S6
Δx	−0.129	0.000	0.250	0.500	0.750	0.871
			New weights			
C1	0.1134	0.0987	0.0704	0.0420	0.0137	0.0000
C2	0.0629	0.0548	0.0391	0.0233	0.0076	0.0000
C3	0.1150	0.1001	0.0714	0.0426	0.0139	0.0000
C4	0.1132	0.0986	0.0703	0.0420	0.0137	0.0000
C5	0.0618	0.0538	0.0384	0.0229	0.0075	0.0000
C6	0.0778	0.0677	0.0483	0.0288	0.0094	0.0000
C7	0.0717	0.0624	0.0445	0.0266	0.0087	0.0000
C8	0.1238	0.1078	0.0768	0.0459	0.0149	0.0000
C9	0.0667	0.0581	0.0414	0.0247	0.0081	0.0000
C10	0.0000	0.1293	0.3793	0.6293	0.8793	1.0000
C11	0.0997	0.0868	0.0619	0.0370	0.0120	0.0000
C12	0.0528	0.0460	0.0328	0.0196	0.0064	0.0000
C13	0.0412	0.0359	0.0256	0.0153	0.0050	0.0000

Figures 3 and 4 show changes to the rankings using different criteria weights when performing the TOPSIS and VIKOR techniques. As shown in Figures 3 and 4, very few alternatives changed when assigning different weights. Based on the results, the most important and the most sensitive criterion is the C10. It is also concluded that the first three rankings (metro, Marmaray, and metrobus) are the same for all scenarios when performing TOPSIS and VIKOR techniques. So, the stability and robustness results are provided for the three top-ranked alternatives.

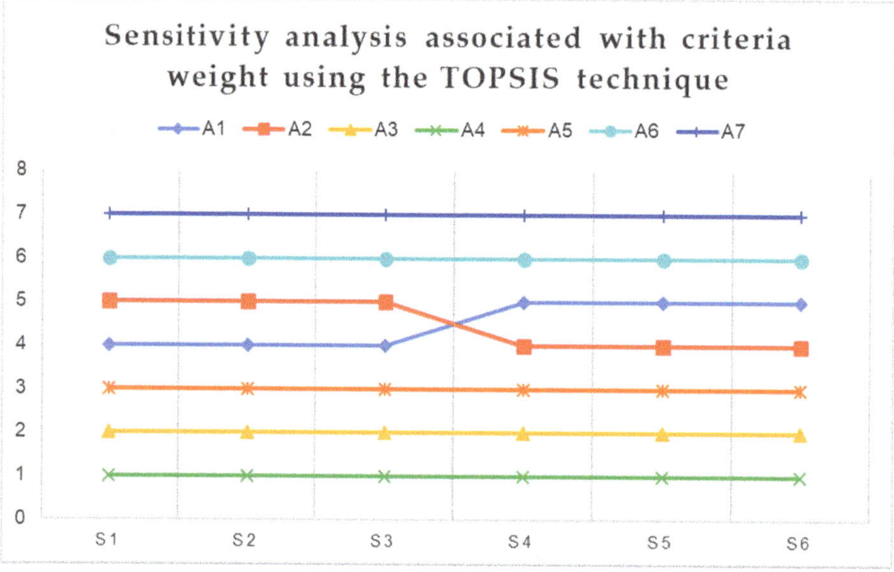

Figure 3. Changes to the rankings of the seven alternatives with the TOPSIS technique.

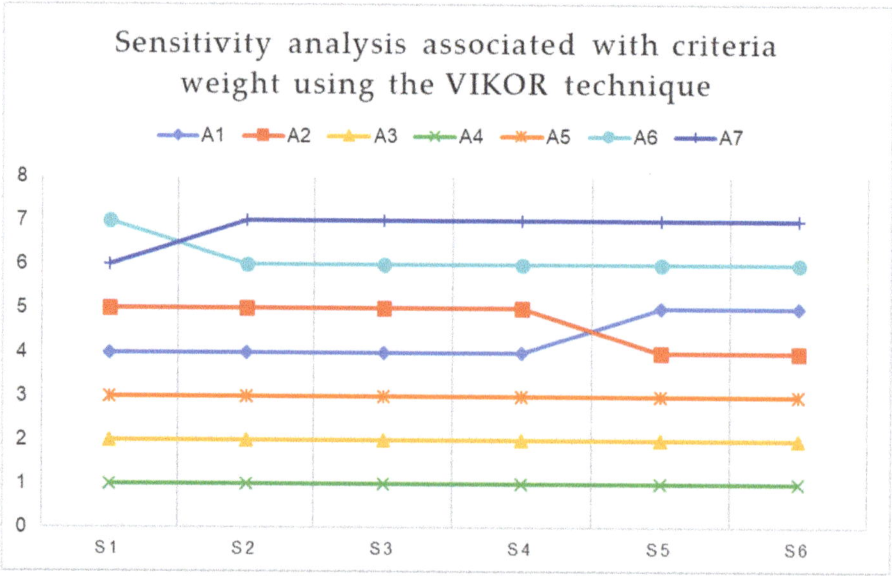

Figure 4. Changes to the rankings of the seven alternatives with the VIKOR technique.

4. Managerial Recommendations

Managerial recommendations are provided for policymakers associated with the results of this study when assessing sustainable and traditional public transportation modes. Based on the AHP results, the top three criteria are service frequency (C10), vehicle type and its mechanism (C8), and ease of accessibility (C3). These results indicate that transportation planning activities are crucial to meeting passengers' expectations. The policymakers may focus on short-term and long-term planning activities for the public transportation system. Policymakers may plan daily operations based on the passengers' behaviors, needs, and demands when increasing the frequency of service and accessibility of the public transportation system. Policymakers may also determine a pricing strategy when dealing with the number of transfers. When analyzing the TOPSIS and VIKOR methods, the top three rankings are metro (A4), Marmaray (A3), and metrobus (A5). The number of metro lines can be increased to satisfy passengers' comfort and needs, so policymakers could make an investment program as a long-term strategy for building new metro lines.

Based on the results, passengers prefer sustainable public transportation alternatives. Indeed, the sustainable public transportation system may improve the economic growth and accessibility when negative environmental impacts are decreased and cities are provided with resilience. Sustainable transport-related sustainable development goals (SDGs) introduced by the UN [42] can be achieved based on the results of this study. Remarkably, the sustainable development goals (SDGs) 11.2 and 11.b can be achieved. Thus, an accessible, affordable, safe, and sustainable public transportation system can be designed in terms of the travel behavior decisions of passengers.

Theoretical recommendations are discussed as follows. The AHP method may suffer from the inconsistency ratio and may not be appropriate for some complex decision analysis problems. In this paper, there is no inconsistency concern, and the decision-making problem is suitable for generating PCMs when asking about notions of decision-makers. Many AHP applications use 7 ± 2 criteria in the literature [43]. However, the DMs are not confused or encounter any problems when generating PCMs for the thirteen criteria in this paper. Next, weights in the AHP method are used in TOPSIS. In addition, consistency is not checked using the TOPSIS method [43]. On the other hand, these situations are not concerns for the TOPSIS method. Further, the TOPSIS results are verified by using the VIKOR technique, as well as published reports and papers. The results are verified to rank the seven alternatives. Moreover, the stability and robustness of the results are provided by employing the sensitivity analysis regarding the variation of criteria weight.

5. Conclusions and Further Studies

A public transportation system is a complex and comprehensive process, including many parameters, and needs planning activities. In this paper, sustainable and traditional public transportation modes are evaluated in relation to the travel behavior decisions of passengers using the proposed integrated decision analysis framework. The findings of this study are summarized as follows:

A novel integrated five-phased decision analysis framework is introduced to prioritize the weights and rank the alternatives.

The thirteen criteria are specified using the five experts' notions, published reports, and papers when dealing with the travel behavior decisions of passengers. Based on the AHP results, service frequency (C10), the vehicle type and its mechanism (C8), and ease of accessibility (C3) are the top three significant criteria. The findings are consistent with those of previous research.

The seven public transportation alternatives are determined, including sustainable and traditional ones. The top three alternatives are metro (A4), Marmaray (A3), and metrobus (A5) when applying the TOPSIS and VIKOR methods. Based on the results, passengers have paid attention to sustainable transportation alternatives due to the high service frequency and efficiency. The results also contribute to eliminating adverse environmental effects. Furthermore, the ranking is validated for this study.

Managerial insights are provided for policymakers. First of all, short-term and long-term planning activities may be determined by policymakers for the public transportation system. Second, policymakers may make a plan related to behaviors, needs, and demands when considering the service frequency and accessibility. Third, a suitable pricing strategy may be specified for the number of transfers. Fourth, new metro lines might be constructed by municipalities to satisfy the expectations of passengers.

The theoretical recommendations for the AHP, TOPSIS, and VIKOR methods are presented. In this paper, there are no theoretical concerns about the results. In addition, the sensitivity analysis is performed to check the stability and robustness of the results.

SDGs 11.2 and 11.b can be achieved when designing an accessible, affordable, environmentally friendly, safe, and sustainable public transportation system.

There could be possible extensions of this research. First, other multi-criteria decision analysis techniques could be employed for evaluating public transportation systems. Next, fuzzy set theory could be applied to integrate the proposed decision analysis framework when dealing with uncertainty. Then, the proposed methodology in this paper could be applied to other cities to evaluate their public transportation systems. However, some cities cannot have water transportation systems due to geography. So, in this case, the ferry alternative could be omitted, and others could be evaluated. Moreover, the proposed methodology relies on static expert and survey data. So, dynamic real-time data could be studied for further research.

Author Contributions: P.B.Ş. and A.Ö.: methodology; P.B.Ş. and A.Ö.: software; P.B.Ş., A.Ö., S.K., and T.I.: validation; P.B.Ş. and A.Ö.: investigation; P.B.Ş., A.Ö., S.K. and T.I.: writing—original draft preparation; P.B.Ş., A.Ö., S.K. and T.I.: writing—review and editing; P.B.Ş., A.Ö., S.K. and T.I.: visualization. All authors have read and agreed to the published version of the manuscript.

Funding: This study is partially financed by the European Union-NextGenerationEU through the National Recovery and Resilience Plan of the Republic of Bulgaria, project № BG-RRP-2.013-0001.

Institutional Review Board Statement: Not applicable.

Informed Consent Statement: Not applicable.

Data Availability Statement: Dataset available on request from the authors.

Conflicts of Interest: The authors declare no conflicts of interest.

References

1. TUIK. Address-Based Population Registration System Results. 2025. Available online: https://data.tuik.gov.tr/Bulten/Index?p=Adrese-Dayali-Nufus-Kayit-Sistemi-Sonuclari-2024-53783 (accessed on 31 May 2025).
2. Yedla, S.; Shrestha, R.M. Multi-criteria approach for the selection of alternative options for environmentally sustainable transport system in Delhi. *Transp. Res. Part A Policy Pract.* **2003**, *37*, 717–729. [CrossRef]
3. Aydın, S.; Kahraman, C. Vehicle selection for public transportation using an integrated multi criteria decision making approach: A case of Ankara. *J. Intell. Fuzzy Syst.* **2014**, *26*, 2467–2481. [CrossRef]
4. Camargo Pérez, J.; Carrillo, M.H.; Montoya-Torres, J.R. Multi-criteria approaches for urban passenger transport systems: A literature review. *Ann. Oper. Res.* **2015**, *226*, 69–87. [CrossRef]
5. Bai, C.; Fahimnia, B.; Sarkis, J. Sustainable transport fleet appraisal using a hybrid multi-objective decision making approach. *Ann. Oper. Res.* **2017**, *250*, 309–340. [CrossRef]
6. Barbosa, S.B.; Ferreira, M.G.G.; Nickel, E.M.; Cruz, J.A.; Forcellini, F.A.; Garcia, J.; de Andrade, J.B.S.O. Multi-criteria analysis model to evaluate transport systems: An application in Florianópolis, Brazil. *Transp. Res. Part A Policy Pract.* **2017**, *96*, 1–13. [CrossRef]
7. Ercan, T.; Onat, N.C.; Tatari, O.; Mathias, J.D. Public transportation adoption requires a paradigm shift in urban development structure. *J. Clean. Prod.* **2017**, *142*, 1789–1799. [CrossRef]

8. Büyüközkan, G.; Feyzioğlu, O.; Göçer, F. Selection of sustainable urban transportation alternatives using an integrated intuitionistic fuzzy Choquet integral approach. *Transp. Res. Part D Transp. Environ.* **2018**, *58*, 186–207. [CrossRef]
9. Saplıoğlu, M.; Aydın, M.M. Choosing safe and suitable bicycle routes to integrate cycling and public transport systems. *J. Transp. Health* **2018**, *10*, 236–252. [CrossRef]
10. Lee, D.J. A multi-criteria approach for prioritizing advanced public transport modes (APTM) considering urban types in Korea. *Transp. Res. Part A Policy Pract.* **2018**, *111*, 148–161. [CrossRef]
11. Jasti, P.C.; Vinayaka Ram, V. Integrated and sustainable benchmarking of metro rail system using analytic hierarchy process and fuzzy logic: A case study of Mumbai. *Urban Rail Transit* **2019**, *5*, 155–171. [CrossRef]
12. Nalmpantis, D.; Roukouni, A.; Genitsaris, E.; Stamelou, A.; Naniopoulos, A. Evaluation of innovative ideas for Public Transport proposed by citizens using Multi-Criteria Decision Analysis (MCDA). *Eur. Transp. Res. Rev.* **2019**, *11*, 22. [CrossRef]
13. Errampalli, M.; Patil, K.S.; Prasad, C.S.R.K. Evaluation of integration between public transportation modes by developing sustainability index for Indian cities. *Case Stud. Transp. Policy* **2020**, *8*, 180–187. [CrossRef]
14. Seker, S.; Aydin, N. Sustainable public transportation system evaluation: A novel two-stage hybrid method based on IVIF-AHP and CODAS. *Int. J. Fuzzy Syst.* **2020**, *22*, 257–272. [CrossRef]
15. Alkharabsheh, A.; Moslem, S.; Oubahman, L.; Duleba, S. An integrated approach of multi-criteria decision-making and grey theory for evaluating urban public transportation systems. *Sustainability* **2021**, *13*, 2740. [CrossRef]
16. Dahlgren, S.; Ammenberg, J. Sustainability assessment of public transport, part ii—Applying a multi-criteria assessment method to compare different bus technologies. *Sustainability* **2021**, *13*, 1273. [CrossRef]
17. Görçün, Ö.F. Evaluation of the selection of proper metro and tram vehicle for urban transportation by using a novel integrated MCDM approach. *Sci. Prog.* **2021**, *104*, 0036850420950120. [CrossRef]
18. Romero-Ania, A.; Rivero Gutiérrez, L.; De Vicente Oliva, M.A. Multiple criteria decision analysis of sustainable urban public transport systems. *Mathematics* **2021**, *9*, 1844. [CrossRef]
19. Canbulut, G.; Köse, E.; Arik, O.A. Public transportation vehicle selection by the grey relational analysis method. *Public Transp.* **2022**, *14*, 367–384. [CrossRef]
20. Çelikbilek, Y.; Moslem, S.; Duleba, S. A combined grey multi criteria decision making model to evaluate public transportation systems. *Evol. Syst.* **2023**, *14*, 1–15. [CrossRef]
21. Derbel, A.; Boujelbene, Y. Performance classification of Tunisian public transport operators. *Public Transp.* **2023**, *15*, 535–574. [CrossRef]
22. Zhang, L.; Liu, Y.; Li, S.; Chen, H.; Liu, Y. Evaluating public transport performance to guide public transport planning: A unified prior method. *Travel Behav. Soc.* **2023**, *33*, 100605. [CrossRef]
23. Borghetti, F.; Carra, M.; Besson, C.; Matarrese, E.; Maja, R.; Barabino, B. Evaluating alternative fuels for a bus fleet: An Italian case. *Transp. Policy* **2024**, *154*, 1–15. [CrossRef]
24. Dilian, O.; Beckers, C.; Witlox, F.; Davidovitch, N.; Martens, K. Examining the health effects of public transport use on older adults: A systematic review. *J. Transp. Health* **2024**, *39*, 101931. [CrossRef]
25. Kundu, P.; Görçün, Ö.F.; Garg, C.P.; Küçükönder, H.; Çanakçıoğlu, M. Evaluation of public transportation systems for sustainable cities using an integrated fuzzy multi-criteria group decision-making model. *Environ. Dev. Sustain.* **2024**, *26*, 27655–27684. [CrossRef]
26. Zehra, S.N.; Wong, S.D. Systematic review and research gaps on wildfire evacuations: Infrastructure, transportation modes, networks, and planning. *Transp. Plan. Technol.* **2024**, *47*, 1364–1398. [CrossRef]
27. Aghaabbasi, M.; Sabri, S. Potentials of digital twin system for analyzing travel behavior decisions. *Travel Behav. Soc.* **2025**, *38*, 100902. [CrossRef]
28. Alhassan, J.A.K.; Anciaes, P. Public transport investments as generators of economic and social activity. *J. Transp. Health* **2025**, *41*, 101989. [CrossRef]
29. Lyu, S.; Huang, Y.; Sun, T. Urban sprawl, public transportation efficiency and carbon emissions. *J. Clean. Prod.* **2025**, *489*, 144652. [CrossRef]
30. Nilsson, J.; Jansson, J.; Nicholas, K.; Zhao, C. Examining senior citizens' in public transport: The role of digitalization, environmental concern, and traveler satisfaction. *Travel Behav. Soc.* **2025**, *38*, 100924. [CrossRef]
31. Forman, E.; Peniwati, K. Aggregating individual judgments and priorities with the analytic hierarchy process. *Eur. J. Oper. Res.* **1998**, *108*, 165–169. [CrossRef]
32. Tavana, M.; Soltanifar, M.; Santos-Arteaga, F.J. Analytical hierarchy process: Revolution and evolution. *Ann. Oper. Res.* **2023**, *326*, 879–907. [CrossRef]
33. Google. The Google Maps. 2025. Available online: https://www.google.com/maps/place/%C4%B0stanbul/@41.1061466,28.936207,10.48z/ (accessed on 26 May 2025).
34. Mouwen, A. Drivers of customer satisfaction with public transport services. *Transp. Res. Part A Policy Pract.* **2015**, *78*, 1–20. [CrossRef]

35. Brechan, I. Effect of price reduction and increased service frequency on public transport travel. *J. Public Transp.* **2017**, *20*, 139–156. [CrossRef]
36. Friman, M.; Lättman, K.; Olsson, L.E. Public transport quality, safety, and perceived accessibility. *Sustainability* **2020**, *12*, 3563. [CrossRef]
37. Güzel, T.D.; Alp, K. Modeling of greenhouse gas emissions from the transportation sector in Istanbul by 2050. *Atmos. Pollut. Res.* **2020**, *11*, 2190–2201. [CrossRef]
38. IETT. Railway Station Transportation in Istanbul. 2024. Available online: https://iett.istanbul/icerik/Railway-Station-Transportation (accessed on 2 January 2025).
39. IETT. Metrobus in Istanbul. 2024. Available online: https://iett.istanbul/icerik/Metrobus (accessed on 2 January 2025).
40. Yazdani, M.; Chatterjee, P.; Pamucar, D.; Abad, M.D. A risk-based integrated decision-making model for green supplier selection: A case study of a construction company in Spain. *Kybernetes* **2020**, *49*, 1229–1252. [CrossRef]
41. Więckowski, J.; Sałabun, W. Sensitivity analysis approaches in multi-criteria decision analysis: A systematic review. *Appl. Soft Comput.* **2023**, *148*, 110915. [CrossRef]
42. UN. Sustainable Transport. 2024. Available online: https://sdgs.un.org/topics/sustainable-transport (accessed on 22 December 2024).
43. Shih, H.S.; Shyur, H.J.; Lee, E.S. An extension of TOPSIS for group decision making. *Math. Comput. Model.* **2007**, *45*, 801–813. [CrossRef]

sustainability

MDPI

Article

Assessing Tractors' Active Safety in Serbia: A Driving Simulator Study

Sreten Simović [1], Aleksandar Trifunović [2,*], Tijana Ivanišević [3], Vaidas Lukoševičius [4,*] and Larysa Neduzha [5]

1 Faculty of Mechanical Engineering, University of Montenegro, 81000 Podgorica, Montenegro; sretens@ucg.ac.me
2 Faculty of Transport and Traffic Engineering, University of Belgrade, 11000 Belgrade, Serbia
3 Academy of Professional Studies Sumadija, 34000 Kragujevac, Serbia; tivanisevic@asss.edu.rs
4 Faculty of Mechanical Engineering and Design, Kaunas University of Technology, Studentų Str. 56, 44249 Kaunas, Lithuania
5 Department of Technical Mechanics, Ukrainian State University of Science and Technologies, Lazaryan 2, 49010 Dnipro, Ukraine; nlorhen@i.ua
* Correspondence: a.trifunovic@sf.bg.ac.rs (A.T.); vaidas.lukosevicius@ktu.lt (V.L.)

Abstract

The active safety of tractors remains a major concern in rural road environments, where tractor drivers face high crash risks due to limited vehicle visibility. In Serbia, 1.4% of crashes involve tractors, mainly due to poor visibility (64.3%), lack of beacon lights, unsafe overtaking, and unmarked stopped tractors (14.3% each). These issues reduce safety, increase fuel consumption and emissions, and cause economic losses. A driving simulator study with 117 drivers examined how visibility equipment affects speed perception. The results showed that 20 km/h was best estimated with all visibility aids, while 10 km/h was most accurately judged with only the slow-moving vehicle emblem. These findings emphasize the potential for simple, cost-effective visibility measures to enhance the active safety of tractors in mixed rural traffic conditions. By enhancing tractor visibility, these measures reduce crash risks, minimize unnecessary acceleration and deceleration, and lower fuel consumption and emissions associated with traffic disturbances. Furthermore, by preventing crashes, these solutions contribute to reducing resource consumption in crash-related medical care, vehicle repairs, and infrastructure damage. Integrating improved visibility equipment into rural traffic policy can significantly enhance tractors' active safety and reduce the risk of crashes in agricultural regions.

Keywords: tractor visibility; speed perception; driving simulator; active safety; sustainability

1. Introduction

Modern sustainable transportation requires not only the reduction of emissions and energy consumption [1,2] but also effective management of traffic safety, particularly the active safety of slow-moving vehicles, such as tractors, in rural traffic environments [3,4]. In Serbia, tractors are involved in approximately 1.4% of all road crashes [5], often with severe consequences due to limited vehicle visibility. Contributing factors include poor lighting, lack of standardized warning signs, and unsafe overtaking practices. As the presence of agricultural machinery on public roads increases, evaluating and improving their active safety becomes essential for preventing collisions and protecting all road users.

Despite the widespread recognition of visibility-related risks, there is limited empirical research examining how specific visibility aids, such as daytime running lights (DRLs),

https://doi.org/10.3390/su17136144

beacon lights, and the slow-moving vehicle (SMV) emblem, affect drivers' perceptual accuracy. Most previous studies have focused on passenger vehicles and motorcycles, leaving a knowledge gap concerning tractors, particularly under real-world conditions simulated through driving environments. Given the unique operational characteristics and vulnerability of tractors, this gap represents a critical limitation in advancing evidence-based improvements in their active safety. This study seeks to bridge this research gap by systematically evaluating the influence of various visibility equipment configurations on drivers' speed perception of tractors within a controlled driving simulator environment. By focusing on perceptual accuracy as a key determinant of safe interaction with tractors, the study directly contributes to the assessment of their active safety in rural traffic contexts.

The remainder of this paper is structured as follows: Section 2 provides a review of relevant literature on vehicle visibility and speed perception. Section 3 details the experimental design, including participant characteristics and simulation conditions. Section 4 presents the results of the statistical analysis. Section 5 discusses the implications of the findings, and Section 6 concludes with a summary and recommendations for future research.

2. Literature Review

Sustainable transportation research increasingly recognizes the need to integrate traffic safety considerations into broader environmental and socio-economic frameworks. Within this context, vehicle visibility, particularly for vulnerable road users, such as agricultural machinery, plays a dual role: it enhances safety and contributes to the efficiency and resilience of transport systems. The subsequent subsection examines the integration of vehicle visibility measures within the broader framework of sustainable transportation, prior to delineating their empirically established contributions to traffic safety and driver perception research.

2.1. The Role of Vehicle Visibility in the Context of Sustainable Transportation

While the enhancement of vehicle visibility is primarily viewed through the lens of traffic safety, its broader implications for sustainable transportation are increasingly recognized. Sustainable transportation encompasses more than environmental efficiency, it also integrates social and economic dimensions, such as the reduction of traffic injuries, optimization of energy use, and minimization of infrastructure strain [6–8]. From this perspective, improved vehicle visibility, particularly for vulnerable and slow-moving vehicles, such as tractors, can be seen as a critical factor contributing to long-term sustainability goals.

Firstly, increasing the visibility of agricultural vehicles has a direct impact on reducing crash frequency and severity. Fewer road crashes translate into lower demand for emergency services, medical treatment, vehicle repair, and road infrastructure maintenance, all of which consume substantial resources and generate emissions [9,10]. It has been estimated that road traffic crashes contribute to national carbon footprints, both through direct emissions (e.g., vehicle leaks, fire, fuel loss) and indirect impacts, such as road reconstruction and hospital energy use [11–13]. Moreover, the deployment of driving simulators for research, as used in this study, supports sustainability by eliminating the need for physical vehicle testing [14]. Real-world experiments typically involve the use of fuel, road space, and controlled environments, each of which bears environmental costs. In contrast, simulators offer a reproducible, low-emission, and resource-efficient alternative for studying driver behavior, perception, and safety-related variables [15,16].

However, visibility-enhancing technologies themselves, particularly DRLs, warrant a nuanced consideration. While DRLs improve perceptual conspicuity and reduce crash risk, they also contribute to marginally increased fuel consumption and CO_2 emissions, especially in conventional internal combustion engine vehicles. Estimates suggest that

DRLs can increase fuel use by approximately 0.3–1.5%, depending on the technology employed (e.g., halogen vs. LED) [17,18]. Nonetheless, these environmental costs are typically offset by the broader societal benefits of crash reduction, especially when modern energy-efficient lighting systems, such as LED DRLs, are utilized [19,20]. Furthermore, the economic costs of road crashes in agricultural areas are disproportionately high when considering the limited transport infrastructure and availability of emergency response services. Visibility interventions that help reduce such incidents thus support economic resilience in rural areas [21–23]. In summary, vehicle visibility measures, especially those targeting high-risk vehicle types, like tractors, should be considered not only as safety improvements but also as integral components of a sustainable transportation framework.

2.2. Vehicle Visibility Equipment and Perceptual Factors in Traffic Safety

The role of visibility-enhancing equipment, such as daytime running lights (DRLs), beacon lights, and slow-moving vehicle (SMV) emblems, in improving road safety remains insufficiently explored. These devices are primarily intended to support timely and effective interactions between road users, acting as a fundamental element for maintaining safe and predictable traffic flow [19,24–26].

Between 2019 and 2021 in Serbia, a total of 99 fatal road crashes and 704 injury-related crashes involved tractors. These incidents resulted in 101 fatalities and 1079 injuries. On average, approximately 34 individuals lose their lives and 360 are injured each year in tractor-related crashes. Tractors are involved in about 1.4% of all road crashes, yet they account for roughly 6.5% of total fatalities and about 2% of all injuries in traffic accidents in Serbia [27–30].

A study by Marković et al. [31] further highlights the nature of these crashes, noting that 33.4% of collisions with tractors occur at the rear-left or front-left sections of the vehicle. In the case of tractors towing trailers, as many as 58.6% of collisions involve the rear-left section of the trailer (Figure 1). These patterns are most often linked to overtaking maneuvers under nighttime conditions, where poor visibility makes it difficult for other road users to perceive the tractor in time to avoid a collision [32–37].

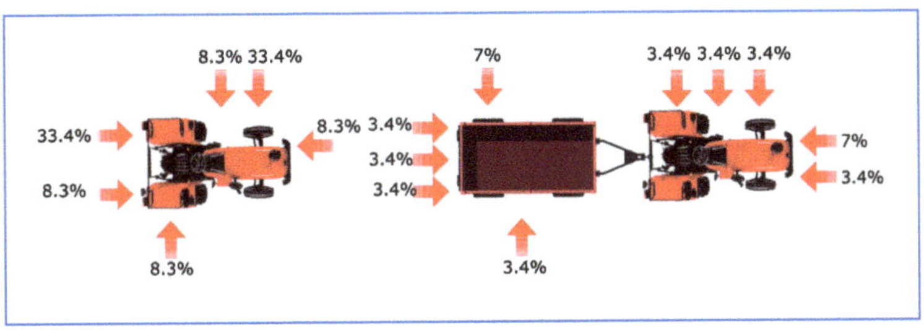

Figure 1. Representation of damage on a tractor and a tractor with a trailer in road crashes [31].

Marković et al. [31] emphasize that a predominant share of road crashes involving tractors—up to 64.3%—can be attributed to insufficient visibility of the vehicle. Specific contributing factors include the absence or malfunctioning of beacon lights, as well as unsafe overtaking practices and the failure to mark stationary tractors or trailers, each accounting for 14.3% of such incidents. An additional 7.1% of cases involve undetermined or ambiguous causes, often due to limitations in post-crash investigation data.

These findings reinforce the crucial role of vehicle visibility in accident prevention. Road safety improvements largely depend on drivers' perceptual abilities, including early detection of an approaching vehicle, accurate estimation of its speed, and correct assessment of the time required to perform a safe maneuver, such as overtaking or yielding [19,

37–40]. Theories of visual perception, such as the illumination contrast theory, further underline that a vehicle's visibility is strongly influenced by its luminance contrast with the background environment [40]. In response to such insights, most developed countries have mandated the use of daytime running lights (DRLs) across all vehicle categories as a cost-effective means of increasing vehicle conspicuity and enhancing traffic safety [19,28,41,42]. In Serbia, by contrast, current legislation requires only the use of a yellow beacon light on tractors and the display of the "slow-moving vehicle (SMV) emblem" as warning indicators [29].

Empirical research supports the effectiveness of DRLs in improving traffic safety outcomes. According to aggregated data, passenger vehicles equipped with DRLs show a 24.6% reduction in fatalities, a 20% reduction in injuries, and a 12.4% decrease in the frequency of multi-vehicle crashes during daytime conditions [19,43]. Historical findings corroborate this impact, with a 1996 study reporting that cars with DRLs were involved in 10–15% fewer multi-vehicle collisions than those without them [44].

Beyond safety statistics, DRLs also influence drivers' perceptual judgments. Several studies have demonstrated that the presence or absence of DRLs significantly affects the accuracy of vehicle speed estimation [19,24,43]. For instance, Pešić et al. [24] revealed that drivers consistently misjudge the speed of oncoming vehicles depending on whether DRLs are turned on or off. These discrepancies become particularly pronounced at higher speed ranges, notably 70 and 90 km/h. The authors concluded that DRLs significantly improve perceptual accuracy on roads with higher speed limits, thereby potentially reducing the likelihood of high-speed overtaking errors [32,45–47]. Similarly, Ivanišević et al. [19] found a statistically significant difference in the perception of motorcycle speed depending on the DRL configuration—whether turned off, turned on, or using an advanced LED DRL system. Their results indicated that speed estimation errors are substantially lower when motorcycles are equipped with LED DRLs, especially at higher speeds (70 and 90 km/h), across both urban and rural road environments. These findings suggest that DRLs—particularly those using LED technology—play a crucial role in enhancing a vehicle's perceptibility and aiding drivers in making accurate perceptual judgments at various speed limits [48–51].

While much of the existing research has focused on passenger vehicles and motorcycles, little is known about the perceptual implications of visibility equipment for tractors. The specific effect of DRLs, beacon lights, and the SMV emblem on the estimation of tractor speed has not yet been systematically investigated [52–54]. This gap in the literature highlights the need for further study, particularly considering the distinct operating speeds and physical characteristics of tractors. To address this gap, the present research aims to explore how different combinations of visibility equipment influence the accuracy of drivers' speed estimation for tractors. By simulating various visibility scenarios using a driving simulator, this study seeks to identify which equipment configurations contribute most effectively to perceptual clarity and improved road safety outcomes.

To provide context for the present research, Table 1 summarizes several key studies that have examined the impact of daytime running lights and vehicle visibility on speed perception and traffic safety. These studies highlight diverse methodological approaches and consistently underscore the role of visibility equipment in enhancing perceptual accuracy and reducing crash risks.

Table 1. Summary of selected studies on vehicle visibility and speed perception.

Authors and Year	Title	Methodology	Key Findings
Pešić et al. (2019) [24]	Evaluation of the effects of daytime running lights for passenger cars	Driving simulator experiment; comparison of DRL on/off conditions	DRLs significantly improve speed estimation accuracy, especially at higher speeds (70–90 km/h)
Ivanišević et al. (2022) [19]	The impact of daytime running (LED) lights on motorcycles speed estimation	Driving simulator; assessment of motorcycle speed with and without LED DRLs	LED DRLs enhance speed perception accuracy at higher speeds on both urban and rural roads
Cavallo & Pinto (2012) [25]	Are car daytime running lights detrimental to motorcycle conspicuity?	Laboratory-based experimental study	DRLs on cars can reduce motorcycle visibility, especially when both vehicles use DRLs
Peña-García et al. (2010) [26]	Influence of daytime running lamps on visual reaction time of pedestrians	Psychophysical tests with pedestrians in controlled environments	DRLs shorten pedestrian reaction time when detecting vehicle turn signals
Elvik (1996) [44]	A meta-analysis of studies concerning the safety effects of daytime running lights on cars	Meta-analysis of multiple studies	DRLs reduce fatalities by ~25%, injuries by ~20%, and daytime crashes by ~12%
Koornstra et al. (1997) [43]	The Safety Effects of Daytime Running Lights	Analytical study using international data	DRLs improve safety, particularly in frontal and side collisions

3. Materials and Methods

For the purposes of this paper, an experiment was carried out on a driving simulator, with the aim of examining the differences in the estimation of the tractor's speed, in situations where the tractor's daytime running lights, beacon lights, and a "slow-moving vehicle (SMV) emblem" were installed.

3.1. Experimental Design Overview

For the purposes of this experiment, the test subjects were presented with 12 different conditions of tractor movement on the driving simulator: 2 conditions when a "slow-moving vehicle (SMV) emblem" was installed on the tractor, 2 conditions when the tractor's daytime running lights were on, 2 conditions when the tractor had the beacon light on, 2 conditions when the beacon light was on AND a "slow-moving vehicle (SMV) emblem" was installed, 2 conditions when the daytime light was on and a beacon light and "slow-moving vehicle (SMV) emblem" were installed, as well as 2 conditions when there was no equipment on the tractor. The tractor moved at speeds of 10 and 20 km/h in all conditions.

The driving simulator was used to simulate traffic on a two-lane, undivided road during a sunny day for the participants [24]. The study focused on analyzing the characteristics of the trajectory during free-flow driving, without any roadside interruptions. The simulation environment included typical traffic signs and vegetation, but no additional objects were introduced in the traffic scenes to prevent influencing the participants' expectations about the movement of the visual targets, as well as avoiding any potential distractions [19,24,55].

In the experiment, the participants were tasked with estimating the speed of a tractor approaching them under all the conditions described above [19]. The participants provided their estimates orally, while an assistant entered the spoken values into the appropriate field of an online questionnaire [19,32,45,55–59]. The questionnaire also included questions about demographic characteristics (gender and age), place of residence, level of education, possession of a driving license (including the category of license held and the number of years the participant had held it), frequency of tractor driving, frequency of driving, and participation in road crashes [19,24,55].

3.2. *Stimuli and Experimental Conditions*

In this study, the tractor was selected as a representative example of slow-moving agricultural vehicles that frequently operate on rural roads and interact with faster traffic streams. While tractors, as slow-moving vehicles in rural areas, are an important subject of active safety analysis, they constitute a critical component of rural mobility systems, especially in countries with strong agricultural sectors. Improving the active safety of such vehicles supports multiple sustainability goals, including injury prevention and resource conservation.

For this study, a 3D model of the "IMT" tractor, model "539" in a red color, was utilized. The tractor's dimensions are 2972 mm in length and 1800 mm in width, and it has a wheelbase of 1830 mm. This model was selected due to its widespread use in agricultural regions of Southeast Europe, ensuring ecological and practical relevance. Two tractor speeds (10 km/h and 20 km/h), were selected for the experiment, reflecting typical real-world operational conditions for tractors on public roads [47].

The visibility equipment selected for this experiment reflects legally required and commonly implemented configurations for agricultural vehicles. The DRL (manufacturer Hella GmbH & Co. KGaA, Lippstadt, Germany) used was a standardized halogen-based daytime running light, designed to provide excellent visibility for the tractor in daylight. In this configuration, the tractor's primary headlight is represented using halogen lighting [24,54].

A "slow-moving vehicle (SMV) emblem" (Figure 2) was placed on the front side of the tractor, while the beacon light (Figure 3) was placed on the rear side of the tractor. The rotating amber beacon light, designed to alert surrounding traffic of slow or obstructive movement, was placed at the highest structural point of the vehicle to maximize visibility. This selection of equipment allowed for systematic testing of six visibility configurations, enabling the evaluation of both isolated and combined effects on driver perception.

Figure 2. Slow-moving vehicle (SMV) emblem.

Figure 3. Beacon light.

3.3. Driving Simulator Setup

Driving simulators are widely recognized as effective tools for studying perceptual and cognitive aspects of driving behavior, especially in situations that would be difficult or unsafe to replicate in real-world conditions [19,24,55].

The driving simulator is equipped with three 42-inch plasma screens (Panasonic TH-42PWD6 plasma screen, manufacturer Panasonic Corporation, Osaka, Japan.), providing participants with a 180° horizontal and 50° vertical field of view of the simulated environment. Each screen has a resolution of 1360 × 768 pixels and a refresh rate of 60 Hz [24,55]. Research has shown that a minimum horizontal field of view of 120° is essential for accurate speed perception [55]. The simulation environment was designed to replicate a two-lane undivided rural road under sunny daylight conditions, incorporating standard road signs and natural roadside vegetation. No additional vehicles or environmental distractions were introduced to ensure experimental control and focus on the target vehicle. Along with visual stimuli, participants received spatially consistent ambient traffic sounds, enhancing immersion and realism. Before beginning the trials, each participant received a standardized briefing and a practice run to familiarize themselves with the equipment and procedures.

3.4. Experimental Protocol

The study was carried out in the Republic of Serbia during October and November 2022. Participants did not receive any form of compensation for their involvement in the research [55]. Each participant was assessed individually and underwent preliminary trials [55]. At the start of the experiment, every participant was assigned a unique sequence of experimental stimuli, determined using a random number generator [19]. This approach aimed to mitigate the anchoring effect by implementing counterbalancing, achieved by randomizing the sequence in which the test stimuli were presented [24]. Respondents were asked to estimate the tractor's speed under 12 distinct visibility conditions: (1) with only the SMV emblem installed, (2) with daytime running lights (DRLs) activated, (3) with a beacon light activated, (4) with both the beacon light and SMV emblem, (5) with the full set of visibility equipment (DRLs, beacon light, and SMV emblem), and (6) with no visibility equipment installed. Each of these six configurations was presented at two different speeds (10 km/h and 20 km/h), resulting in a total of 12 experimental scenarios. Participants observed each scenario within the driving simulator and verbally reported their speed estimates, which were recorded by an assistant. No predefined response options were offered, ensuring open-ended and unbiased input.

3.5. Data Collection and Statistical Processing

The data were gathered through an online questionnaire and subsequently imported into the MS Excel software package, Office 2019 (Microsoft Corporation, Redmond, WA, USA) [60–62]. After importing, the data were thoroughly reviewed and validated [55]. Statistical analysis of the collected data was then performed using IBM SPSS Statistics version 22.0 (IBM Corp., Armonk, NY, USA) [33,63–66]. The normality of the distribution was assessed by examining histograms and performing the Kolmogorov–Smirnov test. Since the distribution of data for all measured variables was found to be normal, parametric methods were applied. To evaluate the significance of differences, the independent samples *t*-test, one sample test, and one-way ANOVA were utilized [24,67–70].

The null hypothesis (H_0) stated that there is no statistically significant difference in tractor speed perceptions. The alternative hypothesis (Ha) is that there are statistically significant differences in tractor speed perceptions.

The statistical significance threshold (α) was set at 5%. Therefore, if the probability (p) is less than or equal to 0.05, the null hypothesis (H_0) is rejected, and the alternative hypothesis (H_a) is accepted. Conversely, if p is greater than 0.05, H_0 is not rejected.

The DRL is a specialized daytime running light designed to provide excellent visibility for the tractor in daylight. In this configuration, the tractor's primary headlight is represented using halogen lighting [24,55]. A "slow-moving vehicle (SMV) emblem" was placed on the front side of the tractor, while the beacon light was placed on the most prominent point of the tractor.

4. Results

4.1. Demographic Data

A total of 117 respondents participated in the study [38,39], with an average age of 28.2 years. The youngest respondent was 17 years old (X_{min}), and the oldest was 55 years old (X_{max}). The sample consisted of 57.3% male respondents and 42.7% female respondents. The highest percentage of respondents, 41.9%, completed high school, followed by 18.8% who completed basic professional studies, 15.4% completed basic academic studies, and 11.1% of respondents completed master's academic studies. Additionally, 12.8% of respondents completed vocational secondary education or other forms of non-academic training. 41.4% of the respondents reported living in rural areas, 44% in urban areas, 10.3% in suburban areas, and 4.3% in small towns. Many respondents, 72.6%, hold a driver's license for passenger vehicles, while 16.2% have a driver's license for trucks. The remaining 11.2% of respondents either hold restricted licenses applicable to motorcycles or agricultural machinery or have no valid driving license at the time of the study. In terms of driving experience, 28.2% of respondents have held a driver's license for more than 10 years, 19.7% for 5 to 10 years, and 21.4% for 3 to 5 years. Additionally, 19.5% reported having less than 3 years of driving experience, while 11.2% did not possess a valid driver's license at the time of the study. The largest percentage of respondents, 66.7%, reported participating in traffic daily as motor vehicle drivers. An additional 20.5% stated that they participate occasionally as motor vehicle drivers. Furthermore, 12.8% reported being regular traffic participants, but not as drivers of motor vehicles (for example, as passengers, cyclists, or users of public transport). Additionally, 38.8% of all respondents reported occasionally driving a tractor, and among them, 11.1% do so on a daily basis, while the rest operate tractors less frequently. Conversely, 61.2% of respondents reported that they rarely or never drive a tractor.

4.2. Tractor Speed Estimate

Table 2 presents the descriptive statistics of the estimated tractor speed across all described conditions and tested speeds. The column labeled t refers to the t-value obtained from the one-sample *t*-test, while p denotes the p-value indicating the level of statistical significance for each tested condition.

In all tested conditions and speeds, respondents tended to overestimate the tractor's speed (Figures 4 and 5). Figure 4 illustrates the mean estimated speed across all six visibility configurations at the actual speed of 10 km/h. It can be observed that the presence of any visibility aid (SMV emblem, DRLs, or beacon light) consistently reduced the degree of overestimation compared to the no-equipment condition. The lowest estimation error occurred when only the SMV emblem was installed, while the absence of any visibility equipment led to the most pronounced overestimation.

Table 2. Descriptive statistics and analysis of the tractor speed estimation at all tested conditions and speeds.

Speed	Conditions	Mean	Str. Deviation	t	*p*
10 km/h	SMV emblem	14.30	9.00	17.18	0.000
	Lights	14.32	8.21	18.87	0.000
	Beacon light	14.82	9.47	16.92	0.000
	Beacon light and SMV emblem	14.74	10.11	15.77	0.000
	Everything from the equipment	14.54	10.41	15.10	0.000
	None of the equipment	15.21	10.08	16.32	0.000
20 km/h	SMV emblem	24.03	11.46	22.69	0.000
	Lights	23.54	11.48	22.17	0.000
20 km/h	Beacon light	25.15	11.01	24.72	0.000
	Beacon light and SMV emblem	24.88	10.34	26.04	0.000
	Everything from the equipment	22.42	10.22	23.74	0.000
	None of the equipment	24.19	11.52	22.72	0.000

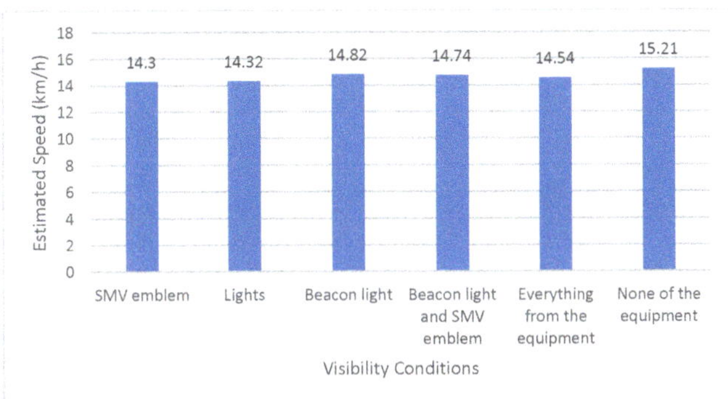

Figure 4. Descriptive statistics of the tractor's speed estimation under all tested conditions at a speed of 10 km/h.

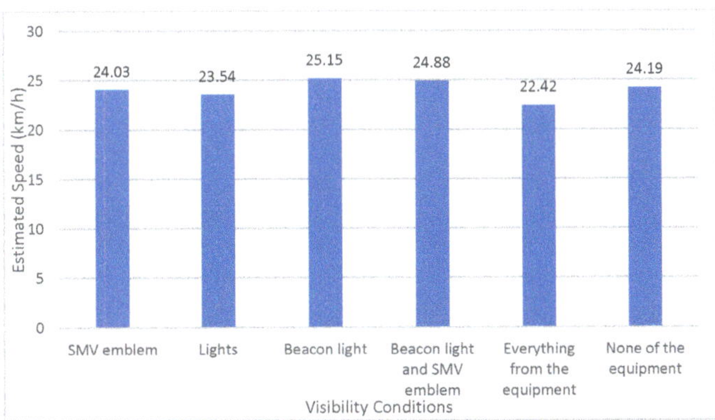

Figure 5. Descriptive statistics of the tractor's speed estimation under all tested conditions at a speed of 20 km/h.

Figure 5 presents the same pattern at the actual speed of 20 km/h. Notably, the configuration with all three visibility aids (DRLs, beacon light, and SMV emblem) resulted in the most accurate perception, with a mean estimated speed of 22.42 km/h. The largest overestimation was recorded when only the beacon light was active or when no equipment was installed, both resulting in perceived speeds exceeding 25 km/h.

Figure 6 provides a comparative overview of the average absolute estimation errors across all visibility conditions and speed levels. It clearly shows that combined visibility configurations generally yield smaller perceptual errors. The data suggest that the addition of even a single visibility feature improves perceptual accuracy, while the simultaneous use of all three yields the best results. This supports the conclusion that multi-modal visibility enhancements can significantly improve drivers' estimation performance and thus promote safer interactions.

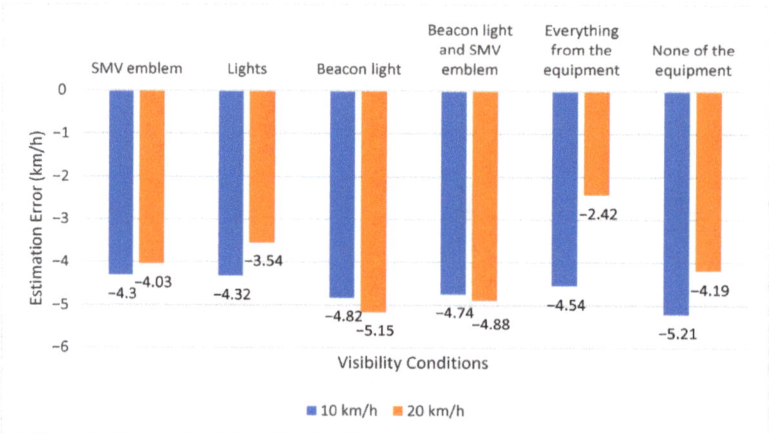

Figure 6. The average error in tractor speed estimation across all speeds and conditions.

In the condition where the tractor's daytime running light and beacon light were on, along with the SMV emblem installed, respondents estimated a speed of 20 km/h with the smallest error (M = 22.42; SD = 10.22). Meanwhile, the speed of 10 km/h was estimated with the smallest error when only the SMV emblem was installed on the tractor (M = 14.30; SD = 9.00). With the largest error, the respondents estimated the speed of 10 km/h in a situation when there was no equipment on the tractor (M = 15.21; SD = 10.08), while they estimated the speed of 20 km/h with the largest error when there was no equipment on and when the beacon light was on (M = 25.15; SD = 11.01).

The one sample test results indicate a statistically significant difference for all tested speeds and conditions, as shown in Table 2 (Hypothesis 1).

4.3. Gender Differences in Tractor Speed Estimation

The independent samples *t*-test was conducted to analyze differences in tractor speed estimation based on the respondents' gender (Hypothesis 2). The results (Table 3) reveal statistically significant differences for all tested speeds and conditions. In Table 3, the column labeled t refers to the t-value derived from the independent samples *t*-test, while *p* indicates the *p*-value representing the probability of obtaining the observed difference by chance.

Table 3. The relationship between the tractor speed estimation for all the tested speeds, for all tested conditions, and half of the respondents.

Speed	Conditions	Gender	Mean	Std. Deviation	t	p	Eta Square	Magnitude of Impacts
10 km/h	SMV emblem Lights	Man	11.687	5.721	−3.525	0.001	0.097	Large
		Woman	17.800	11.223				
	Beacon light	Man	11.701	5.681	−3.980	0.000	0.121	Large
		Woman	17.820	9.701				
	SMV emblem Lights	Man	11.836	6.273	−3.900	0.000	0.117	Large
		Woman	18.820	11.444				
	Beacon light	Man	12.030	8.730	−3.407	0.001	0.091	Large
		Woman	18.360	10.757				
	Everything from the equipment	Man	11.761	7.768	−3.284	0.002	0.086	Large
		Woman	18.260	12.279				
	None of the equipment	Man	12.687	7.097	−3.024	0.003	0.074	Moderate
		Woman	18.580	12.342				
20 km/h	SMV emblem Lights	Man	21.537	8.418	−2.622	0.011	0.056	Moderate
		Woman	27.380	13.975				
	Beacon light	Man	21.104	9.154	−2.584	0.012	0.054	Moderate
		Woman	26.800	13.430				
	SMV emblem Lights	Man	21.940	8.978	−3.711	0.000	0.107	Large
		Woman	29.460	12.048				
	Beacon light	Man	21.896	7.945	−3.617	0.001	0.102	Large
		Woman	28.880	11.804				
	Everything from the equipment	Man	19.851	8.576	−3.152	0.002	0.079	Moderate
		Woman	25.860	11.265				
	None of the equipment	Man	21.731	9.398	−2.615	0.011	0.056	Moderate
		Woman	27.480	13.253				

4.4. The Respondent's Place of Residence and Tractor Speed Estimation

The analysis of variance (ANOVA) revealed statistically significant differences in speed estimation accuracy depending on the respondents' place of residence, particularly in the scenario involving a tractor traveling at 10 km/h with only the SMV emblem installed ($F = 2.581$; $p = 0.041$). Participants residing in rural areas and small towns provided the most accurate estimations, with mean values closest to the actual speed, and the lowest standard deviations. This likely reflects their frequent exposure to slow-moving agricultural vehicles and more intuitive understanding of their movement. In contrast, respondents from urban environments, especially those living in the city center and inner-city areas, exhibited a clear tendency to overestimate tractor speed, with higher average values and greater variability. These findings emphasize the role of environmental familiarity in the development of accurate perceptual models, suggesting that limited real-life interaction with tractors in high-density urban zones may impair estimation accuracy.

4.5. Driving License Category and Tractor Speed Estimation

Statistically significant differences were also observed across all tested visibility conditions when respondents were grouped according to driving license category. Truck drivers consistently demonstrated the highest accuracy in estimating tractor speed at both 10 km/h and 20 km/h, regardless of the installed visibility equipment. Their estimations were closest to the true values, with low variability, indicating a stable and calibrated perceptual response, likely developed through regular professional interaction with large and slow-moving vehicles.

Passenger car drivers showed moderate accuracy overall but were notably less precise than truck drivers, particularly in conditions with no visibility aids or only partial equipment. Respondents without a driving license, as well as those currently in the process of training at a driving school, showed a consistent pattern of overestimation, with the most extreme deviations observed in conditions involving beacon lights or minimal signaling. This group also displayed wide standard deviations, indicating inconsistency in perceptual judgment.

A particularly concerning result emerged for motorcycle license holders, albeit a small subsample, who drastically overestimated speed in nearly all scenarios—sometimes by two to three times the actual value. Their responses suggest a perceptual bias possibly linked to limited exposure to slow-moving vehicles or differing spatial heuristics acquired through motorcycle operation.

Overall, these findings highlight that the category of driving license is a strong predictor of perceptual performance, with professional experience (e.g., truck driving) correlating with lower estimation error. This suggests that the perceptual calibration gained through regular exposure to large vehicles can significantly improve judgment accuracy. The results also underscore the need to reinforce perception-related training—especially concerning agricultural vehicles—in driver education programs across all license categories, including novice drivers and those with limited driving experience.

4.6. Years of Possession of the Driver's License and Tractor Speed Estimation

One-way ANOVA was used to analyze differences in tractor speed estimation based on the years of driver's license ownership (Hypothesis 5). The results show no statistically significant differences in tractor speed perception, regardless of the years of license ownership, for all tested speeds and conditions.

4.7. Frequency of Driving a Motor Vehicle and Tractor Speed Estimation

One-way ANOVA was used to analyze the relationship between the frequency of driving a motor vehicle (daily, less than 3 times a week, 3 to 5 times a week, less than 3 times a month, and less than three times a year) and tractor speed estimation (Hypothesis 6). The results show significant statistical differences in the estimation of 10 km/h with the tractor's daytime running lights on ($F = 2.560$; $p = 0.031$), 20 km/h with the SMV emblem installed ($F = 2.394$; $p = 0.042$), and 20 km/h with both the beacon light and SMV emblem installed ($F = 3.112$; $p = 0.011$).

Respondents who drive a vehicle 3 to 5 times a week provided the most accurate estimates with the smallest error when estimating the tractor's speed at 10 km/h with the tractor's daytime running lights on (M = 10.00; SD = 5.873), as well as when estimating the tractor's speed at 20 km/h with the SMV emblem installed (M = 16.00; SD = 6.633). For estimating the speed of 20 km/h in the condition where both the beacon light and SMV emblem were installed on the tractor, the most accurate estimate (smallest error) was provided by respondents who drive a motor vehicle less than 3 times a year (M = 16.667; SD = 7.637).

4.8. Impact of Tractor Driving Experience on Tractor Speed Estimation

The relationship between the frequency of tractor driving and the accuracy of tractor speed estimation was first examined using a one-way ANOVA. The results indicated no statistically significant differences across all tested speeds and visibility conditions, suggesting that how often a respondent drives a tractor does not substantially influence their ability to estimate its speed.

However, a more pronounced effect was observed when comparing respondents who reported driving a tractor in traffic with those who do not. An independent samples *t*-test revealed statistically significant differences in speed estimation for all tested speeds and conditions. Across nearly all scenarios, respondents with actual experience operating tractors in traffic consistently demonstrated lower estimation errors compared to those without such experience. These differences were particularly notable in configurations where visibility equipment was either partially present or completely absent.

The largest perceptual gap was observed in the condition with only the beacon light activated at 20 km/h, where drivers without tractor experience overestimated the speed to a significantly greater extent than those who regularly operate tractors. Effect size analyses confirmed that these differences ranged from small to large, depending on the specific visibility configuration, with the strongest effects appearing at higher speeds and in conditions involving incomplete or minimal visibility equipment.

These findings underscore the importance of practical exposure to slow-moving agricultural vehicles in developing accurate perceptual judgments. Familiarity with the operational dynamics and typical appearance of tractors under real traffic conditions appears to contribute significantly to more accurate estimation of their speed, which in turn has direct implications for overtaking behavior and overall traffic safety.

4.9. The Relationship Between Road Crashes and Tractor Speed Estimation

The potential connection between respondents' participation in road crashes and tractor speed estimation was examined using the independent samples *t*-test (Hypothesis 8). The results show statistically significant differences in the estimation of tractor speed at 10 km/h when the tractor has its daytime running light, beacon light, and SMV emblem installed ($F = -2.031$; $p = 0.045$).

4.10. Cluster Analysis of Drivers Based on Speed Perception Patterns

To further explore the variability in drivers' speed estimation accuracy, a cluster analysis was performed using k-means clustering. This unsupervised machine learning technique was applied to identify distinct subgroups among the respondents based on both their demographic attributes and behavioral patterns in speed estimation. Variables included in the clustering model were gender, age, place of residence, driving license category, frequency of general vehicle use, frequency of tractor use, and average estimation error across the 12 visibility conditions. Prior to clustering, all variables were normalized to ensure comparability. The optimal number of clusters was determined using the Elbow Method, which suggested a three-cluster solution.

Cluster 1 ("Experienced Rural Estimators") consisted predominantly of older, rural-based respondents with frequent exposure to tractors. This group exhibited the lowest average error in speed estimation, particularly in scenarios involving visibility equipment.

Cluster 2 ("Urban Overestimators") was characterized by younger drivers living in urban areas, most of whom reported limited contact with tractors. They showed consistently higher estimation errors, especially in low-visibility conditions.

Cluster 3 ("Mixed Experience Respondents") included participants from various environments and driving experience levels. Their estimation errors varied considerably across visibility conditions, indicating a lack of consistent estimation patterns.

A one-way ANOVA confirmed significant differences between clusters in average estimation error ($F = 6.41$; $p < 0.01$). Post hoc analysis using Tukey's HSD test indicated that Cluster 1 had significantly lower errors than Cluster 2 ($p < 0.01$), while differences between Clusters 1 and 3 and Clusters 2 and 3 were marginally significant.

These findings suggest that perception accuracy is not solely determined by individual factors, such as gender or license type, but emerges from a combination of demographic and behavioral characteristics. The results further highlight the potential value of tailored safety campaigns, particularly aimed at urban drivers who are less familiar with slow-moving agricultural vehicles.

5. Discussion

The research that was conducted has significant implications for understanding the impact of vehicle visibility equipment on the safety of all road users, especially tractor drivers. The results point to several key conclusions that can contribute to the improvement of road safety and adjust the legal regulations. This discussion highlights several key aspects based on the results obtained.

The data on the number of road crashes [32,33,35,47] with tractors in Serbia emphasize the seriousness of the safety challenges that these vehicles represent. Especially since tractors account for about 1.4% of all road crashes, with a significant number of deaths and injuries [31]. This raises the question of the effectiveness of current safety measures and the need for further innovation. Based on the results obtained and presented in the paper, it can be concluded that vehicle visibility equipment significantly contributes to a better perception of the tractor's speed and thereby improves the tractor's active safety. It should be emphasized that the use of only a beacon light is not sufficient for the accurate perception of the tractor speed. The perception of the tractor, as well as the accuracy of the estimation of the tractor's speed, has a great influence on the occurrence and severity of the consequences of road crashes involving tractors.

An analysis of the types of road crashes, especially collisions with the rear and front of tractors, as well as with trailers, provides additional information on the risks [8,31,71–76]. This type of road crash most often occurs outside the settlement. The mentioned fact only confirms the results obtained and shown in the paper about the importance of the perception of the tractor's speed. The accuracy of the perception of tractor speed is also influenced by the demographic characteristics of the respondents (gender, place of residence, frequency of driving, driver's license category). While for a detailed explanation of the existence of gender differences in the perception of tractor speed, additional research must be done (primarily a personality test and other tests on male and female subjects, other differences can be explained more easily. The place of residence has an influence on the accuracy of the estimation of tractor speed, so that respondents from rural areas, who encounter tractors more often (or can operate a tractor more often), have a more accurate estimation of tractor speed, in contrast to respondents from urban areas. The frequency of driving [4,57,77,78] also influences the accuracy of the estimation of the tractor's speed, so drivers with more experience have a more accurate estimation of the tractor's speed. These findings indicate the importance of experience for the accuracy of the estimation of tractor speed. On the other hand, the driver's license category has an impact on the accuracy of the tractor speed estimate. Like experience, this finding only indicates the importance of experience for the accuracy of tractor speed estimation. The above facts should be used to create an educational program [4,77,78] in driving schools, but also in other ways (campaigns on traffic, school competitions, tribunes, etc.), in order to draw attention to the importance of using vehicle visibility equipment for the accuracy of estimating tractor speed, but also the importance of other factors for the improvement of all road users. Also, the importance of the data obtained and presented and analyzed in the work should also be a task for the well-founded and science-based attitudes of the state legislative authority for the creation of legal solutions regarding tractor drivers [79,80].

The results of the cluster analysis provide an important extension to the previously reported findings, offering a more nuanced understanding of how various driver profiles relate to perceptual accuracy. The identification of the "Experienced Rural Estimators" cluster, with significantly lower estimation errors, reinforces the idea that both exposure to agricultural traffic and routine familiarity with tractors contribute to more accurate speed judgments. In contrast, the "Urban Overestimators" cluster illustrates a clear vulnerability among drivers with limited interaction with tractors, who consistently overestimated speed—especially in scenarios lacking visibility aids. These insights suggest that perceptual biases are not uniformly distributed across the driver population but instead follow patterns shaped by contextual and experiential factors. From a policy and education standpoint, these findings underscore the importance of targeted interventions—such as simulator-based training or awareness campaigns—focused on urban and less-experienced drivers, who may benefit most from increased exposure to realistic scenarios involving slow-moving agricultural vehicles.

Beyond academic relevance, the findings of this study have important practical implications for traffic safety policy, vehicle design, and educational initiatives. For example, the demonstrated impact of specific visibility equipment configurations on perceptual accuracy can inform legislation mandating the use of DRLs and SMV emblems on agricultural vehicles. In addition, the results may be used to improve driver education curricula, particularly by integrating simulator-based training to help urban drivers better estimate the speed of slow-moving vehicles. The findings can also guide manufacturers in designing more conspicuous lighting and signaling systems for tractors and similar vehicles operating on public roads.

Limitations and Future Research

Despite the methodological rigor of this study, several limitations should be acknowledged. First, the use of a driving simulator, although effective for experimental control, may not fully replicate the complexity of real-world driving, particularly under dynamic environmental conditions (e.g., variable lighting, weather, or road distractions). Second, the study focused on a specific tractor model and two selected speeds, which may limit the generalizability of findings to other types of agricultural vehicles or higher-speed contexts. Third, the sample consisted of participants from one country, and cultural or regional familiarity with agricultural traffic might influence perceptual accuracy. Additionally, while this study analyzed some demographic influences, such as gender and residence, it did not explore deeper psychological or cognitive variables (e.g., risk perception, attention span, or personality traits), which may also play a role in speed estimation. Future research should expand to include more diverse vehicle types, broader speed ranges, and real-world conditions, as well as a more comprehensive psychological profiling of drivers [28,81,82].

6. Conclusions

The findings of this study confirm that the perception of tractor speed is significantly influenced by the presence and configuration of vehicle visibility equipment. The most accurate estimations occurred when tractors were equipped with a full set of visibility aids, including daytime running lights, beacon lights, and the SMV emblem, particularly at higher operational speeds. Conversely, the absence of such equipment led to a notable increase in perceptual error, emphasizing the crucial role of visual cues in enhancing driver awareness and response. Beyond the equipment itself, individual differences among drivers, shaped by gender, place of residence, driving habits, and license category, further contributed to variability in speed perception. These differences were particularly evident among respondents from urban environments and those with limited exposure to agricultural vehicles, who demonstrated a tendency toward overestimation. Notably, the frequency of general vehicle use, rather than tractor-specific experience, emerged as a more consistent predictor of estimation accuracy. While some factors, such as years of license possession or frequency of tractor driving, did not show a statistically significant influence, others, such as crash involvement history, proved relevant under specific conditions, highlighting the multifaceted nature of perceptual performance in traffic contexts.

The inclusion of cluster analysis in this study significantly enhanced the understanding of individual differences in speed perception related to tractor visibility. By identifying distinct driver profiles based on demographic and behavioral attributes, the analysis demonstrated that estimation accuracy is not uniformly distributed across the population. Instead, it is shaped by factors, such as driving experience, place of residence, and frequency of interaction with agricultural vehicles. These findings provide a solid basis for the development of more effective, evidence-based road safety strategies tailored to the needs of specific driver groups. Incorporating this knowledge into educational programs and legislative planning may contribute to a more inclusive and efficient approach to enhancing rural traffic safety and promoting sustainable mobility.

Overall, the study offers empirically grounded insights into how low-cost visibility measures can reduce perceptual errors and improve the active safety of tractors in rural road environments. These conclusions are especially relevant for safety policies involving slow-moving agricultural vehicles. As such, the practical application of these findings supports the development of targeted educational interventions, especially for urban drivers, and encourages the integration of visibility standards into national traffic safety regulations. The results may also serve as a foundation for future revisions to road traffic legislation, aimed at standardizing the use of visual warning systems on slow-moving vehicles.

Future research should build on these insights by examining a broader range of visibility technologies and their interaction with human cognitive factors. Potential directions include the integration of smart lighting systems, real-time warning interfaces, and adaptive perception-based technologies. Moreover, cross-cultural comparisons could shed light on perceptual and behavioral variations in international contexts, contributing to the development of harmonized global safety standards. In parallel, interdisciplinary studies that combine psychological profiling, risk perception analysis, and neurocognitive testing would offer a deeper understanding of the mechanisms underlying speed misjudgment. Altogether, this research paves the way toward a safer, more perceptually aligned traffic system, particularly in agricultural and mixed rural environments.

Author Contributions: Conceptualization, A.T. and T.I.; methodology, S.S.; software, T.I.; validation, A.T., T.I., and S.S.; formal analysis, T.I.; investigation, S.S.; resources, A.T.; data curation, T.I.; writing—original draft preparation, T.I.; writing—review and editing, T.I., V.L., and L.N.; visualization, S.S.; supervision, T.I.; project administration, T.I.; funding acquisition, L.N. All authors have read and agreed to the published version of the manuscript.

Funding: This research received no external funding.

Institutional Review Board Statement: Not applicable.

Informed Consent Statement: Not applicable.

Data Availability Statement: Data are contained within the article.

Conflicts of Interest: The authors declare no conflicts of interest.

References

1. Pitka, P.; Miličić, M.; Kovačević, T.; Majstorović, M. Development of information systems in road transportation. *J. Road Traffic Eng.* **2024**, *70*, 41–45. [CrossRef]
2. Hasanov, R.I.; Giyasova, Z.; Azizova, R.; Huseynova, S.; Zemri, B.E. Long-Term Dynamics Between Human Development and Environmental Sustainability: An Empirical Analysis of CO_2 Emissions in Azerbaijan. *Chall. Sustain.* **2024**, *12*, 273–280. [CrossRef]

3. Trifunović, A.; Senić, A.; Čičević, S.; Ivanišević, T.; Vukšić, V.; Simović, S. Evaluating the Road Environment Through the Lens of Professional Drivers: A Traffic Safety Perspective. *Mechatron. Intell. Transp. Syst.* **2024**, *3*, 31–38. [CrossRef]
4. Khan, D. A Functional Energy Minimization Framework for the Detection of Crash Stones on Road Surfaces in Intelligent Transportation Systems. *Mechatron. Intell. Transp. Syst.* **2025**, *4*, 81–91. [CrossRef]
5. Integrated Database on Road Safety Features. Available online: https://bazaabs.abs.gov.rs/absPortal/ (accessed on 17 February 2025).
6. Feroz Khan, A.B.; Ivan, P. Integrating Machine Learning and Deep Learning in Smart Cities for Enhanced Traffic Congestion Management: An Empirical Review. *J. Urban Dev. Manag.* **2023**, *2*, 211–221. [CrossRef]
7. Hussein, A.M.; Osman, B.M. The Impact of Rapid Urbanization on Poverty Levels in the Context of Climate Change: Empirical Evidence from Somalia. *Chall. Sustain.* **2024**, *12*, 281–291. [CrossRef]
8. Senić, A.; Dobrodolac, M.; Stojadinović, Z. Development of Risk Quantification Models in Road Infrastructure Projects. *Sustainability* **2024**, *16*, 7694. [CrossRef]
9. Kalem, A.; Tadić, S.; Krstić, M.; Čabrić, N.; Medić, A.; Branković, N. Evaluation of Railway Infrastructure Managers' Efficiency Using a Pearson's Correlation-Based DEA Method Model. *Oppor. Chall. Sustain.* **2024**, *3*, 256–268. [CrossRef]
10. Stević, Ž.; Subotić, M.; Softić, E.; Božić, B. Multi-Criteria Decision-Making Model for Evaluating Safety of Road Sections. *J. Intell. Manag. Decis.* **2022**, *1*, 78–87. [CrossRef]
11. Liao, X.; Wu, G.; Yang, L.; Barth, M.J. A Real-World Data-Driven approach for estimating environmental impacts of traffic accidents. *Transp. Res. Part. D Transp. Environ.* **2023**, *117*, 103664. [CrossRef]
12. Lankarani, K.B.; Heydari, S.T.; Aghabeigi, M.R.; Moafian, G.; Hoseinzadeh, A.; Vossoughi, M. The impact of environmental factors on traffic accidents in Iran. *J. Inj. Violence Res.* **2014**, *6*, 64. [CrossRef]
13. Adshead, F.; Salman, R.A.-S.; Aumonier, S.; Collins, M.; Hood, K.; McNamara, C.; Moore, K.; Smith, R.; Sydes, M.R.; Williamson, P.R. A strategy to reduce the carbon footprint of clinical trials. *Lancet* **2021**, *398*, 281–282. [CrossRef]
14. Ivanišević, T.; Trifunović, A.; Marković, N.; Simović, S. Analysis of the Influence of Vehicle Color on Speed Perception and Estimation. *J. Road Traffic Eng.* **2025**, *71*, 47–51. [CrossRef]
15. Calsavara, F.; Kabbach Junior, F.I.; Larocca, A.P.C. Effects of Fog in a Brazilian Road Seg-ment Analyzed by a Driving Simulator for Sustainable Transport: Drivers' Visual Profile. *Sustainability* **2021**, *13*, 9448. [CrossRef]
16. Christoforou, Z.; Kallianiotis, A.; Farhi, N. Design, Development, and Validation of Driving Simulators for Enhancing the Safety and Sustainability of Electric Microvehicles. *Sustainability* **2025**, *17*, 3260. [CrossRef]
17. D'Elia, A.; Newstead, S. Evaluation of the effectiveness of daytime running lights (DRLs). *J. Saf. Res.* **2023**, *85*, 95–100. [CrossRef] [PubMed]
18. Knight, I.; Sexton, B.; Bartlett, R.; Barlow, T.; Latham, S.; McCrae, I. *Daytime Running Lights (DRL): A Review of the Reports from the European Commission*; TRL Limited: Wokingham, UK, 2006.
19. Ivanišević, T.; Ivković, I.; Čičević, S.; Trifunović, A.; Pešić, D.; Vukšić, V.; Simović, S. The impact of daytime running (LED) lights on motorcycles speed estimation: A driving simulator study. *Transp. Res. Part. F Traffic Psychol. Behav.* **2022**, *90*, 47–57. [CrossRef]
20. Ferreira, M.; d'Orey, P.M. On the impact of virtual traffic lights on carbon emissions mitigation. *IEEE Trans. Intell. Transp. Syst.* **2011**, *13*, 284–295. [CrossRef]
21. Czech, S.; Shakeshaft, A.P.; Byrnes, J.M.; Doran, C.M. Comparing the cost of alcohol related traffic crashes in rural and urban environments. *Accid. Anal. Prev.* **2010**, *42*, 1195–1198. [CrossRef]
22. Dhakal, B.; Al-Kaisy, A. An Empirical Evaluation of a New Heuristic Method for Identifying Safety Improvement Sites on Rural Highways: An Oregon Case Study. *Sustainability* **2024**, *16*, 2047. [CrossRef]
23. Bridgelall, R. Spatial Analysis of Middle-Mile Transport for Advanced Air Mobility: A Case Study of Rural North Dakota. *Sustainability* **2024**, *16*, 8949. [CrossRef]
24. Pešić, D.; Trifunović, A.; Ivković, I.; Čičević, S.; Žunjić, A. Evaluation of the effects of daytime running lights for passenger cars. *Transp. Res. Part. F Traffic Psychol. Behav.* **2019**, *66*, 252–261. [CrossRef]
25. Cavallo, V.; Pinto, M. Are car daytime running lights detrimental to motorcycle conspicuity? *Accid. Anal. Prev.* **2012**, *49*, 78–85. [CrossRef] [PubMed]
26. Peña-García, A.; de Oña Lopez, R.; Espín Estrella, A.; Aznar Dols, F.; Calvo Poyo, F.J.; Molero Mesa, E.; de Oña López, J. Influence of daytime running lamps on visual reaction time of pedestrians when detecting turn indicators. *J. Saf. Res.* **2010**, *41*, 385–389. [CrossRef]
27. Road Safety Agency. Overview Report. Safety of Tractor Drivers in Traffic. 2022. Available online: https://www.abs.gov.rs/admin/upload/documents/20221107135715-bezbednost-traktorista-u-saobracaju_2019_2021.pdf (accessed on 17 February 2025).
28. Pešić, A.; Stephens, A.N.; Newnam, S.; Čičević, S.; Pešić, D.; Trifunović, A. Youth Perceptions and Attitudes towards Road Safety in Serbia. *Systems* **2022**, *10*, 191. [CrossRef]
29. Law on Traffic Safety on the Roads. In *Official Gazette*; Ministry of the Republic of Serbia: Belgrade, Serbia, 2021.
30. Stanojević, P.; Jovanović, D.; Lajunen, T. Influence of Traffic Enforcement on the Attitudes and Behavior of Drivers. *Accid. Anal. Prev.* **2013**, *52*, 29–38. [CrossRef]

31. Marković, N.; Ivanišević, T.; Petrović, T.; Vukšić, V. Analysis of the Causes of Traffic Accidents with the Participation of Tractors. In Proceedings of the 2nd Expert Seminar with International Participation "Traffic Safety in the Local Community"; Ministry of Transport and Communications of Republika Srpska; Republic of Srpska Traffic Safety Agency; Auto-Moto Association of Republika Srpska, Banja Luka, Bosnia and Herzegovina, 31 October–1 November 2013; pp. 205–213, ISBN 978-99938-615-3-9.
32. Truelove, V.; Freeman, J.; Kaye, S.A.; Watson, B.; Mills, L.; Davey, J. A Unified Deterrence-Based Model of Legal and Non-Legal Factors That Influence Young Driver Speeding Behaviour. *Accid. Anal. Prev.* **2021**, *160*, 106327. [CrossRef]
33. Soleymani Kermani, M.; Namazian Jam, A. Modifying PIARC's Linear Model of Accident Severity Index to Identify Roads' Accident Prone Spots to Rehabilitate Pavements Considering Nonlinear Effects of the Traffic Volume. *J. Rehabil. Civ. Eng.* **2016**, *4*, 45–51. [CrossRef]
34. Trifunović, A.; Pešić, D.; Čičević, S.; Antić, B. The Importance of Spatial Orientation and Knowledge of Traffic Signs for Children's Traffic Safety. *Accid. Anal. Prev.* **2017**, *102*, 81–92. [CrossRef]
35. Orriols, L.; Salmi, L.R.; Philip, P.; Moore, N.; Delorme, B.; Castot, A.; Lagarde, E. The Impact of Medicinal Drugs on Traffic Safety: A Systematic Review of Epidemiological Studies. *Pharmacoepidemiol. Drug Saf.* **2009**, *18*, 647. [CrossRef]
36. Gerberich, S.G.; Robertson, L.S.; Gibson, R.W.; Renier, C. An Epidemiological Study of Roadway Fatalities Related to Farm Vehicles: United States, 1988 to 1993. *J. Occup. Environ. Med.* **1996**, *38*, 1135–1140. [CrossRef]
37. Sundet, J.M. Effects of Colour on Perceived Depth. Review of Experiments and Evaluation of Theories. *Scand. J. Psychol.* **1978**, *19*, 133–143. [CrossRef] [PubMed]
38. Glavić, D.; Trpković, A.; Milenković, M.; Jevremović, S. The E-Scooter Potential to Change Urban Mobility—Belgrade Case Study. *Sustainability* **2021**, *13*, 5948. [CrossRef]
39. Hussain, I. An Adaptive Multi-Stage Fuzzy Logic Framework for Accurate Detection and Structural Analysis of Road Cracks. *Mechatron. Intell. Transp. Syst.* **2024**, *3*, 190–202. [CrossRef]
40. Lee, Y.M.; Sheppard, E. The effect of lighting conditions and use of headlights on drivers' perception and appraisal of approaching vehicles at junctions. *Ergonomics* **2018**, *61*, 444–455. [CrossRef]
41. Davoodi, S.R.; Hossayni, S.M. Role of Motorcycle Running Lights in Reducing Motorcycle Crashes during Daytime; A Review of the Current Literature. *Bull. Emerg. Trauma* **2015**, *3*, 73–78. [PubMed]
42. Al-Awar Smither, J.; Torrez, L.I. Motorcycle conspicuity: Effects of age and daytime running lights. *Human Factors* **2010**, *52*, 355–369. [CrossRef]
43. Koornstra, M.; Bijleveld, F.; Hagenzieker, M. *The Safety Effects of Daytime Running Lights*; SWOV Institute for Road Safety Research: Hague, The Netherlands, 1997.
44. Elvik, R. A meta-analysis of studies concerning the safety effects of daytime running lights on cars. *Accid. Anal. Prev.* **1996**, *28*, 685–694. [CrossRef]
45. Vankov, D.; Schroeter, R.; Twisk, D. Understanding the Predictors of Young Drivers' Speeding Intention and Behaviour in a Three-Month Longitudinal Study. *Accid. Anal. Prev.* **2021**, *151*, 105859. [CrossRef]
46. Elvik, R.; Høye, A.; Vaa, T.; Sørensen, M. *The Handbook of Road Safety Measures*; Emerald Publishing Limited: Leeds, UK, 2009. [CrossRef]
47. Greenan, M.; Toussaint, M.; Peek-Asa, C.; Rohlman, D.; Ramirez, M.R. The Effects of Roadway Characteristics on Farm Equipment Crashes: A Geographic Information Systems Approach. *Inj. Epidemiol.* **2016**, *3*, 31. [CrossRef]
48. Logan, E.; Kaye, S.A.; Lewis, I. The Influence of the Revised Reinforcement Sensitivity Theory on Risk Perception and Intentions to Speed in Young Male and Female Drivers. *Accid. Anal. Prev.* **2019**, *132*, 105291. [CrossRef] [PubMed]
49. Himes, S.C.; Donnell, E.T.; Porter, R.J. Posted Speed Limit: To Include or Not to Include in Operating Speed Models. *Transp. Res. Part. A Policy Pract.* **2013**, *52*, 23–33. [CrossRef]
50. van Schagen, I.; Commandeur, J.J.F.; Goldenbeld, C.; Stipdonk, H. Monitoring Speed before and during a Speed Publicity Campaign. *Accid. Anal. Prev.* **2016**, *97*, 326–334. [CrossRef]
51. Lin, M.R.; Kraus, J.F. A Review of Risk Factors and Patterns of Motorcycle Injuries. *Accid. Anal. Prev.* **2009**, *41*, 710–722. [CrossRef] [PubMed]
52. Chellappa, S.L.; Steiner, R.; Blattner, P.; Oelhafen, P.; Götz, T.; Cajochen, C. Non-Visual Effects of Light on Melatonin, Alertness and Cognitive Performance: Can Blue-Enriched Light Keep Us Alert? *PLoS ONE* **2011**, *6*, e16429. [CrossRef] [PubMed]
53. Allen, M.J.; Clark, J.R. Automobile Running Lights—A Research Report. *Clin. Exp. Optom.* **2021**, *47*, 329–345. [CrossRef]
54. Tofflemire, T.C.; Whitehead, P.C. An Evaluation of the Impact of Daytime Running Lights on Traffic Safety in Canada. *J. Saf. Res.* **1997**, *28*, 257–272. [CrossRef]
55. Simović, S.; Ivanišević, T.; Trifunović, A.; Čičević, S.; Taranović, D. What affects the e-bicycle speed perception in the era of eco-sustainable mobility: A driving simulator study. *Sustainability* **2021**, *13*, 5252. [CrossRef]
56. Geber, S.; Baumann, E.; Klimmt, C. Tailoring in Risk Communication by Linking Risk Profiles and Communication Preferences: The Case of Speeding of Young Car Drivers. *Accid. Anal. Prev.* **2016**, *97*, 315–325. [CrossRef]

57. Fierro, I.; Gómez-Talegón, T.; Alvarez, F.J. The Spanish Pictogram on Medicines and Driving: The Population's Comprehension of and Attitudes towards Its Use on Medication Packaging. *Accid. Anal. Prev.* **2013**, *50*, 1056–1061. [CrossRef]
58. Bina, M.; Graziano, F.; Bonino, S. Risky Driving and Lifestyles in Adolescence. *Accid. Anal. Prev.* **2006**, *38*, 472–481. [CrossRef]
59. Rezaee Arjroody, A.; Hosseini, S.a.; Akhbari, M.; Safa, E.; Asadpour, J. Accurate Estimation of Cost and Time Utilizing isk Analysis and Simulation (Case Study: Road Construction Projects in Iran). *Int. J. Constr. Manag.* **2024**, *24*, 19–30. [CrossRef]
60. Hameed, H.; Faheem, S.; Paiva-Santos, A.C.; Sarwar, H.S.; Jamshaid, M. A Comprehensive Review of Hydrogel-Based Drug Delivery Systems: Classification, Properties, Recent Trends, and Applications. *AAPS PharmSciTech* **2024**, *25*, 64. [CrossRef] [PubMed]
61. Kopeček, J. Hydrophilic Biomaterials: From Crosslinked and Self-Assembled Hydrogels to Polymer-Drug Conjugates and Drug-Free Macromolecular Therapeutics. *J. Control. Release* **2024**, *373*, 1–22. [CrossRef]
62. Wu, H.; Eungpinichpong, W.; Ruan, H.; Zhang, X.; Dong, X. Relationship between Motor Fitness, Fundamental Movement Skills, and Quality of Movement Patterns in Primary School Children. *PLoS ONE* **2021**, *16*, e0237760. [CrossRef] [PubMed]
63. Barzegar, A.; Ghadipasha, M.; Forouzesh, M.; Valiyari, S.; Khademi, A. Epidemiologic Study of Traffic Crash Mortality among Motorcycle Users in Iran (2011–2017). *Chin. J. Traumatol.* **2020**, *23*, 219–223. [CrossRef]
64. Kapatsa, C.; Kavishe, N.; Maro, G.; Zulu, S. The Identification of Sustainability Assessment Indicators for Road Infrastructure Projects in Tanzania. *Sustainability* **2023**, *15*, 14840. [CrossRef]
65. Pallant, J. *SPSS Survival Manual*, 3rd ed.; McGraw-Hill Open University Press: Maidenhead, UK, 2007.
66. IBM Corporation. *IBM SPSS Statistics 28 Brief Guide*; IBM Corporation: Armonk, NY, USA, 2021.
67. Singh, N.; Katiyar, S.K. Application of Geographical Information System (GIS) in Reducing Accident Blackspots and in Planning of a Safer Urban Road Network: A Review. *Ecol. Inform.* **2021**, *66*, 101436. [CrossRef]
68. Kuo, Y.C.; Lu, S.T. Using Fuzzy Multiple Criteria Decision Making Approach to Enhance Risk Assessment for Metropolitan Construction Projects. *Int. J. Proj. Manag.* **2013**, *31*, 602–614. [CrossRef]
69. Flyvbjerg, B.; Bester, D.W. The Cost-Benefit Fallacy: Why Cost-Benefit Analysis Is Broken and How to Fix It. *J. Benefit-Cost. Anal.* **2021**, *12*, 395–419. [CrossRef]
70. Luttinen, R.T. *Statistical Analysis of Vehicle Time Headways*; Aalto University: Espoo, Finland, 1996.
71. Merola, F.; Bernardeschi, C.; Lami, G. A Risk Assessment Framework Based on Fuzzy Logic for Automotive Systems. *Safety* **2024**, *10*, 41. [CrossRef]
72. Love, P.E.D.; Ahiaga-Dagbui, D.D.; Irani, Z. Cost Overruns in Transportation Infrastructure Projects: Sowing the Seeds for a Probabilistic Theory of Causation. *Transp. Res. Part. A Policy Pract.* **2016**, *92*, 184–194. [CrossRef]
73. Zhang, X.; Mohandes, S.R. Occupational Health and Safety in Green Building Construction Projects: A Holistic Z-Numbers-Based Risk Management Framework. *J. Clean. Prod.* **2020**, *275*, 122788. [CrossRef]
74. Shahbodaghlou, F.; Samani, B.A. A Fuzzy Systematic Approach to Construction Risk Analysis. *J. Risk Anal. Cris. Response* **2012**, *2*, 275–284. [CrossRef]
75. Han, S.H.; Kim, D.Y.; Kim, H.; Jang, W.S. A Web-Based Integrated System for International Project Risk Management. *Autom. Constr.* **2008**, *17*, 342–356. [CrossRef]
76. Dikmen, I.; Birgonul, M.T.; Han, S. Using Fuzzy Risk Assessment to Rate Cost Overrun Risk in International Construction Projects. *Int. J. Proj. Manag.* **2007**, *25*, 494–505. [CrossRef]
77. Kodepogu, K.; Manjeti, V.B.; Siriki, A.B. Machine Learning for Road Accident Severity Prediction. *Mechatron. Intell. Transp. Syst.* **2023**, *2*, 211–226. [CrossRef]
78. Trifunović, A.; Cicevic, S.; Pesic, D.; Samčović, A.; Markovic, V. Surveying Disadvantaged Children's Traffic Safety Education in a Comparison between Paper and Electronic Methods: A Case Example for the Expanded Use of Educational Technology. *Transp. Res. Rec.* **2023**, *2677*, 401–417. [CrossRef]
79. Ali, R. Intelligent Road Crack Detection Using Fuzzy Logic and Multi-Scale Optimization. *Inf. Dyn. Appl.* **2025**, *4*, 1–11. [CrossRef]
80. Badi, I.; Bouraima, M.B.; Kiptum, C.K. Integrating Cultural Norms and Behav-ioral Risk Factors into Traffic Accident Mitigation: A Hybrid MCDM Approach for Libya. *Mechatron. Intell. Transp. Syst.* **2025**, *4*, 41–48. [CrossRef]
81. Li, Y.; Liu, B.; Zhang, W. Driving-Related Cognitive Abilities Prediction Based on Trans-former's Multimodal Fusion Framework. *Sensors* **2025**, *25*, 174. [CrossRef]
82. Trifunović, A.; Senić, A. Pyramid of Contribution Review: A Structured Model for Functional Literature Integration in Scientific Writing. *Educ. Sci. Manag.* **2025**, *3*, 40–56. [CrossRef]

sustainability

MDPI

Review

Crowdfunding in Transport Innovation and Sustainability: A Literature Review and Future Directions

Marta Mańkowska [1], Dominika Kordela [2] and Monika Pettersen-Sobczyk [3,*]

1 Institute of Management, University of Szczecin, 71-004 Szczecin, Poland; marta.mankowska@usz.edu.pl
2 Institute of Economics and Finance, University of Szczecin, 71-101 Szczecin, Poland; dominika.kordela@usz.edu.pl
3 Institute of Spatial Management and Socio-Economic Geography, University of Szczecin, 71-101 Szczecin, Poland
* Correspondence: monika.pettersen-sobczyk@usz.edu.pl; Tel.: +48-605-696-262

Abstract

Sustainable transport innovation often faces funding gaps, as traditional public and private sources rarely support early-stage or high-risk initiatives. Crowdfunding, enabled by digital transformation, is emerging as a complementary financing mechanism for this sector. This study presents a literature review combined with bibliometric mapping to examine the evolving research landscape on crowdfunding in transport. Three research questions guide the analysis: RQ1—What are the dominant research areas at the intersection of crowdfunding and transport? RQ2—What types of transport projects are financed via crowdfunding? RQ3—What research gaps and future directions emerge for transport innovation financing? Findings reveal three core research areas: (1) Sustainability and finance, (2) Fintech and blockchain, and (3) Management and consumer behavior. We propose a typology of crowdfunded transport projects comprising five categories: (1) Large-scale transport infrastructure, (2) Sustainable local mobility, (3) Innovative start-ups, (4) New business models, and (5) Advanced systems and technologies. This demonstrates crowdfunding's versatility beyond traditional infrastructure, supporting high-risk innovations critical for decarbonization and technological transformation. The study highlights domain-specific challenges—such as integrating PPP models with digital finance and ensuring investor protection—and emphasizes crowdfunding's role as an enabler of low-carbon transition aligned with global climate strategies (EU Green Deal, SDGs). Despite its potential, investor safety remains a major concern. Policy implications include sandbox regulation, standardized risk assessment, and operationalizing PPP–crowdfunding hybrids to unlock large-scale and innovative transport projects.

Keywords: crowdfunding; transport innovation; decarbonisation

1. Introduction

Information and Communication Technologies (ICT) contribute to the digital transformation of transport by enabling platform-based financing and novel business models. One recognized financial innovation resulting from this development is crowdfunding. As both an alternative financial instrument and an internet-native business model, crowdfunding leverages digital payments and online communities to transfer capital at low coordination cost [1].

In recent years, crowdfunding has gained recognition as a vital form of financing across various industries with high demand for external funding and limited access to

https://doi.org/10.3390/su18020576

traditional financial sources. Simultaneously, it has become the subject of extensive scientific study, exploring its mechanism, entities' motives, investor behavior, platform development, and regulation dilemmas [2–13]. Research also focuses on crowdfunding's impact on supporting innovation, the influence of the concept of openness, and the role of platforms' innovation in campaign success [14,15].

However, while the crowdfunding literature is extensive, the transport-specific evidence remains fragmented. This is even though the importance and potential of crowdfunding for the transport sector have been highlighted in key documents, such as the 2017 OECD report, given its indication as one of the innovative financing mechanisms proper for sustainable transport and mobility [16]. Data from this report confirm this scarcity, showing transport projects account for a very small share (e.g., 1.0% of projects) but frequently require the highest average funding goals, suggesting a specific, capital-intensive niche.

This makes the consideration of crowdfunding even more relevant when viewed against the backdrop of global and regional climate strategies—including the EU Green Deal and the United Nations Sustainable Development Goals (SDGs)—which face significant challenges in financing the decarbonization of transport and mobility systems. Current research highlights the critical need for innovative financing approaches beyond traditional public funding to achieve net-zero objectives [17]. Studies indicate that while policy momentum is strong, investment decisions often lag behind net-zero scenarios [18]. Investors remain cautious due to risk expectations and profitability uncertainties in green technologies. Green climate funds are emerging as potential solutions, with evidence suggesting they can enhance capacity to channel both public and private funding [19]. Technological innovations are seen as key enablers, and scholars recommend a favorable regulatory environment to generate positive short-term effects and create significant environmental, economic, and social impacts [20]. The research underscores that no single solution exists, and a combination of financial, technological, and policy innovations will be crucial to achieving climate objectives. Within this landscape, crowdfunding emerges as a promising complementary mechanism—yet remains underexplored in transport and mobility research—underscoring the need for deeper studies on its potential role in advancing sustainable, innovative, and climate-aligned transport solutions.

The current research evidence is insufficient to determine the extent to which crowdfunding can address funding gaps for high-risk transport and mobility innovations with decarbonization potential. This review addresses that gap and examines how crowdfunding can support the technological transformation and decarbonisation of the transport sector, including early-stage, high-risk projects. To achieve this, we employed a methodology that combined literature review and bibliometric keyword co-occurrence mapping to map current research, classify crowdfunded transport projects, and identify future research directions for sustainable mobility.

2. Research Method and Tools

Three research questions (RQ) guide this study:

RQ1—What are the dominant research areas at the intersection of crowdfunding and transport?

RQ2—What types of transport projects are financed via crowdfunding?

RQ3—What research gaps and future directions emerge for transport innovation financing?

To answer these questions, we adopted elements of the systematic literature review method [21,22], integrating a scoping review technique [23–26] to effectively map concepts, evidence, and research gaps. The review protocol was comprehensive, covering search strategy, screening, data extraction, and synthesis. The analysis was complemented by bibliometric keyword co-occurrence mapping using the VOSviewer IT tool (version 1.6.20) [27,28].

https://doi.org/10.3390/su18020576

This technique was specifically employed to address RQ1 by visualizing the main research areas and their interrelationships within the corpus. Figure 1 shows the three distinct stages of the SLR process.

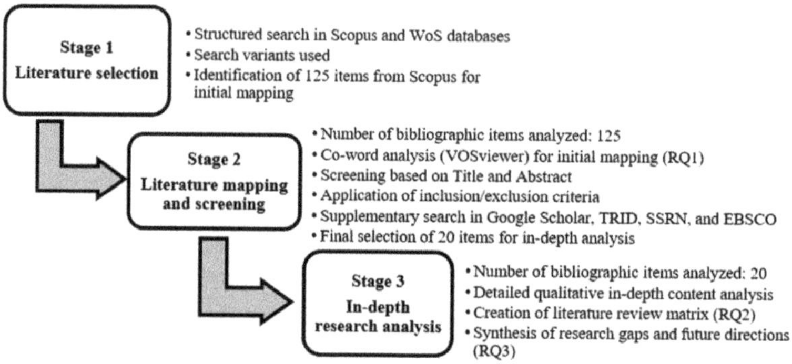

Figure 1. Stages of the literature review process on crowdfunding in transport.

Stage 1: literature selection was conducted using the Scopus and Web of Science (WoS) databases (as of 13 January 2025). These databases were selected because they are comprehensive and widely recognized sources for management, transport, and innovation studies, providing robust metadata for bibliometrics. Given the recent emergence of the topic, we applied an open time window up to the search date.

We used structured queries combining terms for crowdfunding and transport (e.g., 'crowdfunding AND transport'). To address term variability and maximize coverage, the search was conducted in two primary variants: 'crowdfunding refined by transport' and 'transport refined by crowdfunding'. Table 1 summarizes the full queries and corresponding initial item counts.

Table 1. Search queries and initial results for crowdfunding in transport.

Database	Number of Bibliographic Items by Search Variants (Query)		
	Crowdfunding and Transport	Crowdfunding Refined by Transport	Transport Refined by Crowdfunding
Scopus	16	125	116
WoS	8	30	22

The search for the query 'crowdfunding refined by transport' in Scopus yielded the largest number of results (125). Since these Scopus results included items found in the WoS search, the 125 articles from the Scopus database were selected for the initial literature mapping (Stage 2).

In Stage 2, a co-occurrence analysis of keywords was conducted using the VOSviewer IT tool on the 125 items exported in RIS format. This bibliometric analysis primarily addressed RQ1 by visualizing the main research areas in crowdfunding for transport projects and their interrelationships within the corpus. This full dataset of 125 items was intentionally used for bibliometric mapping (RQ1) because a preliminary analysis using only highly relevant items produced insufficient density and a low-quality visualization.

Following the bibliometric mapping, the 125 items were subjected to a rigorous screening based on titles and abstracts, applying standard inclusion/exclusion criteria:

1. Inclusion: peer-reviewed articles, conference papers, and book chapters in English that examine crowdfunding applied to transport projects, organizations, or policies.

https://doi.org/10.3390/su18020576

2. Exclusion: studies focusing solely on crowdsourcing without financing, purely conceptual pieces without transport relevance, non-scholarly sources, and duplicates.

Due to the narrow scope of the topic and the limited number of highly relevant studies, the screening and selection of the final corpus were performed independently by the three co-authors to ensure maximum coverage and minimize the risk of omitting valuable items.

Given the severely limited number of highly relevant results (just a dozen) from the primary databases, the search was supplemented by a targeted review using Google Scholar to identify valuable studies not indexed in Scopus or WoS. The top 100 literature items (based on relevance for the keyword 'crowdfunding and transport') were reviewed. After discarding repetitions from earlier stages, 4 additional literature items were identified.

Following suggestions made during the first round of article review, in December 2025, we further expanded our search using the same criteria to include additional databases: TRID (Transport Research International Documentation), SSRN (Social Science Research Network), and EBSCO. After removing duplicates from previous results in WoS, Scopus, and Google Scholar, we identified two additional references—one peer-reviewed paper from EBSCO and one preprint from SSRN. This brought the final corpus to 20 references for in-depth content analysis in Stage 3 (14 from primary databases, 4 from Google Scholar, 1 from SSRN, and 1 from EBSCO).

Additionally, to strengthen the conclusions drawn from the scientific literature review, we used EBSCO to access its non-academic content (including press reports, descriptive articles, and project summaries) to identify six examples of crowdfunding projects not covered in scholarly research but directly demonstrating the application of this financing mechanism in the transport sector. This complementary analysis allowed us to better illustrate the practical implementation of the discussed funding models and to contextualize the research findings with a broader set of illustrative cases.

In Stage 3, the final corpus of 20 literature items was subjected to in-depth qualitative content analysis. Data extraction and synthesis were performed by coding the study context, crowdfunding model, methodological approach, and key findings. This analysis was complemented by the illustrative examples from EBSCO, which provided additional insights into how the identified models operate in practice. The analysis primarily addressed RQ2 through the creation of a literature review matrix and RQ3 through the synthesis of emerging research gaps and future directions.

The keyword co-occurrence mapping based on 125 items served as an exploratory step to visualize the broad conceptual landscape (RQ1), while the in-depth qualitative analysis focused on a refined corpus of 20 highly relevant studies to address RQ2 and RQ3. This two-tiered approach combines bibliometric mapping, which provides a comprehensive overview of thematic clusters—even including peripheral sources—with targeted content analysis to extract detailed insights.

3. Results and Findings

3.1. Literature Mapping (RQ1)

Keyword co-occurrence analysis was carried out using the VOSviewer IT tool to address RQ1 by identifying the dominant research areas. The minimal number of occurrences of the keyword was set at level 5, which yielded 12 keywords that met the threshold for inclusion in the network analysis. The total strength of the co-occurrence links was calculated for each of these 12 keywords, which were subsequently divided into 3 distinct clusters (Figure 2). The co-occurrence map clearly demonstrates the central importance of the terms 'crowdfunding' and 'crowdsourcing' within the network, indicated by their large node size and the density of their links. This confirms that these concepts form the core methodological and conceptual foundation of the transport financing research.

https://doi.org/10.3390/su18020576

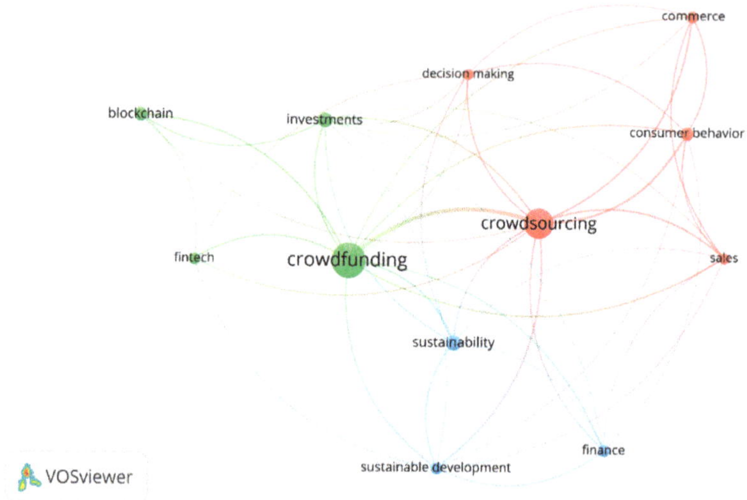

Figure 2. Co-occurrence network map of keywords related to crowdfunding in transport (network visualization).

The resulting network visualization revealed three primary, interconnected research areas constituting the current landscape of crowdfunding in transport:

Cluster 1 (blue): Sustainability and finance shows links with sustainability, sustainable development, and finance. In general studies on crowdfunding, researchers focus on the influence of sustainable orientation on the success of crowdfunding campaigns and society's engagement in sustainable projects. However, there is an emerging interest in issues regarding sustainable transport financing.

Cluster 2 (green): Fintech and blockchain covers crowdfunding, blockchain, fintech, and investments. This cluster links to the innovation perspective, focusing on financial services as a technology for faster and easier capital transfer and access to funding, rather than the core product itself. Cluster 2 represents a fast-growing area.

Cluster 3 (red): Management and consumer behavior includes research on crowdsourcing, commerce, consumer behavior, decision-making, and sales. These publications are more management-oriented, focusing on consumer decision-making and shifting business models toward the sharing economy, co-production, and co-creation. The research concentrates on modern systems and applications that are user-friendly, easily accessible, and mostly digital.

The mapping analysis confirms a close relationship between crowdfunding and crowdsourcing, which is understandable as both are based on harnessing the power of the internet community. Both terms demonstrate a nexus to sustainability, sustainable development, and finance (C1). Moreover, the study area is closely linked to management (C3) and financial innovations (C2).

The distribution of the publication over time provides deeper insights into the field's evolution (Figure 3).

https://doi.org/10.3390/su18020576

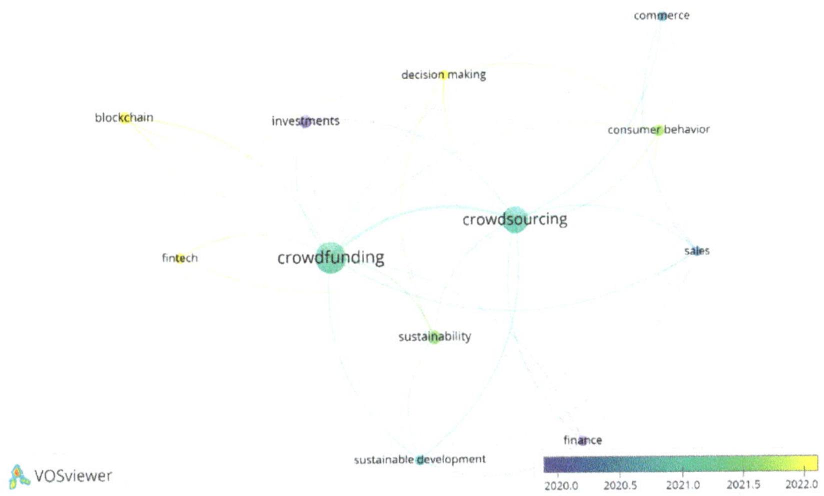

Figure 3. Co-occurrence network map of keywords related to crowdfunding in transport (overlay visualization).

Mapping shows that the intersection of crowdfunding and transport is a relatively new area of study. While the oldest publications are from 2017 and study infrastructure projects, the growth has been visible since 2020. The overlay visualization reveals specific temporal trends:

Emerging frontiers (newest): keywords related to the fintech/blockchain (C2) and sustainability (C1) clusters, such as blockchain, fintech, sustainability, and sustainable development, represent the newest research focus (average publication year 2021–2022, indicating these are the emerging frontiers of the field.

Initial focus (older): terms in the management cluster (commerce, sales, consumer behavior) (C3) show slightly earlier activity (cooler colors), suggesting they formed the initial layer of research interest, focusing on the commercial and behavioral aspects of using crowdfunding.

The overall trend shows rising attention to fintech/blockchain and decision-making, confirming the high innovative potential of this research area, driven by IT-backed decision-making linked strongly to innovative services and products and digitalization.

3.2. In-Depth Research Analysis (RQ2, RQ3)

The mapping of 125 literature items revealed general research areas on crowdfunding and transport. To identify the types of transport projects financed through crowdfunding, we conducted a detailed review of 20 selected articles (Stage 3). We focused on studies that describe real-transport projects funded via crowdfunding. The results are presented in Table 2.

We identify five types of projects: large-scale transport infrastructure; sustainable local mobility; innovative start-ups; innovative business models; and innovative systems and technologies. Definitions are provided to clarify boundaries. Categories may overlap, and classification reflects the dominant focus.

https://doi.org/10.3390/su18020576

Table 2. Categorization of transport projects financed via crowdfunding.

Study	Project Type				
	Large-Scale Transport Infrastructure	Sustainable Local Mobility	Innovative Start-Ups	New Business Models	Advanced Systems and Technology
Arifin & Wisudanto, 2017 [29]	v				
Corona-Treviño, 2017 [30]				v	
Soto, 2017 [31]	v				
Zhu et al., 2017 [32]		v			v
Miller & Coutts, 2018 [33]		v			
Kelemen et al., 2019 [34]			v		
Polishchuk et al., 2019 [35]			v		
Berawi et al., 2020 [36]	v				
Fadhil et al. 2020 [37]		v			
Klein et al., 2020 [38]				v	
Harris & Wonglimpiyarat, 2021 [39]					v
Zhang et al., 2021 [40]				v	
De Crescenzo et al., 2022 [41]		v			
Kosheleva et al., 2022 [42]					v
Ling et al., 2022 [43]				v	
Pranata et al., 2022 [44]	v				
Shukla et al., 2022 [45]	v		v		
Tavmen, 2024 [46]					v
Erfani et al., 2024 [47]		v			
Yin et al., 2024 [48]				v	

Legend: The symbol 'v' indicates that the respective study addresses the given project type.

The studies on crowdfunding in transport infrastructure projects are the earliest and can be divided into large-scale infrastructure projects and smaller sustainable mobility projects. The studies on large-scale transport infrastructure development focus on using crowdfunding in financing initial project stages [31] and also explore its integration with the public–private partnership (PPP) model [36]. They highlight the potential of implementing this financing model in countries with less developed infrastructure and limited public funds, such as those in the Asian region [29,44]. These studies also indicate significant barriers, including the risk of insufficient funds and the lack of experience among communities or crowdfunding organizations in managing large-scale transport projects, which are often considered unbankable, highly risky, and characterized by long return periods. They underline the need for appropriate regulations to ensure investment security. It is worth mentioning that in the case of debt crowdfunding for public transportation projects, the risk is lower because the local government or municipal company often serves as the guarantor [37]. A practical illustration of integrating crowdfunding into the PPP model is the InfraShares platform, launched in the United States in 2015 by Brian Ross. InfraShares enables citizens to invest directly in large-scale transport assets, such as toll roads and transit hubs, through equity crowdfunding [49]. This approach transforms residents from passive taxpayers into active stakeholders, bridging funding gaps while strengthening transparency and local accountability. Although still at an early stage, the

https://doi.org/10.3390/su18020576

concept demonstrates how crowdfunding can scale to major infrastructure projects and democratize transport financing.

Research on the application of crowdfunding in sustainable local mobility includes projects like bicycle infrastructure [33] and public transportation networks in smaller communities [41]. Crowdfunded projects in this sector are often smaller in scale, community-driven, and tailored to the specific needs of local populations. Recent studies also explore crowdfunding for equitable electric vehicle (EV) charging infrastructure, emphasizing social equity and access for underserved communities [47]. The findings suggest that crowdfunding can be an effective tool when combined with other financing mechanisms, even in cases where public budgets are limited. For illustration, practical examples of sustainable local mobility initiatives financed through crowdfunding include projects promoting micromobility and clean transport solutions. Early cases feature equity crowdfunding for innovative e-bike concepts, such as the Freygeist Bikes campaign on the Companisto platform in 2015, which raised approximately €1.5 million from over 1000 investors to launch an ultralight electric bicycle with a range of up to 100 km [50]. More recent initiatives leverage global platforms like Indiegogo to introduce electric scooters, as in Yadea's ElitePrime campaign in 2023, offering a high-performance "Urban SUV" model with a range of 120 km and super early bird pricing starting at $1299 [51]. These examples illustrate how crowdfunding supports the diffusion of sustainable micromobility technologies and complements traditional financing mechanisms.

Crowdfunding is a new and desirable research direction in implementing innovative transport projects, such as start-ups, business models, systems, and technologies.

The area of innovative start-up research primarily focuses on transportation projects that do not meet the funding criteria of traditional financial institutions, particularly in the context of seed funding. Like infrastructure project studies, they highlight that crowdfunding is especially vulnerable to risks such as fraud, unethical practices, and lack of transparency, which may discourage potential investors, particularly those with limited expertise in the transportation industry. Therefore, these studies concern supporting decision-making processes and propose quantitative risk assessment models of start-up projects in the transport sector, aimed at increasing the security of crowdfunding platforms and minimizing investment risks [34,35]. An example of such a project is Hackrod, a California-based start-up founded in 2018, which launched an equity crowdfunding campaign on StartEngine to raise $1 million for a custom 3D-printed car platform [52]. The concept integrates VR design, AI optimization, and additive manufacturing, with its La Bandita speedster featuring a recycled Tesla powertrain and a 3D-printed chassis. This case demonstrates how crowdfunding accelerates disruptive automotive technologies and democratizes vehicle manufacturing.

Research on innovative business models in transport is grounded in the concept of the sharing economy, also known as uberization, whose main idea is to connect owners of underused assets with clients via digital platforms. Early studies analyze services such as Uber and BlaBlaCar [30] and crowdfunding platforms like GoFundMe.com [38], focusing on technological management of capital assets, market risks, and revenue distribution. Complementing these multi-campaign analyses, recent work examines individual campaigns to understand how personalized narratives and transparent cost framing influence donor engagement [48]. GoFundMe campaign demonstrates how emotional appeals and clear cost breakdowns can mobilize donors for individual mobility needs. The newest research topic is the concept of flight crowdfunding, where individuals or organizations raise funds to cover the cost of flights using online platforms. Zhang et al. [40] analyze airline crowdfunding as a novel airline sales strategy to optimize ticket pricing and reduce costs. They highlight its potential to enhance airline profits while generating social benefits,

https://doi.org/10.3390/su18020576

such as reduced greenhouse gas emissions. Ling et al. [43] examined the collaboration between airlines and online travel agencies (OTAs) in airline crowdfunding. The study finds crowdfunding is more likely to be adopted in highly competitive markets or when fixed flight costs are high. The findings suggest that, under optimal market conditions, crowdfunding can create a win-win scenario for airlines, OTAs, and consumers. A practical illustration of this type of project is also the Crowdfunding-Rail project initiated by Locomore in Germany in 2015 [53]. The railway operator pioneered a hybrid financing model combining ticket pre-sales with equity crowdfunding to launch its Stuttgart–Berlin open-access route. The campaign raised approximately €460,000 from over 1000 supporters, who could pre-purchase tickets at discounted rates or invest in company shares. This approach not only secured start-up capital but also created strong customer loyalty and brand identity before operations began.

These projects show how crowdfunding enables new business models in passenger transport, fostering community engagement and offering alternatives to incumbent operators.

A very timely topic is using crowdfunding to implement innovative technical and technological advances in transportation, such as electric personal vehicles (e-scooters, e-bikes), smart city solutions (intelligent traffic management systems, transport/logistics-information clusters), and large-scale R&D projects, such as hyperloop or autonomous vehicle projects. The research of Zhu et al. [32] on charging piles for electric vehicles indicates that crowdfunding has similar effects to subsidies and is more efficient and straightforward in promoting the expansion of new technological solutions in transport. Harris & Wonglimpiyarat [39] analyze the potential for large-scale development of Hyperloop technology, using the case study of SpaceX Hyperloop. They showed crowdfunding as a key component of the financial infrastructure in innovative projects alongside corporate venture funding, initial coin offerings, and traditional financing. The survey by Kosheleva et al. [42] presented crowdfunding as one of the models that enable the development of the intelligent city system. The earliest studies by Tavmen [46] analyzed the role of open data in the development of smart cities based on the case study of the Citymapper transportation app in London (TfL), which became the fastest fundraising campaign in the platform's history. The campaign attracted 9000 individual investors, raising £6.7 million within 24 h, despite concerns regarding the company's financial projections and long-term sustainability. The study highlights that Citymapper's success in the crowdfunding campaign reflects the app's strong public appeal rather than a demonstration of a sustainable and profitable business model. The potential for financing complex technical assets is further exemplified by the APOC Aviation case. APOC Aviation, a Netherlands-based aviation aftermarket company specializing in aircraft part-out and component trading, launched an equity crowdfunding campaign in 2019 to finance the acquisition of two Boeing 737–700 airframes. The campaign successfully raised USD 2.6 million from private investors, enabling the purchase of low-cycle airframes previously configured as Boeing Business Jets [54]. This innovative financing approach illustrates how crowdfunding can support high-value, capital-intensive operations in the aviation sector, particularly in niche markets such as asset recovery and sustainable reuse of aircraft components. The case involves complex technical assets and demonstrates the potential of alternative funding models for specialized transport technologies.

The reviewed studies reveal five distinct categories of transport projects financed through crowdfunding, each with specific characteristics and risk profiles. Large-scale infrastructure initiatives typically employ early-phase crowdfunding and, in some cases, community bonds, but face substantial challenges related to construction risk, demand uncertainty, and long payback periods. Local sustainable mobility projects are small-scale and community-driven, aiming to improve accessibility and encourage modal shift

https://doi.org/10.3390/su18020576

toward greener transport options. Innovative start-ups often rely on seed-stage crowdfunding to support unbankable concepts, yet remain exposed to technological, execution, and fraud risks, highlighting the need for decision-support tools. New business models, such as sharing-economy platforms and service innovations, raise governance and fairness concerns in platform design and market regulation. Finally, advanced systems and technologies—including e-mobility solutions, smart-city applications, and deep-tech projects—are characterized by high techno-economic uncertainty and interoperability challenges, requiring robust frameworks for risk assessment and investor protection.

Evidence from the reviewed studies and project cases indicates that campaign design and governance strongly influence outcomes across project types. Community-scale initiatives, such as cycling routes, benefit from reward-based structures and transparent information disclosure, which increase pledging likelihood. In contrast, airline crowdfunding performs well only under specific market conditions—high fixed costs or intense competition—while start-ups and deep-tech projects remain highly exposed to fraud and execution risk, reinforcing the need for platform-level risk-rating and advanced assessment models. At a systemic level, weak or absent regulation consistently emerges as a critical barrier, amplifying uncertainty and limiting adoption.

The analysis of these studies also indicates clear regional variation. In Asia, PPP–crowdfunding hybrids and blockchain-based models are explored as mechanisms to bridge public funding gaps, while North America demonstrates policy-driven adoption through local mobility projects, equity-based EV initiatives, and airline crowdfunding strategies. Europe combines strong retail participation with concerns over the long-term viability of platform-based mobility models. These context-specific factors—market competition, fiscal constraints, equity policies, and regulatory maturity—shape success dynamics across project types.

The five identified project types align closely with the three thematic clusters revealed in the keyword mapping (Section 3.1). Large-scale transport infrastructure projects show strong links to Sustainability and Finance (C1) and Fintech and Blockchain (C2), reflecting debates on regulatory frameworks and PPP integration. Sustainable local mobility projects primarily connect to Sustainability and Finance (C1), emphasizing environmental goals and community-driven funding. Innovative start-ups, new business models, and advanced systems and technologies exhibit multi-label associations with Fintech and Blockchain (C2) and Management and Consumer Behavior (C3), particularly in discussions on risk assessment, platform governance, and user engagement. To capture these relationships without introducing composite labels, we map each project type (RQ2) onto the original research areas (RQ1) and indicate primary (•) and secondary (○) linkages. This alignment is summarized in Table 3.

Table 3. Mapping of crowdfunded transport project types onto research areas.

Type of Projects/ Research Areas	(1) Sustainability and Finance	(2) Fintech and Blockchain	(3) Management and Consumer Behavior
Large-scale transport infrastructure	•	•	○
Sustainable local mobility	•	○	○
Innovative start-ups	○	•	○
New business models	○	○	•
Advanced systems and technologies	○	•	○

Legend: The symbol "•" indicates a primary linkage while the symbol "○" indicates a secondary linkage between project types and research areas.

https://doi.org/10.3390/su18020576

To sum up, the analysis of RQ2 confirms that crowdfunding is moving beyond traditional sustainable infrastructure projects to encompass innovative and intelligent transport solutions with a high level of investment risk. This mapping illustrates that project types often span multiple research areas (RQ1), reflecting diverse technological, financial, and behavioral dimensions. Across all categories, the overarching common denominator is the critical need to strengthen investor safety and security—through appropriate regulations, risk assessment models, and platform governance—given the significant barriers and high-risk profile of emerging transport innovations. These findings set the stage for RQ3, which addresses research gaps and future directions.

4. Discussion and Future Research Directions

The three clusters identified through keyword mapping provide an interpretive framework for the five project types analyzed in RQ2. These relationships are multi-layered: infrastructure and local mobility align with sustainability-oriented finance, while start-ups and advanced technologies connect financial innovation to user behavior and platform governance. Building on this alignment, RQ3 synthesizes three emerging research streams: (S1) risk, regulation, and hybrid finance; (S2) behavioral and geographic expansion; and (S3) methodological and technological evolution.

Our study indicates that although crowdfunding is well described in the literature (several thousand records in WoS and Scopus), its application to transport projects remains rarely addressed. The literature review conducted here, complemented by an analysis of illustrative crowdfunding projects retrieved from EBSCO's non-academic content (such as press reports and descriptive sources), reveals that this emergent field is highly dynamic, covering a wide spectrum of projects—from large-scale infrastructure [29,31,36,44,45], to local mobility initiatives [33,37,41], innovative start-ups [34,35], new business models [30,38,40,43], and advanced systems and technologies [39,42,46].

Despite this diversity, the field faces a critical tension: while research signals strong innovative potential (e.g., fintech, Hyperloop, smart city), practical implementation remains constrained by significant barriers and low awareness among decision-makers [45]. This gap underpins the need for future research directions outlined in Table 4, aimed at addressing operational challenges and unlocking the innovative potential of crowdfunding in transport.

Table 4. Research gaps and future directions for crowdfunded transport projects.

Project Type	Key Gaps	Future Research Directions
Large-scale transport infrastructure	- unbankability - long payback periods - unclear PPP–CF integration	- develop hybrid PPP–crowdfunding models - regulatory pilots - risk-sharing contracts - viability assessments
Sustainable local mobility	- lack of standardized impact metrics (GHG, modal shift) - dispersed benefits	- create standardized GHG and energy metrics - implement civic CF governance - match-funding schemes
Innovative start-ups	- high failure risk - weak investor protection - limited screening mechanisms	- decision-support tools - sandbox regulation - comparative studies across platforms and jurisdictions

https://doi.org/10.3390/su18020576

Table 4. *Cont.*

Project Type	Key Gaps	Future Research Directions
New business models	- user-investor heterogeneity - fairness and ethics in platform design	- behavioral studies on motives - equity and fairness frameworks - market design for competition
Advanced systems and technologies	- techno-economic uncertainty - interoperability - data scarcity	- open-data protocols - life-cycle and techno-economic analysis - trials for e-mobility and smart systems

Across all project types, three cross-cutting research streams emerge:

(S1) Risk, regulation and hybrid finance;
(S2) Behavioral and geographic expansion;
(S3) Methodological and technological evolution, including energy and emissions assessment.

Stream 1 focuses on institutional and market risks, investor protection, and integration with traditional financing (e.g., PPP models [29,31,36,44,45], municipal bonds [33,37,41]). Stream 2 emphasizes expanding geographic scope beyond Europe and Asia and exploring user-investor heterogeneity, including motivations and demographic differences [34,35]. Stream 3 calls for new analytical approaches to assess market dynamics and platform governance, examine technological innovations such as blockchain and ICOs [39,42,46] and evaluate environmental impacts through life-cycle and techno-economic assessments.

5. Conclusions

The primary aim of this literature review was to comprehensively map the emerging field of crowdfunding in the transport sector and to propose a structured research agenda. The findings underscore that while crowdfunding is a mature funding instrument across various sectors, its application in transport remains nascent, yet highly versatile—covering municipal infrastructure, innovative start-ups, and deep-tech solutions.

This study contributes to the literature on transport financing and crowdfunding in three primary, verifiable ways:

First, by identifying and delineating three main thematic research areas: (1) Sustainability and finance, (2) Fintech and blockchain, and (3) Management and consumer behavior (RQ1).

Second, by classifying transport projects utilizing crowdfunding into five distinct types: (1) Large-scale transport infrastructure, (2) Sustainable local mobility, (3) Innovative start-ups, (4) New business models, and (5) Advanced systems and technologies (RQ2).

Finally, by proposing three research streams—(1) Risk, regulation, and hybrid finance; (2) Behavioral and geographic expansion; and (3) Methodological innovation and technological evolution—offering a roadmap for future scholarly work (RQ3).

Unlike prior reviews that examine crowdfunding in general, this study embeds these thematic areas within the specific context of transport innovation and sustainability. By mapping the interplay between research clusters and project types, we reveal domain-specific linkages—such as PPP–crowdfunding integration for infrastructure, fintech-driven models for mobility start-ups, and governance challenges in platform-based transport services. These intersections provide novel theoretical and practical insights that underpin the three research streams proposed above.

Crowdfunding, as an internet-based financial innovation, reinforces the digital transformation of transport by integrating fintech and blockchain solutions.

https://doi.org/10.3390/su18020576

Transport crowdfunding introduces unique regulatory, behavioral, and technological challenges—such as integrating PPP models with digital finance and ensuring investor protection in high-risk mobility innovations—issues not systematically explored in prior research.

The analysis suggests several recommendations for policymakers and practitioners. Policymakers should clarify the legal and regulatory frameworks that enable hybrid funding models (e.g., match-funding or PPP–crowdfunding hybrids) to mitigate risk and unlock large-scale infrastructure projects. Future research should operationalize PPP–crowdfunding hybrids through pilot frameworks and risk-sharing mechanisms, as indicated in Table 4. Given the capital-intensive nature of transport projects and investor risk aversion, crowdfunding offers a mechanism to mobilize resources for early-stage, high-risk innovations. Furthermore, crowdfunding should be promoted as an alternative source for innovative and environmentally friendly transport projects where traditional funding is limited, alongside efforts to foster adoption in under-studied regions (e.g., Latin America and Africa) to diversify the evidence base.

Crowdfunding thus emerges as a critical enabler of the low-carbon transition in transport by de-risking and funding projects—such as early-stage EV infrastructure or sustainable mobility concepts—that traditional financial institutions often avoid. These findings align with global climate strategies, such as the EU Green Deal and SDGs, emphasizing crowdfunding's role in bridging financing gaps for sustainable mobility. Policy frameworks should incorporate sandbox regulation and standardized risk assessment models to enhance investor protection.

Our analysis, while robust, is subject to methodological and empirical limitations that inform future research. Although the evidence base was expanded beyond WoS, Scopus, and Google Scholar to include TRID, SSRN, and EBSCO, and complemented by illustrative non-academic examples, the overall corpus remains small and unevenly distributed, concentrated mainly in Europe and Asia. To achieve a broad overview, our analysis adopted a general perspective without deeply considering country-specific regulatory contexts. We acknowledge that keyword variability and indexing limitations across databases may have led to omissions, and the categorization process, while transparently reported, involved subjective judgment. Further studies should examine success and failure factors using platform-level data and quantitative models to inform decision-support tools. Integrating transport-specific frameworks, such as sustainable mobility and innovation diffusion theory, could deepen the domain specificity of future research.

These limitations pave the way for future work: researchers should conduct detailed comparative case studies on regional regulatory frameworks, expand longitudinal diffusion studies to assess long-term viability, and develop qualitative research on user-investor heterogeneity and behavioral drivers of pledging decisions.

In sum, this review establishes a foundation for systematic research on transport crowdfunding, highlighting critical gaps and directions for future inquiry.

Author Contributions: Conceptualization, M.M., M.P.-S. and D.K.; methodology, M.M.; validation, M.M. and M.P.-S.; writing—original draft preparation, M.M., M.P.-S. and D.K.; writing—review and editing, M.M., M.P.-S. and D.K.; visualization, M.M.; supervision, M.M. All authors have read and agreed to the published version of the manuscript.

Funding: Co-financed by the Minister of Science under the "Regional Excellence Initiative".

Institutional Review Board Statement: Not Applicable.

Informed Consent Statement: Not Applicable.

Data Availability Statement: The original contributions presented in this study are included in the article. Further inquiries can be directed to the corresponding author.

https://doi.org/10.3390/su18020576

Conflicts of Interest: The authors declare no conflicts of interest.

References

1. Xie, P.; Zou, C.; Liu, H. Research on Internet financial model. *Financ. Res.* **2012**, *12*, 11–22.
2. Belleflamme, P.; Lambert, T.; Schwienbacher, A. Crowdfunding: Tapping the right crowd. *J. Bus. Ventur.* **2014**, *29*, 585–609. [CrossRef]
3. Cumming, D.J.; Johan, S.A.; Zhang, Y. The role of due diligence in crowdfunding platforms. *J. Bank. Financ.* **2019**, *108*, 105661. [CrossRef]
4. Motylska-Kuzma, A. Crowdfunding and sustainable development. *Sustainability* **2018**, *10*, 4650. [CrossRef]
5. Yin, L.Y.; Khalil, F.C.; Khalil, L.J.; Kaspard, J.A. The Role of Project Description in the Success of Sustainable Crowdfunding Projects. *J. Risk Financ. Manag.* **2025**, *18*, 200.
6. Hervé, F.; Manthé, E.; Sannajust, A.; Schwienbacher, A. Determinants of individual investment decisions in investment-based crowdfunding. *J. Bus. Financ. Account.* **2019**, *46*, 762–783. [CrossRef]
7. Lukkarinen, A.; Shneor, R.; Wallenius, J. Growing pains and blessings: Manifestations and implications of equity crowdfunding industry maturation. *Decis. Support Syst.* **2022**, *157*, 113768. [CrossRef]
8. Nitani, M.; Riding, A.; He, B. On equity crowdfunding: Investor rationality and success factors. *Ventur. Cap.* **2019**, *21*, 243–272. [CrossRef]
9. Rossi, A.; Vismara, S. What do crowdfunding platforms do? A comparison between investment-based platforms in Europe. *Eurasian Bus. Rev.* **2018**, *8*, 93–118. [CrossRef]
10. Shneor, R.; Vik, A.A. Crowdfunding success: A systematic literature review 2010–2017. *Balt. J. Manag.* **2020**, *15*, 149–182. [CrossRef]
11. Vismara, S.; Wirtz, P. Fundraising, Governance and Environmental Ethics: Evidence from Equity Crowdfunding. *J. Bus. Ethics* **2025**, *200*, 841–865.
12. Szewczyk, A.; Stempnakowski, Z. Social energy as the driving force behind crowdfunding—Analysis and classification of selected attributes. *Energies* **2021**, *14*, 6062. [CrossRef]
13. Gai, L.; Algeri, C.; Ielasi, F.; Manganiello, M. Sustainability-Oriented Equity Crowdfunding: The Role of Proponents, Investors, and Sustainable Development. *Sustainability* **2025**, *17*, 2188. [CrossRef]
14. Bargoni, A.; Ferraris, A.; Bresciani, S.; Camilleri, M.A. Crowdfunding and innovation: A bibliometric review and future research agenda. *Eur. J. Innov. Manag.* **2024**, *27*, 1109–1133. [CrossRef]
15. Chu, C.C.; Cheng, Y.F.; Tsai, F.S.; Tsai, S.B.; Lu, K.H. Open innovation in crowdfunding context: Diversity, knowledge, and networks. *Sustainability* **2019**, *11*, 180. [CrossRef]
16. Charbit, C.; Desmoulins, G. *Civic Crowdfunding: A Collective Option for Local Public Goods?* OECD Publishing: Paris, France, 2017; Volume 2017.
17. Kurekova, L.; Cermakova, K.; Hromada, E.; Kaderabkova, B. Public funding in R&D and R&D outcome sustainable development: Analysis of Member States EU. *Int. J. Econ. Sci.* **2023**, *12*, 40–62.
18. Hunt, O.B.; Tilsted, J.P. 'Risk on steroids': Investing in the hydrogen economy. *Environ. Plan. A* **2025**, *57*, 1131–1150. [CrossRef]
19. Mohsin, M.; Dilanchiev, A.; Umair, M. The Impact of Green Climate Fund Portfolio Structure on Green Finance: Empirical Evidence from EU Countries. *Ekonomika* **2023**, *102*, 130–144. [CrossRef]
20. Brahmi, M.; Esposito, L.; Parziale, A.; Dhayal, K.S.; Agrawal, S.; Giri, A.K.; Loan, N.T. The Role of Greener Innovations in Promoting Financial Inclusion to Achieve Carbon Neutrality: An Integrative Review. *Economies* **2023**, *11*, 194. [CrossRef]
21. Hensel, P. *Systematyczny Przegląd Literatury w Naukach o Zarządzaniu i Jakości*; Wydawnictwo Naukowe Wydziału Zarządzania Uniwersytetu Warszawskiego: Warszawa, Poland, 2020; ISBN 9788366282193.
22. Munn, Z.; Peters, M.D.J.; Stern, C.; Tufanaru, C.; McArthur, A.; Aromataris, E. Systematic review or scoping review? Guidance for authors when choosing between a systematic or scoping review approach. *BMC Med. Res. Methodol.* **2018**, *18*, 143. [CrossRef]
23. Colquhoun, H.L.; Levac, D.; O'Brien, K.K.; Straus, S.; Tricco, A.C.; Perrier, L.; Kastner, M.; Moher, D. Scoping reviews: Time for clarity in definition, methods, and reporting. *J. Clin. Epidemiol.* **2014**, *67*, 1291–1294.
24. Tricco, A.C.; Lillie, E.; Zarin, W.; O'Brien, K.; Colquhoun, H.; Kastner, M.; Levac, D.; Ng, C.; Sharpe, J.P.; Wilson, K.; et al. A scoping review on the conduct and reporting of scoping reviews. *BMC Med. Res. Methodol.* **2016**, *16*, 15. [CrossRef]
25. Pettersen-Sobczyk, M.; Mańkowska, M. Artificial Intelligence in Optimizing the Selection of Incoterms Rules in Cross-Border Trade. State of Knowledge and Needs for Further Research. In Proceedings of the Proceedings of the 32nd International Conference on Information Systems Development, Gdańsk, Poland, 26–28 August 2024; University of Gdańsk: Gdańsk, Poland, 2024.
26. Cooper, S.; Cant, R.; Kelly, M.; Levett-Jones, T.; McKenna, L.; Seaton, P.; Bogossian, F. An Evidence-Based Checklist for Improving Scoping Review Quality. *Clin. Nurs. Res.* **2021**, *30*, 230–240. [CrossRef]
27. van Eck, N.J.; Waltman, L. Software survey: VOSviewer, a computer program for bibliometric mapping. *Scientometrics* **2010**, *84*, 523–538.

https://doi.org/10.3390/su18020576

28. Bukar, U.A.; Sayeed, M.S.; Razak, S.F.A.; Yogarayan, S.; Amodu, O.A.; Mahmood, R.A.R. A method for analyzing text using VOSviewer. *MethodsX* **2023**, *11*, 102339. [CrossRef]
29. Arifin, S.R.; Wisudanto, W. Crowdfunding Sebagai Alternatif Pembiayaan Pembangunan Infrastruktur. In Proceedings of the Simposium II UNIID 2017, Palembang, Indonesia, 19–20 September 2017; Volume 2, pp. 309–314.
30. Corona-Treviño, L. Technology management of capital assets and risks in the service sharing economy: Cases of uberization of crowdfunding and transportation in Mexico. In Proceedings of the 2017 Portland International Conference on Management of Engineering and Technology (PICMET), Portland, OR, USA, 9–13 July 2017; Volume 6, pp. 346–357. [CrossRef]
31. Soto, E.G. *Built By the Crowd, Crowdfunding Transport Infrastructure Projects*; Tilburg University: Tilburg, The Netherlands, 2017.
32. Zhu, L.; Zhang, Q.; Lu, H.; Li, H.; Li, Y.; McLellan, B.; Pan, X. Study on crowdfunding's promoting effect on the expansion of electric vehicle charging piles based on game theory analysis. *Appl. Energy* **2017**, *196*, 238–248. [CrossRef]
33. Miller, S.; Coutts, C. A multiple case study of local & creative financing of bicycle and pedestrian infrastructure. *Case Stud. Transp. Policy* **2018**, *6*, 257–264. [CrossRef]
34. Kelemen, M.; Polishchuk, V.; Gavurová, B.; Szabo, S.; Rozenberg, R.; Gera, M.; Kozuba, J.; Andoga, R.; Divoková, A.; Blišťan, P. Fuzzy Model for Quantitative Assessment of Environmental Start-up Projects in Air Transport. *Int. J. Environ. Res. Public Health* **2019**, *16*, 3585.
35. Polishchuk, V.; Kelemen, M.; Kozuba, J. Technology Improving Safety of Crowdfunding Platforms Functioning in the Context of the Protection of the Start-Up Investors in the Financial and Transport Sectors. *J. Konbin* **2019**, *49*, 313–330. [CrossRef]
36. Berawi, M.A.; Reyes, G.; Sari, M.; Saroji, G. Developing public-private partnership model with blockchain-based crowdfunding concept for port city project. In ACM International Conference Proceeding Series, Proceedings of the 2nd International Scientific Conference on Innovations in Digital Economy (SPBPU IDE '20), St. Petersburg, Russia, 22–23 October 2020; Association for Computing Machinery: New York, NY, USA, 2020.
37. Fadhil, D.N.; Jinca, A.; Prakoso, P.; Hoegeng, J. Alternative Approach to Sustainable Urban Mobility in Developing Countries. *J. Infrastruct. Policy Manag.* **2021**, *4*, 19–25. [CrossRef]
38. Klein, N.J.; Tran, M.; Riley, S. "Desperately Need a Car": Analyzing crowdfunding campaigns for car purchases and repairs on Gofundme.com. *Travel Behav. Soc.* **2020**, *21*, 247–256. [CrossRef]
39. Harris, W.L.; Wonglimpiyarat, J. SpaceX Hyperloop: A future technology disruption of the transportation system? *Foresight* **2021**, *23*, 528–543. [CrossRef]
40. Zhang, Z.; Ling, L.; Yang, F. Pricing strategy and campaign design in flight crowdfunding: A creative way to sell flight tickets. *Nav. Res. Logist.* **2021**, *68*, 795–809.
41. De Crescenzo, V.; Simeoni, F.; Ulrich, K.; Ribeiro Navarrete, S. Searching for the crowd in sustainable tourism and leisure projects. *Eur. J. Innov. Manag.* **2022**, *26*, 110–126. [CrossRef]
42. Kosheleva, T.; Ksenofontova, T.; Yue, W. Chinese Experience in the Formation of Transport-Information Clusters on a Digital Basis. *Lect. Notes Netw. Syst.* **2022**, *402 LNNS*, 719–725.
43. Ling, L.; Bao, X.; Zhang, Z.; Yang, F. Mechanisms of cooperation between airlines and online travel agencies for flight crowdfunding. *Int. Trans. Oper. Res.* **2022**, *29*, 3537–3564. [CrossRef]
44. Pranata, N.; Firdaus, N.; Mychelisda, E.; Hidayatina, A. Crowdfunding for Infrastructure Project Financing: Lessons Learned for Asian Countries. In *Handbook of Research on Big Data, Green Growth, and Technology Disruption in Asian Companies and Societies*; IGI Global: Hershey, PA, USA, 2022; pp. 276–300.
45. Shukla, G.P.; Kumar, S.; Kumar, M.; Kumar, A.; Chhetri, M. A Fuzzy Ecosystem Benchmarking for Crowdfunding in Transport Sector. *J. Adv. Transp.* **2022**, *2022*, 7071279. [CrossRef]
46. Tavmen, G. "Open data means business": Infrastructural and economic implications of opening up data in smart London. *Digit. Geogr. Soc.* **2024**, *7*, 100098. [CrossRef]
47. Erfani, A.; Cui, Q.; Decorla-Souza, P. Crowdfunding for Equitable EV Charging Infrastructure. *arXiv* **2024**, arXiv:2406.14295. [CrossRef]
48. Yin, Z.; Huang, G.; Zhao, R.; Wang, S.; Shang, W.L.; Han, C.; Yang, M. Information disclosure and funding success of green crowdfunding campaigns: A study on GoFundMe. *Financ. Innov.* **2024**, *10*, 147. [CrossRef]
49. Cho, A. Entrepreneurs Hope to Bring Crowdfunding to P3 Projects. Available online: https://www.enr.com/articles/2146-entrepreneurs-hope-to-bring-crowdfunding-to-p3-projects (accessed on 19 December 2025).
50. Hobey, E. Freygeist Redefines E-Bike: Rides with Companisto for Its Sustainable Fun Equity Crowdfunding Campaign. Available online: https://www.crowdfundinsider.com/2015/03/64248-freygeist-redefines-e-bike-rides-with-companisto-for-its-sustainable-fun-equity-crowdfunding-campaign/ (accessed on 19 December 2025).
51. Yadea. Yadea Scooter ElitePrime Launches CrowdFunding Campaign on Indiegogo: The Urban SUV. Available online: https://www.prnewswire.com/apac/news-releases/yadea-scooter-eliteprime-launches-crowdfunding-campaign-on-indiegogo-the-urban-suv-301796285.html (accessed on 19 December 2025).

https://doi.org/10.3390/su18020576

52. Jackson, B. Hackrod Launches Million Dollar Crowdfund for Custom 3D Printed Car Platform. Available online: https://3dprintingindustry.com/news/hackrod-launches-million-dollar-crowdfund-custom-3d-printed-car-platform-128997/ (accessed on 19 December 2025).
53. Herbstreuth, M. Konkurrenz zur Deutschen Bahn-Mit dem Crowdfunding-Zug von Berlin nach Stuttgart. Available online: https://www.deutschlandfunkkultur.de/konkurrenz-zur-deutschen-bahn-mit-dem-crowdfunding-zug-von-100.html (accessed on 19 December 2025).
54. AviTrader Aviation News APOC Aviation's Crowdfunding Campaign Raises US$2.6 Million. Available online: https://avitrader.com/2019/04/28/apoc-aviations-crowdfunding-campaign-raises-us2-6-million/ (accessed on 19 December 2025).

https://doi.org/10.3390/su18020576

sustainability

MDPI

Article

Achieving Port Sustainability by Harnessing the Potential of Port Operations, Ships and Tugboats

Vytautas Paulauskas [1,*], Donatas Paulauskas [1] and Vytas Paulauskas [2]

1 Marine Engineering Department, Klaipeda University, H. Manto Street 84, LT-92294 Klaipeda, Lithuania; paulauskasd75@gmail.com
2 Faculty of Electrical and Electronics Engineering, Kaunas Technological University, Studentu Street 48, LT-51368 Kaunas, Lithuania; vytaspaulauskas1@gmail.com
* Correspondence: vytautaskltc@gmail.com; Tel.: +370-685-60884

Abstract

Port sustainability focuses on balancing economic growth by focusing on environmental protection and social well-being, involving green infrastructure, renewable energy, efficient operations, and community engagement to reduce pollution, cut emissions, manage waste, and support sustainable supply chains, guided by global goals. The port is one of the most important parts of the transport system; therefore, when assessing the sustainability of the port, it is necessary to assess that the port could receive ships characterized by maximum possible parameters with minimal delays when they enter the port due to hydro meteorological and other adverse conditions. One of the conditions for a sustainable port is to optimally use the port's capabilities, provide high-quality and fast passenger services and cargo processing, and reduce the negative impact on the environment due to the generation of pollutants emitted by ship and terminal activities. As ship parameters in ports increase and port sustainability is achieved, appropriate and accurate assessment of ship maneuvering capabilities and port tugboat capacities, as well as optimal use of port infrastructure and superstructure, becomes increasingly important. Ensuring shipping safety in ports, optimizing port loading operations, reducing environmental impact and achieving port sustainability are very important parts of maritime transport activities. Since port sustainability includes technical, technological, environmental, social and other aspects, this article focuses mainly on technical, technological and environmental aspects. This article presents a methodology for assessing port infrastructure related to shipping safety in the port, port (terminal) activities and reducing environmental impacts, based on a method expressed as a comparative port sustainability index. The developed methodology can be applied in various ports, port terminals and other locations to achieve port and maritime transport sustainability.

Keywords: sustainable water transport; sustainable port infrastructure; ship navigational safety; optimal use of port infrastructure and superstructure; environmental impact assessment; comparative port sustainability index

1. Introduction

Port sustainability is one of the main features of modern port operations and is related to the planning of port infrastructure and superstructure and the navigational safety of ships [1]. The sustainability of the port is analyzed in this article, primarily focusing on the safety of port navigation (port entrances and internal navigation channels, ship turning basins and approaches to the quays) and the capabilities of port tugs to ensure the safety of port navigation. This article also focuses on the reduction of environmental impact and the efficient operation of the port and terminals in serving passengers and handling cargo (including high qualification and work culture of port and terminal employees, decarburization and digitalization of the port and terminal, and other factors).

The optimal development of port superstructure, reducing the time of ship handling in the port, plays an important role in reducing the environmental impact of ships and port terminals [2]. Safe maneuvering of ships during entry and maneuvering in the port, while reducing the fuel consumption of ships and, if tugs are used, their fuel consumption, reduces the amount of emissions from port facilities and ships and ultimately reduces the environmental impact [3]. The sustainability of ports from a technical, technological and environmental impact point of view is directly related to optimal conditions for ship entry and maneuvering in ports, convenient ship mooring options at quays and optimal passenger service, and cargo handling technologies.

The navigation safety of ships entering ports and maneuvering in them independently or with the help of port tugs is important not only from the point of view of shipping safety and ship arrival time, but also from the point of view of environmental protection, as it reduces the time of ships entering the port and maneuvering in it, fuel consumption during such operations and the resulting emissions [4]. Individual port areas, such as port entrances and internal navigation channels, ship turning basins, approaches to quays, often have complex configurations and require many and very precise ship maneuvers.

Modern ports are trying to accommodate as many large ships as possible, developing port infrastructure and superstructure, using tugboats and trying to minimize downtime. In some ports, when ships enter and maneuver in ports, it is mandatory to order tugboats, although some large ships, such as large cruise ships, can maneuver and moor independently. In some countries and ports, port navigation rules require that during autonomous mooring operations, for example, for cruise ships, one or two tugboats be on standby next to such mooring ships. Such rules apply in some ports in Japan, where one tug must be on standby next to a ship mooring independently, and in many ports in South Korea, for example, two tugs must be ordered and on standby next to a ship mooring independently [5,6]. Tugs may be used compulsorily based on port regulations or at the decision of the ship's captain and pilot if the ship does not have sufficient maneuverability or due to difficult hydro meteorological and hydrological conditions.

Accurate knowledge and assessment of a ship's maneuvering characteristics allows reducing the number of maneuvers when maneuvering in the port, reducing fuel consumption and at the same time minimizing the resulting emissions [4,7], which is important in developing sustainable ports.

Optimizing the ability of ships to maneuver independently or using tugboats in ports is related to the good development of sustainable port infrastructure and is important from a theoretical and practical point of view, as it allows for the assessment of optimal ship maneuvering conditions in advance and the timely implementation of necessary additional factors. Additional actions are related to the optimal planning of port infrastructure, the assessment and ordering of the required tugboat traction force (quantity and power) in order to avoid unforeseen situations during ship maneuvering. It is necessary to consider that once ship maneuvers in the port have begun, there are often very limited conditions and time to attract additional capacity (tugboats) if during the maneuver it turns out that the initial decisions made are insufficient.

The superstructure of port terminals is very important in the development of sustainable ports, so that ships are processed in the shortest possible time, without causing additional environmental impact. The superstructure of port terminals must correspond

https://doi.org/10.3390/su18041916

as closely as possible to the optimal ship processing capabilities, since, for example, when loading bulk cargo onto ships at a particularly high intensity, it is not always possible to avoid significant dispersion of solid particles beyond the ship's loading areas.

The development of sustainable ports is a very important task from a maritime safety, economic and environmental perspective; therefore research in this area is important for many countries and ports.

This article analyses the following key factors of port sustainability: optimality of port entrances and internal navigation channels (maneuverability and environmental impact); sustainability of port turning basins (maneuverability and environmental impact); sustainability of port tugs (capacity and environmental impact); optimality of quayside (area and environmental impact); sustainability of terminal (port) activities (efficiency, intensity, energy demand for port operations and environmental impact); emissions; additional actions (port organization, staff quality, fuel quantities, etc.) that influence port sustainability and environmental impact.

The main research objectives are as follows: port (terminal) sustainability (optimality), based on ensuring shipping safety in ports, using the necessary power capabilities of port tugs; improving port sustainability, based on reducing environmental impact, using more friendly energy sources; sustainability of port terminal operations, assessing optimal passenger service and cargo handling.

The novelty, innovation and contribution to the development of science in the field of port sustainability of this article is manifested in the fact that a comprehensive study of the sustainability of ports (terminals) has been carried out, supplemented by factors of shipping safety and tugboat capabilities, which were not assessed in previous port sustainability studies (most were limited to port economic and environmental factors). This article studies a comprehensive assessment of port sustainability, including (but not limited to) the following port sustainability factors: ensuring shipping safety in ports, using the necessary port tugboat capacities; reducing environmental impact by using more friendly energy sources; ensuring the sustainability of port terminal operations, assessing optimal passenger service and cargo handling. Based on the results of the study, new methods for assessing sustainable port elements and their applicability in various ports have been developed. The developed sustainable port benchmark index allows for assessing the port's capacity and comparing the port's sustainability situation with other ports, which is important when implementing elements of sustainable port good practice in any port.

In order to achieve innovation and innovativeness, the following structure of this article was adopted: first, an analysis of the current situation and literature of ports was performed, a mathematical model of methods was created, a case study was conducted with real ships and ports. Based on the case study, the created mathematical model was tested and its correction was made, and a discussion section was prepared, which discusses the achieved results and anticipates possible directions for further research, as well as conclusions.

2. Review of Port Sustainability Conditions and the Literature

Port sustainability is one of the main conditions for successful port operations, as it allows for the optimization of many areas of port operations, ensuring shipping safety in the port and minimal environmental impact. Many ports around the world are trying to optimally develop port infrastructure, installing the latest navigation systems, using modern tugs to ensure shipping safety, and accepting the largest possible ships, including new passenger service and cargo handling technologies [8]. When optimizing port operations, efforts are made to ensure that ships spend as little time in the port as possible, performing passenger service and cargo handling operations [9,10]. By optimizing ship

https://doi.org/10.3390/su18041916

entry into the port and port operating conditions, fuel and energy consumption of ships and port facilities in the port is reduced and at the same time emissions generated from ships and port facilities are minimized [11,12].

The water and air resistance of ships, studied from various aspects, and their accurate assessment are important in order to ensure navigation safety in the port, optimize ship utilization and increase port sustainability [7,13,14]. Many researchers have limited themselves to studying ship handling in individual port locations. The handling of ships due to the effects of wind, currents, waves and shoals can vary significantly in the port entrance channels, which in most cases are affected by all of the above-mentioned effects, in the inner harbor channels and ship turning basins, where waves are usually absent, and in the approaches to the quays, where current parameters often change [15].

Port infrastructure, i.e., port entrances and internal navigation channels, ship turning basins, approach to quays, piers, port navigation equipment and other elements of port infrastructure are very important in creating a optimally (sustainable) port, because a well-planned and developed port infrastructure can increase the parameters of ships entering the port and ensure their safe navigation [1,10,16].

Port tugs are very important in ensuring shipping safety in the port, accepting the largest ships possible, because they help to ensure a safe entry for ships into the port, maneuvering and mooring in the port and departure from the ports [17]. Modern tugs are usually equipped with "azipod" systems, their maneuverability is significantly better compared to traditional (classic) tugs; therefore, with the use of new tugs, additional opportunities for safe entry of larger ships into the port appear [17,18]. Thus, research into ship maneuvering in ports using tugs is important in ensuring shipping safety, reducing emissions and achieving the potential for sustainable port development. The entry of larger ships into the port not only improves the economic indicators of the port and the image of the port, but also makes the port sustainable, because there are opportunities to develop a wider range of port activities, increase navigation safety and reduce environmental impact. Reducing environmental impact is manifested in the fact that using larger ships reduces relative fuel consumption and at the same time the amount of emissions per one conditional passenger or cargo unit [19].

The use of port tugs to ensure the safety of port navigation is related to the port use or navigation rules and the decisions of the ship's captain or harbor master, if in specific conditions there are no sufficient ship control capabilities and there is a reasonable risk to the safety of the ship [20,21]. Port tugs usually have powerful engines, in some cases up to 4000–5000 kW; therefore, when carrying out ship towing operations in ports under difficult conditions, their need (number of tugs and their traction force) must be assessed as accurately as possible. Inaccurate preliminary preparation of maneuvers and their execution in ports, using high-power engines of ships and tugs, consumes a lot of fuel and generates large amounts of emissions, which negatively affects the sustainability of the port [22].

In some cases, navigation conditions in ports can change in a very short time. If such a possibility exists, for example, the probability of squalls, ship and port masters, aware of the possibility of a change in navigation conditions, usually take precautionary measures in advance, using additional port tug capacities to safely maneuver the ship or looking for other options to ensure the safety of the ship. If there is a significant possibility of a change in navigation conditions, the ship's entry into or departure from the port is often postponed, additional mooring lines are used when the ship is moored at the quay, and so on [23]. The use of additional measures to increase the safety of the ship also requires additional costs; therefore, the most accurate assessment of the ship's capabilities becomes very important both from the economic and port sustainability point of view [24].

https://doi.org/10.3390/su18041916

The port terminal superstructure is also very important from the perspective of a sustainable (optimal) port, as a well-designed and operated port terminal superstructure shortens the processing time of ships in the port, improves the economic condition of ships and port terminals. Due to the shorter time of ships in the port, the environmental impact of the ship in the port is reduced, which is important for achieving sustainable port conditions [9,10].

Port operations related to passenger service and cargo handling are important from an environmental impact perspective, as such operations consume a lot of energy and emit a lot of pollutants [11,25]. Optimizing port operations is one of the most important factors in reducing environmental impact and at the same time improving port sustainability. Port cargo handling operations, such as grain drying and loading, require high energy inputs, using high-power cranes or other handling machines. Optimizing loading operations, if correctly and accurately assessed, planned and implemented, significantly saves energy requirements and reduces the amount of pollutants generated in port operations [25].

By properly planning and developing port infrastructure (especially when creating new ports or individual parts thereof), optimizing the well-prepared arrival of ships to and departure from ports and port terminal activities, it is possible to reduce energy consumption and emissions, improve the economic performance of the port and terminals, and at the same time improve the sustainability of the port.

Port sustainability has so far been largely focused on the economic and environmental indicators of the port. Transport is associated with increased risk; therefore, maritime safety is of particular importance in the development of sustainable ports. Maritime safety hearings are regulated by PIANC, IMO and other recommendations and guidelines [26,27], which provide very general results and only partially cover the elements of port sustainability. Therefore, it is necessary to assess the provision of maritime safety as one of the most important factors of port sustainability as accurately as possible. Port tugs play an important role in providing maritime safety in modern ports; therefore, an accurate assessment of their capabilities is important as one of the factors of sustainable port development.

As can be seen from the analysis of the port situation and literature from the perspective of port sustainability, there are many scientific and practical problems to be solved in order to increase the sustainability of ports (terminals).

3. Mathematical Basis for Assessing Port Sustainability

In order to achieve port sustainability by improving port efficiency and increasing shipping safety in ports, reducing energy consumption and emissions of incoming ships, port tugs and port operations, it is very important to create a mathematical model of port sustainability. It is also important to verify the created mathematical model based on real experimental data, make corrections to it and, after creating the final mathematical model, conduct port (terminal) sustainability studies.

3.1. Research Progress and Principle of Mathematical Model Development

When developing a mathematical model for assessing port sustainability, it is first necessary to collect a port database based on statistics and literature analysis of port operations of ships, port tugs and environmental impact. Based on the collected database and acceptable mathematical research methods, a principle for developing a mathematical model is created. Using the principle for developing a mathematical model and the collected database, a primary mathematical model for assessing port sustainability is created. The created primary mathematical model is verified using the results of experiments related to real port activities, ships, port tugs and port operations and environmental impact. Based

https://doi.org/10.3390/su18041916

on the results of the verification of the primary mathematical model, the created primary mathematical model is corrected, a final mathematical model is created and with its help the obtained port sustainability assessment results are verified (Figure 1).

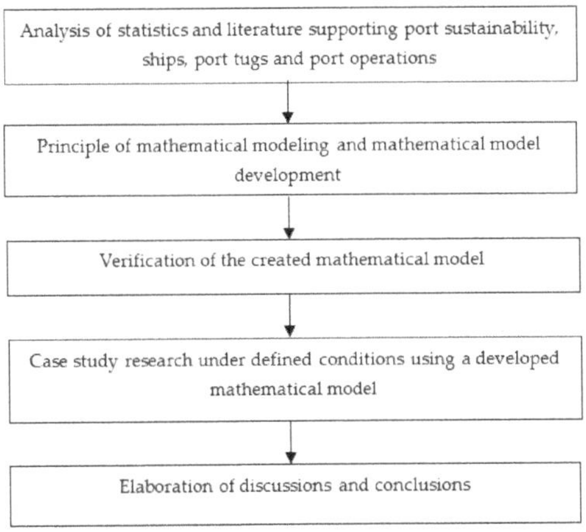

Figure 1. The algorithm for the sustainable port research methodology.

During the verification of the mathematical model, the mathematical model for assessing the sustainability of the port is adjusted and, using the created and adjusted mathematical model, the sustainability of the port is assessed and options for improving the sustainability of the port are sought. Based on the obtained assessment results, a discussion part is prepared, in which the obtained results, possible further studies of the sustainability of the port are discussed and conclusions are prepared and presented.

3.2. Mathematical Model for Assessing Port Sustainability Based on the Comparative Index Principle

For the assessment of port sustainability, it is appropriate to create a generalized assessment system (model) that would combine the impact of port infrastructure and superstructure on port sustainability, as well as the ship or ships entering the port and their controllability, the capabilities of port tugs, the assessment of port operations and other possible factors that have or may have an impact on port sustainability. For the assessment of the generalized sustainable port impact, it is appropriate to use the principle of the comparative method, expressed in a comparative index (E_S), which can be expressed in the following functional dependence:

$$E_S = f\left(B_{ChE},\ B_{ChN},\ D_{TB}, F_{Tb},\ L_{QE}, Q_{TER},\ E_E, A_{FT}\right), \quad (1)$$

where: B_{ChE}—port entrance channel optimality (sustainability) (maneuverability and environmental impact); B_{ChN}—port inside navigational channels optimality (sustainability) (maneuverability and environmental impact); B_{ChE}—port ships turning basins optimality (sustainability) (maneuverability and environmental impact); F_{Tb}—port tugboats sustainability (capability and environmental impact); L_{QE}—optimality (sustainability) of the berth (area and environmental impact); Q_{TER}—terminal (port) operation sustainability (productivity, intensively, energy requirements for port operations and environmental impact); E_E—quantities of generated emissions; A_{FT}—additional actions (port organization, employ quality, fuel quantities and so on), affecting the port sustainability and environmental impact.

https://doi.org/10.3390/su18041916

To assess the sustainability of a port, it is necessary to adopt either ideal conditions for port sustainability or to use a comparison of ports from a sustainability perspective. In general, comparisons between ports are usually inaccurate, since it is practically impossible to find ports that are absolutely similar in terms of their areas of activity, parameters, etc. In many cases, it is appropriate to compare port terminals with each other, since their activities are similar, for example, container, bulk cargo, Ro-Ro or other specific terminals are similar at least in terms of their activities and in many cases in terms of basic parameters. Therefore, when assessing the sustainability of ports, it is appropriate to limit ourselves to comparing individual port terminals or their groups.

Given the large number of different factors that affect the sustainability of ports or terminals, the comparative index is often based on deterministic assumptions [26]. Given the large number of factors, which is typical for assessing the sustainability of ports, according to the centile limit theorem, a normal distribution is approached and in this case Gauss's law can be used [28]. By using factor weight coefficients, nonlinear influences of factors are also taken into account when assessing the comparative sustainability index of ports. In such a case, the comparative port sustainability index can be written as follows:

$$E_S = \frac{1}{\eta_k}\left(k_{ChE}\frac{B_{ChEi}}{B_{ChE0}} + k_{ChN}\frac{B_{ChNi}}{B_{ChN0}} + k_{TB}\frac{D_{TBi}}{D_{TB0}} + k_{Tb}\frac{F_{Tbi}}{F_{Tb0}} + k_{QE}\frac{L_{QEi}}{L_{QE0}} + k_{TER}\frac{Q_{TERi}}{Q_{TER0}} + k_E\frac{E_{E0}}{E_{Ei}} + k_{GT}\frac{A_{FTi}}{A_{GF0}}\right), \quad (2)$$

Here: η_k—correlation coefficient, depending on the number of similar elements, can be from 0.95 to 1.0 (1.0 will be if there are no similar factors or elements). The correlation coefficient is adopted when evaluating similar factors; in this study, the interaction of port infrastructure, ship maneuverability, and tugboat capabilities was evaluated; k_i—factors weight coefficients; $B_{ChEi}, B_{ChNi}, D_{TBi}, F_{Tbi}, L_{QEi}, Q_{TERi}, E_{Ei}, A_{FTi}$—elements (factors) of the analyzed port (terminal), can be expressed as coefficients or absolute values; $B_{ChE0}, B_{ChN0}, D_{TB0}, F_{Tb0}, L_{QE0}, Q_{TER0}, E_{E0}, A_{FT0}$—elements of an ideal sustainable or comparable port, expressed in the same parameters as the elements of the analyzed port. A relatively ideal port can be recognized as a port that meets all international and national requirements in terms of shipping safety (in the broad sense), has the minimum possible impact on the environment at any given moment, and meets the best technical, technological, organizational and legal international and national requirements. In reality, such ports are difficult to find; therefore, for the purpose of assessing the comparative port sustainability index, a comparable port may be adopted, the above indicators of which, or some of them, are better than the port under study.

It is worth noting that the developed sustainable port comparative index allows the use of all or part of the factors, as well as expanding the number of factors. In all cases, it is necessary to evaluate the weight coefficients of the factors; their sum in all cases must be equal to one.

3.3. Mathematical Model for Assessing Navigation Safety Based on Ship Trajectory and Water Area Parameters

Port infrastructure optimality (sustainability) (approach and inside port navigational channels, ships turning basins configuration and main parameters, influence on navigational safety and environmental impact), could be evaluated via channels parameters and ship's maneuverability, that means whether a ship can independently or with the help of tugboats enter the port, maneuver in it and moor, performing a minimum number of maneuvers.

Port infrastructure: port entrances and internal navigation channels, ship turning basins, quays, etc. should be designed so that the trajectories of the largest possible ships entering the port "fit" into the planned channels, i.e., so that ships can safely enter the port and maneuver in it. The ship trajectory, i.e., the coordinates of the trajectory in the absence of external forces (wind, current, waves, shallows), X_{0i} and Y_{0i} can be calculated as follows [7]:

$$X_{0i} = \int_0^t v_i dt \cdot \cos\left(\int_0^t \omega_i dt - \beta_i\right); \tag{3}$$

$$Y_{0i} = \int_0^t v_i dt \cdot \sin\left(\int_0^t \omega_i dt - \beta_i\right), \tag{4}$$

where v_i is ship's speed module (can be taken from ship's navigational equipment or could be calculated [7,13]); ω_i is turning velocity, rad/s (can be taken from ships navigational equipment and calculate using methodology, presented in [7,13]; β_i is drift angle, rad; v_i, ω_i and β_i can be calculated depending on the sailing conditions using methods presented in [7,13].

In most cases, when ships enter and maneuver in ports, they are affected by aerodynamic and hydrodynamic forces, i.e., wind, current and waves (waves mainly affect the ship's trajectory in the port's entrance channels). In such situations, when the ship's trajectory is affected by current, waves and wind, and the ship's propulsion system is also used, the ship's trajectory coordinates can be calculated as follows [7,13,14]:

$$X_{0i} = \int_0^t v_x dt + \int_0^t v_{xsr} dt + \int_0^t v_{xd} dt + \int_0^t v_{xw} dt, \tag{5}$$

$$Y_{0i} = \int_0^t v_y dt + \int_0^t v_{ysr} dt + \int_0^t v_{yd} dt + \int_0^t v_{yw} dt. \tag{6}$$

where v_{xsr}, v_{ysr} are the current speeds in the X and Y directions; v_{xd}, v_{xd} are the ship's drift speeds in the X and Y directions, which can be calculated using the methodology given in [7,13,14]; v_{xw}, v_{yw} are the ship's acquired speeds in the X and Y directions due to the impact of waves on the ship, which can be calculated using the methodology given in [13,14].

Taking into account the calculation method presented in Equations (5) and (6) and after considering the effect of shallowing in the port entrance channel and other parts of the port, Equations (5) and (6) can be written as follows:

$$X_{0i} = \int_0^t v_{is} \cos \int_0^t (\omega_{is} dt - \beta_{is}) dt + \int_0^t v_{xsr} dt + \int_0^t v_{xd} dt + \int_0^t v_{xw} dt, \tag{7}$$

$$Y_{0i} = \int_0^t v_{is} \sin \int_0^t (\omega_{is} dt - \beta_{is}) dt + \int_0^t v_{ysr} dt + \int_0^t v_{yd} dt + \int_0^t v_{yw} dt. \tag{8}$$

Here: v_{is} the speed of the ship when moving along a curvilinear trajectory at shallow depth can be calculated using the methodology presented in [7,13]; ω_{is} the angular speed of rotation of the ship when moving along a curvilinear trajectory at shallow depth can be calculated using the methodology presented in [7,13]; β_{is} the angle of drift of the ship when moving along a curvilinear trajectory at shallow depth can be calculated using the methodology presented in [7,13]. At the same time, when considering specific port conditions and the relatively short sailing time of a ship through the port entrance and internal navigation channels, in many cases it is possible to assume constant wind, current and depth parameters and then simplify the calculation of the ship's trajectory.

The width of the port entrance and internal navigation channels, so that the largest planned ships can safely enter the port, can be calculated according to the following formula, taking into account the coordinates of the ship's navigation trajectory, from which the navigation lane occupied by the ship is calculated [7,14]:

$$B_C = L sin\beta_{is} + B cos\beta_{is} + L sin\Delta K + P' \sigma_y + b_n, \tag{9}$$

https://doi.org/10.3390/su18041916

Here: L—length of the ship between perpendiculars (maximum ship's length between perpendiculars is assumed); B—width of the ship (width of the widest ship is assumed); ΔK—angle of the ship's course inclination when navigating the channel (for large ships it should not exceed 2 degrees); P' probability supply coefficient, 95% supply is assumed in navigation; therefore P' applicable in this case is about 2; LNG ships, oil tankers, chemical tankers have specific requirements; therefore a coefficient of up to 3 is applied, i.e., assuming up to 99.7% probability supply [7,26,27]; σ_y—accuracy of determining the ship's location relative to the channel axis, for example, the sensitivity of the guide; b_n—navigation margin, which depends on the accuracy of the channel slope position and the stability of the channel slopes (in depth, it is calculated based on the geotechnical properties of the soil in which the channel is dug) and amounts to (0.5–1.0) B [7,13,26,27].

After evaluating the ship's trajectory and the navigation lane occupied by the ship (at each point in time, the width of the navigation lane occupied by the ship must be added to the ship's navigation trajectory), the required minimum channel width and its configuration (entry and internal navigation channels) are obtained.

The minimum width of the navigation channel (entry and internal) must not be less than the navigation lane occupied by the vessel, calculated from the vessel's navigation trajectory using Formulas (7) and (8), and estimating the width of the vessel's navigation lane according to Formula (9). In individual cases, the width of the vessel's navigation lane may be reduced by using additional vessel steering devices (thrusters) or tugs, or by reducing the permissible wind and current parameters.

For practical purposes, when calculating the minimum width of a port entrance or internal navigation channels, it is recommended to use a ship's rudder angle of no greater than 20° in order to have a margin of controllability in case external forces suddenly change (wind gusts) or the ship's controllability is not accurately assessed. It should be emphasized that the presented methodology for calculating channel parameters can be applied in any navigation area.

If the channel width is smaller than that obtained using the presented calculation methodology, additional ship steering devices (thrusters) or tugs can be used to improve the ship's maneuverability. At the same time, it is necessary to take into account that if, under specific hydro meteorological and hydrological conditions, the ship cannot independently safely navigate the port's entrance and internal navigation channels, the ship's navigation is restricted; therefore, the ports set maximum permissible boundary conditions for entering and leaving the port (wind and current parameters). The minimum channel width for typical ships, taking into account the ship's drift angle and other ship movement parameters, can be calculated in advance using Formula (9) or other methods, for example, using PIANC [26] or other recommendations [27]. Assuming that the ship can reach a drift angle of up to 15° during a turn [7,13,14], depending on the main geometric parameters of the ship, i.e., The preliminary results of calculating the width of the channels, assuming a probability of about 95%, are presented in Figure 2 as an example, based on the ratio of the length to the width of the ship (this ratio is taken to be approximately 6.5).

If the channel width is insufficient during the turn, depending on the length of the ship and the specified basic parameters (drift angle and accuracy of ship location determination), ship thrusters, tugs or tugs can be used to increase the angular speed of the ship and at the same time reduce the channel width required during the turn. When a ship moves through channels, the best result is achieved by using a stern thruster or a stern tug, because the ship's pole of rotation, when the ship moves forward, is located in the forward part of the ship and in this case a larger turning moment is created. The abscissa of the ship's pole of rotation, when moving forward of the ship (from the middle of the ship, i.e., from the

https://doi.org/10.3390/su18041916

intersection point of the waterline and midship planes), can be calculated according to the following formula [7,13]:

$$x_0 = L(0.4 + 11.5\frac{T_{as} - T_{bo}}{L} - 0.23\alpha), \quad (10)$$

Here: T_{as}—ship's stern draft; T_{bo}—ship's bow draft; α—rudder plate rotation angle, rad.

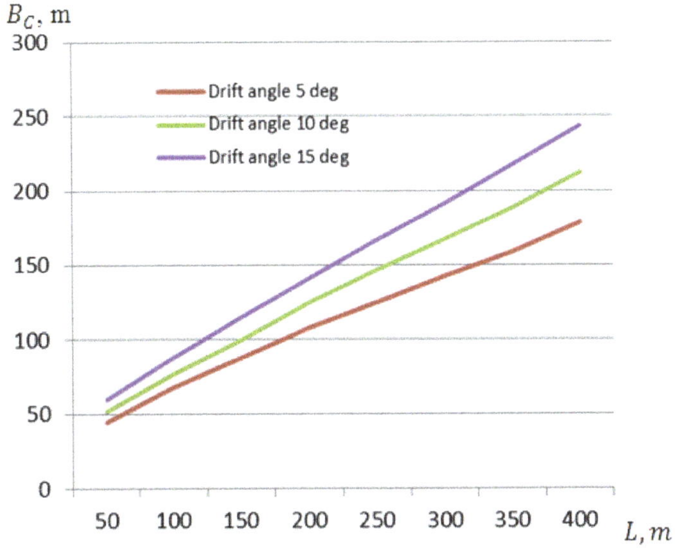

Figure 2. Required width of navigation channels depending on the length of the ship and the angle of drift.

In this way, using stern thruster(s) or a tugboat at the stern can significantly increase the ship's maneuverability when navigating through straits and channels, which is very important for the development of sustainable ports.

When assessing the sustainability of a port from the point of view of the approach and internal navigational channels, it is very important that the minimum width of the channels is as close as possible to the calculated one using Formula (9). In the event that the actual minimum width of the port channels is greater than that calculated using Formula (2), the ratio B_{ChEi}/B_{ChE0} or B_{ChNi}/B_{ChN0}, is taken equal to unity.

Ship turning basins are one of the important elements of port infrastructure, and for a optimality (sustainable) port it is very important that the ship turning basin(s) meet the needs of shipping safety. In ship turning basins, tugboats are used to improve the maneuverability of large ships that do not have high-power propulsion systems (thrusters). For example, cruise ships, whose additional control systems (thrusters) sometimes have a power of up to 12–14 MW, (greater than the power of 3–4 port tugs), usually do not use tugs.

The recommended diameter of ship turning basins, i.e., about 1.6 of the maximum possible ship length [13,14,26,27], is difficult to achieve in many ports, since the length of modern ships sometimes reaches 350–400 m. Preparing turning basins of the specified diameter of about 1.6 of the maximum ship length, especially in old ports, is often difficult due to geographical conditions; therefore, in many ports the diameter of ship turning basins is smaller than specified in the recommendations. In many ports, turning basins currently being prepared for the longest ships currently in operation (length up to 400 m) are about 500 m in diameter [20,21], i.e., the ratio of the turning basin diameter to

https://doi.org/10.3390/su18041916

the longest ship length is about 1.25. With a similar ratio of turning basin to ship length (about 1.25), an accurate preliminary estimate of the number of tugs and their power or pulling force is very important. The moments created by the ship and external forces must be accurately estimated.

3.4. Mathematical Model to Estimate the Required Pulling Power of Port Tugs and Their Capabilities for Navigation Safety

The moments acting on the ship when entering and maneuvering in the port must be compensated by the forces and moments created by the ship's rudder or, if the ship uses the assistance of steering devices or tugboats, by the forces and moments created by additional steering devices (thrusters) and/or tugboats.

Large ships in ports usually use port tugboats, especially if the ship does not have its own additional steering devices (thrusters). Considering that there are no waves in the ships turning basins of ports, the current changes little when the ship turns (due to the relatively small time interval), and the ship has a minimum longitudinal speed, using D' Lambert's principle [7,13,14], the equation of the ship's turning moments can be written as follows:

$$M_{in} + M_k + M_a + M_{sh} + M_N + M_T + M_{tug} = 0, \tag{11}$$

where M_{in} is the inertia moment; M_k is the moment created by the ship's hull, which could be calculated by using the methodology stated at [7,13,14]; M_N is the moment created by thrusters [7,13,14]; M_a is aerodynamic moment, which could be calculated using the methodology stated at [7,13,14]; M_{sh} is the moment created by shallow water effect [7] (in port conditions, this parameter is very important, especially when the ratio of the ship's draft to the depth of the turning basin is greater than 0.9); M_T is the moment created by ship's propeller (propellers) and rudder(s), which could be calculated using the methodology stated in [7,13,14]; and M_{tug} is the moment created by tugboats. Additional moments could be created by anchor or mooring ropes or other factors.

The moment of inertia of a ship moving through canal bends and turning the ship in a turning basin can be expressed as follows [7,13]:

$$M_{in} = (I_z + \lambda_{66})\frac{d\omega}{dt}, \tag{12}$$

where I_z is the moment of inertia of the ship; λ_{66} is added moment of inertia of the ship turning in water; $\frac{d\omega}{dt}$ is the acceleration of the ship's rotational angular velocity.

The moment of inertia of a ship, assuming the ship as a rigid body, can be calculated using the principles of theoretical mechanics, as follows [29,30].

$$I_z = \rho V L^2/12, \tag{13}$$

where ρ is water density; V is ships displacement; L is ship's length.

When evaluating the rotational moment of the connected fluid mass, the total moment of inertia can be calculated as follows: [7,13,29,30]:

$$M_{in} = ((1 + k_{66})\rho V L^2/12)(\frac{d\omega}{dt}\ (exp(-3t/T_{in})), \tag{14}$$

where k_{66}—added moment coefficient, which for the analyzed situation (ship turning in port turning basin in case of T/H = 0.90–0.95), is equal to 3 [7,13,14]; T_{in} is ship's inertia period, can be taken as $T_{in} \approx L/3$ [7,14].

https://doi.org/10.3390/su18041916

The ship's hull moment (M_k) can be calculated as the hull's resistance to lateral acceleration ($\frac{dv_y}{dt}$) [7,13,14]:

$$M_k = (k_{22}\rho VL\frac{dv_y}{dt}(1 - exp\left(-\frac{3t}{T_{in}}\right)))(1 + 4.95(T/H)^2)\,, \quad (15)$$

where k_{22} is the coefficient of the added water mass when the ship moves in the transverse direction; v_y is the speed of the ship's movement in the transverse direction; T is the average draft of the ship; H is the depth of the port ships turning basin.

The aerodynamic moment when the ship is turning can be calculated as follows [13,14]:

$$M_a = C_a\frac{\rho_1}{2}\sqrt{S_x^2 + S_y^2 v_a^2 x_a sinq_a}, \quad (16)$$

where: C_a is the aerodynamic coefficient of the above-water part of the ship; ρ_1 here is air density, 1.25 kg/m^3 can be accepted for calculations; S_x and S_y here are the areas of the projections of the above-water part of the ship to the middle and transverse planes; v_a here is the wind speed; x_a is the abscissa of the aerodynamic force addition point with respect to the middle plane of the ship; q_a is the wind heading angle at the start of the maneuver.

The ship will start turning when the moment created by the thrusters, or tugboats is greater than the moments of inertia and other external forces. A lot of big conventional ships have not thrusters and in this case, tugs created moment can be calculate as follows:

$$M_{tug} > M_{in} + M_k + M_a \quad (17)$$

In this way, the part of the moment created by the tugboats, which will act to turn the ship, will be calculate as follows [31]:

$$\Delta M_{tug} = \sum M_{tug} - M_{in} + M_k + M_a, \quad (18)$$

Port tugboats sustainability (capability and environmental impact) (F_{Tbi}), means having a sufficient number and the required pulling power of tugboats in the port to provide safe entry of vessels into the port and maneuvering in it up to the permissible hydro meteorological and hydrological conditions. When mooring ships and turning them in the ship turning basin or other port area, the pulling force of tugs is required to create the necessary torque, which would be greater than the torques created by other external forces. The minimum torque created by tugs depends on the pulling force of tugs and the moment arm must be greater as forces created by external factors (wind, current, waves, shallow water effect and so on. The necessary lateral force created by tugs (taking into account the ship's course) together with ships propulsion mechanisms can be calculated as follows [7]:

$$Y_{tug} + Y_{Sp} = Y_{Sin} + Y_{Sk} + +Y_{Sa} + Y_{Ssh}, \quad (19)$$

where: Y_{Sp}—forces are generated by the ship's propulsion mechanisms (thrusters); Y_{Sin}—ship's inertia force; Y_{Sk}—ship's developed hull force; Y_{Sa}—aerodynamic force generate by ship; Y_{Sc}—the force is created by the current on ship; Y_{Ssh}—the force is created by the impact of shallow water on ship.

Considering the short-term necessary traction force of tugs (pushers) and the relatively low speed of the ship, when the ship moves through the port channels or in the port ship turning basin and when it is moored or unmoored, the inertia force, taking into account the resistance of the ship's lateral hull, can be considered to be approximately 1.3–1.5 times greater than the lateral resistance force of the ship. Then, regardless of the forces generated

https://doi.org/10.3390/su18041916

by the engine(s) (if the ship has one) (they remain as an additional safety margin of the ship), Equation (19) can be written as follows [7,14]:

$$Y_{tug} = 1.3C\frac{\rho}{2}LTv_y^2(1 + 4.95(\frac{T}{H})^2) + C_a\frac{\rho_1}{2}S_x v_a^2 sinq_a + C\frac{\rho}{2}LTv_c^2 sinq_c, \quad (20)$$

where: C—the hydrodynamic coefficient of the hull of the ship, for marine ships during mooring or turning the ship, at relatively low speeds in the Y direction, can be taken as about 1.3–1.5 (as a plate with rounded edges placed across the flow) [7,13]; ρ—water density; L—ship length; T—mean draft of the ship; H—depth; S_x—the area of the projection of the above-water part of the ship to the middle plane; C_a—aerodynamic coefficient [7,13]; v_y—the speed of the ship moving on Y dirrection to the ship's course during sailing in channels, turning or during moving to quay or from quay [13,14,32]; v_a—wind velocity; v_c—current speed; q_a—the angle of the wind to the ship's course; q_c—the angle of the current to the ship's course.

The ability of tugboats in the port to help turn a ship around in extreme hydro meteorological and hydrological conditions is an important factor in a sustainable port. In many ports, maximum wind and current speeds are set, up to which ships can enter the port (enter, maneuver and moor). In the case when the power of the ship's engines or tugboats (traction forces) is equal to or greater than the forces created by external forces, calculated according to Formula (20), the ratio F_{Tbi}/F_{Tb0} is considered equal to one.

Entrance to quay sustainability (area and environmental impact) (L_{QE}) means the comfortable approach of a ship to the quay, without the need for special maneuvers. In most cases, this corresponds to the required depths for access to the quay, which are located at the quay at a distance of not less than 0.1 L on both sides of the largest moored ship and at an angle of not less than 45° (Figure 3). Under better conditions than those specified above, the ratio L_{QEi}/L_{QE0} is taken equal to unity.

Figure 3. Dredged water area near the embankment (quay) (delimited from the embankment to the navigation channel by yellow lines).

https://doi.org/10.3390/su18041916

3.5. Mathematical Model for Assessing Port (Terminal) Activity

Port sustainability (activity intensity and energy demand for port operations and environmental impact) (Q_{TER}) refers to the loading capacity and energy demand of port facilities for passenger and cargo handling, for example per passenger or per tonne of cargo, and the corresponding emissions [7,33,34]. This parameter depends on the superstructure and equipment present in the port and can be estimated using terminal intensity, energy consumption and emissions [35–37].

The optimal terminal intensity (one of the conditions for terminal sustainability) depends on the type of cargo and the planned ship sizes. As examples, the capacity of container, car, bulk, liquid bulk and Ro-Ro terminals evaluation formulas presented below. Container terminal capacity can be evaluated using the terminal area utilization factor, the vertical loading of containers and the average container storage time at the terminal. Studies conducted at various terminals allow for a more accurate assessment of the volume and area utilization coefficients of sustainable ports and terminals, which often differ from the design norms of various countries [25,27,38]. The annual terminal capacity (N_{KT}) can be calculated using the following formula [7,38,39]:

$$N_{KT} = k_{TS}\frac{S_T}{S_K}n_{Kv}\frac{T_T}{t_{Kv}}, \quad (21)$$

Here: k_{TS}—terminal area utilization factor, applicable to many container terminals, taking into account the total area of the terminal and the area required for container handling (access roads, gaps between container storage racks, area required between quay(s) and storage areas, etc.). Based on studies of terminal area utilization factors in several container terminals, it has been found that this indicator varies from 0.2 to 0.5 in different terminals. The application of optimization principles has shown that in sustainable container terminals this factor can range from 0.25 to 0.35; S_T—total terminal area; S_K—area occupied by one container, for a 20-foot container (TEU—twenty equalent unit) it is about 15 m^2; n_{Kv}—average vertical loading of containers, this value in many container terminals is from 2 to 4; T_T—total operating time of the container terminal per year, if the terminal operates all year round without breaks, this time can be taken as 365 days; t_{Kv}—average storage time of containers in the terminal, in many container terminals this time is up to 10 days.

Terminals for cars and other wheeled vehicles are very important and many ports around the world have such terminals. The annual capacity of a sustainable car terminal (N_{AT}) can be calculated using the following formula [7,40]:

$$N_{AT} = k_{AS}\frac{S_A}{S_{1A}}n_{Av}\frac{T_A}{t_{Av}}. \quad (22)$$

Here: k_{AS}—terminal area utilization factor. Studies conducted in several car terminals (single and multi-level) have shown a very wide range of this coefficient, i.e., from 0.3 in multi-level terminals (access roads, other infrastructure required for the structure are required), to 0.85 in single-level car terminals. For a sustainable car terminal, the optimal terminal area utilization factor, considering the total terminal area and the area required for car servicing, can be accepted as 0.5 to 0.8; S_A—total terminal area; S_{1A}—area occupied by one car, for passenger cars it is about 10 m^2; n_{Av}—number of levels (storeys) of the car warehouse, if multi-storey terminals (warehouses) are used, this value usually amounts to 3–4 levels; T_A—total working time of the car terminal per year, in most terminals of this type work all year round without breaks, then this time can be taken as 365 days; t_{Av}—average car storage time at the terminal, calculated in days. In most car terminals, this time is about 10–20 days.

https://doi.org/10.3390/su18041916

The annual capacity of a bulk cargo terminal (N_{BT}) (Formula (23)) and closed-type liquid cargo storage facilities (N_{SKT}) (Formula (24)), located in many ports around the world, can be calculated as follows: [7,41,42]:

$$N_{BT} = k_{BT} V_{BT} n_{BT}. \quad (23)$$

$$N_{SKT} = k_{SKT} V_{SKT} n_{SKT}, \quad (24)$$

Here: k_{BS}—bulk terminal area utilization factor. Based on research conducted at bulk terminals (research conducted at 5 bulk cargo terminals), depending on the design, it can be from 0.7 to 1.0, for example, (1.0 could be for silo terminals); V_{BT}—bulk cargo terminal volume in cubic meters or tons; n_{BT}—terminal turnover per year, in many terminals it is up to 6–15, depending on the type of bulk cargo and market conditions; k_{SKT}—liquid cargo terminal area or volume utilization coefficient. Based on research conducted at 6 liquid cargo terminals (research conducted at oil, oil products and LNG cargo terminals) in most cases it is about 0.90–0.95, i.e., from the maximum filling of the tanks; V_{SKT}—liquid cargo terminal volume in cubic meters or tons; n_{SKT}—terminal turnover per year, in many terminals it is up to 30–40, depending on the type of liquid cargo and market conditions.

Ro-Ro transport is very important in supplying the activities of many countries; therefore Ro-Ro terminals are being developed in many ports. The capacity of Ro-Ro terminals (N_{RT}) can be calculated as follows [7,43]:

$$N_{RT} = k_{RT} \frac{S_{RT}}{S_R} \frac{T_R}{t_{Rv}}, \quad (25)$$

Here: k_{RT}—terminal area utilization factor. Based on the research conducted on Ro-Ro terminals (5 terminals of this type in 3 ports were studied), it can be assumed that for many Ro-Ro terminals, when assessing the total terminal area and the necessary area for servicing Ro-Ro vehicles (approach roads, spaces between Ro-Ro vehicle storage rows, other additional areas necessary for vehicles to exit or enter without moving others), it can be taken from 0.4 to 0.6; S_{RT}—total terminal area, m^2; S_R—area occupied by one Ro-Ro unit (truck or trailer, on average for Ro-Ro vehicles it is about 45 m^2; T_R—total working time of the Ro-Ro terminal per year, in most terminals of this type work all year round without breaks, then this time can be taken as 365 days; t_{Rv}—average storage time of Ro-Ro vehicles at the terminal, calculated in days. In most Ro-Ro terminals this time is from 0.5 to 1.0 days.

The presented formulas for assessing the annual turnover of sustainable terminals can be used in any port by adjusting the conditions of the specific port, in particular, such as customs procedures, cargo arrival and departure from the terminal, and other conditions.

For further research, as example, it is appropriate to present a detailed evaluation methodology for one terminal. In modern ports, container terminals have the most equipment and mechanisms; therefore, as an example, it is appropriate to analyze this type of (container) terminal. The main equipment of container terminals includes: STS (shore to ship) cranes, RTG cranes, terminal tugs (tractors), reach stackers, vertical lifts (forklifts) and other auxiliary equipment [7,38,39].

The average number of container transshipments from ship to quay or vice versa by one STS crane is up to 30–35 per hour. In many feeder ports, container ships carry out loading operations for no longer than 12 h, and in hub ports—for no longer than 24 h. Thus, knowing or assuming the planned number of container transshipments per ship, the required number of STS cranes (n_{STS}) for the hub or feeder port can be calculated according to the formula [7,38,39]:

$$n_{STS} = n_L \frac{Q_{KL}}{T_L k_{STS}}, \quad (26)$$

https://doi.org/10.3390/su18041916

where: n_L—number of ships scheduled at the terminal at the same time; Q_{KL}—average number of containers brought to the port by container ship; T_L—time of ship's stay in the port (in hours); k_{STS}—hourly capacity of one STS crane.

The amount of equipment operating in a container terminal is calculated based on the planned annual container traffic at the terminal. The main equipment of a container terminal consists of: rubber tyred gantry cranes (RTG), terminal tractors, reach stackers and forklifts. In addition, frame lifts and other additional equipment may be used.

The number of RTG cranes (n_{RTG} (Formula (27)), terminal tractors (n_{TV}) (Formula (28)), reach stackers (n_{SK}) (Formula (29)) and forklifts (n_{VK}) (Formula (30)) depends on the annual container flow and can be calculated as follows [7,39]:

$$n_{RTG} = \frac{Q_T}{k_{RTG}}, \tag{27}$$

$$n_{TV} = \frac{Q_T}{k_{TV}}, \tag{28}$$

$$n_{SK} = \frac{Q_T}{k_{SK}}, \tag{29}$$

$$n_{VK} = \frac{Q_T}{k_{VK}}, \tag{30}$$

where: k_{RTG}—RTG coefficient corresponding to the average number of containers processed by one RTG per year, in modern sustainable container terminals it is accepted on average from 40 thousand to 60 thousand; k_{TV}—terminal tug coefficient, corresponding to the average number of containers handled by one tug per year, for modern container terminals it is taken from 30 thousand to 50 thousand; k_{SK}—terminal reach stackers coefficient, corresponding to the average number of containers processed by the side lift per year, for modern container terminals it is accepted from 90 thousand to 110 thousand; k_{VK}—forklifts (vertical lift) coefficient, corresponding to the average number of containers processed by one vertical lift per year, for modern sustainable container terminals it is accepted from 190 thousand to 210 thousand.

The coefficients of the necessary equipment of a container terminal depend significantly on the container handling technology used, i.e., currently, STS crane grabs are widely used, which pick up one 40-foot container or two 20-foot containers at a time and are accompanied by other terminal handling equipment. Some container terminals have started to use a block container handling system, when STS cranes load two 40-foot containers or four 20-foot containers at a time. When assessing the rapid change in technology, it is important to examine the real situation of the available container terminal equipment in each specific case and adjust the coefficients for calculating the necessary equipment accordingly.

Some equipment typically used in container terminals is often used in other terminals, especially general cargo and multi-purpose terminals, so the methods provided for calculating the amount of equipment can be applied to other types of terminals.

The majority of the port superstructure equipment, such as RTG cranes, terminal tractors, reach stackers, forklifts and other equipment and mechanisms with internal combustion engines, consume a significant amount of fuel, especially diesel. The annual fuel consumption of RTGs, terminal tugs, side and vertical lifts can be calculated by assessing the number of equipment and mechanisms in the terminal, the relative fuel consumption of the engines of equipment and mechanisms, the power of the engines of equipment and mechanisms, the power utilization factor of the engines of equipment and mechanisms and the average coefficient of their use time.

https://doi.org/10.3390/su18041916

3.6. Mathematical Model for Estimating Fuel Consumption and Environmental Impact of Ships and Port Equipment

Thus, the amount of fuel consumed per year in the terminal (q_f) can be calculated according to the formula [44,45]:

$$q_f = 8760 n_i q_i' k_{Ni} k_{Ti} N_i, \quad (31)$$

where: n_i—number of specific equipment (mechanisms) in the terminal; q_i'—relative fuel consumption of specific equipment (mechanisms), the relative fuel consumption of diesel engines of many terminal equipment (mechanisms) is from 0.25 to 0.35 kg/kWh (more accurate data is provided in the specifications of specific engines); k_{Ni}—power utilization factor of specific equipment (mechanisms) engines; k_{Ti}—time coefficient of specific equipment (mechanisms) utilization; N_{RTG}—engine power of specific equipment (mechanisms).

The fuel consumption of another port terminal or other facility, such as a warehouse, equipment, per year (or other period of time) (q_{kU}) can be calculated according to the general formula [7,44]:

$$q_{kU} = 8760 n_U q_U' k_{NU} k_{TU} N_{VU}, \quad (32)$$

where: n_U—the amount of equipment or mechanisms of a terminal or other facility; q_U'—the relative fuel consumption of the equipment or mechanisms of a terminal or other facility, the relative fuel consumption of diesel or other engines of many port or other facility equipment or mechanisms is from 0.12 to 0.40 kg/kWh (more precise data is provided in the specifications of the engines of specific equipment or mechanisms); k_{NU}—the power utilization factor of the engines of the equipment or mechanisms of a terminal or other facility; k_{TU}—the time coefficient of the use of the equipment or mechanisms of the terminal or other facility; N_{VU}—the power of the engines of the equipment or mechanisms of a terminal or other facility.

The total fuel consumption at the terminal per year or other period ($\sum q_{fT}$) can be calculated using the formula [7,45]:

$$\sum q_{fT} = q_{fRTG} + q_{fTV} + q_{fSK} + q_{fVK} + q_{fU}, \quad (33)$$

Here: q_{fRTG}—the amount of fuel consumed by the terminal's RTG cranes per year; q_{fTV}—the amount of fuel consumed by the terminal's tugs per year; q_{fSK}—the amount of fuel consumed by the terminal's side lifts per year; q_{fVK}—the amount of fuel consumed by the terminal's vertical lifts per year; q_{fU}—the amount of fuel consumed by other terminal equipment per year.

The methodology obtained in this way allows for the assessment of fuel consumption of port superstructure equipment and mechanisms. By assessing the fuel consumption of port superstructure equipment and mechanisms, the average operating time of equipment and mechanisms (engine hours) and the average power used by equipment and mechanisms, the amount of emissions generated can be estimated.

A sustainable port should have minimal environmental impact, but it employs a wide range of machinery and equipment. Many of the vehicles operating in the port have powerful engines and consume a lot of fuel. Ships consume a lot of fuel when sailing into the port, maneuvering in it, and slightly less when they are moored at quays or anchored at anchorages, i.e., while waiting to enter the port. The fuel consumption of vehicles and other machinery operating in the port is calculated during sailing or at another time. The

https://doi.org/10.3390/su18041916

fuel consumption of vehicles and machinery engines if they use fossil fuels, such as diesel, LNG or other fuels, (Q_f) can be calculated as follows [44,45]:

$$Q_f = \int_0^t k_f q'_f N_{TM} dt, \quad (34)$$

where: k_f is the coefficient, which depends on the type of engine; q_f is the consumption of fuel for the definite engine; N_{TM} is the engine's average power during the working period, which can be calculated using Equation (35):

$$N_{TM} = \frac{\int_0^t N_{TMi} dt}{t}, \quad (35)$$

where: N_{TMi}—is the instantaneous transport means (equipment) engine power; t—is the transport means working time in hours.

Equipment engine power (N) and the amount of fuel consumed (q_f) over a given period of equipment working, during which equipment working, time (t), e.g., an hour, and the relative fuel consumption (q'_f) link as [44,45]:

$$N = q_f / (q'_f t) \quad (36)$$

The amount of fuel consumed by equipment when it works can be calculated as:

$$q_f = \int_0^t q'_f N_{av} dt \quad (37)$$

Here, N_{av} is the average engines power of the equipment during it work.

The amount of pollutants emitted by ships and port equipment and port machinery during operation directly depends on the amount and quality of fuel used, engine power and engine operating time [7,37,46–48]. The main types of pollutants emitted by engines are: carbon dioxide (CO_2), nitrogen oxides (NO_x), carbon monoxide (CO), sulfur oxides (SO_x) and particulate matter (PM) [37,46,48]. Thus, the amount of carbon dioxide, Sulphur oxide, carbon monoxide, nitrogen oxides and particulate matter emissions could be calculated according to the Formulas (38)–(42) [38,46,47]:

$$CO_2 = k_{CO_2} q_f \quad (38)$$

$$SO_x = k_{SO_x} q_f \quad (39)$$

$$CO = \int_0^t N_{av} k_{CO} dt \quad (40)$$

$$NO_x = \int_0^t N_{av} k_{NO_x} dt \quad (41)$$

$$PM = \int_0^t N_{av} k_{PM} dt \quad (42)$$

Here, k_{CO_2} is carbon dioxide coefficient for petroleum products (diesel, fuel oil) is between 3.0 and 3.5, for LNG between 2.5 and 2.9 [7,37,47]; k_{SO_x} is the Sulphur oxide coefficient, which depends on the type of fuel: for petroleum products it ranges from 0.001 to 0.035, for LNG it is around zero [37,47]; k_{CO} is carbon monoxide coefficient, which depends on the type of engine [7,37,46]; k_{NO_x} is nitrogen oxide coefficient, depending on engine type [37,46]; k_{PM} is the particulate matter coefficient, which depends on the type of engine and the type of fuel, up to 10 g/kWh for petroleum products and close to zero for LNG fuels [37,46,48].

https://doi.org/10.3390/su18041916

Additional actions affecting the sustainability and environmental impact of the port (A_S) refer to the necessary amount of energy and correspondingly generated emissions used to perform additional operations at the port, including daily maintenance of the port (terminal), not related to passenger service and cargo processing.

The developed methodology for assessing a sustainable port (terminal) using the sustainable port comparative index allows revealing the real situation of the technical, technological, organizational and other factors of the port (terminal) from the perspective of a sustainable port. Knowing the real situation of the port (terminal) allows you to effectively seek opportunities to optimally and sustainably develop the technical, technological, organizational and other capabilities of the port (terminal) and at the same time, by assessing the weaknesses of the port (terminal) according to individual factors of the sustainable port benchmark index, achieve better port performance results.

4. Case Study Assessing the Sustainability Situation of a Port (Terminal)

For the case study, as an example, a port container terminal was selected to assess the sustainability of a port (terminal), located next to the port's inner navigation channel and ship turning basin.

Depending on the type of port terminal (container, Ro-Ro, LNG, etc.) and its location in the port, the probability parameters of shipping safety must be assessed. For LNG, crude oil and petroleum product terminals, a probability of 99.7% is assumed, for other terminals, under simple navigational conditions, the standard shipping probability of 95% is assumed [7,26]. For terminals located far from the port entrance channel and with a complex channel configuration, a probability between the above-mentioned probabilities is assumed, after additional navigational safety assessments have been performed [7,26].

The port's entrance channel is 150 m wide, its depth is 15.5 m, and the diameter of the ship turning basin is 410 m. The terminal has 5 STS cranes, 16 RTG cranes, 20 terminal tugs, 8 side lifts, and 4 vertical lifts. The container terminal's capacity is approximately 800,000 TEU per year. The NEW PANAMAX container ship was selected for the assessment of the sustainable port terminal, with a length between perpendiculars of 355 m, a width of 48 m, a draft of 14 m, and a container capacity of approximately 14,000 TEU. The ship handles an average of approximately 4000 TEU at the terminal, and the average ship's parking time at the terminal is up to 24 h. The area of the projection of the above-water part of the ship onto the midplane is approximately 14,000 m^2, and onto the midship plane—approximately 2200 m^2. The area of the projection of the underwater part of the ship onto the midship plane, at a draft of 14 m, is approximately 5100 m^2. The sailing of a G-class container ship (length 400 m, width 61 m) to the port was also partially analyzed.

Experimental data on the arrival, maneuvering and departure of real ships to the port were used for the case study. Experimental studies of the arrival, maneuvering and departure of various types of real ships from ports were carried out by the authors (over 50 ships) in 20 ports, as well as using AIS data [7,24,37]. To confirm the theoretical results, it is very important to conduct experiments with real objects (ships) using highly accurate navigation equipment [49–51], and according to the results obtained, adjust the mathematical models so that they can be applied to any dynamic objects (ships) and in various locations in the ports. i.e., ship motion parameters: trajectories, linear and angular velocities, drift angles, obtained parameters of the main engine of the ships, etc.

Currently, simulators are often used, with the help of which various situations can be created. At the same time, it is not always possible to accurately enter all possible parameters into simulators, especially for dynamic objects (ships); therefore it is very important that the ship parameters in the simulator are as close as possible to real ships. Based on the results of experiments with real ships, calibration coefficients were calculated

https://doi.org/10.3390/su18041916

for the simulator, as the differences between the results obtained in analogous conditions on real ships and in the simulator. Using calibration coefficients, with the help of simulators it is possible to create various research situations, the necessary parameters of ship movement and maneuvering [7,24,52].

The calibrated simulator "Simflex Navigator" [53] was used for the research, for the calibration of which experimental results of real ships with analogous parameters were used.

After performing calculations of ship trajectories and other parameters using the methodology presented in Section 3 and comparing them with experimental results obtained on real ships and with the help of a calibrated simulator, it was determined that the ship motion parameters (calculated and experimental) differed from the values of the parameters themselves by no more than 10% [7,24,37,51,52].

When manoeuvring the ship, i.e., entering the port and turning it around in the port's ship turning basin, approaching the terminal quay and towing the ship away from the terminal quay, and when the ship leaves the port, 2 tugboats were used, each with a maximum pulling force of up to 520 kN.

The real ship's approach to the port, maneuvering in it, mooring and sailing from the quay and sailing from the port were carried out in a southwesterly wind with a speed of about 12 m/s, and the current in the channel direction towards the port gate reached up to 0.5 knots. The same external conditions were used in the calibrated simulator. The simulator was also used to collect statistical data.

The ship's movement trajectories in the port's entrance and internal navigation channels and the required channel width were calculated using the methodology presented in Section 3. The ship's movement trajectories obtained on real ships and with the help of a simulator during experiments were also used and analyzed. During the ship's navigation in the port's entrance and internal navigation channel, the wind speed was 12 m/s, the wind direction angle in the port's entrance channel was about 120° from the starboard side, in the internal navigation channel when entering the port—about 90° from the starboard side, and when sailing towards the quay and sailing away from the quay and when sailing from the port was from 90° to 60° from the port side. The current speed in the entrance channel (at the port gate) and in the internal navigation channel reached up to 0.4–0.5 knots. The direction of the current coincided with the direction of the channels (inlet and internal). The maximum drift angle of the ship in the port entrance channel reached up to 4°, in the inner navigation channel—up to 5°, in the turn, when sailing in the inner navigation channel—up to 10°. The angle of the ship's yaw about the course reached up to 1–2° (Figures 4–7).

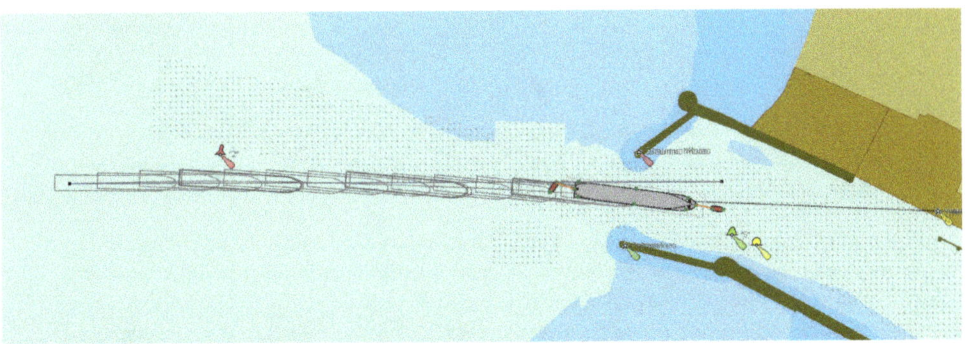

Figure 4. NEW PANAMAX container ship navigation through the port entrance channel during the vessel's entry into the port [53].

https://doi.org/10.3390/su18041916

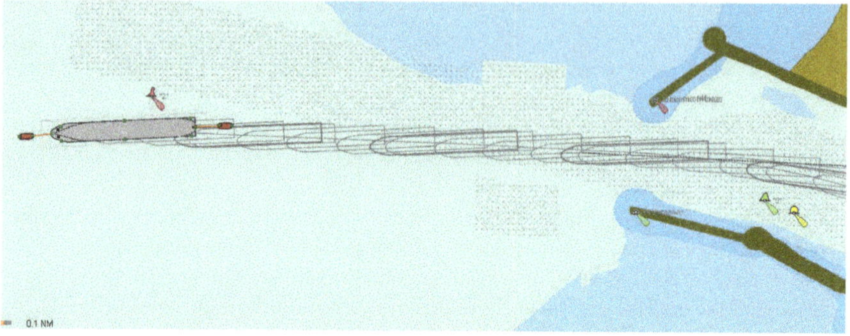

Figure 5. NEW PANAMAX container ship navigation through the port entrance channel as the vessel departs the port [53].

Figure 6. NEW PANAMAX container ship lanes in the port entrance channel under rough sea conditions [53].

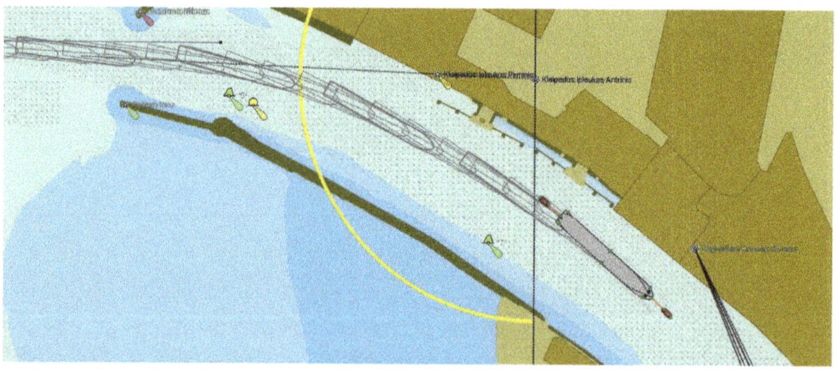

Figure 7. NEW PANAMAX container ship navigation in the inner port navigation channel during a turn after passing through the port gate as the vessel enters the port (yellow line show lighthouse sector) [53].

The port entrance channel is very important for achieving a sustainable port; therefore its parameters must ensure not only the navigation safety of incoming and outgoing ships, but also its configuration is important, allowing ships to navigate easily without additional special ship maneuvers. Examples of ship navigation lanes when a ship navigates through the port entrance channel, entering the port and leaving the port are presented in Figures 4 and 5.

When assessing the sustainability of the port entrance channel, the necessary channel width was estimated using the methodology presented in Section 3 (using the results of

https://doi.org/10.3390/su18041916

calculations and experiments) and assuming that the ship's drift angle reached up to 4°, the ship's yaw angle reached up to 2°, the accuracy of determining the ship's location with the help of a lead line was about 8 m. Assuming a probability of 95% and a navigation margin of 0.5 B, the calculated width of the port entrance channel was obtained to be about 125 m.

In rough sea conditions, the width of the port entrance channel must correspond to the calculated width, as the effects of external forces in individual cases require additional maneuvers to keep the ship in the channel (Figure 6).

The actual width of the port entrance channel is about 150 m. Thus, considering similar drift angles, the sustainable width of the port entrance channel is sufficient for ships up to 400 m in length and 62 m in width (class G container ship), for which the calculated channel width is about 149 m. Thus, for both analyzed ships, the port entrance channel comparative port sustainability index can be assumed to be equal to one.

The inner navigation channels of the port, especially if they contain large turns, are very important for the safe passage of ships in the port. As an example, Figures 7 and 8 show the lanes of ships in the inner navigation channel when a ship enters and leaves the port and makes a turn in the inner navigation channel of the port.

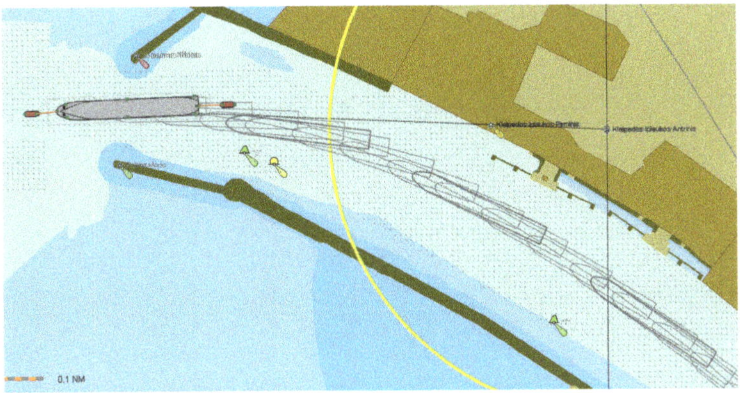

Figure 8. NEW PANAMAX container ship navigation in the inner port navigation channel and turning before approaching the port gate, when the vessel leaves the port (yellow line show lighthouse sector) [53].

In order to assess the sustainability of the inner navigation channel, maximum ship drift angles were adopted, which reached up to 10° when turning at the port gate. The width of the inner navigation channel, the above-mentioned ship sailing conditions, for a 355 m long ship (NEW PANAMAX container ship), is about 161 m.

The width of the inner navigation channels is also important due to possible additional effects, such as hydrodynamic effects between the ship and the quay walls or channel slopes, when ships pass ships moored at the quays, etc. Ship navigation lanes in the inner navigation channel, as an example, are shown in Figure 9.

For a ship with a length of 400 m (class G container ship), the calculated channel width, when the drift angle reaches 10°, is about 192 m. The actual width at the channel bend is about 200 m. Thus, for both analyzed ships, the port internal navigation channel comparative port sustainability index can be taken equal to one.

Port ship turning basins are essential for the safe turning of ships in the port. As an example, Figure 10 shows the ship turning trajectory when a NEW PANAMAX container ship is turned in the ship turning basin with the help of two tugboats.

To assess the sustainability of the port's ship turning basin, a NEW PANAMAX container ship was used, which was turned around with the help of two tugboats, each

https://doi.org/10.3390/su18041916

with a pulling force of up to 520 kN. Using the methodology for assessing the sustainable diameter of the port's ship turning basin presented in Section 3, the recommended diameter of the port's ship turning basin for a sustainable port was calculated. The actual diameter of the port's ship turning basin is 410 m, the length of the ship is 355 m, so the ratio of the diameter of the ship turning basin to the length of the ship is 1.15, which does not provide a 95% probability that the ship will be able to turn around safely. In this case, the sustainability coefficient of the port's ship turning basin is about 0.87.

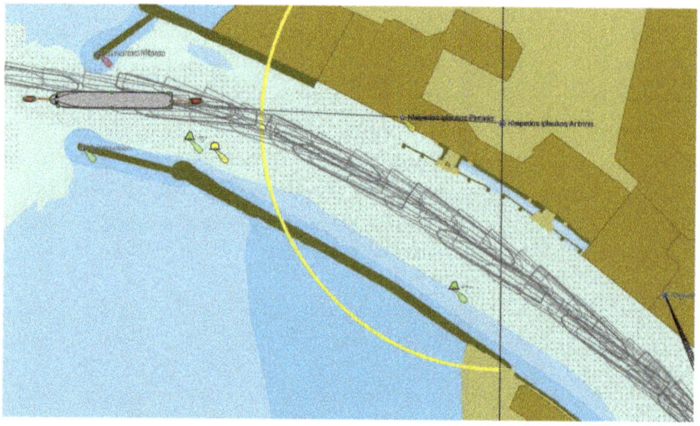

Figure 9. NEW PANAMAX container ship lanes in the inner navigation channel (yellow line show lighthouse sector) [53].

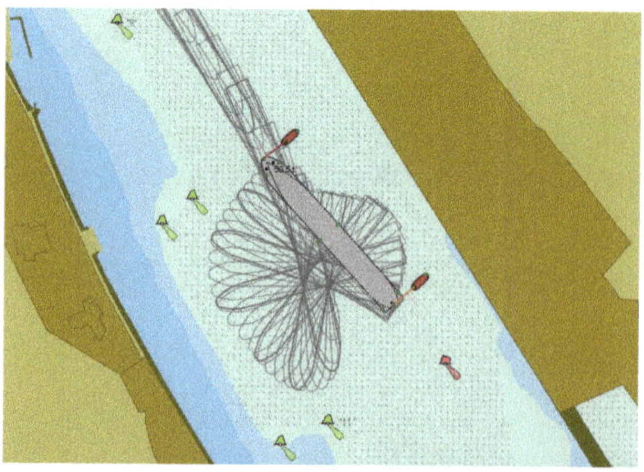

Figure 10. Turning a NEW PANAMAX container ship in a turning basin using two tugboats [53].

Access to the quay is very important in cases where there is limited depth at the approach to the quay. Using the methodology for assessing the sustainability of quay access presented in Section 3. The depths of other quays near the specified quay are similar to the depths at which the ship is moored, which means that there is sufficient space for the ship to manoeuvre. Figure 11 shows the movement lane of a ship as it approaches the quay. The same conditions apply when the ship is being driven away from the quay (Figure 12).

When assessing the sustainability of the ship's arrival and departure from the quay, the depth of the nearby water area and the power (traction force) of the tugs used are very important. In the case where there are enough tugs in the port and their traction force is high, optimal solutions can be found, but it is always very important to assess the required traction forces of the tugboats in advance.

https://doi.org/10.3390/su18041916

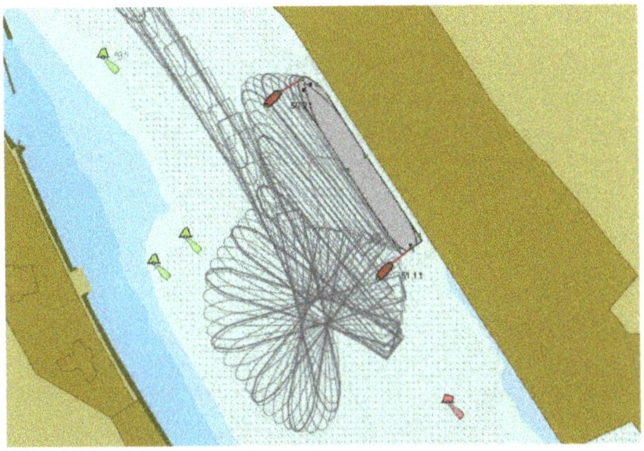

Figure 11. Bringing the NEW PANAMAX container ship to the quay (motion lane) [53].

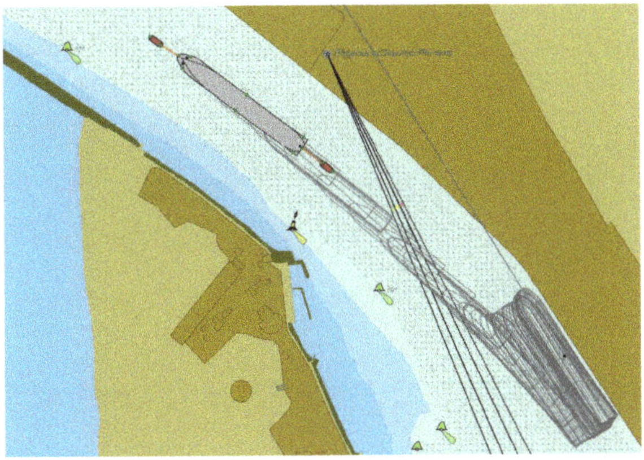

Figure 12. Lane for guiding a NEW PANAMAX container ship away from the quay and navigating in the internal navigation channel [53].

Using the methodology presented in Section 3, the required tugboat traction force was estimated and verified with similar real ships under similar conditions. The tugboat traction force required to turn and moor to and unmoor from the berth of the NEW PANAMAX ship (the calculation results are presented in Figure 13). The tugboat traction force required for mooring operations of the NEW PANAMAX ship, depending on the current and wind parameters and depth at the maneuvering site, was obtained by calculation and by conducting experiments with real ships and using a calibrated simulator. The difference between the obtained calculation and experimental results did not exceed 10%.

The tugboats pulling force, obtained by calculation and experimentally, depending on the current and wind parameters and the effect of the shallowing effect, is important when planning ship maneuvers in the port. As can be seen from the results obtained, at wind speeds up to 12 m/s and current speeds up to 0.5 knots, when its direction corresponds to the direction of the channel, the use of two tugboats, each with a pulling force of up to 520 kN, is sustainable. The tugboat pulling forces and pulling directions when turning the NEW PANAMAX ship and when mooring and unmooring it under the above conditions, obtained experimentally with a real ship and using a calibrated simulator, are presented in Figures 14 and 15. At higher wind and current speeds and in other directions, for the above-mentioned type of ship (NEW PANAMAX container ship) could be necessary to have higher tugboats pulling forces, i.e., three or more tugboats may be used and this

https://doi.org/10.3390/su18041916

information is very important to have when planning the entry or exit of a ship into or out of port.

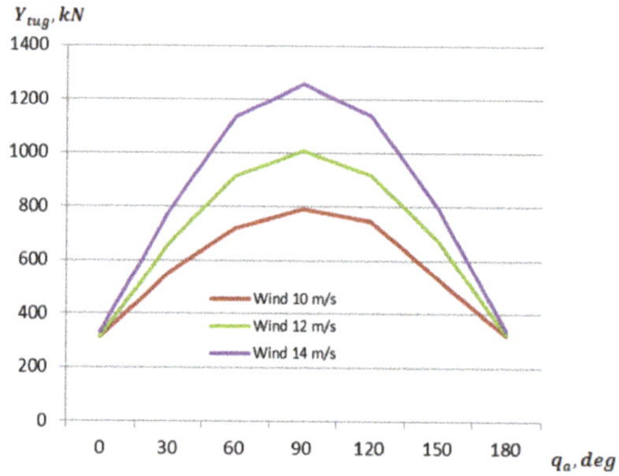

Figure 13. Required tugboats pulling force when mooring a NEW PANAMAX vessel at a current of 0.5 knots, depending on wind speed and direction.

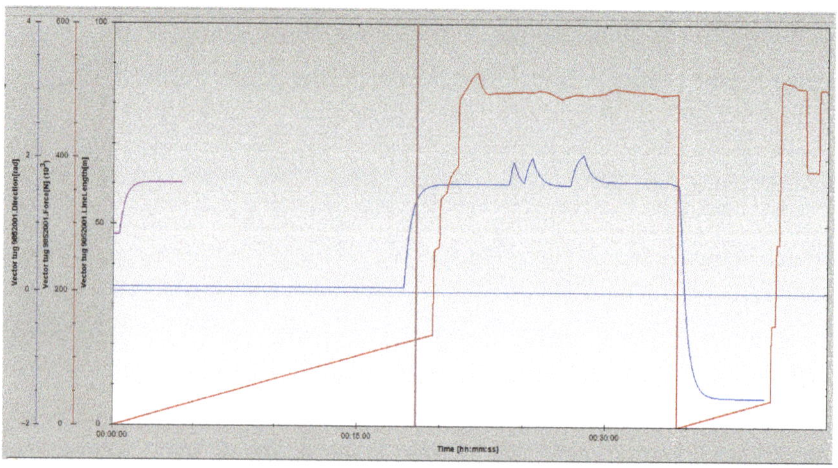

Figure 14. The pulling force of the first tugboat's rope, the length of the towing rope and the pulling directions when turning the NEW PANAMAX container ship and mooring the ship to the quay [53].

The sustainable use of port tugs in the port is important to ensure the safety of shipping navigation and to optimize the number of maneuvers of ships with high main engine power, while reducing the environmental impact, which is important for sustainable port factors.

The impact on the environment in the port is caused by the equipment and mechanisms of the incoming vessels, tugs and terminals. During the experiments on real analyzed ships (NEW PANAMAX), it was determined that the sailing time of the ship entering the port was about 45 min (from the entrance channel to the start of the mooring operation), the average power of the main engine was about 7000 kW and the ship consumed about 1100 kg of diesel fuel in 45 min (Figure 16).

When reloading from a container ship at the terminal, an average of about 4000 TEU, the ship stays at the terminal quay for an average of about 24 h, the average power of the auxiliary engines while standing at the quay is about 2000 kW, and the ship consumes about 9500 kg of diesel fuel during the stay. The work of the tugboats towing the ship into the port and mooring it takes about 1.5 h, the average power of the engines (of both

https://doi.org/10.3390/su18041916

tugboats together) is about 4500 kW, and during this time they consume about 1600 kg of diesel fuel.

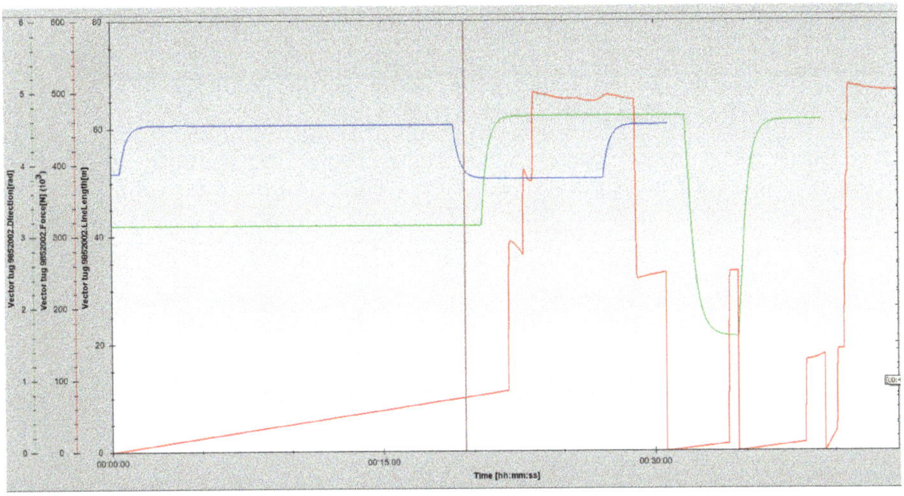

Figure 15. The pulling force of the second tugboat's rope, the length of the towing rope and the pulling directions when turning the NEW PANAMAX container ship and mooring the ship to the quay [53].

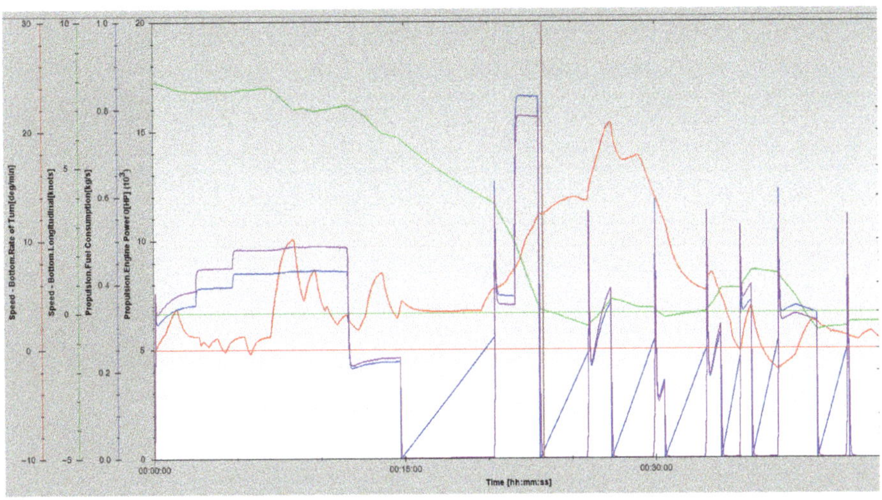

Figure 16. Parameters of ship motion and energy consumption when the ship enters the port: rate of turn (deg/min); longitudinal speed (knots); fuel consumption (kg/s); engine power (HP) [53].

The energy required to process one container (TEU) using STS cranes (which use electricity), RTG cranes, terminal tugs, skateboards and forklifts, using the methodology presented in Section 3 and the information provided in [54–57], is about 4 kWh per TEU. Converted into fuel consumption, it can be stated that about 4800–5000 kg of diesel fuel and about 16,000–18,000 kWh of energy are consumed for processing one container cargo of the analyzed ship (on average about 4000 TEU) at the terminal.

https://doi.org/10.3390/su18041916

The ship's departure from the port (from the terminal quay to the departure beyond the port's entrance channel) takes about 32 min, the average ship's engine power is about 4000 kW and during this time the ship consumes about 500 kg of diesel fuel (Figure 17).

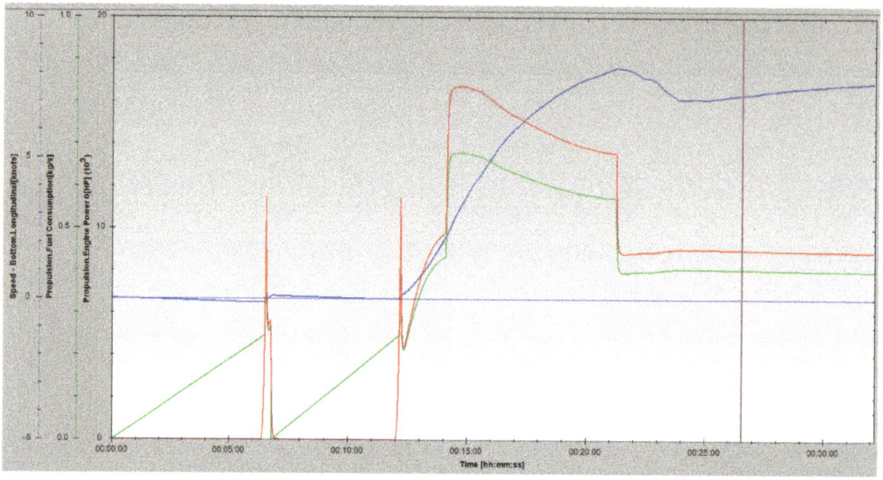

Figure 17. Parameters of NEW PANAMAX container ship motion and energy consumption when leaving the port: longitudinal speed (knots); consumption (kg/s); engine power (HP) [53].

The port tugs, when unmooring and taking a ship out of the port (from the terminal quay to the exit from the port entrance channel), work for about 32 min, the average engine power used (both together) is about 4000 kW, and the diesel consumption is about 640 kg (Figures 18 and 19).

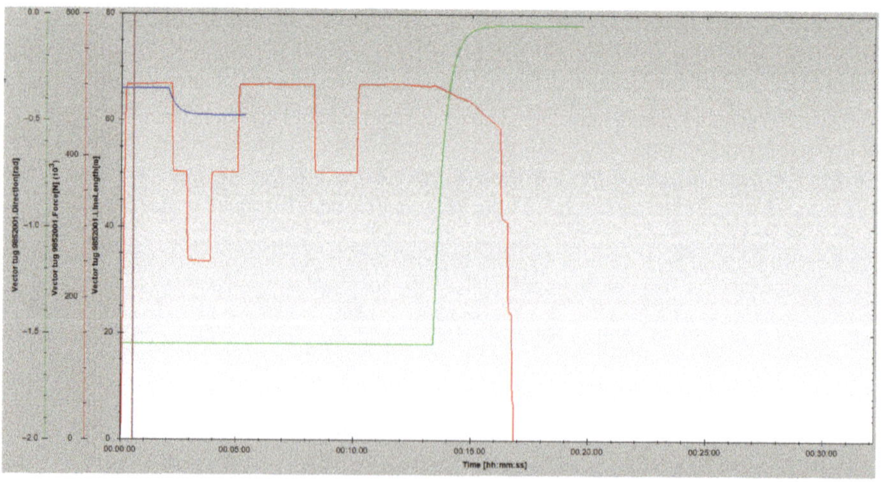

Figure 18. The pulling force and pulling directions of the first tugboat's rope, the length of the towing rope when unmooring the ship from the terminal quay and leaving the port [53].

In this way, a NEW PANAMAX ship consumes about 3200–3500 kg of diesel fuel during one port call, the average power of the main and auxiliary engines used is about 3200 kW, and the ship's stay in the port (from entering the port's entrance channel to leaving the port's entrance channel) is about 26 h on average.

https://doi.org/10.3390/su18041916

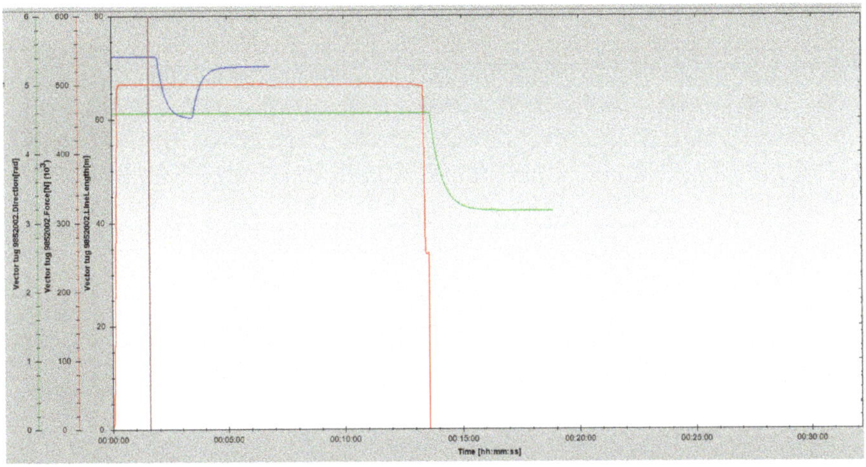

Figure 19. The pulling force and pulling directions of the second tugboat's rope, the length of the towing rope when unmooring the ship from the terminal quay and leaving the port [53].

Port tugboats, servicing the NEW PANAMAX vessel entering the port, consume about 2250 kg of diesel fuel, worked for about 2 h in total and the average total power used by the engines (of both tugs) during the aforementioned 2 h was about 4300 kW.

The terminal equipment and mechanisms operate for about 20 h during the arrival of one analyzed vessel, consume about 18,000 kWh of energy and about 5400 kg of diesel fuel.

Sustainable port development requires minimal environmental impact, which can be achieved by using greener fuels. The likelihood that ships, tugs and terminal equipment and machinery arriving at the port at the same time will start using, for example, LNG fuel is very low. Ships, port tugs and terminal equipment and machinery are gradually switching to LNG fuel. The colorimetric value of LNG fuel is on average about 15% higher than that of diesel [53]; therefore, about 15% less fuel mass is required to obtain the same amount of energy. The results of different fuel consumption (diesel and LNG) and the emission levels generated, using the methodology presented in Section 3, are presented in Table 1.

Table 1. Consumption and emissions generated by different fuels (diesel and LNG) of the NEW PANAMAX vessel, two tugboats and terminal equipment and mechanisms (one vessel call).

Object	Diesel Fuil, kg	CO_2, kg	SO_x, kg	CO, kg	NO_x, kg	PM, kg	LNG, kg	CO_2, kg	CO, kg	NO_x, kg
Ship	3500	11,200	3.5	400	800	41	2975	8033	250	333
Tugboats	2250	7200	2.25	43	86	5	1912	5162	26	35
Terminal equipent	5400	17,280	5.4	30	70	4	4590	12,393	25	33
Total	11,150	35,680	11.15	473	956	50	9477	25,588	301	401

Note: It is assumed that ship engines comply with Euro0, terminal equipment and mechanisms comply with Euro4 standards [55].

In the event that the ship, tugboats and terminal equipment and mechanisms had used LNG fuel, LNG fuel consumption would be about 15% less than when using diesel fuel, i.e., instead of about 11,150 kg of diesel fuel consumed, about 9477 kg of LNG fuel would have been consumed.

The results of emissions from ships, port tugs and terminal equipment and mechanisms clearly demonstrate the benefits of using more environmentally friendly energy

https://doi.org/10.3390/su18041916

sources for environmental impact. In the studied cases, the use of LNG fuel reduces the amount of CO_2 by about 28%, CO by about 35%, NO_x by about 58%. When using LNG fuel, SO_x and PM emissions are not generated. Recently, a significant part of port terminals are developing or connecting to alternative energy source generations (solar, wind, hydroelectric power plants) and are starting to use other energy sources, such as hydrogen, some ships are starting to use ammonium fuels. By using new environmentally friendly energy sources, emissions can be reduced by up to 60–80% or more.

After evaluating the results obtained, the comparative index (E_S) for the analyzed ship and port terminal is presented in Table 2.

Table 2. Comparative index of sustainable port factors.

Port (Terminal) Benchmark Factor	Port (Terminal) Under Analysis	Ideal (Comparable) Port (Terminal) (Using LNG Fuel)	The Ratio of the Analyzed and Ideal (Comparable) Ports Factors
Port entrance channel, m	125	150	1
Port inside navigational channels, m	161	200	1
Port ships turning basins, m	410	440	0.93
Port tugboats, kN	1040	1020	1
Entrance to quay, deg	More than 45°	More than 45°	1
Terminal operation, TEU/h	180	180	1
Quantities of generated CO_2/ CO/ NO_x	35,680/473/956	25,588/301/401	0.71/0.64/0.42
Additional actions (fuel consumption), kg	11,150	9477	0.85

The weighting coefficients of sustainable port factors were obtained using an expert method, evaluating the port data [56]. To determine the factor weighting coefficients, 20 ship masters, 10 terminal management representatives, 12 port pilots and 5 tugboat masters from 3 ports were interviewed. The survey results were processed using the maximum distribution methods [7,24,28,56,57].

In the case of the port under study, when compared with an ideal (comparable) port, assessing the parameters of the port infrastructure (port entrance and internal navigation channels, ship turning basin), the port terminal operational capabilities, container handling, and the environmental impact of using diesel and LNG fuels, it can be stated that achieving a sustainable port (terminal) even with individual factors allows improving the image of the port, increasing shipping safety conditions in the port, using the necessary tugboat capabilities, and reducing environmental impact. For example, the obtained ratio of the diameter of the port's ship turning basin (0.93) shows that when accepting ships of similar parameters, it is necessary to accurately assess the capabilities of the port's tugboats and order the required number of tugboats with the necessary pulling force to avoid non-standard situations during ship maneuvers.

By adopting the weighting coefficients of sustainable port factors and factor values presented in Table 2, the sustainable port comparative index was obtained, presented in Table 3.

The resulting sustainable port benchmark index (compared to an ideal port) is not so important in its absolute value, but at the same time it shows individual sustainable port factors that need to be improved in order to achieve better sustainable port results. From the obtained sustainable port benchmark index, it is clear which of them should be developed in order to increase shipping safety in the port, reduce environmental impact,

https://doi.org/10.3390/su18041916

which should positively affect working conditions in the port, improve the quality of life in the port region, by reducing generated emissions.

Table 3. Sustainable port comparative of the analyzed port (terminal).

Factor	Conditional Value of the Factor (from Table 2)	Factor Weight Coefficient	Comparative Index of the Factor	Notes
Port entrance channel	1	0.15	0.15	The actual channel width is larger than the calculated one.
Port inside navigational channels	1	0.15	0.15	The actual channel width is larger than the calculated one.
Port ships turning basins	0.93	0.1	0.093	Actual diameter is less than calculated
Port tugboats	1	0.1	0.1	Real tugboat capacities greater than calculated
Entrance to quay	1	0.1	0.1	The actual depth at the approaches to the quays is sufficient
Terminal operation	1	0.15	0.15	The terminal equipment is sufficient for the current flow
Quantities of generated emissions	0.71	0.1	0.071	Emissions are higher than when using more environmentally friendly fuel
Additional actions (fuel consumption)	0.85	0.15	0.1275	More environmentally friendly fuel can be used
TOTAL		1.0	0.9415	

The developed sustainable port comparative index methodology is important in that it can be applied not only to the assessment of the overall sustainable port index, but also to the assessment of individual sustainable port factors, and it can be applied to any port or terminal after adapting it to local conditions.

5. Discussion

The creation of a sustainable port is a long process and it includes many factors; therefore, the development of new technologies and opportunities, as well as changing international and local requirements, especially in the field of environmental protection, force us to look for new solutions. The creation of a sustainable port is related not only to technical and technological aspects, but the human factor is very important, since the level of education and practical training of people determines which technical, technological, organizational and legal factors will be adopted and developed for the creation of a sustainable port.

The assessment of a sustainable port (terminal) is very important; therefore future research in this direction is very important, so that as many people (specialists) as possible have a common understanding of similar phenomena in the process of sustainable port development. In this article, using the sustainable port comparative index, an attempt is made to pay more attention to technical and technological factors. Currently, in different countries and even in different ports of one country, the elements of a sustainable port are more often assessed differently, because some pay more attention to technical and technological factors, while others are more focused on environmental impact processes; therefore, complex research on sustainable port development is very important [9,12,46,58,59].

Comparative research methods are often successfully applied in sustainable port development studies, using several or a dozen factors that are very important, but the

https://doi.org/10.3390/su18041916

weight of the factors is not currently very clearly resolved. In many cases, the currently applied expert assessment of factor weight coefficients allows determining the general situation of a sustainable port or terminal, but it is not always possible to determine clear priorities for sustainable port development due to not always clearly justified factor weight coefficients [56]. Therefore, research on the weight (coefficients) of sustainable port factors is promising and important from a theoretical and practical point of view.

Many different coefficients are used in the assessment of a sustainable port (terminal), such as factor weights, area utilization, amount of equipment, etc. The aforementioned coefficients often change due to the implementation of new technologies, changes in working conditions and occupational safety requirements, etc.; therefore further research in this area is promising and important in the processes of implementing sustainability in ports (terminals).

Sustainable port development research related to reducing environmental impact is important not only for the ports themselves, but also for the surrounding residential areas and suburbs. One of the goals of sustainable port development must be to reduce environmental impact, since ports are often also industrial zones and not only passenger service and cargo processing take place there, but also various production and logistics activities are carried out [33,35,60]. As a result of these activities, large land transport (cars, railways) flows are formed, which affect not only the environment, but also directly affect the standard of living of people; therefore this type of research is important and promising.

6. Conclusions

Studies on the assessment and development of sustainable ports have been carried out, allowing us to assess the current situation of ports and develop appropriate solutions for achieving port sustainability in the areas of navigational safety, port operations, and environmental impact reduction. Studies on the assessment of sustainable port development include technical, technological, and environmental aspects, and therefore a number of important conclusions can be drawn: 1. ports strive to achieve sustainability by developing port entrances and internal navigation channels, port ship turning basins, and access to quays, thereby improving navigational safety; 2. port terminals, by implementing the latest passenger service and cargo processing technologies, increase terminal productivity and reduce environmental impact, thereby achieving the prospects for creating a sustainable port; 3. a methodology for assessing port sustainability has been developed, using a sustainable port comparative index, allowing us to identify weak factors of a port or terminal and seek specific solutions to improve the situation and increase port sustainability; 4. a methodology for assessing port sustainability has been developed using a sustainable port comparative index, which can be adapted and implemented in any port and the most effective methods and measures for increasing port sustainability have been sought.

The developed methodology for the comparative port (terminal) sustainability index requires a relatively large amount of data; therefore it is recommended that ports, based on the methodology, prepare practical calculation programs and use them when developing the port, accepting ships that are not common in the port, and installing alternative energy sources.

Author Contributions: Conceptualization, V.P. (Vytautas Paulauskas); methodology, V.P. (Vytautas Paulauskas) and D.P.; software, V.P. (Vytautas Paulauskas) and D.P.; validation, V.P. (Vytautas Paulauskas), D.P. and V.P. (Vytas Paulauskas); formal analysis, V.P. (Vytautas Paulauskas); investigation, V.P. (Vytautas Paulauskas) and V.P. (Vytas Paulauskas); resources, V.P. (Vytautas Paulauskas) and D.P.; data curation, D.P. and V.P. (Vytas Paulauskas); writing—original draft preparation, V.P. (Vytautas Paulauskas) and D.P.; writing—review and editing, D.P. and V.P. (Vytas Paulauskas); visualization, D.P.; supervision, V.P. (Vytautas Paulauskas); project administration, V.P. (Vytautas Paulauskas);

https://doi.org/10.3390/su18041916

funding acquisition, V.P. (Vytautas Paulauskas) All authors have read and agreed to the published version of the manuscript.

Funding: This research received no external funding.

Institutional Review Board Statement: Not applicable.

Informed Consent Statement: Not applicable.

Data Availability Statement: The original contributions presented in this study are included in the article. Further inquiries can be directed to the corresponding author.

Acknowledgments: This article is based on the research conducted in the Maritime Engineering Department of the Klaipeda University.

Conflicts of Interest: The authors declare no conflicts of interest.

References

1. Wagner, N.; Kotowska, I.; Pluciński, M. The Impact of Improving the Quality of the Port's Infrastructure on the Shippers' Decisions. *Sustainability* **2022**, *14*, 6255. [CrossRef]
2. Konstantinos, K.; Nikas, A.; Daniil, V.; Kanellou, E.; Doukas, H. A multi-criteria decision support framework for assessing seaport sustainability planning: The case of Piraeus. *Marit. Policy Manag.* **2022**, *50*, 1030–1056. [CrossRef]
3. Glavinovi'c, R.; Krčum, M.; Vuki'c, L.; Karin, I. Cold Ironing Implementation Overview in European Ports—Case Study—Croatian Ports. *Sustainability* **2023**, *15*, 8472. [CrossRef]
4. Mersin, K. Review of total emission of transit Ships in the Dardanelle which including possible CO_2 emission of 1915 Canakkale Bridge. *Therm. Sci.* **2020**, *24*, 391–398. [CrossRef]
5. Yokohama Port Regulations. Available online: https://www.scribd.com/document/591569779/Port-Entry-Manual-Port-of-Yokohama (accessed on 1 December 2025).
6. Doha Port Regulations. Available online: https://visitqatar.com/intl-en/plan-your-trip/doha-port (accessed on 1 December 2025).
7. Paulauskas, V.; Paulauskas, D.; Paulauskas, V. *Smart and Green Sea Transport and Port*; Klaipeda University Publish House: Klaipeda, Lithuania, 2025; 500p. (In Lithuanian)
8. Beskovnik, B.; Bajec, P. Strategies and approach for smart city–port ecosystems development supported by the internet of things. *Transport* **2021**, *36*, 433–443. [CrossRef]
9. Rao, A.R.; Wang, H.; Gupta, C. Predictive Analysis for Optimizing Port Operations. *Appl. Sci.* **2025**, *15*, 2877. [CrossRef]
10. Xu, L.; Chen, Y. Overview of Sustainable Maritime Transport Optimization and Operations. *Sustainability* **2025**, *17*, 6460. [CrossRef]
11. Rehmatulla, N.; Smith, T. The impact of split incentives on energy efficiency technology investments in maritime transport. *Energy Policy* **2020**, *147*, 111721. [CrossRef]
12. PORTUS Green and Effective Operations at Terminals and in Ports "GREEN EFFORTS". Available online: https://portusonline.org/green-and-effective-operations-at-terminals-and-in-ports-green-efforts-2/?pdf=11253 (accessed on 30 November 2025).
13. Rawson, K.J.; Tupper, E.C. *Basic Ship Theory*, 5th ed.; Elsever: Amsterdam, The Netherlands, 2001; 727p.
14. Čerka, J. *Laivo Eigumas*; Klaipeda University Publish House: Klaipeda, Lithuania, 2005; 200p. (In Lithuanian)
15. Kurt, I.; Aymelek, M. Operational adaptation of ports with maritime autonomous surface ships. *Transp. Policy* **2024**, *145*, 1–10. [CrossRef]
16. Wawruch, R. Ships' Routes to Polish Ports of the Gulf of Gdańsk. *TransNav Int. J. Mar. Navig. Saf. Sea Transp.* **2021**, *15*, 583–590. [CrossRef]
17. Choi, J.-H.; Jang, J.-Y.; Woo, J. A Review of Autonomous Tugboat Operations for Efficient and Safe Ship Berthing. *J. Mar. Sci. Eng.* **2023**, *11*, 1155. [CrossRef]
18. Aydin, C.; Karabulut, U.C.; Ünal, U.O.; Sariöz, K. Practical computational procedures for predicting steering and braking forces of escort tugs. *Gemive Deniz Teknol.* **2017**, *21*, 21–36.
19. Kalam, M.A.; Husnawan, M.; Masjuki, H.H. Exhaust emission and combustion evaluation of coconut oil-powered indirect injection diesel engine. *Renew. Energy* **2003**, *28*, 2405–2415. [CrossRef]
20. Port of Roterdam Ongoing Projects. 2025. Available online: https://www.portofrotterdam.com/en/building-port/ongoing-projects (accessed on 28 February 2025).
21. Port of Zeebruge. 2025. Available online: https://www.portofantwerpbruges.com/en/location/zeebrugge (accessed on 25 February 2025).
22. Zou, Y.; Chen, Y.; He, J.; Pang, G.; Zhang, K. 4D Time Density of Trajectories: Discovering Spatiotemporal Patterns in Movement Data. *ISPRS Int. J. Geo-Inf.* **2018**, *7*, 212. [CrossRef]

https://doi.org/10.3390/su18041916

23. Perkovič, M. Advances in Navigability and Mooring. *J. Mar. Sci. Eng.* **2024**, *12*, 1601. [CrossRef]
24. Paulauskas, V.; Paulauskas, D.; Simutis, M. Estimation of Tug Pulling Power (Bollard Pull) and Number of Tugs Required During Ship Mooring Operations. *J. Mar. Sci. Eng.* **2024**, *12*, 1959. [CrossRef]
25. Tang, G.; Li, M.; Fang, Z.; Shen, C.; Qi, Y.; Zhen, Z.; Zhao, Z. Efficiency and Energy Consumption of the Automated Container Yard with Twin Rail-Mounted Gantry Cranes Considering Crane Scheduling Strategies and Handshake Area Designs. *Simul. Model. Pract. Theory* **2024**, *135*, 102981. [CrossRef]
26. PIANC. Harbour Approach Channels and Design Guidelines—Report No. 121-2014, s.l.: The World Association for Waterborne Transport Infrastructure. 2014. Available online: https://files.pca-cpa.org/pcadocs/ua-ru/02.%20UA%20Observations%20on%20Jurisdiction/01.%20Exhibits/UA-88.pdf (accessed on 1 December 2025).
27. Puertos del Estado. *ROM 3.1-99: Recommendation for Design of the Maritime Configuration of Ports, Approach Channels and Harbour Basins*; Puertos del Estado [Ports of the State]: Madrid, Spain, 1999; 382p. Available online: https://www.puertos.es/en/services/publications/recommendation-maritime-works (accessed on 1 December 2025).
28. Knill, O. *Probability Theory and Stochastic Processes with Applications*; Overseas Press: New York, NY, USA, 2009; 382p.
29. Raymond, A. Serway. In *Physics for Scientists and Engineers*, 2nd ed.; Saunders College Publishing: Philadelphia, PA, USA, 1986; p. 202.
30. Kianejad, S.S.; Enshaei, H.; Duffy, J.; Ansarifard, N. Prediction of a ship roll added mass moment of inertia using numerical simulation. *Ocean Eng.* **2019**, *173*, 77–89. [CrossRef]
31. Piaggio, B.; Viviani, M.; Martelli, M.; Figari, M. Z-Drive Escort Tug maneuverability model and simulation. *Ocean Eng.* **2019**, *191*, 106461. [CrossRef]
32. PIANC Fender Guidelines 2024. Guidelines for the Design, Manufacturing and Testing of Fender Systems. 274p. Available online: https://www.pianc.org/publication/pianc-fender-guidelines-2024/ (accessed on 25 November 2025).
33. Matuszak-Flejszman, A.; Preisner, A.; Banach, J.K. Transport Related Emissions and Transition Strategies for Sustainability—A Case Study of the Fast Fashion Industry. *Sustainability* **2024**, *16*, 7749. [CrossRef]
34. Johansson, L.; Jalkanen, J.P.; Kukkonen, J. Global assessment of shipping emissions in 2015 on a high spatial and temporal resolution. *Atmos. Environ.* **2017**, *167*, 403–415. [CrossRef]
35. Styhre, L.; Winnes, H.; Black, J.; Lee, J.; Le-Griffin, H. Greenhouse gas emissions from ships in ports—Case studies in four continents. *Transp. Res. Part D Transp. Environ.* **2017**, *54*, 212–224. [CrossRef]
36. Jiang, M.; Wang, B.; Hao, Y.; Chen, S.; Wen, Y.; Yang, Z. Quantification of CO_2 emissions in transportation: An empirical analysis by modal shift from road to waterway transport in Zhejiang, China. *Transp. Policy* **2023**, *145*, 177. [CrossRef]
37. Paulauskas, V.; Filina-Dawidowicz, L.; Paulauskas, D. The Method to Decrease Emissions from Ships in Port Areas. *Sustainability* **2020**, *12*, 4374. [CrossRef]
38. Saini, M.; Lerher, T. Assessing the factors impacting shipping container dwell time: A multi-port optimization study. *Bus. Theory Pract.* **2024**, *25*, 51–60. [CrossRef]
39. Owusu-Oware, E.; Effah, J.; Adam, I.O.; Amankwah-Sarfo, F. Actualizing the affordances of seaport smart container terminal system in a developing country. *J. Inf. Technol. Case Appl. Res.* **2023**, *25*, 339–367. [CrossRef]
40. Lee, B.K.; Low, J.M.W. Resource Capacity Requirement for Multi-Terminal Cooperation in Container Ports. *Appl. Sci.* **2021**, *11*, 9156. [CrossRef]
41. Cao, Z.; Sun, Q.; Wang, S. Dry bulk terminal operations: A critical review of models and optimization methods. *Transp. Res. Part C Emerg. Technol.* **2025**, *178*, 105221. [CrossRef]
42. Alzbutas, R.; Vaisnoras, M.; Saruniene, I.; Krikstolaitis, R.; Valincius, M.; Babilas, E.; Augutis, J.; Rimkevicius, S.; Iesmantas, T.; Anusauskas, F.; et al. Aggregated Risk Assessment and Survey for Risk Reduction in Oil Terminals. *Sustainability* **2021**, *13*, 12169. [CrossRef]
43. Malandri, C.; Mantecchini, L.; Paganelli, F. Towards Green and Smart Ro–Ro Port Terminal Operations: A Comparative Analysis of ICE, BET and e-AGT Tractors. *Future Transp.* **2025**, *5*, 121. [CrossRef]
44. Heinrich, L.; Koschinsky, A.; Markus, T.; Singh, P. Quantifying the fuel consumption, greenhouse gas emissions and air pollution of a potential commercial manganese nodule mining operation. *Mar. Policy* **2020**, *114*, 103678. [CrossRef]
45. Brzezinski, M.; Pyza, D.; Archutowska, J.; Budzik, M. Method of Estimating Energy Consumption for Intermodal Terminal Loading System Design. *Energies* **2024**, *17*, 6409. [CrossRef]
46. Endresen, Ø.; Sørgård, E.; Sundet, J.K.; Dalsøren, S.B.; Isaksen, I.S.A.; Berglen, T.F.; Gravir, G. Emission from international sea transportation and environmental impact. *J. Geophys. Res. D Atmos.* **2003**, *108*, ACH 14-1–ACH 14-22. [CrossRef]
47. Safety4Sea. Cold Ironing: The Role of Ports in Reducing Shipping Emissions. Available online: https://safety4sea.com/cm-cold-ironing-the-role-of-ports-in-reducing-shipping-emissions/ (accessed on 30 March 2023).
48. Lack, D.A.; Corbett, J.J. Black carbon from ships: A review of the effects of ship speed, fuel quality and exhaust gas scrubbing. *Atmos. Chem. Phys.* **2012**, *12*, 3985–4000. [CrossRef]

https://doi.org/10.3390/su18041916

49. Specht, C. Maritime DGPS System Positioning Accuracy as a Function of the HDOP in the Context of Hydrographic Survey Performance. *Remote Sens.* **2023**, *15*, 10. [CrossRef]
50. Alissa, S.; Håkansson, M.; Henkel, P.; Mittmann, U.; Hüffmeier, J.; Rylander, R. Low bandwidth network-RTK correction dissemination for high accuracy maritime navigation. *TransNav—Int. J. Mar. Navig. Saf. Sea Transp.* **2021**, *15*, 171–179. [CrossRef]
51. Gucma, L.; Montewka, J. Land borne laser rangefinder measurements for navigation safety assessment. *Eur. J. Navig.* **2005**, 3, 1–6.
52. Gucma, L. The Risk Assessment of Ships Maneuvering on the Waterways Based on Generalized Simulation. *DataWIT Trans. Built Environ.* **2007**, *94*, 58–69.
53. SimFlex "SimFlex Navigator" Simulator. 2016. Available online: https://forcetechnology.com/-/media/force-technology-media/pdf-files/0-to-4000/1922-maritime-engineering-brochure.pdf (accessed on 30 November 2025).
54. Roales, J.; Moscoso, F.G.; Vargas, A.P.; Lopes-Costa, T.; Pedrosa, J.M. Colorimetric Gas Detection Using Molecular Devices and an RGB Sensor. *Chemosensors* **2023**, *11*, 92. [CrossRef]
55. Jokubyniene, V.; Liebuviene, J. Research in vehicle exhaust gas compliance with Euro standard. In Proceedings of the 22nd International Scientific Conference Engineering for Rural Development, Jelgava, Latvia, 24–26 May 2023; 15p. [CrossRef]
56. Johansen, J.; Fischer-Hübner, S. Expert Opinions as a Method of Validating Ideas: Applied to Making GDPR Usable. In *Human Factors in Privacy Research*; Gerber, N., Stöver, A., Marky, K., Eds.; Springer: Cham, Switzerland, 2023. [CrossRef]
57. Jin, H.; Peng, S. Optimal unbiased estimation for maximal distribution. *Am. Inst. Math. Sci.* **2021**, *6*, 189–198. [CrossRef]
58. He, J. Berth allocation and quay crane assignmentin a container terminal for the trade-off between time-saving and energy-saving. *Adv. Eng. Inform.* **2016**, *30*, 390–405. [CrossRef]
59. Energy Consumption and Container Terminal Efficiency. Available online: https://www.researchgate.net/publication/313405671_Energy_consumption_and_container_terminal_efficiency (accessed on 24 November 2025).
60. Li, X.; Zhou, R.; Zhu, L. The Influence of Operation Platform on the Energy Consumption of Container Handling. *Sustainability* **2023**, *15*, 385. [CrossRef]

https://doi.org/10.3390/su18041916

sustainability

MDPI

Article

Harmonisation of Navigation Signs on Inland Waterways in Poland with European Regulations to Improve Navigation Safety

Łukasz Pieron [1,2,*], Kasper Jędrzychowski [1], Stefan Iwicki [3] and Piotr Durajczyk [4,*]

[1] National Water Management Authority, State Water Holding Polish Waters, 00-613 Warsaw, Poland; kasper.jedrzychowski@wody.gov.pl
[2] Faculty of Geographical Sciences, Kazimierz Wielki University, 85-033 Bydgoszcz, Poland
[3] Regional Water Management Authority in Szczecin, State Water Holding Polish Waters, 70-030 Szczecin, Poland; stefan.iwicki@wody.gov.pl
[4] Faculty of Engineering and Economics of Transport, Maritime University of Szczecin, 70-500 Szczecin, Poland
* Correspondence: lukasz.pieron@wody.gov.pl (Ł.P.); p.durajczyk@pm.szczecin.pl (P.D.)

Abstract

Navigation marking is an essential element of inland waterways. The visibility, location, and quality of navigation signs determine navigation conditions. Therefore, this study aimed to identify opportunities to improve navigation safety on inland waterways in Poland. Particular attention was paid to Polish legislation and international regulations in order to identify discrepancies and propose amendments to individual legal acts. These recommendations were preceded by field studies of selected navigation signs, including assessments of their legibility at night and in conditions of limited visibility. On this basis, and drawing on many years of practical experience on navigable waterways in Poland and Western European countries, the authors developed original proposals for the use of navigation signs, including designs for floating navigation signs and bridge clearance signs, as well as concepts for navigation marking plans for selected infrastructure components related to inland waterways. The proposed measures are expected to improve the recognisability and consistency of navigation markings, which may contribute to better operating conditions for inland waterway users, including both professional inland navigation operators (freight and passenger) and individual recreational users.

Keywords: inland waterways; sustainable infrastructure; navigation marks; navigation safety; navigation regulations; sustainable transport

1. Introduction

Inland waterways are extensive systems consisting of natural and regulated sections of rivers and lakes, as well as artificial canals, used for the transport of passengers and cargo [1]. Individual sections are often subject to inconsistent standards across waterway classes, which affects safety levels [2]. The diversity of inland waterway sections also influences the navigation signs used [3,4]. Proper navigation signs are one of the key factors affecting the safety of users, including both professional inland waterway transport operators (freight and passenger) and individual tourist users [5]. Recommendations for improving safety conditions on inland waterways, including reducing collisions, emphasise the need to install a sufficient number of navigation signs (cardinal and lateral marks, buoys, etc.) at the required locations for both day and night navigation [6].

Research on inland waterway navigation safety has traditionally focused on traffic management [7], hydrological constraints [8], and vessel operations [9], while navigation marking systems have received comparatively less systematic academic attention. Nevertheless, several studies have highlighted the importance of clear and standardised navigation aids for safe vessel manoeuvring and traffic organisation on inland waterways, including research on automatic navigational buoy recognition [10] and buoy behaviour and stability in the Automatic Identification System [11]. Other works analysing accident databases and operational safety in inland navigation also emphasise the role of different navigational systems [12–14].

In Poland, attention is also drawn to the need for the systematic modernisation of signage, which is in line with European trends [15], and the need to harmonise regulations and implement modern technical solutions [16]. The European Commission also confirms the need to harmonise many aspects of inland navigation and inland waterways, including the implementation by Member States of European regulations into national legislation [17], and in this respect, the application of the "European Code for Signs and Signals on Inland Waterways" (SIGNI) and the "European Code for Inland Waterways" (CEVNI) may contribute to reducing the number of incidents on inland waterways.

Notwithstanding the above, it should be remembered that inland waterway transport is considered sustainable and the safest mode of transport in Europe due to the low number of fatalities and casualties on inland waterways [18–20]. Nevertheless, in an era of challenges related to climate change, we can expect a deterioration in navigation conditions on individual inland waterways, e.g., in terms of water level fluctuations and the resulting effects on costs and reliability [21]. The safe manoeuvrability of a vessel depends on the available water depth in the navigation area in relation to the draught of the vessel [22]. As in other countries, droughts can seriously disrupt inland waterway transport services, reducing water levels to completely unnavigable levels or to levels that force operators to reduce the amount of cargo on board, which directly affects the profitability of transport [23]. The changes taking place may also necessitate increased research to ensure safe navigation in poorer navigational conditions, in particular the detection of navigational obstacles in waterways with extremely shallow depths [24]. Already, one of the causes of shipping accidents is the inadequate maintenance of inland waterways, which is related to unmarked navigational obstacles in the form of drifting branches, underwater objects or shoals, which should be removed [25]. In addition to professional inland navigation, many inland waterways in Europe are intensively used by recreational users such as motorboat operators, sailors, kayakers and other tourism-related users [26]. These groups often have varying levels of navigational training and experience, which increases the importance of clear and intuitive navigation marking systems [27]. In Poland, small vessels up to 10 m in length may operate without a licence [28], which further emphasises the need for navigation signs that are easily recognisable and consistent across different waterways.

In Poland, the maintenance of inland waterways in a condition ensuring safe navigation is the responsibility of the waterway administration. This role is performed by the State Water Holding Polish Waters (Państwowe Gospodarstwo Wodne Wody Polskie), which erects and maintains navigation signs on navigable waterways and on its facilities (locks and slipways), issues navigational notices, and publishes current navigational conditions. Navigation signs in ports and on facilities, structures, and transmission lines crossing waterways are installed by their owners in agreement with the waterway administration [28]. The inland waterway administration authorities are the minister responsible for inland waterway transport and the directors of the local inland waterway offices. The minister is the head of the inland waterway administration and, among other things, supervises the directors' activities and is responsible for legislation. The responsibilities of the directors of inland waterway administration offices include supervising the safety of inland

https://doi.org/10.3390/su18083844

waterway transport and inspecting the condition of navigable waterway markings, locks, slipways, bridges, waterway structures, and port entrances. The division of competences and institutional arrangements in Poland is similar to, for example, the administrative structure in the Federal Republic of Germany, where the German Federal Waterways and Shipping Administration (Wasserstraßen- und Schifffahrtsverwaltung des Bundes, WSV) is responsible for the maintenance of inland waterways [29]. In other regions, international organisations such as the Central Commission for the Navigation of the Rhine [30] and the Danube Commission [31] coordinate navigation standards among several countries.

Despite the practical importance of navigation marking systems, the issue of harmonisation between national and international regulations has not been extensively analysed in the academic literature in Poland. This creates a research gap, particularly in the context of countries where international standards such as SIGNI and CEVNI have not been fully incorporated into national legislation.

Therefore, the objective of this study was to analyse the current system of navigation marking on inland waterways in Poland and to identify possibilities for its harmonisation with European regulatory frameworks. The study aimed to identify discrepancies between Polish regulations and international standards, assess selected practical aspects of navigation mark visibility through field observations, and propose design and regulatory recommendations that could improve the recognisability and consistency of navigation marking systems. To this end, a comprehensive analysis of national and international regulations was carried out, making it possible to identify the most significant discrepancies in the legislation currently in force. On this basis, proposals were developed for the implementation of modern navigation sign solutions in Poland for use both during the day and at night, including in conditions of limited visibility.

2. Materials and Methods

2.1. Study Area

The analysis covered navigation markings used on inland waterways in Poland (Figure 1) and related legal conditions at the national and international level. The length of navigable waterways in Poland exceeds 3700 km [32], of which 3549 km was used for navigation in 2023 [33]. This is the fifth longest network in Europe, surpassed only by France (8507 km), Finland (8086 km), the Federal Republic of Germany (7675 km) and the Netherlands (7040 km) [34]. Therefore, Poland is second only to countries (except Finland) that make much greater use of inland navigation for sustainable development and transport.

Due to the diversity of navigational conditions in Poland, the study focused on selected waterway sections representing different types of inland navigation environments, including international waterways, regional waterways and urban navigation nodes. Field observations were conducted in three main locations:

- The border section of the Oder River (international waterway, cross-border navigation conditions);
- The Wrocław Water Junction (complex urban navigation system with intensive infrastructure);
- The Vistula River in Kraków (regional waterway with significant tourism traffic).

These locations were deliberately chosen to represent different functional types of inland waterways, intended for professional navigation and individual users. At the same time, these were the only sections of waterways in Poland adapted to 24 h navigation, including at night, thanks to the use of reflective foil-covered markings, illuminated buoys, and radar screens on individual sections. Although the selected sites do not represent the entire Polish inland waterway network, they provide examples of typical marking problems occurring on waterways with different operational characteristics, which are known to the authors of the article due to their participation in inspection cruises and preparation of navigational marking plans on the remaining sections.

The analysed waterway sections include a total of 32 bridges (the border section of the Oder River—8, the Wrocław Water Junction—12, and the Vistula River in Kraków—12) and numerous ports and harbours, which were also included in the analysis of navigation signage.

Inland waterways in Poland vary in terms of operating parameters, which is related to their natural conditions and the degree of use of hydrotechnical infrastructure [35]. Less than 6% of them meet international requirements (classes IV–Vb), while 94% have regional operating parameters (classes Ia–III) [32]. The sections with international designation are located in different parts of the country and do not form a coherent network of waterways, which means that their potential is not being exploited [36].

At the same time, the greatest challenge to achieving proper navigation conditions is ensuring adequate transit depths for the navigable route [37]. Maintaining the parameters of inland waterways specified for a given class is also a demanding task [36].

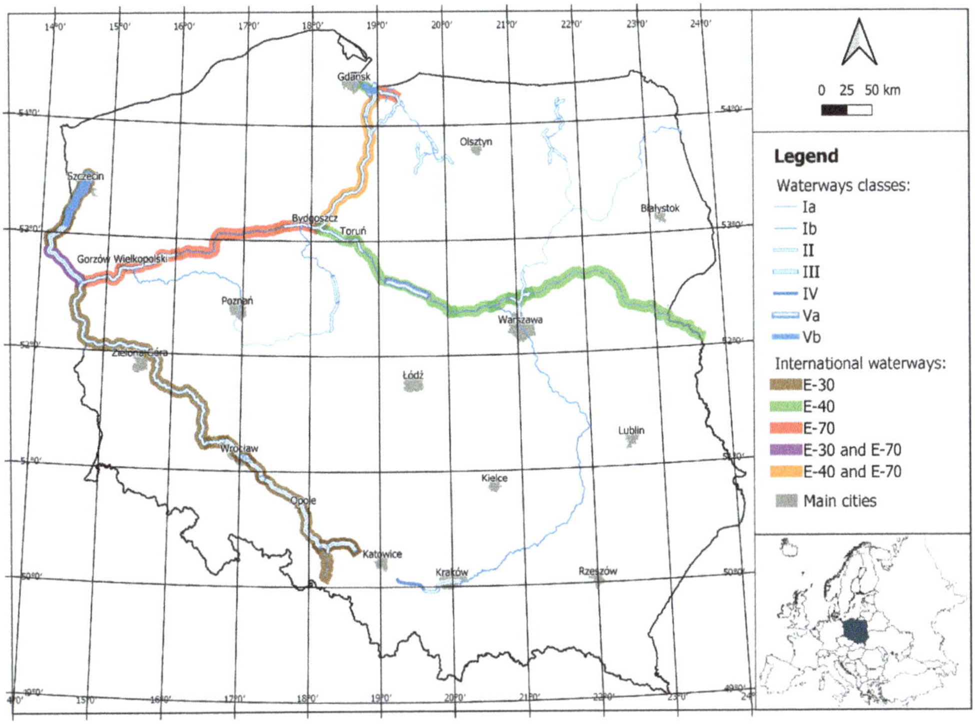

Figure 1. Inland waterways in Poland. Source: own study based on [32,38].

Waterways of international importance have been designated in Poland, as specified in the European Agreement on Main Inland Waterways of International Importance (AGN), which constitute parts of European connections [39]:

- E30—the Oder Waterway together with the Gliwice Canal (E30-01) and the Western Oder, as well as the missing section of the Danube–Oder–Elbe connection;
- E40—the Vistula waterway from Gdańsk to Warsaw, including the missing link to Brest;

https://doi.org/10.3390/su18083844

- E70—section of the Oder Waterway, the Vistula–Oder waterway, section of the Vistula River waterway and the Nogat or Szkarpawa River (Figure 1).

There are 129 navigation locks and 5 slipways on inland waterways in Poland, which are operated by the waterway administration [40]. It is worth noting that in the case of the Oder Waterway, most of the barrages have two locks (including twin locks).

To ensure safe navigation, the State Water Holding Polish Waters (Państwowe Gospodarstwo Wodne Wody Polskie) has approximately 14,000 shore (permanent markings placed on the shore and on hydrotechnical structures) and floating signs (buoys located on the water) [41–50]. Shore signs dominate this list, as they form the basis of navigation signs on most sections of inland waterways in Poland. Among the approx. 10,000 signs of this type, navigational route signs predominate. On the other hand, approx. 4000 navigational signs are buoys, mainly for marking navigational route boundaries and dangerous areas, including navigational obstacles. To ensure safe navigation, almost 66% of shore signs (approx. 6800) and 75% of floating signs (approx. 3000) owned by the waterway administration are displayed (Figure 2), while the remaining signs are spare or worn out and require renovation. At the same time, it should be remembered that the above list only concerns navigation signs within the competence of the waterway administration, which is responsible for the navigable route and navigation infrastructure (locks and slipways) and related water structures and equipment. Point infrastructure and navigation signs intersecting with the waterway are installed by the owners [28].

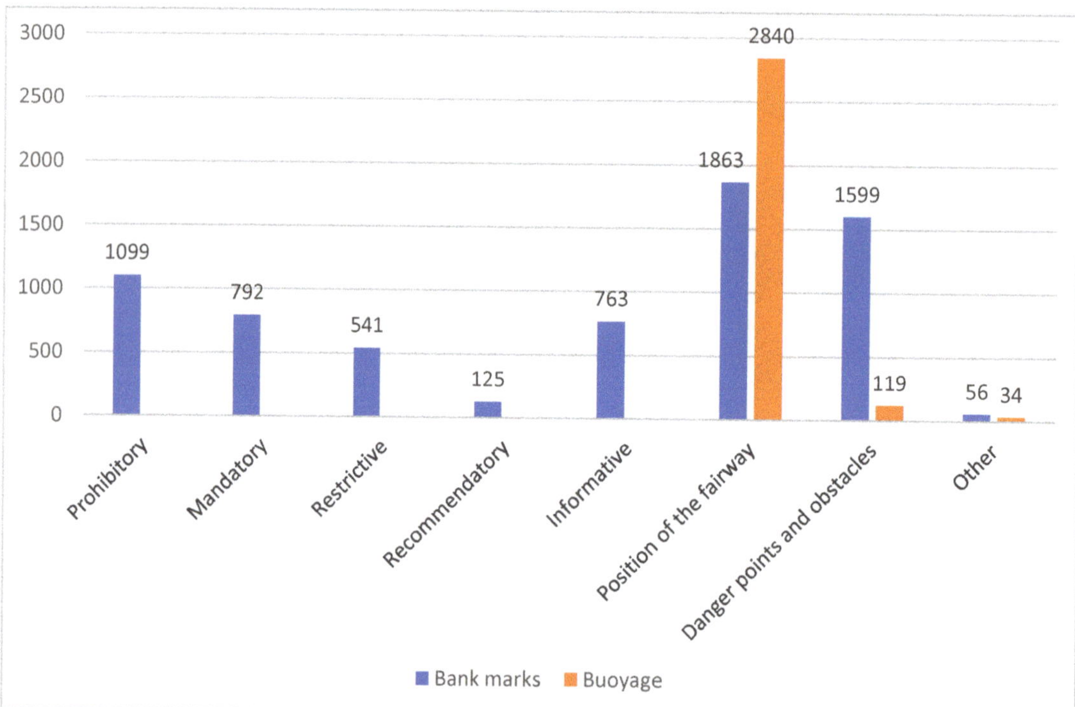

Figure 2. Inventory of navigation signs displayed by the waterway administration. Source: own study based on [41–50].

2.2. Data

A comprehensive approach to the issue of navigation safety on inland waterways in Poland requires both a detailed analysis of the applicable legal provisions and field

https://doi.org/10.3390/su18083844

research, as well as the development of practical solutions aimed at improving the level of navigation safety. Therefore, the research methodology consisted of three complementary components:

(1) Analysis of national and international legal frameworks;
(2) Field observations and visibility tests of navigation marks;
(3) Expert-based evaluation and development of design recommendations.

The combination of these approaches made it possible to compare regulatory frameworks, verify selected issues through field observations and formulate practical recommendations for the improvement of navigation marking systems.

The analysis of possibilities for improving navigation safety on inland waterways in Poland was based on the following legal acts:

- The Act of 21 December 2000 on inland navigation—defines, among other things, the inland navigation administration authorities, their competences, conditions for navigation, safety requirements and rules for the classification and maintenance of inland waterways [28].
- The Act of 20 July 2017—Water Law [51]—regulates ownership rights to surface waters, including inland waterways, and assigns responsibility for their maintenance.
- The Regulation of the Minister of Infrastructure of 28 April 2003 on navigation regulations on inland waterways [52]—specifies, among other things, the method of marking navigable waterways, hydrotechnical structures and facilities, navigational obstacles, and the designs of navigational signs and signals.
- The Regulation of the Minister of Transport and Maritime Economy of 10 September 1998 on technical conditions to be met by railway structures and their location [53]—contains guidelines for the marking of railway bridges crossing waterways.

In addition, local laws specifying detailed safety conditions for the movement and mooring of vessels on selected sections of waterways have been taken into account, including:

- Order No. 1/2020 of the Director of the Inland Navigation Office in Szczecin of 18 February 2020 [54];
- Order No. 1/2021 of the Director of the Inland Navigation Office in Wrocław of 25 May 2021 [55];
- Order No. 1/2022 of the Director of the Inland Navigation Authority in Szczecin of 10 March 2022 [56];
- Order No. 7/2024 of the Director of the Inland Navigation Office in Bydgoszcz of 26 June 2024 [57].

The analysis of international legislation took into account:

- The European Code for Signs and Signals on Inland Waterways (SIGNI) [58]—Implemented by the European Economic Commission, this act contains recommendations on the installation and use of navigation signs on European inland waterways.
- The European Code for Inland Waterways (CEVNI) [59]—Introduced by the United Nations Economic Commission for Europe, this document specifies, among other things, navigation signs for inland waterways.

At the same time, it was confirmed that Poland, as a member of the United Nations, has not ratified the above-mentioned international regulations and that they therefore do not constitute binding national law.

The analysis of international legislation was supplemented by a review of selected aspects of the shipping regulations in force in the Federal Republic of Germany

https://doi.org/10.3390/su18083844

(Binnenschifffahrtsstraßen-Ordnung) [60], which is particularly important due to the cross-border nature of the Oder Waterway and the Western Oder.

The objective of the regulatory analysis was to identify discrepancies between Polish legislation and international navigation standards, particularly in relation to the technical parameters of navigation signs, their visibility requirements and the marking of infrastructure intersecting waterways. In the first stage of the research, the regulations in force in Poland were reviewed in terms of navigation marking standards; then, solutions were proposed to increase the safety of inland waterway users. In the next stage, the differences between Polish regulations and those of the United Nations were identified.

An inventory of the navigation signs used on inland waterways in Poland was also carried out. For this purpose, a structured questionnaire survey was conducted among regional waterway management units responsible for the installation and maintenance of navigation signs, which made it possible to determine the quantity of signs broken down by type (land and floating). The questionnaire collected information on:

- The number of shore navigation signs;
- The number of floating navigation signs (buoys);
- The operational status of signs (in use, reserve, or damaged);
- The distribution of navigation signs across different waterway sections.

The next stage of the research involved field observations and visibility tests of navigation signs on selected inland waterways. These activities were informed by the review of national and international regulations, but field research was also necessary. The visibility of shore-based navigation signs was tested during regular checks of the readiness of waterways and navigation routes for night-time navigation on the border section of the Oder River, the Wrocław Water Junction, and the Vistula River in Kraków (Figure 3). These locations were deliberately selected to represent different functional types of inland waterways, including an international waterway section (the border section of the Oder River), an urban navigation node with complex infrastructure (the Wrocław Water Junction), and a regional waterway with significant tourism traffic (the Vistula River in Kraków). Although these sites do not represent the entire inland waterway network in Poland, they illustrate typical navigation marking challenges occurring on waterways with different operational characteristics.

To assess the situation comprehensively, a comparative analysis was also carried out during the day-time. The fieldwork involved the use of vessels, a searchlight, binoculars with a rangefinder, and a lux metre. Radar displays were also an important element of the study (Figure 3), as the images obtained from them provided a clear basis for determining the position of a vessel [61], as well as the location of navigation signs and radar beacons on inland waterways. This work made it possible to determine the distances from which signs of various sizes and signs covered with different types of reflective film were visible, and to develop solutions for the marking of inland waterways for navigation at night and in conditions of limited visibility.

For the purpose of this study, visibility and recognisability were assessed using two operational criteria:

- Detection distance—the distance at which a navigation sign could first be visually detected by the observer;
- Recognition distance—the distance at which the observer could recognise the meaning of the sign.

Distances were estimated using range-finding binoculars and radar observations where applicable (Figure 3). The study covered all navigation signs located on the analysed sections of the navigable waterway, thus providing a comprehensive verification of naviga-

https://doi.org/10.3390/su18083844

tion conditions during the day-time and night-time. Measurements on individual sections were conducted under comparable and favourable navigational conditions, understood here as excellent visibility, calm water, and low wind strength.

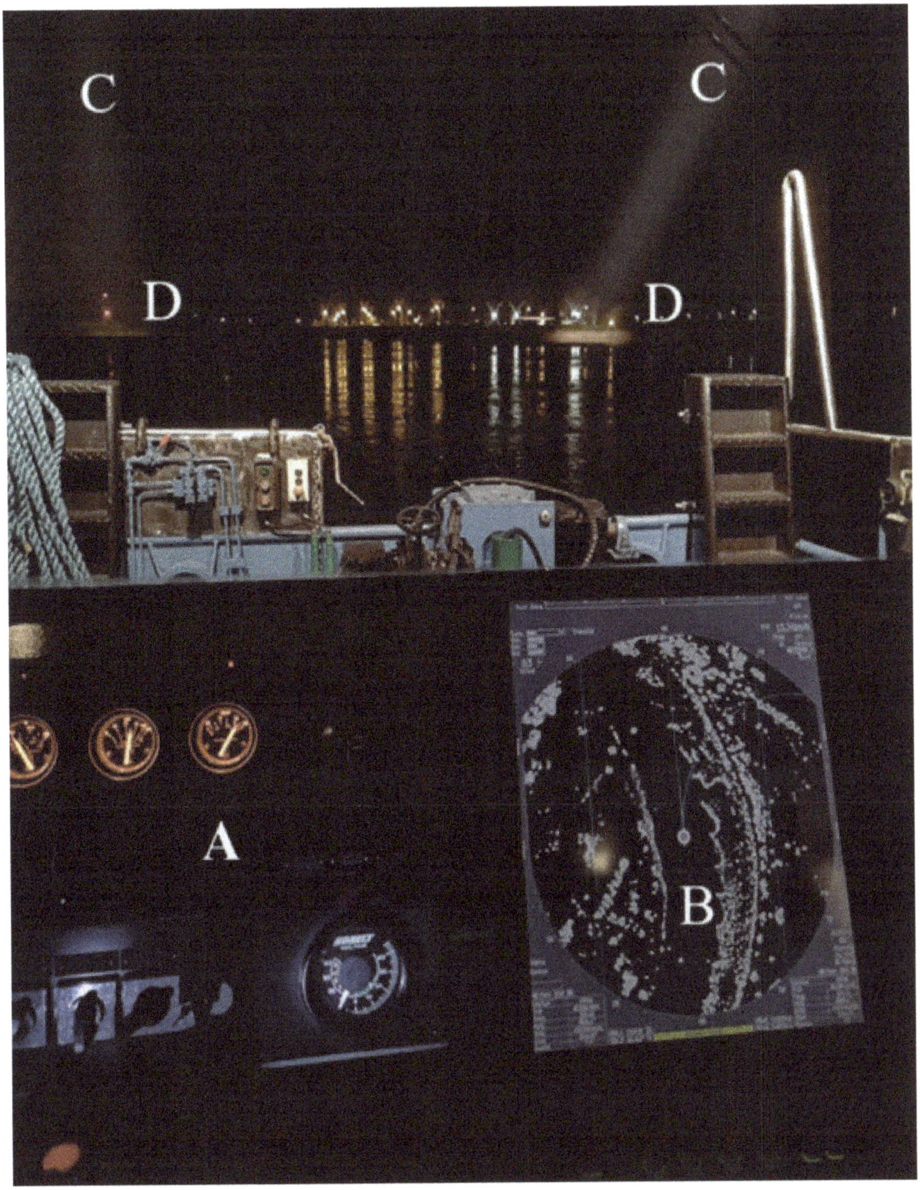

Figure 3. Testing solutions to improve the safety of navigation at night and in conditions of limited visibility. Explanation: (A)—ship, (B)—radar screen, (C)—searchlight, (D)—navigation signs covered with reflective film. Source: own study.

Based on the above research, several measures were proposed. These included introducing different size groups of signs into Polish legislation, preparing guidelines and proposed navigation marking plans for administrators of point infrastructure and the infrastructure intersecting waterways, and identifying navigation signs that are missing from

https://doi.org/10.3390/su18083844

Polish regulations in comparison with international standards. Another important element was the development, using design software, of conceptual designs for floating navigation signs representing individual types of signs. The study also found that information on the current vertical clearance under bridges is an element that increases inland waterway safety. Solutions used on bridges in Germany, the Netherlands, and Belgium were compared with Polish installations in terms of clearance gauges (inverted water level gauges). On this basis, and in cooperation with the Institute of Meteorology and Water Management, Poznań Branch [62], standardised designs recommended for use on inland waterways in Poland were developed.

It should be emphasised that the research was also based on expert knowledge and experience gained during inspections of navigable routes in 2021–2023, covering over 2500 km of waterways of international and regional importance. The expert evaluation involved specialists responsible for inland waterway management and navigation safety within the Polish waterway administration. These experts included waterway inspectors and navigation safety specialists with long-term professional experience in maintaining navigation markings and supervising navigable routes. This made it possible to verify the condition of navigational markings on the navigable routes and points and intersecting infrastructure. This was supplemented by observations and analysis of solutions used on selected waterways in Western Europe, from which good practices were adapted in this study.

3. Results

3.1. Analysis of Polish Regulations

An analysis of Polish legal acts and practical experience in work carried out to ensure safe navigation has identified several problematic issues concerning navigation marks:

(1) Dimensions of navigational signs—Navigation regulations [52] specify the designs of navigational signs, but do not define their dimensions. As a result, there are discrepancies between different waterways in this respect. It has also been established that the internal regulations of the State Water Management Authority [63] specify uniform sizes for signs throughout the country, based on European standards contained in the SIGNI study. However, it should be remembered that this is not generally applicable law and appropriate legislative changes should be introduced to clarify the issue of navigation sign sizes throughout Poland. In practice, this regulatory gap leads to inconsistent sign dimensions across different waterways, which may reduce the uniformity and recognisability of navigation markings for waterway users.

(2) Navigation marking of point infrastructure and infrastructure intersecting with inland waterways—The display of navigation signs in ports and of devices, structures and transmission lines intersecting with waterways is the responsibility of their owners [28], but not all entities follow the guidelines of the waterway administration, and missing or incorrect navigation signs are noticeable. Polish regulations do not include tools to enforce the mandatory use of appropriate signage. A good solution would be to oblige the owners of the designated facilities to agree with the waterway administration on navigation marking plans and their subsequent compliance. The lack of clear enforcement mechanisms results in the inconsistent marking of bridges, ports, pipelines and other infrastructure intersecting inland waterways, which may affect navigation safety and traffic organisation.

(3) Areas for water skiing or similar activities—Water skiing or similar activities and the towing of aircraft behind inland waterway vessels are only permitted during daylight hours, in good visibility and in water areas designated by the director of the inland navigation authority, in consultation with the waterway administration, and marked

https://doi.org/10.3390/su18083844

with information signs such as E.17, E.18, E.19, E.20, E.21 and E.24 [52]. However, the regulations do not specify who is responsible for marking the designated zones, and this responsibility should be assigned to the entity that has an interest in designating such a zone and leases the relevant water area for this purpose.

(4) Installation of clearance signs on bridge structures for navigation purposes—The regulation on the technical conditions to be met by railway structures and their location [53] recommends the installation of water level gauges. However, neither the Inland Navigation Act [28] nor the navigation regulations [52] specify the appearance of the gauge or where it should be located. At the same time, a review of the solutions used on inland waterways in Poland and Western Europe shows that there are no uniform designs. Therefore, it is recommended that an obligation be introduced to place clearance signs on bridge structures intersecting inland waterways, or in front of them with sufficient advance notice (e.g., 100–200 m). In this way, they would provide users of navigable routes with information about the current vertical clearance. This information may improve the situational awareness of vessel operators and reduce the risk of navigation errors related to the incorrect estimation of bridge clearance. A vessel's master would then be able to verify, before reaching the bridge, whether the current navigation parameters are sufficient for the vessel to pass safely through the structure.

(5) Incomplete guidelines for the marking of waterways where night navigation is permitted—Navigation regulations [52] only mention the need to cover navigation signs with reflective material. The distance from which the signs should be visible needs to be specified. In addition, in some places, lamps should be used to illuminate signs at night, and it may even be necessary to use additional navigation lamps if the regulations allow for interchangeable day and night markings. The lack of clearly defined visibility parameters may lead to differences in the practical implementation of night navigation marking systems on various inland waterways.

(6) Incomplete guidelines for navigation in conditions of limited visibility—The navigation regulations [52] are not precise in terms of supplementing navigation signs with radar reflectors enabling the display of signs on the radar screen during navigation at night or in fog or heavy rain. Therefore, it should be clarified on which navigation signs radar reflectors are mandatory in order to allow radar navigation in a given area. The introduction of such requirements could contribute to greater consistency in navigation marking systems used in conditions of reduced visibility.

(7) Differences in local regulations—Not all directors of inland waterway authorities have laid down guidelines in their local regulations [55,57] on the adaptation of waterways for night-time navigation and navigation in conditions of reduced visibility. Such provisions are extremely important for ensuring safe navigation in the circumstances indicated.

At the same time, it was found that the Water Law [51] does not address the issue of navigational markings.

3.2. Differences Between Polish Regulations and United Nations Regulations

A comparative analysis of national and international regulations (SIGNI [58] and CEVNI [59]) revealed a number of significant discrepancies that affect the safety and efficiency of inland navigation in Poland. In order to present the differences clearly, key issues have been summarised to enable readers to quickly identify areas requiring harmonisation and implementation in Polish legislation (Table 1).

This situation should primarily be attributed to the fact that international regulations have not been implemented into Polish law. It should be noted, however, that such

https://doi.org/10.3390/su18083844

action is not mandatory and each country is free to choose its own approach in this regard. Regardless of this, Polish navigation regulations [52] date back to 2003. Since then, numerous updates and technical clarifications concerning navigation marking systems have been introduced in international regulations such as SIGNI and CEVNI. As a result, significant differences exist between the detailed requirements of international regulations and the current Polish regulatory framework. These differences concern, in particular:

- Technical parameters and dimensions of navigation signs;
- Requirements for lighting and visibility distances;
- The use of radar reflectors and modern navigation technologies;
- The introduction of new types of navigation signs responding to emerging operational needs.

Table 1. Differences between Polish and international regulations. Source: own study based on SIGNI [58] and CEVNI [59].

No	Issue	Polish Regulations	SIGNI/CEVNI
1	Dimensions of navigation signs	No specific dimensions, only sign designs	Detailed dimensions and proportions of signs
2	Sign lighting	Optional, imprecise	Detailed guidelines for sign lighting
3	Radar reflectors	Optional	Recommended on floating signs and bridges
4	Marking of lake exits	Special signs used on lakes,	No such sign system, standard signage for side roads and junctions used
5	No motorised vessels	No dedicated sign	Dedicated sign A.1.1.
6	Characteristics of navigation lights on signs	Colours and rhythm in most cases described as arbitrary—not very precise	Detailed cycles and characteristics as well as colours of lights—distinguishing each sign
7	Visibility of signs (distances)	No requirements	Minimum visibility distances specified
8	Dynamic (digital) signs	No regulations	Described in SIGNI appendices
9	Standardisation of floating signs	No detailed parameters	No detailed parameters
10	Clearance patches	No specific guidelines	No detailed guidelines
11	Wreck float	No regulations	Introduced in recent amendments

Therefore, international experience should be taken into account, at least in part, when amending national regulations. Such harmonisation does not necessarily require the full legal adoption of international conventions, but may involve the gradual adaptation of selected technical standards that improve the consistency and recognisability of navigation marking systems.

3.3. Guidelines for Navigational Marking in Poland

Based on the analysis of Polish legal acts, the comparison with international regulations and field observations of navigation markings, a set of recommendations for improving navigation marking systems on inland waterways in Poland has been developed. The proposed solutions address both regulatory gaps and practical problems observed during inspections of navigable waterways.

The recommendations presented below concern several key aspects of navigation marking systems, including the dimensions of navigation signs, the marking of infrastructure intersecting waterways, the introduction of missing signs, and the improvement of marking systems used for night navigation and navigation in conditions of limited visibility.

https://doi.org/10.3390/su18083844

3.3.1. Introduction of Standards for the Size of Navigation Signs

In accordance with SIGNI standards [58], a single size of shore navigation signs is used on European inland waterways. These standards specify detailed dimensions and proportions, which facilitate the recognition and standardisation of signs. However, the study has shown that this solution can be adapted to local conditions, which include not only waterways of international importance but also regional waterways with lower navigability, often used exclusively for water tourism.

The parameters of navigation signs indicated in SIGNI [58] are mostly based on sign dimensions of 1.0 m × 1.0 m and 1.2 m × 1.0 m. The use of such signs significantly improves the consistency and recognisability of navigation marking on inland waterways. When appropriate reflective foil is used, this type of marking is highly visible to waterway users. The use of uniform sign dimensions also means that waterway users have no doubt as to which regulation indicated on the sign applies to a given section of the waterway. The same standard dimensions for sign boards also reduce the risk that vessel operators will overlook orders, prohibitions, or other information when smaller signs are placed between boards compliant with SIGNI.

Notwithstanding the above, experience gained during inspections and consultations with waterway administration staff responsible for displaying navigation signs also indicates several disadvantages of using uniform sign sizes:

- SIGNI-compliant navigation signs, when displayed in large numbers next to each other, may interfere with perception, in particular when important navigation signs (e.g., prohibition or mandatory signs) are located near signs of lesser importance (e.g., information signs)—with the current sizes, the legibility of the information presented is already impaired (Figure 4).
- Higher production costs.
- Higher repair costs due to the use of more consumables.
- The larger surface area of the navigation sign increases its susceptibility to damage caused by strong winds.

Due to potential difficulties in introducing uniform sizes for shore marks, but also taking into account the specific nature of navigable waterways in Poland, an alternative method of navigational marking has been developed. It involves adjusting the size of the signs depending on the inland waterway class and, at the same time, adjusting their visibility to ensure safe navigation. A similar solution is used in road engineering, where five groups of sign sizes have been established, mainly based on the category of the road: warning, prohibition, mandatory, information, and direction and location [64,65].

Based on the research conducted, it was proposed to use three groups of sizes for shore navigation signs (Figure 5), based on the proportions specified in SIGNI:

- Group I—signage corresponding to 100% of the size of SIGNI signs;
- Group II—signage corresponding to 80% of the size of SIGNI signs (with navigation symbols reduced proportionally to 80% of the original size);
- Group III—signage corresponding to 60% of the size of SIGNI signs (with navigation symbols reduced proportionally to 60% of the original size).

The proposed proportions of 100%, 80% and 60% were adopted as practical scaling variants derived from the base dimensions specified in SIGNI. Their purpose is to preserve the proportions and recognisability of sign symbols while allowing the dimensions of sign boards to be adapted to waterways with different operational characteristics, traffic intensity and installation constraints.

In particular, the largest size group is recommended for waterways of higher importance and longer viewing distances, whereas the reduced groups may be applied on

https://doi.org/10.3390/su18083844

regional waterways and in locations where the installation of full-size signs would create excessive visual clutter or technical difficulties. The proposed thresholds therefore result from operational experience, field observations and the need to maintain proportional symbol geometry rather than from arbitrary dimensional reduction.

Figure 4. Examples of grouped navigation signs. Source: own study.

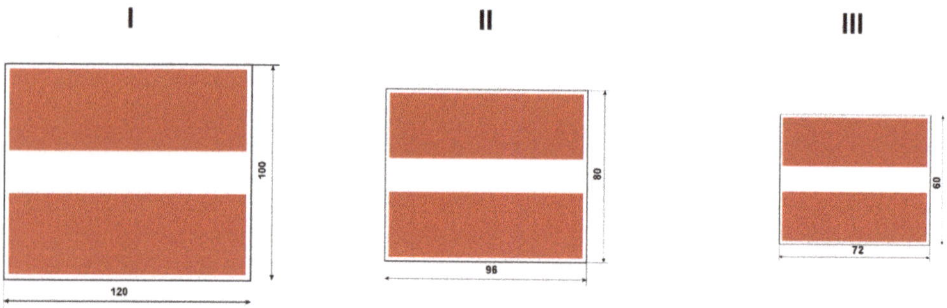

Figure 5. Proposal to introduce different sizes of navigation signs based on sign A.1. Explanation: Arabic numerals indicate the numbers of sign groups: (**I**)—group I, (**II**)—group II, (**III**)—group III. Source: own study based on SIGNI [58].

When determining signs from groups II and III (Figure 5), it is not necessary to introduce all schemes in Polish regulations relating to each type of marking. It is sufficient to add the appropriate notation indicating multipliers of 0.8 and 0.6 of the base value of the reference signs from group I.

The analyses carried out have shown the possibility of using different sizes of navigation signs depending on:

- Waterway class—In this solution, the principle of displaying navigation signs according to the waterway class has been adopted, assuming the need to ensure standards compliant with SIGNI, in particular on waterways constituting supra-regional connections. Therefore, the following principles are proposed:
    - Waterways of international importance (class IV-Vb)—use of signs from group I;

https://doi.org/10.3390/su18083844

    - Waterways of regional importance in classes II and III—use of signs from groups I and II;
    - Waterways of regional importance in classes Ia and Ib—use of signs from groups II and III.
- Type of signs—The proposed variant takes into account the importance of a given group of signs for ensuring safe navigation. On this basis, the following have been adopted:
    - Prohibition signs—use of signs from group I;
    - Mandatory signs—use of signs from group I;
    - Restriction signs—use of signs from group I and group II;
    - Recommendation signs—use of signs from groups I and II;
    - Marking of dangerous places—use of signs from groups I and II;
    - Information signs—use of signs from groups I, II and III;
    - Auxiliary signs—use of signs from groups I, II and III;
    - Other signs—use of signs from groups I, II and III.
- Visibility for the vessel's operator—SIGNI includes provisions specifying the distance from which signs should be visible to waterway users. The application of this criterion may also be crucial for the display of navigation signs, and therefore the possibility of displaying signs from individual groups has been proposed (Figure 6).

It should be remembered that navigational safety plays a key role in the use of appropriate navigation signs. In addition to the examples discussed above, other conditions may arise depending on the specific water area and inland waterway. For example, on some water areas night navigation is permitted, and in such situations navigation signs smaller than those specified in SIGNI, i.e., those outside group I (Figure 6), should be excluded because reduced-size signs may not provide a sufficient recognition distance under such conditions.

3.3.2. Obligation of Owners of Point Infrastructure and Infrastructure Intersecting with Inland Waterways to Agree with the Waterway Administration on Navigation Marking Plans and Their Subsequent Compliance

Guidelines have been prepared that administrators of individual facilities should follow when planning and installing navigation markings:

- Bridges and footbridges (Figure 7)—Each span must be marked as appropriate to the situation: if it is impassable, closed, or traffic is permitted only from the opposite direction, it must be marked with sign A.1; each navigable span must be marked with signs A.10, D.1a or D.1b, and C.2 (sign A.10 may be replaced by sign D.2 only in places where the depths outside the channel limits also correspond to the parameters of the waterway and there are no underwater obstacles, and such a change must be approved by the waterway administration); and if the navigable channel under the span is narrower than that specified for the inland waterway class, the span must also be marked with sign C.3. The signs must not protrude beyond the bridge clearance and, on inland waterways adapted for night navigation, they must also be illuminated at night and preferably equipped with a radar reflector.
- Power lines (Figure 7)—Each power line should be marked with sign E.2, supplemented in the lower right corner with a value indicating the height of the overhead line above the highest navigable water level (determined on the warmest days, owing to the dependence of line length on weather conditions).
- Ferry crossings—Each crossing should be marked as follows: a moored ferry with sign E.4a and a freely moving ferry with sign E.4b; in addition, each ferry should be marked with signs B.7 and B.8.

https://doi.org/10.3390/su18083844

- Pipelines and underwater lines—These should be marked with sign A.6b, which, unless supplemented with additional signs in the form of an arrow indicating the direction and distance to which it applies, must be cancelled by sign E.11.
- Marking of port and harbour entrances (Figure 7)—For this purpose, the following should be used: sign IV.A.1 on the left and sign IV.A.2 on the right, and on inland waterways suitable for night navigation, these signs should additionally be equipped with radar reflectors and illuminated, i.e., sign IV.A.1 with a red flashing light of any rhythm and sign IV.A.2 with a green flashing light of any rhythm.

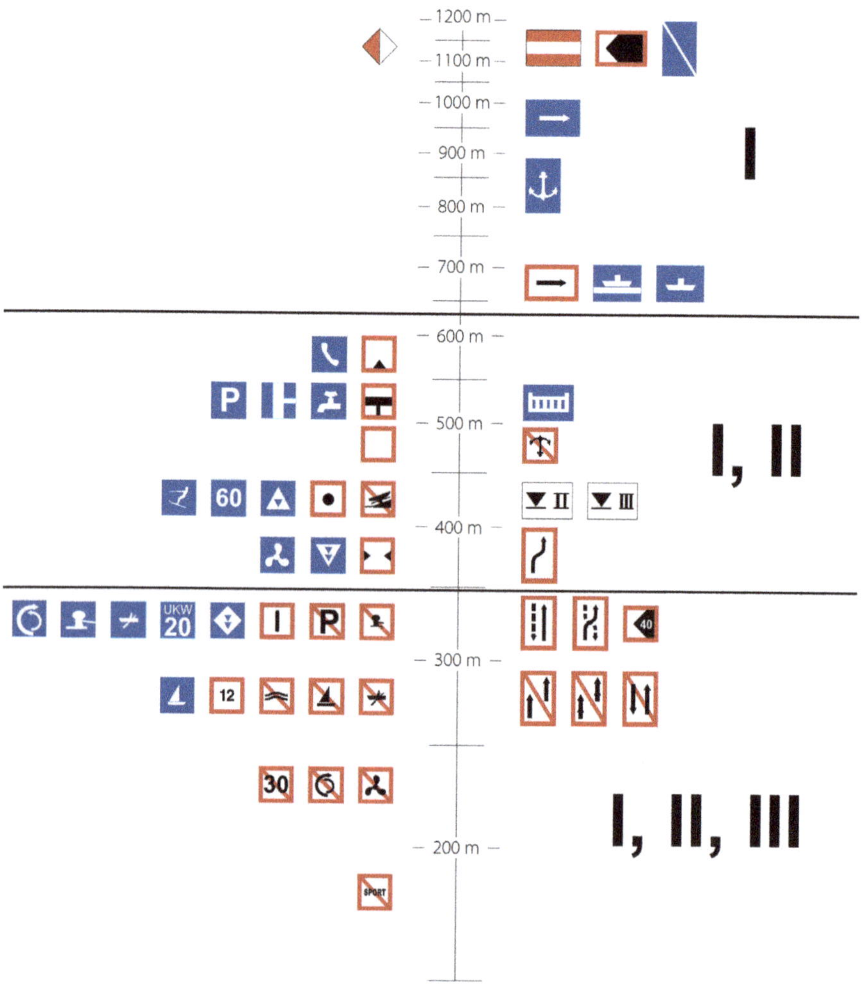

Figure 6. Division of sign sizes according to visibility for the vessel operator. Explanation: Arabic numerals indicate the numbers of sign groups: I—group I, II—group II, III—group III. Source: own work based on SIGNI [58].

The introduction of a formal requirement to agree upon navigation marking plans with the waterway administration would make it possible to standardise the marking of infrastructure intersecting waterways and reduce the occurrence of incomplete or inconsistent marking solutions.

https://doi.org/10.3390/su18083844

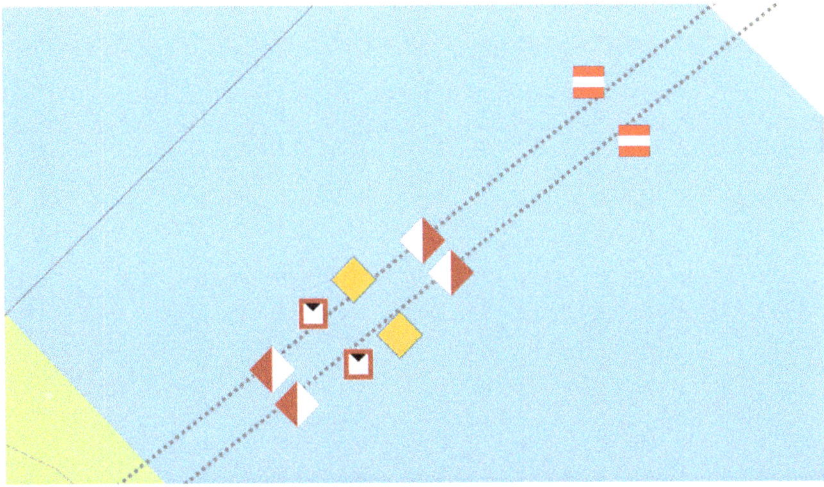

Figure 7. Proposed navigation marking plan as guidelines for owners of infrastructure intersecting with inland waterways. Explanation: The symbols used are consistent with those contained in SIGNI [58]. Source: [66] based on these guidelines.

3.3.3. Clarification of the Provisions Concerning Water Skiing or Similar Equipment

It is proposed to add to the navigation regulations [52] a provision specifying that responsibility for marking areas designated for water skiing or similar activities lies with the entity that leases the water area and is responsible for ensuring safety therein. At the same time, the location of such a place and the method of marking it with information signs E.17 (Figure 8) should be approved by the waterway administration.

3.3.4. Installation of Clearance Markers on Bridge Structures for Navigation Purposes

In view of the identified need to improve safety in the vicinity of bridge structures, which constitute navigation obstacles, it has been proposed that provisions be introduced into the navigation regulations [52] concerning the use of clearance signs on the piers of bridges crossing navigable waterways (in both directions, with a recommendation to place them on the right-hand side). This would make it possible to read the current vertical clearance and improve the situational awareness of vessel operators approaching the structure. It is worth noting that neither the Polish navigation regulations [52] nor the international regulations SIGNI [58] and CEVNI [59] contain any official guidelines on the use of clearance signs. Solutions of this type are implemented locally, based on the practice and experience of waterway managers rather than on formal regulations. The introduction of clear standards in this area could contribute to the harmonisation of the information system for waterway users, improve navigational consistency, and reduce the risk of misinterpreting bridge-clearance conditions.

It is therefore reasonable to introduce definitions of clearance signs in Polish and international regulations, detailed requirements for their design, colour and scale, as well as standardisation of their location on both sides of the bridge in a place visible to the crew. This should also include an obligation for the owner of a structure crossing a waterway to regularly inspect and maintain the equipment. At the same time, a template for a device to be installed on bridge piers or, in exceptional cases, in front of the structure has been developed with the following parameters: the size of the board with the number is 1.5 m × 1.0 m and, additionally, the board with the scale is 0.5 m × 1.0 m (scale every 10 cm with a clear marking of half a metre) (Figure 9).

https://doi.org/10.3390/su18083844

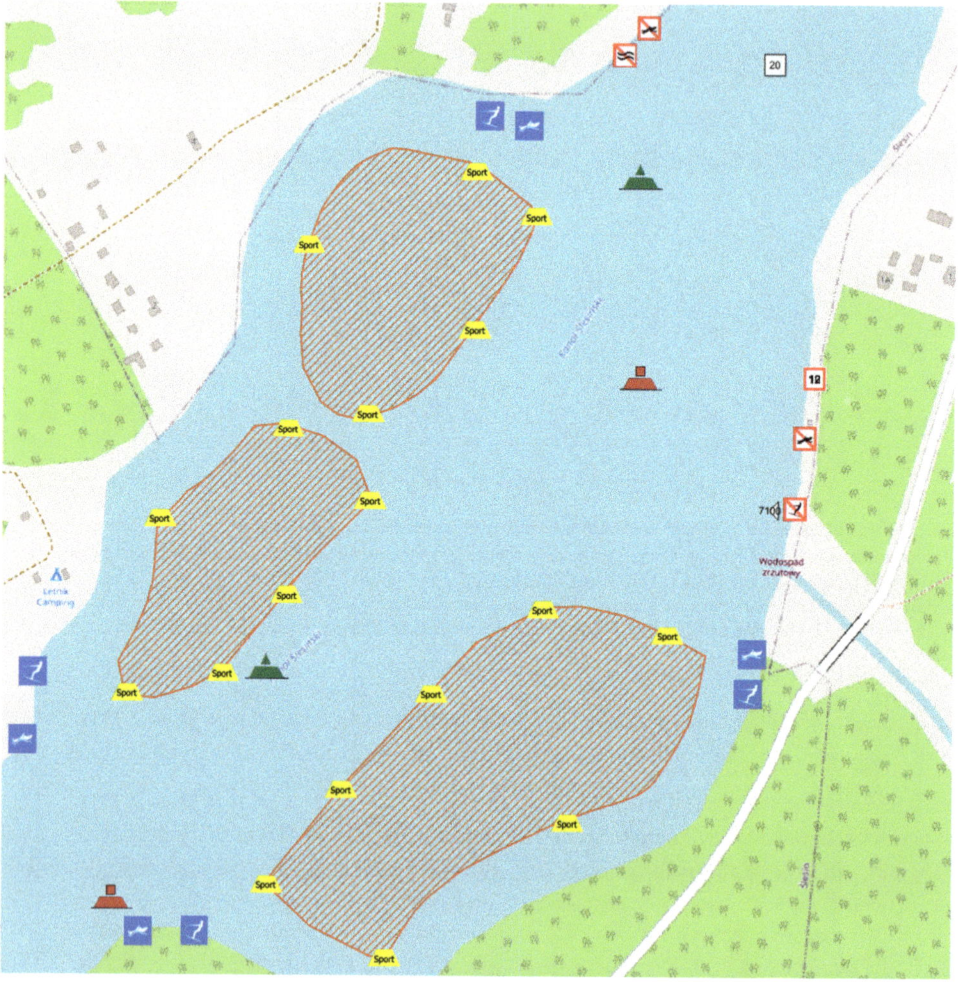

Figure 8. Example of marking zones for water skiing or similar equipment. Explanation: Sports zones are marked in red and their boundaries are marked by yellow buoys with the word "sport". Source: own work based on [52].

At the same time, it should be noted that a more advantageous solution would undoubtedly be the use of sensors on bridges and associated automatic dynamic (digital) signs displaying the current clearance, as found on selected inland waterways in Western Europe. Therefore, the pilot implementation of such a solution on selected sections should be considered, especially on bridges with clearance lower than that required by regulations on inland waterways used for tourism. However, this method is much more expensive, both at the implementation stage and during subsequent operation, and is not required given the current volume of shipping traffic in Poland. National regulations, unlike SIGNI, do not regulate the use of dynamic signs.

https://doi.org/10.3390/su18083844

Figure 9. Designed model of a clearance sign for navigation purposes in various variants. Source: Institute of Meteorology and Water Management, Poznań Branch [62].

3.3.5. Marking of Lake Exits

The special marking of lake route exits used in Poland (mainly on water bodies such as Lake Dąbie, the Great Masurian Lakes System and the Elbląg Canal) is not provided for in SIGNI and CEVNI. Therefore, it is recommended to use standard signs IV.A.1 and IV.A.2. This will harmonise the marking of lake exits with the European system, limiting the use of special signs. The proposed system will be consistent, and all locations should be marked in the same way, as there is no difference between an entrance to a port, a side canal or a branch on a lake.

3.3.6. Addition of Missing Signs

Polish navigation regulations do not include a prohibition sign for motorised vessels. In SIGNI [58] and German regulations [60], there is a sign, A.1.1, which clearly prohibits motorised vessels from entering. Therefore, it is recommended that sign A.1.1 be introduced into national regulations, as this would facilitate traffic management in recreational waters.

The situation is similar with the new navigation mark for wreck floats. SIGNI has introduced a sign for a newly discovered and marked hazard—a buoy with yellow and blue stripes and a yellow cross at the top. It was developed on the basis of practices used in maritime areas and is compatible with the IALA system [67]. The temporary wreck buoy has been designed as an additional navigation mark for use in both visual and radar navigation. It should be displayed as close to the navigational obstacle as possible. It is recommended that such marking be introduced into Polish navigation regulations.

3.3.7. Implementation of Solutions for the Marking of Inland Waterways for Night-Time Navigation and in Conditions of Reduced Visibility

In accordance with the proposed introduction of standards for the size of navigation signs suitable for night-time navigation and in conditions of limited visibility, it is essential to use shore-based navigation signs with the parameters specified in SIGNI, i.e., those classified in group I (Figure 6). Navigation signs should be covered with reflective film,

https://doi.org/10.3390/su18083844

and tests have shown that third-generation film has the most favourable properties in this respect (Figure 10).

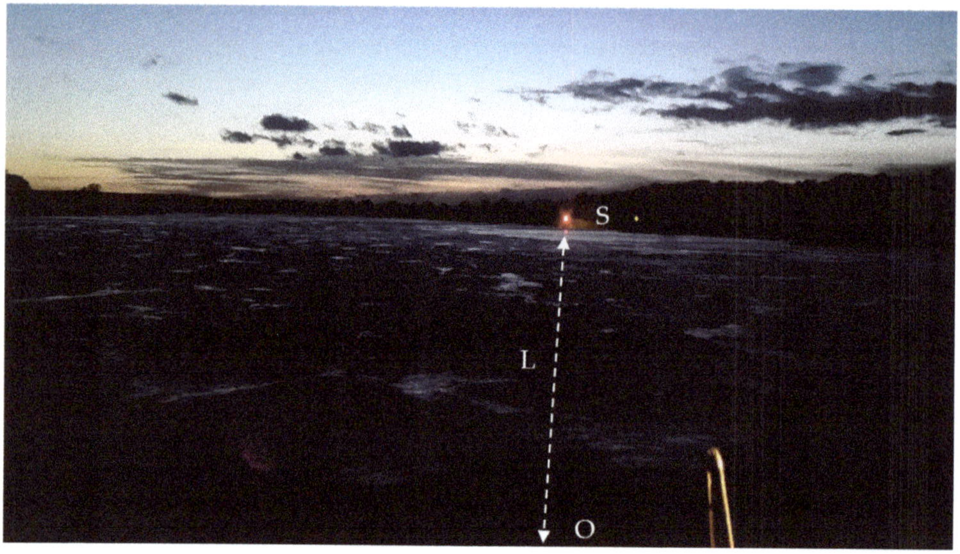

Figure 10. Visibility of navigation signs at night. Explanation: L—visibility distance of 3 km; O—observer/radar screen; S—navigation signs covered with reflective film. Source: own work.

Observations (Figure 11) show that navigation markings covered with third-generation reflective film are visible from a distance of up to 2–3 km. In contrast, navigation signs covered with second-generation reflective film can be seen from a distance of 1.2 km, and those covered with first-generation film only from approximately 0.5 km. Of course, a person without optical instruments is unable to read the information on the sign but can see the light reflected from these signs and is therefore aware of the direction of travel and the need to familiarise themselves with the signage. Only when approaching the sign at a distance of approximately 0.8–0.3 km, depending on individual abilities, are they able to read the information on the sign without the use of additional devices. Field observations therefore indicate that higher classes of reflective film may substantially improve the detection distance of navigation signs during night navigation.

In addition, point infrastructure and infrastructure intersecting with inland waterways must be additionally illuminated at night. On waterways intensively used at night, especially by passenger ship operators and individual users, it is also recommended that buoys with autonomous rhythmic white lights be used at branches, and that appropriately rhythmic red lights be used on the right-hand side of the route and rhythmic green lights on the left-hand side at entrances to ports and harbours. In this regard, Polish regulations should be supplemented with detailed light characteristics, which would improve the consistency of night-time marking and harmonise national regulations with international standards. At present, the lighting of signs is treated as optional and is described imprecisely, which may lead to poor visibility at night or in conditions of limited visibility. In addition, navigation signs, particularly on bridges, at port entrances, and on floating signs on navigable waterways, should be equipped with radar reflectors (Figure 11), which improve radar conspicuity and facilitate navigation in reduced-visibility conditions.

At the same time, it is recommended that appropriate regulations be introduced into local law to clarify the rules for navigation at night and in conditions of limited visibility.

https://doi.org/10.3390/su18083844

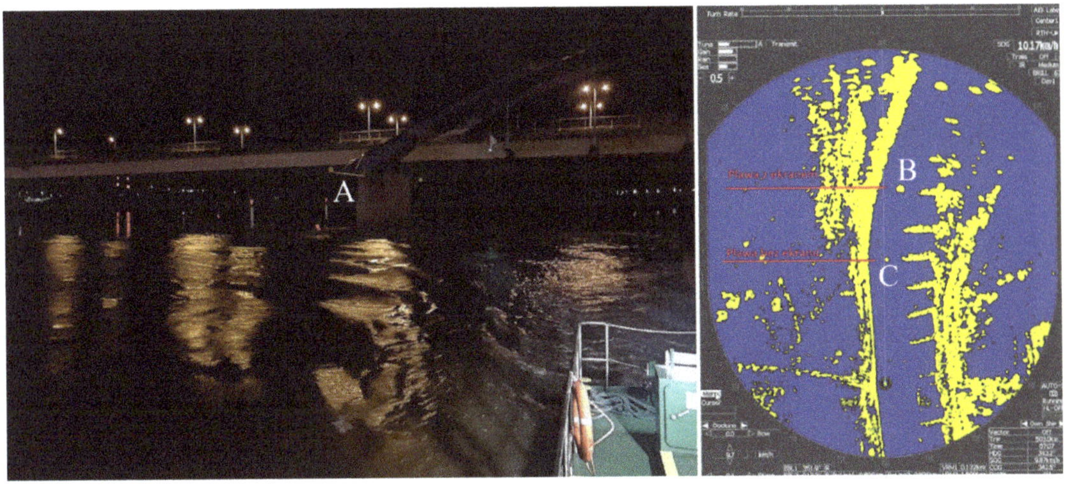

Figure 11. Practical application of radar reflectors on infrastructure intersecting with a waterway and floating navigation signs. Explanation: A—radar reflectors on a bridge structure; B—buoy equipped with a radar reflector; C—buoy without a radar reflector. Source: own work.

3.3.8. Development of Floating Navigation Mark Designs

Due to the lack of clear parameters for floating navigation signs in shipping regulations and SIGNI, and additionally due to the different designs in the first of the aforementioned legal acts, there is a need to clarify this issue.

Floating navigation signs adapted for night-time navigation and in conditions of limited visibility are being introduced on many inland waterways in Poland (Figures 3 and 10–12). However, due to the experience gained so far with their operation and the need to standardise their parameters, conceptual designs for floating navigation signs have been developed (Figure 13).

Figure 12. Example of floating navigation marks adapted for night-time navigation and conditions of limited visibility. Source: own work.

When developing guidelines for buoys to be used in the future on Polish and other countries' inland waters, the following groups were established (similarly to land-based signs):

- Marking of main shipping lanes—use of larger sizes;

https://doi.org/10.3390/su18083844

- Marking of secondary navigable routes and closed waters—use of smaller sizes;
- Marking of hydrotechnical structures and branches—use of larger sizes (Figure 13).

Type of sign	Designs according to navigation regulations	Conceptual drawing
Right-hand boundary of the navigable route	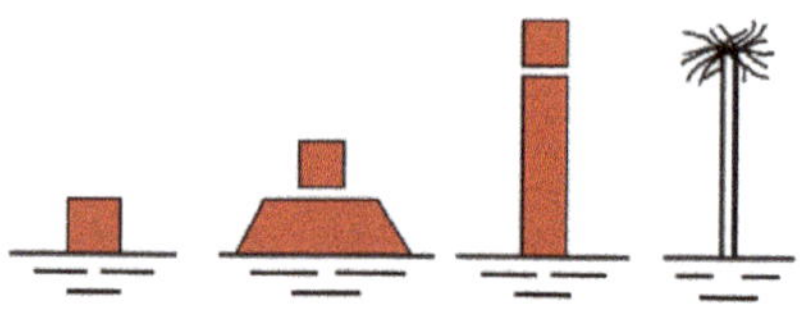	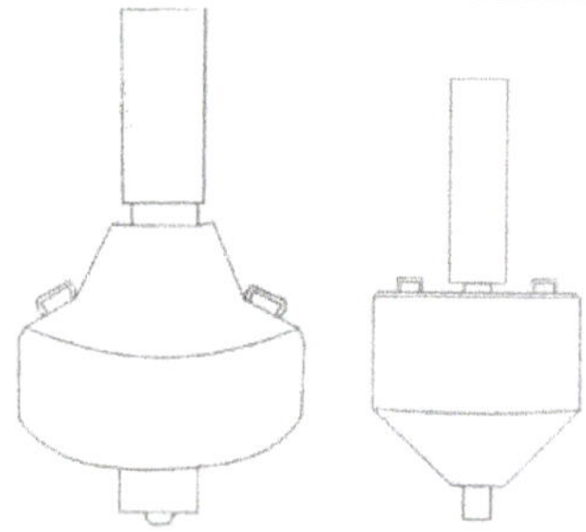
Left boundary of the navigable route	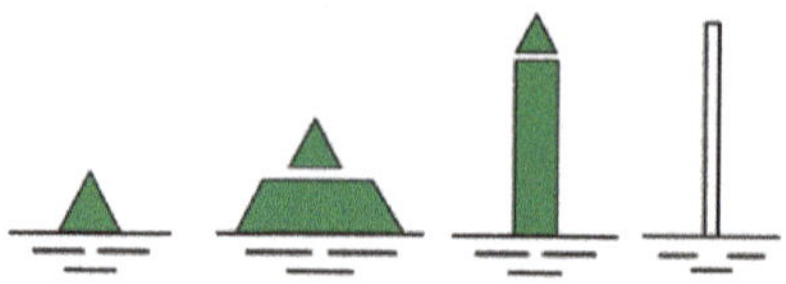	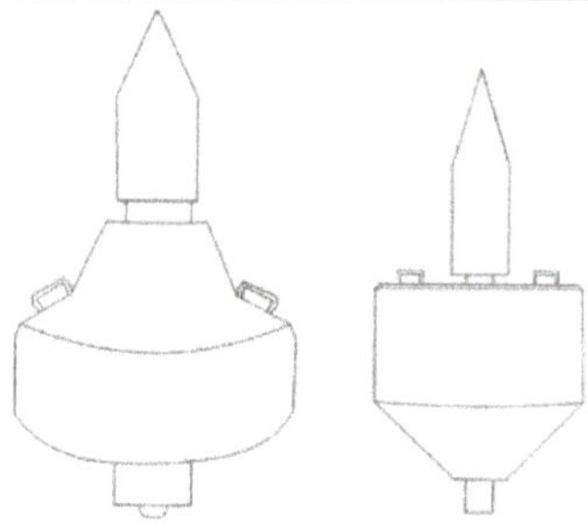
Branching of the navigable route		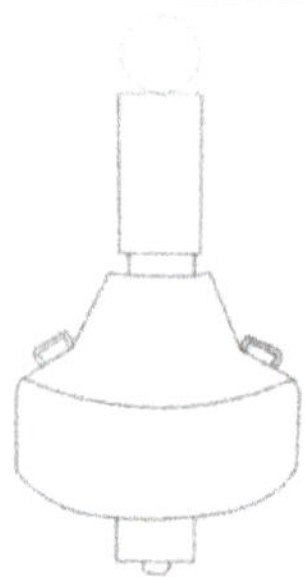
Water area closed to shipping traffic	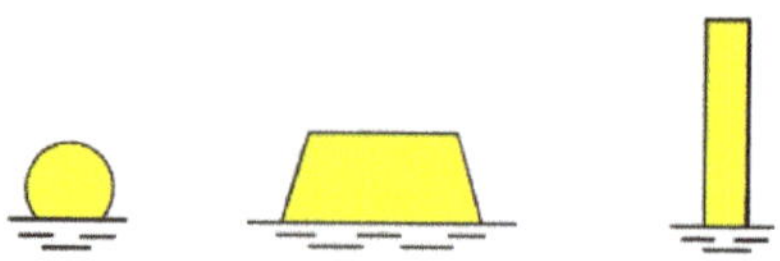	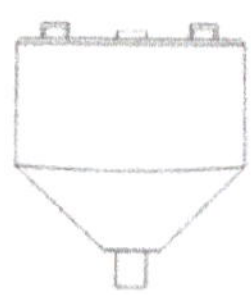

Figure 13. Conceptual proposal for standardising floating markings on inland waterways. Source: own work based on [52].

Therefore, larger buoys should be used for main navigable routes than for side navigable routes or hydrotechnical structures. Their technical parameters are the subject of further work, and at this stage, prototypes have been prepared and tested in real-world conditions.

However, it should be emphasised that the proposed solutions should be tested using pilot buoys in order to assess their structural correctness and effective functioning in real conditions. In this respect, the authors plan to carry out further research leading to the implementation of the described buoys.

https://doi.org/10.3390/su18083844

It should also be noted that the above recommendations are based on a comparison of national and international regulations (3.3.1, 3.3.2, 3.3.5, 3.3.6, and 3.3.7), field observations (3.3.1, 3.3.3, 3.3.4, 3.3.7, and 3.3.8), and design recommendations (3.3.1, 3.3.2, 3.3.3, 3.3.4, 3.3.5, 3.3.6, 3.3.7, and 3.3.8). Most of them are the result of a convergence of several factors.

4. Discussion

This study made it possible to compare national and international regulations concerning navigation marking. The navigation regulations applied in Poland [52] are, in several respects, very general and relatively old (having been in force for more than two decades). They have not yet been updated, whereas in Europe many modifications and clarifications to the original provisions have already been introduced in SIGNI [58] and CEVNI [59]. As a result, there are differences between the detailed regulations and the navigation signs used, the most significant of which are listed in this study. Some of the main problems are the dimensions of navigation signs, unclear provisions concerning navigation at night and in conditions of limited visibility, the absence of clearance signs, and the lack of standard designs for floating signs. In line with international practice, the proposed changes to navigation marking systems may improve the recognisability and consistency of navigation signs and contribute to more coherent marking practices on inland waterways. Such improvements may support safer navigation conditions [68], particularly during night navigation or in reduced visibility.

After analysing both Polish and European legal acts and taking into account expert experience, it should be concluded that there are two standards for the demarcation and marking of inland waterways in Europe, with the Elbe River in eastern Germany forming the boundary between them. On the Elbe itself and to the east of it, there are signs in the form of frames and crosses that regulate the course of the shipping route and comply with the regulations in force in this area [60]. To the west of the Elbe, these signs have their equivalents in the form of boards. These differences are partly the result of the historical development of navigation systems and institutional frameworks in different parts of Europe. Western European waterways, particularly those associated with the Rhine basin, have developed highly standardised marking systems coordinated by international river commissions, while waterways in Central and Eastern Europe often rely more strongly on national regulatory frameworks. SIGNI also includes sign A.1.1, which is also provided for in the inland waterway regulations of the Federal Republic of Germany. In Western European countries, clearance markers are also used for shipping purposes (e.g., in the Netherlands). At the same time, on inland waterways in Western Europe, various solutions have been introduced to increase user safety, together with numerous improvements resulting from technological advances in navigation signage, such as navigation lighting using photovoltaics, more advanced solutions in the form of sensors and active signs on bridges displaying the current clearance above the highest navigable water level, radar reflectors, reflective materials, and high-visibility colours for signs.

It should be noted that Poland has not ratified the international CEVNI and SIGNI regulations, which means that they are not binding law. Therefore, consideration should be given to taking steps to implement these regulations or—when introducing changes to national legislation and recommended updates to navigation regulations—taking into account European and global standards for navigation signs and user safety, which serve as a kind of roadmap for other developed countries. Some of the provisions have been introduced by the waterway administration in its internal regulations [63], on the basis of which uniform standards for the navigational marking of navigable waterways and point and intersecting infrastructure (e.g., specific dimensions and precise appearance of pictograms) should be applied, in accordance with the regulations in force in Poland.

https://doi.org/10.3390/su18083844

However, this is not a generally binding act and is therefore not mandatory for all entities responsible for issuing navigation signs. Despite this, the highest standards of navigation marking have been introduced, inter alia, on sections of the Oder (including the border section of the Oder, the Szczecin Waterway Junction [69], the Wrocław Waterway Junction [70]) and the Vistula (Kraków [71] and Warsaw [72]), which have thus been adapted for night-time navigation. New navigation signs are also found on shipping routes used for tourist purposes, including the heavily used Augustów [73] and Ślesiński [74] canals.

Unfortunately, this type of navigation marking is very expensive, which significantly limits its comprehensive replacement and use on all inland waterways in Poland (Figure 1). Analyses show that a buoy with lighting and a radar reflector is, on average, approximately EUR 300 more expensive than a buoy not adapted for night navigation. There has also been a systematic increase in the prices of shore markers in recent years, while prices for floating markers have remained relatively stable. Therefore, it is recommended that this type of marking be introduced first on sections with the highest shipping traffic, in particular in large cities and on the Great Masurian Lakes route. From a governance perspective, the implementation of the proposed solutions would involve several stakeholders, including waterway administrations, infrastructure owners (e.g., bridge operators, port authorities, and utility companies), as well as vessel operators and tourism enterprises. The introduction of clearer regulations concerning navigation marking plans for infrastructure intersecting waterways may therefore require institutional coordination and administrative enforcement mechanisms to ensure compliance among different stakeholders. At the same time, it should be remembered that for each party responsible for installing navigation marks on its facilities, and in the case of the waterway administration also on the navigable routes, the introduction of new guidelines regarding navigation marks would involve additional costs. Furthermore, in the case of bridges and other infrastructure crossing waterways, these proposals may be used in the development of a reliable bridge-collision warning system for inland vessel navigation [75]. This is important because, for example, in Serbia most shipping accidents occurred near Belgrade, at the confluence of the Sava and Danube rivers, and near bridges—places characterised by complex navigation conditions [25].

During the study, it was also found that shore markings in the form of frames and crosses should be maintained, due to the costs already incurred for the purchase and maintenance of existing signs, and supplemented with floating markings on the main rivers in Poland. Furthermore, the authors are aware of the legislative difficulties involved in introducing new solutions and the related requirements for navigational signs in order to increase safety on navigable waterways. In 2021, amendments to the Inland Navigation Act [28] were under consideration in Poland, which were intended, among other things, to introduce provisions concerning coordination with the waterway administration on navigation marking plans for point infrastructure and infrastructure intersecting waterways [76]. The guidelines prepared would fit well within the legal provisions under consideration and would greatly facilitate the work of the owners of ports, bridges, power lines, and similar infrastructure. However, these provisions did not enter into force, and no further plans for their implementation are known. The proposed changes could also eliminate another obstacle to improving navigation safety, namely the difficulty of enforcing the use of appropriate navigation signs on water structures and equipment managed by entities other than the waterway administration. Not all owners of point infrastructure and infrastructure intersecting inland waterways recognise the need to use appropriate navigation signs, and there are no effective tools to compel them to take action.

https://doi.org/10.3390/su18083844

Taking into account the results presented and the above limitations, it should be noted that appropriate navigation marking will also be of great importance for the gradual introduction of autonomous inland waterway vessels and the automated detection of navigation signs [77]. Therefore, research on the recognition of navigation signs is being conducted worldwide [78]. However, the wider use of unmanned inland waterway vessels requires the development of new design codes at national and European levels and their continuous updating, although regulations allowing for their testing are already in place [79]. Despite the legal and social difficulties mentioned above, captain-assistance functions could make accurate and safe inland navigation a reality, even in difficult navigational conditions [80]. In Poland, however, the implementation of such solutions will be difficult due to the insufficient operating parameters of inland waterways and the related neglect of the inland navigation sector [81]. As a result, the possible use of autonomous vessels could be limited to selected waterways, primarily those with adequate transit depths and proper navigational markings, in particular sections with at least class IV waterway status, including sections in the immediate vicinity of the seaports of Gdańsk and Szczecin [36].

Digitalisation is also important for inland waterway users, as it can increase the safety of inland waterway transport, as well as its efficiency and competitiveness [82]. Various solutions are being tested, including digital twins [83] and improved algorithms for route planning [84] and ship target detection and tracking [85]. Currently, however, the main tool supporting inland waterway transport is the river information system, which improves its competitiveness and contributes to the overall objective of making inland navigation a safe, efficient and environmentally friendly link in the logistics chain [86]. The RIS therefore facilitates the safe and efficient flow of information, which translates into numerous benefits for water transport, including, above all, the possibility of increasing the safety and efficiency of navigation [87]. In Poland, the River Information System was introduced thanks to two projects implemented by the Inland Navigation Authority in Szczecin, namely "Pilot implementation of RIS Lower Odra" in 2010–2013 and "Full implementation of RIS Lower Odra" in 2016–2023. Thanks to these projects, the RIS covers a total of 240.8 km of waterways E30 and E70 in the hinterland of the Szczecin–Świnoujście seaports [88]. The "Comprehensive implementation of RIS on the Oder Waterway" is planned by 2028 [89]. EUR 22.5 million has been earmarked for this purpose in the current financial perspective of the European Union [90]. A "Pilot implementation of RIS on the Lower Vistula" is also planned, but this is not a priority task and no funding has been secured for it so far [89]. At the same time, in 2024, the State Water Holding Polish Waters (Państwowe Gospodarstwo Wodne Wody Polskie) launched a Virtual River Information System on all inland waterways in Poland, which also contributes to improving safety. In this respect, it offers the possibility of electronic reporting of obstacles (e.g., navigational obstacles, incidents and accidents) on navigable routes, publishes navigational messages and information on navigational signs, and automatically calculates transit depths and clearances under bridges; it is also possible to purchase digital tickets for the use of inland waterways and their infrastructure, including route planning [66]. Previously, this type of information was published in the form of navigational notices issued by the waterway administrator [91].

Furthermore, it was recognised that the analysis of the harmonisation of navigation signs on inland waterways in Poland with European regulations is primarily addressed to national administrations responsible for drafting legislation. However, it may also provide valuable information for inland navigation and inland waterway authorities, entities dealing with technical standards, international river commissions, managers of point infrastructure and infrastructure intersecting navigable routes, as well as ship owners and operators [27].

https://doi.org/10.3390/su18083844

Finally, it is important to emphasise that the demarcation and marking of inland waterways should be clear and unambiguous and should not give rise to any doubts among users. It should be remembered that users include both educated and experienced captains and tourists with motorboat or sailing licences, but there are also people without licences or extensive knowledge. Polish regulations [28] allow small vessels up to 10 m in length to operate without a licence. These include small rowing boats, kayaks and vessels equipped with engines up to 10 HP. Preparing waterways for the latter group of users (without licences) represents an important challenge for waterway administrations. Therefore, the introduction of standardised signs and lighting them at night and simultaneous public education may significantly improve the clarity of navigation marking systems and contribute to better safety awareness among waterway users.

5. Conclusions

The analyses carried out clearly indicate that harmonising navigation signs on inland waterways in Poland with European regulations is an important step towards improving the consistency and clarity of navigation marking systems used on inland waterways in Poland. The discrepancies identified between national navigation regulations and international standards, such as SIGNI and CEVNI, made it possible to identify areas requiring regulatory clarification and to develop recommendations for the use of modern navigation signs and related legislative changes. Field observations indicate that the introduction of clearer parameters for sign dimensions, visibility requirements, and the technical equipment of navigation signs may improve their recognisability, particularly during night navigation and in conditions of limited visibility such as fog or heavy rain. It is also crucial to implement the proposed solutions for point infrastructure and infrastructure intersecting waterways, together with the coordination of navigation marking plans with waterway authorities, as this could reduce inconsistencies in infrastructure marking and improve the overall coherence of navigation marking systems. The implementation of technical solutions, such as clearance signs and radar reflectors, may further improve the detectability of navigation marks and facilitate navigation in conditions of reduced visibility, particularly in the context of the increasing hydrological variability associated with climate change.

It is recommended that new solutions be introduced in stages, first on sections with the highest traffic intensity and in areas with increased navigation risk. Pilot implementations will allow the effectiveness of the proposed changes to be assessed and adapted to the specific characteristics of Polish inland waterways. At the same time, it is also recommended that designs for floating navigation signs be developed on the basis of the proposed guidelines, with implementation planned for the next stage of work. Similarly, in the case of the developed concepts for clearance signs for navigation purposes, it is recommended that sample units be produced and tested under real conditions, including, above all, checking their legibility and the distances from which they are visible. Finally, the results of this study may contribute to future discussions on the harmonisation of inland navigation marking systems in Central and Eastern Europe, particularly in countries where international regulations such as SIGNI and CEVNI have not yet been fully incorporated into national legislation.

Author Contributions: Conceptualization, Ł.P.; methodology, Ł.P. and K.J.; formal analysis, Ł.P., S.I. and P.D.; investigation, K.J. and S.I.; writing—original draft, Ł.P. and K.J.; writing—review and editing, K.J. and P.D.; supervision, Ł.P.; project administration, K.J. All authors have read and agreed to the published version of the manuscript.

https://doi.org/10.3390/su18083844

Funding: This research was funded by the National Centre for Research and Development (NCBR) grant number Hydrostrateg 1/OO1P/2022.

Institutional Review Board Statement: This study is waived for ethical review as the work was carried out as part of official duties, in accordance with agreements signed in the State Water Holding Polish Waters.

Informed Consent Statement: Informed consent for participation was obtained from all subjects involved in the study.

Data Availability Statement: The original contributions presented in this study are included in the article. Further inquiries can be directed to the corresponding authors.

Conflicts of Interest: The authors declare no conflict of interest.

References

1. Segovia, P.; Rajaoarisoa, L.; Nejjari, F.; Duviella, E.; Puig, V. Model predictive control and moving horizon estimation for water level regulation in inland waterways. *J. Process Control* **2019**, *76*, 1–14. [CrossRef]
2. Łozowicka, D.; Kaup, M. Safety aspects of inland waterway transport in Poland. *TTS Tech. Transp. Szyn.* **2015**, *12*, 2016–2020.
3. Police Regulations for the Navigation of the Rhine. Available online: https://www.ccr-zkr.org/13020500-en.html#01 (accessed on 1 July 2025).
4. Wołejsza, P.; Jędrzychowski, H.; Jędrzychowski, K.; Karkos, D.; Kazimierski, W.; Wiśniecki, B. Development and Pilot Testing of a Navigation Support System on the Lower Vistula Section Between Warsaw and Gdańsk (Waterway E-40). INWAPO Project Implemented Through the CENTRAL EUROPE Programme Co-Financed by the ERDF 2014. Available online: https://docplayer.pl/1377624-Modernizacja-wodnych-drog-srodladowych-i-portow-morskich-wp-4-needs-assessment.html (accessed on 1 July 2025).
5. Skupień, E.T. Assessment of Factors Influencing the Development of Inland Navigation in Poland. *Sustainability* **2024**, *16*, 6663. [CrossRef]
6. Uddin, I.; Islam, M.R.; Ibn Awal, Z.; Newaz, K.M.S. An Analysis of Accidents in the Inland Waterways of Bangladesh: Lessons from a Decade (2005–2015). *Procedia Eng.* **2017**, *194*, 291–297. [CrossRef]
7. Steiner, J.; Havlíček, J.; Duša, T.; Heinrichs, G. The Vulnerability of Inland Waterway AIS to GNSS Radio Frequency Interference. *Eng. Proc.* **2023**, *54*, 26. [CrossRef]
8. Kubiak-Wójcicka, K.; Manoiu, V.-M. International Inland Waterways in Poland: Current State and Their Importance in EU Transport Policy. *Water* **2025**, *17*, 3190. [CrossRef]
9. Pulido, J.M.; Bedoya-Maya, F.; van Hassel, E.; Vanelslander, T.; Carlan, V. Improving the visibility of waterway events influencing traffic along the Rhine, Main, and Danube. *J. Transp. Geogr.* **2025**, *128*, 104385. [CrossRef]
10. Wei, Y.; Lu, W.; Chu, C.; Jia, X. Research on Automatic Navigational Buoy Recognition Based on YOLOv11. *Int. J. Transp. Eng. Technol.* **2026**, *12*, 23–30. [CrossRef]
11. Jurkovič, M.; Molnárová Baracková, A.; Prabowo, A.R.; Gorzelanczyk, P.; Melnyk, O. Transformative Technology for Inland Waterway Navigation: AIS AtoNs. *Transp. Res. Procedia* **2026**, *93*, 95–100. [CrossRef]
12. Ding, H.; Weng, J. A robust assessment of inland waterway collision risk based on AIS and visual data fusion. *Ocean. Eng.* **2024**, *307*, 118242. [CrossRef]
13. Liu, J.; Jiang, X.; Huang, W.; He, Y.; Yang, Z. A novel approach for navigational safety evaluation of inland waterway ships under uncertain environment. *Transp. Saf. Environ.* **2022**, *4*, tdab029. [CrossRef]
14. Zhang, J.; Wan, C.; He, A.; Zhang, D.; Soares, C.G. A two-stage black-spot identification model for inland waterway transportation. *Reliab. Eng. Syst. Saf.* **2021**, *213*, 107677. [CrossRef]
15. Woś, K. System of harmonised river information services (RIS) on the lower section of the Odra waterway. *Econ. Probl. Serv.* **2010**, *59*, 268.
16. Ariefjew, I.; Tarchalski, M. Some problems of inland navigation in Poland in comparison with selected countries of the world. *Sci. J. Marit. Univ. Szczec.* **2001**, *62*, 5–20.
17. Communication from The Commission to The European Parliament, The Council, The European Economic And Social Committee and The Committee of The Regions NAIADES III: Boosting Future-Proof European Inland Waterway Transport. Available online: https://eur-lex.europa.eu/legal-content/EN/TXT/?uri=CELEX%3A52021DC0324 (accessed on 1 July 2025).
18. UNECE. White Paper on the Progress, Accomplishment and Future of Sustainable Inland Water Transport. Available online: https://unece.org/transport/publications/white-paper-progress-accomplishment-and-future-sustainable-inland-water (accessed on 29 June 2025).

https://doi.org/10.3390/su18083844

19. Via Donau. Manual on Danube Navigation. Available online: https://rewway.at/en/teaching-materials/manual-danube-navigation-viadonau (accessed on 29 June 2025).
20. European Commission. Communication from the Commission to the European Parliament, the Council, the European Economic and Social Committee and the Committee of the Regions Towards Quality Inland Waterway Transport NAIADES II. Available online: https://eur-lex.europa.eu/legal-content/EN/TXT/?uri=celex:52013DC0623 (accessed on 29 June 2025).
21. Hendrickx, C.; Breemersch, T. The Effect of Climate Change on Inland Waterway Transport. *Procedia—Soc. Behav. Sci.* **2012**, *48*, 1837–1847. [CrossRef]
22. Schoeneich, M.; Habel, M.; Szatten, D.; Absalon, D.; Montewka, J. An Integrated Approach to an Assessment of Bottlenecks for Navigation on Riverine Waterways. *Water* **2023**, *15*, 141. [CrossRef]
23. Christodoulou, A.; Christidis, P.; Bisselink, B. Forecasting the impacts of climate change on inland waterways. *Transp. Res. Part D Transp. Environ.* **2020**, *82*, 102159. [CrossRef]
24. Hu, W. Research on Intelligent Navigation Collision Avoidance Decision-making for Typical Inland Waterway. *J. Eng. Syst.* **2024**, *2*, 12–17. [CrossRef]
25. Bačkalov, I.; Vidić, M.; Rudaković, S. Lessons learned from accidents on some major European inland waterways. *Ocean Eng.* **2023**, *273*, 113918. [CrossRef]
26. Economic Commission for Europe. European Recreational Inland Navigation Network—Resolution No. 52, Revision 2. Available online: https://unece.org/fileadmin/DAM/trans/doc/2013/sc3wp3/ECE-TRANS-SC3-164-Rev1e_01.pdf (accessed on 14 March 2026).
27. Commissioned by Maritime New Zealand 2021. Recreational Boating Literature Review. Available online: https://www.maritimenz.govt.nz/media/wvqnqwjj/recreational-boating-literature-review-2021.pdf (accessed on 14 March 2026).
28. Act of 21 December 2000 on Inland Navigation. Available online: https://isap.sejm.gov.pl/isap.nsf/DocDetails.xsp?id=WDU20010050043 (accessed on 29 June 2025).
29. Sorgatz, J.; Kayser, J. Assessment of maintenance efforts and probabilities of failure at German inland waterways to advance the design of bank revetments. *J. Coast. Hydraul. Struct.* **2022**, *2*, 1–21. [CrossRef]
30. Central Commission for the Navigation of the Rhine. Available online: https://www.ccr-zkr.org/ (accessed on 1 July 2025).
31. Danube Commission. Available online: https://www.danubecommission.org/dc/en/ (accessed on 1 July 2025).
32. Announcement of the Prime Minister of 31 March 2022 on the Announcement of the Uniform text of the Regulation of the Council of Ministers on the Classification of Inland Waterways. Available online: https://isap.sejm.gov.pl/isap.nsf/DocDetails.xsp?id=WDU20220001170 (accessed on 29 June 2025).
33. Statistical Office in Szczecin. Inland Waterways Transport in Poland in 2023. Available online: https://stat.gov.pl/obszary-tematyczne/transport-i-lacznosc/transport/transport-wodny-srodladowy-w-polsce-w-2023-roku,4,14.html (accessed on 29 June 2025).
34. Eurostat. Length of Navigable Inland Waterways by Waterway Type. Available online: https://ec.europa.eu/eurostat/databrowser/view/iww_if_infrastr/default/table?lang=en&category=iww.iww_if (accessed on 27 April 2025).
35. Kulczyk, J.; Winter, J. *Inland Water Transport*; Wrocław University of Technology Publishing House: Wrocław, Poland, 2001.
36. Pieron, Ł. The scale of challenges related to the adoption of Polish inland waterways covered by the AGN Agreement to international standards. *Water Manag.* **2017**, *11*, 348–351.
37. Pieron, Ł.; Woś, K.; Wrzosek, K. Water Reservoirs in Plans to Improve Navigability of the Lower Section of the Vistula. *Water* **2022**, *14*, 4042. [CrossRef]
38. United Nations. *European Agreement on Main Inland Waterways of International Importance (AGN)*; United Nations: Geneva, Switzerland, 1996.
39. Pieron, Ł. Inventory of Components of Inland Waterways of Particular Transport Importance. 2017. Available online: https://www.gov.pl/web/infrastruktura/publikacje-i-materialy-informacyjne68 (accessed on 27 April 2025).
40. National Protection System Data. Available online: https://wody.isok.gov.pl/imap_kzgw/?gpmap=gpSIGW (accessed on 27 April 2025).
41. Regional Water Management Authority in Białystok, State Water Holding Polish Waters data—E-mail dated 21 February 2022.
42. Regional Water Management Authority in Bydgoszcz, State Water Holding Polish Waters data—E-mail dated 21 February 2022.
43. Regional Water Management Authority in Lublin, State Water Holding Polish Waters data—E-mail dated 21 February 2022.
44. Regional Water Management Authority in Gdańsk, State Water Holding Polish Waters data—E-mail dated 21 February 2022.
45. Regional Water Management Authority in Gliwice, State Water Holding Polish Waters, email dated 21 February 2022.
46. Regional Water Management Authority in Kraków, State Water Holding Polish Waters data—E-mail dated 21 February 2022.
47. Regional Water Management Authority in Poznań, State Water Holding Polish Waters data—E-mail dated 21 February 2022.
48. Regional Water Management Authority in Szczecin, State Water Holding Polish Waters data—E-mail dated 21 February 2022.
49. Regional Water Management Authority in Warsaw, State Water Holding Polish Waters data—E-mail dated 21 February 2022.

https://doi.org/10.3390/su18083844

50. Regional Water Management Authority in Wrocław, State Water Holding Polish Waters data—Email dated 21 February 2022.
51. Act of 20 July 2017—Water Law [Journal of Law 2017 Item 1566]. Available online: https://isap.sejm.gov.pl/isap.nsf/download.xsp/WDU20170001566/U/D20171566Lj.pdf (accessed on 27 April 2025).
52. Regulation of the Minister of Infrastructure of 28 April 2003 on Navigation Regulations on Inland Waterways. Available online: https://isap.sejm.gov.pl/isap.nsf/DocDetails.xsp?id=WDU20032122072 (accessed on 27 April 2025).
53. Regulation of the Minister of Transport and Maritime Economy of 10 September 1998 on the Technical Conditions to be Met by Railway Structures and Their Location. Available online: https://isap.sejm.gov.pl/isap.nsf/DocDetails.xsp?id=WDU19981510987 (accessed on 27 April 2025).
54. Order 1/2020 of the Director of the Inland Navigation Office in Szczecin of 17 February 2020 on Detailed Conditions for the Safety of Ship Traffic and Berthing. Available online: https://e-dziennik.szczecin.uw.gov.pl/WDU_Z/2020/906/akt.pdf (accessed on 27 April 2025).
55. Order 1/2021 of the Director of the Inland Navigation Office in Wrocław of 25 May 2021 on Detailed Conditions for the Safety of Ship Traffic and Berthing. Available online: https://edzienniki.duw.pl/WDU_D/2021/2550/akt.pdf (accessed on 27 April 2025).
56. Order 1/2022 of the Director of the Inland Navigation Office in Szczecin of 10 March 2022 on Navigation in the Border Waters of the Odra, Western Oder and Lusatian Neisse Rivers. Available online: https://www.szczecin.uzs.gov.pl/wp-content/uploads/2022/03/zachodniopomorskie-publ..pdf (accessed on 27 April 2025).
57. Order 7/2024 of the Director of the Inland Navigation Office in Bydgoszcz of 4 June 2024 on Detailed Conditions for the Safety of Ship Traffic and Berthing [Order 1/2021 of the Director of the Inland Navigation Office in Bydgoszcz of 4 June 2024 on Detailed Conditions for the Safety of Ship Traffic and Berthing]. Available online: https://edzienniki.bydgoszcz.uw.gov.pl/WDU_C/2024/3920/akt.pdf (accessed on 27 April 2025).
58. UNECE. SIGNI—European Code for Signs and Signals on Inland Waterways—Resolution No. 90. Available online: https://unece.org/DAM/trans/main/sc3/publications/SIGNI_2019_e.pdf (accessed on 27 April 2025).
59. UNECE. CEVNI—European Code for Inland Waterways (Sixth Revised Edition). Available online: https://unece.org/transport/publications/cevni-european-code-inland-waterways-rev6 (accessed on 27 April 2025).
60. Binnenschifffahrtsstraßen-Ordnung (BinSchStrO) [Inland Waterways Regulations]. Available online: https://www.elwis.de/DE/Schifffahrtsrecht/Binnenschifffahrtsrecht/BinSchStrO/BinSchStrO-node.html (accessed on 27 April 2025).
61. Naus, K.; Wąż, M.; Szymak, P.; Gucma, L.; Gucma, M. Assessment of ship position estimation accuracy based on radar navigation mark echoes identified in an Electronic Navigational Chart. *Measurement* **2021**, *169*, 108630. [CrossRef]
62. Szymański, K. Inverted water gauge staff pattern for navigation purposes. 2022.
63. Order No. 10/2021 of the President of the State Water Management Authority Polish Waters of 17 February 2021 on the introduction of the "Instructions for the maintenance of inland waterways".
64. Regulation of the Ministers of Infrastructure and Internal Affairs and Administration of 31 July 2002 on Road Signs and Signals. Available online: https://isap.sejm.gov.pl/isap.nsf/DocDetails.xsp?id=wdu20021701393 (accessed on 27 April 2025).
65. Regulation of the Minister of Infrastructure of 3 July 2003 on Detailed Technical Requirements for Road Signs and Signals and Road Safety Devices and Conditions for Their Placement on Roads. Available online: https://isap.sejm.gov.pl/isap.nsf/DocDetails.xsp?id=wdu20032202181 (accessed on 27 April 2025).
66. State Water Holding Polish Water. Virtual River Guide. Available online: https://wir.wody.gov.pl/ (accessed on 12 March 2026).
67. Naval Hydrographic Office. *Maritime Signalling—IALA System [IALA Maritime Buoyage System]*; Naval Hydrographic Office: Dehradun, India, 2017.
68. Lai, C.; Ismail, M. Navigational Safety of Inland Waterway Transport System (IWTS) in Sarawak: Rajang River. *J. Sustain. Sci. Manag.* **2023**, *18*, 102–118. [CrossRef]
69. Regional Water Management Authority in Szczecin, State Water Holding Polish Waters. Available online: https://szczecin.wody.gov.pl/aktualnosci/1191-parkingi-dla-lodolamaczy-i-nowe-oznakowanie-szlaku-zeglownego-kolejna-inwestycja-na-odrze (accessed on 3 July 2025).
70. Regional Water Management Authority in Wrocław, State Water Holding Polish Waters. Notice to Skippers 34/2021. Available online: https://wroclaw.wody.gov.pl/komunikaty-nawigacyjne/128-zegluga/1156-notices-to-skippers (accessed on 3 July 2025).
71. Regional Water Management Authority in Kraków, State Water Holding Polish Waters. Available online: https://krakow.wody.gov.pl/aktualnosci/2446-otwarcie-odcinkow-szlaku-zeglugowego-w-krakowie-dla-ruchu-nocnego-od-tynca-do-stopnia-wodnego-dabie-mozna-bezpieczne-plywac (accessed on 3 July 2025).
72. Regional Water Management Authority in Warsaw, State Water Holding Polish Waters. Navigation Announcement No. 26/2023. Available online: https://warszawa.wody.gov.pl/komunikaty-nawigacyjne?id=1155:navigation-annoucements&catid=114 (accessed on 3 July 2025).
73. Regional Water Management Authority in Białystok, State Water Holding Polish Waters. Available online: https://bialystok.wody.gov.pl/aktualnosci/1782-nowe-oznakowanie-nawigacyjne-w-200-rocznice-kanalu-augustowskiego (accessed on 3 July 2025).

https://doi.org/10.3390/su18083844

74. Regional Water Management Authority in Poznań, State Water Holding Polish Waters. Available online https://poznan.wody.gov.pl/aktualnosci/1281-nowe-oznakowanie-nawigacyjne-na-jeziorach-kanalu-slesinskiego (accessed on 3 July 2025).
75. Heßelbarth, A.; Ziebold, R.; Sandler, M.; Alberding, J.; Uhlemann, M.; Hoppe, M.; Bröschel, M. Towards a reliable bridge collision warning system for inland vessel navigation based on RTK height determination. In Proceedings of the 30th International Technical Meeting of The Satellite Division of the Institute of Navigation (ION GNSS+ 2017), Portland, OR, USA, 27–29 October 2017.
76. Draft Act Amending the Inland Navigation Act and Certain Other Acts. Available online: https://legislacja.rcl.gov.pl/projekt/12349052 (accessed on 3 July 2025).
77. Du, Y.; Sun, S.; Qiu, S.; Li, S.; Pan, M.; Chen, C. Intelligent recognition system based on contour accentuation for navigation marks. *Wirel. Commun. Mob. Comput.* **2021**, *2021*, 6631074. [CrossRef]
78. Pan, M.; Liu, Y.; Cao, J.; Li, Y.; Li, C.; Chen, C.-H. Visual Recognition Based on Deep Learning for Navigation Mark Classification. *IEEE Access* **2020**, *8*, 32767–32775. [CrossRef]
79. Nzengu, W.; Faivre, J.; Pauwelyn, A.S.; Bolbot, V.; Wennersberg, L.A.L.; Theotokatos, G. Regulatory framework analysis for the unmanned inland waterway vessel. *WMU J. Marit. Aff.* **2021**, *20*, 357–376. [CrossRef]
80. Hesselbarth, A.; Medina, D.; Ziebold, R.; Sandler, M.; Hoppe, M.; Uhlemann, M. Enabling Assistance Functions for the Safe Navigation of Inland Waterways. *IEEE Intell. Transp. Syst. Mag.* **2020**, *12*, 123–135. [CrossRef]
81. Dziubińska, A.; Weintrit, A. Wrocław Waterway Hub. *Logistics* **2014**, *3*, 1602–1627.
82. Tubis, A.A.; Skupień, E.T.; Jędrzychowski, K.K.; Jędrzychowski, H. A Sector-Specific Digital Maturity Model for Inland Waterway Transport. *Systems* **2025**, *13*, 347. [CrossRef]
83. Wu, Z.; Ren, C.; Wu, X.; Wang, L.; Zhu, L.; Lv, Z. Research on Digital Twin Construction and Safety Management Application of Inland Waterway Based on 3D Video Fusion. *IEEE Access* **2021**, *9*, 109144–109156. [CrossRef]
84. Cao, S.; Fan, P.; Yan, T.; Xie, C.; Deng, J.; Xu, F.; Shu, Y. Inland Waterway Ship Path Planning Based on Improved RRT Algorithm. *J. Mar. Sci. Eng.* **2022**, *10*, 1460. [CrossRef]
85. Jie, Y.; Leonidas, L.; Mumtaz, F.; Ali, M. Ship Detection and Tracking in Inland Waterways Using Improved YOLOv3 and Deep SORT. *Symmetry* **2021**, *13*, 308. [CrossRef]
86. PIANC. Guidelines and Recommendations for River Information System. InCom WG Report 125/I, 2019. Available online: https://izw.baw.de/publikationen/pianc/0/InCom-WG-125-1.pdf (accessed on 4 July 2025).
87. Durajczyk, P. The Analysis of the Possibility to Improve the Efficiency of Container Transport Via the Oder Waterway with the Use of the RIS System. *Naše More* **2020**, *67*, 199–208. [CrossRef]
88. Durajczyk, P.; Drop, N.; Niedzielski, P. Possibilities of Using Inland Navigation to Improve the Efficiency of Interurban Freight Transport with the Use of the RIS System—Case Study of the Route Opole–Szczecin. *Sustainability* **2024**, *16*, 10754. [CrossRef]
89. Resolution No. 180/2023 of the Council of Ministers of 3 October 2023 on the Establishment of a Development Programme Called "National Shipping Programme Until 2030". Available online: https://www.gov.pl/web/infrastruktura/krajowy-program-zeglugowy-do-roku-2030 (accessed on 4 July 2025).
90. European Funds for Infrastructure, Climate, Environment 2021–2027 Programme. Available online: https://www.gov.pl/web/funds-regional-policy/the-largest-national-programme-in-the-entire-eu-has-been-launched---european-funds-for-infrastructure-climate-environment-2021-2027 (accessed on 4 July 2025).
91. State Water Holding Polish Water. Notice to Skippers. Available online: https://www.gov.pl/web/wody-polskie/komunikaty-nawigacyjne (accessed on 4 July 2025).

https://doi.org/10.3390/su18083844

MDPI AG
Grosspeteranlage 5
4052 Basel
Switzerland
Tel.: +41 61 683 77 34

Sustainability Editorial Office
E-mail: sustainability@mdpi.com
www.mdpi.com/journal/sustainability

www.ingramcontent.com/pod-product-compliance
Lightning Source LLC
Chambersburg PA
CBHW061551220926
51850CB00005B/304
* 9 7 8 3 7 2 5 8 7 9 8 7 8 *